Heavy Metal Toxicity in Plants

Heavy Metal Toxicity in Plants

Physiological and Molecular Adaptations

Edited by

TARIQ AFTAB AND KHALID REHMAN HAKEEM

CRC Press
Taylor & Francis Group
Boca Raton London New York

CRC Press is an imprint of the
Taylor & Francis Group, an **informa** business

First edition published 2022
by CRC Press
6000 Broken Sound Parkway NW, Suite 300, Boca Raton, FL 33487-2742
and by CRC Press

4 Park Square, Milton Park, Abingdon, Oxon, OX14 4RN

CRC Press is an imprint of Taylor & Francis Group, LLC

Library of Congress Cataloging-in-Publication Data

Names: Aftab, Tariq, editor. | Hakeem, Khalid Rehman, editor.
Title: Heavy metal toxicity in plants : physiological and molecular adaptations / Tariq Aftab, Khalid Rehman Hakeem.
Description: First edition. | Boca Raton : C&H/CRC Press, 2021. | Includes bibliographical references and index.
Identifiers: LCCN 2021019424 (print) | LCCN 2021019425 (ebook) | ISBN 9780367725075 (hardback) |
ISBN 9780367725150 (paperback) | ISBN 9781003155089 (ebook)
Subjects: LCSH: Plants--Effect of heavy metals on. | Heavy metals--Toxicology.
Classification: LCC QK753.H4 H43 2021 (print) | LCC QK753.H4 (ebook) |
DDC 581.7--dc23
LC record available at https://lccn.loc.gov/2021019424
LC ebook record available at https://lccn.loc.gov/2021019425

ISBN: 9780367725075 (hbk)
ISBN: 9780367725150 (pbk)
ISBN: 9781003155089 (ebk)

DOI: 10.1201/9781003155089

Typeset in Times LT Std
by KnowledgeWorks Global Ltd.

Contents

Preface

Soils polluted with heavy metals have become commonplace across the globe due to increases in geologic and anthropogenic activities. Growth reduction as a result of changes in physiological and biochemical processes in plants growing on heavy metal polluted soils has been recorded. Some heavy metals, namely, cobalt (Co), copper (Cu), iron (Fe), manganese (Mn), molybdenum (Mo), nickel (Ni), and zinc (Zn), are considered to be essential for plants, whereas chromium (Cr) and antimony (Sb) are essential for animals. These metal elements can directly influence growth, senescence, and energy-generating processes due to their high reactivity. Their concentration in the soil beyond permissible limits is toxic to plants, either causing oxidative stress through free radicals and/or disrupting the functions of enzymes by replacing essential metals and nutrients.

Some of the direct toxic effects caused by high metal concentration include inhibition of cytoplasmic enzymes and damage to cell structures due to oxidative stress. An example of indirect toxic effect is the replacement of essential nutrients at cation exchange sites of plants. Further, the negative influence heavy metals have on the growth and activities of soil microorganisms may also indirectly affect the growth of plants. The bioaccumulation of heavy metals in excessive concentrations may replace essential metals in pigments or enzymes disrupting their function and causing oxidative stress. Heavy metal toxicity hinders the growth process of the underground and aboveground plant parts and the activity of the photosynthetic apparatus, which is often correlated with progress in senescence.

This book covers a wide range of topics, discussing the heavy metal toxicity impact on plants. Moreover, this is a unique reference on the topic highlighting the various metal-induced impacts on plants and adaptation strategies employed by plants to avoid these stressful conditions. In this volume, we highlight the working solutions as well as open problems and future challenges for heavy metal toxicity in plants.

About the Editors

Dr. Tariq Aftab

Tariq Aftab earned his PhD from the Department of Botany, Aligarh Muslim University, India and is currently an assistant professor there. He is the recipient of the prestigious Leibniz-DAAD fellowship from Germany, Raman Fellowship from the Government of India, and has won the Young Scientist Awards from the State Government of Uttar Pradesh (India) and Government of India. After completing his doctorate, he worked as a research fellow at the National Bureau of Plant Genetic Resources, New Delhi and as a post-doctorate fellow at Jamia Hamdard, New Delhi, India. Dr. Aftab was also a visiting scientist at Leibniz Institute of Plant Genetics and Crop Plant Research (IPK), Gatersleben, Germany and in the Department of Plant Biology, Michigan State University, East Lansing, MI. He is a member of various scientific associations from India and abroad.

He has edited 11 books with international publishers, including Elsevier Inc., Springer Nature, and CRC Press (Taylor & Francis Group), co-authored several book chapters, and published over 60 research papers in peer-reviewed international journals. His research interests include physiological, proteomic, and molecular studies in crop and medicinal plants.

Dr. Khalid Rehman Hakeem

Khalid Rehman Hakeem, PhD, is a professor at King Abdulaziz University, Jeddah, Saudi Arabia. After completing his doctorate (botany; specialization in plant eco-physiology and molecular biology) from Jamia Hamdard, New Delhi, India, in 2011, he worked as a lecturer at the University of Kashmir, Srinagar, for a short period. Later, he joined Universiti Putra Malaysia, Selangor, Malaysia as a post-doctorate fellow in 2012 and fellow researcher (associate professor) from 2013–2016. Dr. Hakeem has more than 10 years of teaching and research experience in plant eco-physiology, biotechnology and molecular biology, medicinal plant research, and plant-microbe-soil interactions, as well as in environmental studies. He is the recipient of several fellowships at both national and international levels; he has also served as the visiting scientist at Jinan University, Guangzhou, China. Currently, he is involved with a number of international research projects with different government organizations.

So far, Dr. Hakeem has authored and edited more than 60 books with international publishers, including Springer Nature, Academic Press (Elsevier), and CRC Press. He also has to his credit more than 125 research publications in peer-reviewed international journals and 55 book chapters in edited volumes with international publishers.

At present, Dr. Hakeem serves as an editorial board member and reviewer of several high-impact international scientific journals from Elsevier, Springer Nature, Taylor & Francis, Cambridge, and John Wiley Publishers. He is included in the advisory board of Cambridge Scholars Publishing, UK.

Prof. Khalid was recently elected as a fellow, Royal Society of Biology, UK. He is also a fellow of the Plantae group of the American Society of Plant Biologists, member of the World Academy of Sciences, member of the International Society for Development and Sustainability, Japan, and member of the Asian Federation of Biotechnology, Korea. Dr. Hakeem has been listed in the Marquis Who's Who in the World from 2014–2019. Currently, Dr. Hakeem is engaged in studying the plant processes at eco-physiological as well as molecular levels.

List of Contributors

Muhammad Adeel
Institute of Soil and Environmental Science
University of Agriculture
Faisalabad, Punjab, Pakistan

Tariq Aftab
Department of Botany
Aligarh Muslim University
Aligarh, India

Puja Agarwal
Constituent College in Purnea University
Purnia, Bihar, India

S.C. Agrawal
Phycology Laboratory
Department of Botany
University of Allahabad
Prayagraj, UP, India

Hamaad Raza Ahmad
Institute of Soil and Environmental Sciences
University of Agriculture
Faisalabad, Punjab, Pakistan

Sajid Rashid Ahmad
College of Earth and Environmental Sciences
University of the Punjab
Lahore, Pakistan

Zahoor Ahmad
Department of Botany
University of Central Punjab
Bahawalpur Campus, Punjab Group of Colleges
Pakistan

Sharif Ahmed
International Rice Research Institute
Bangladesh Office
Dhaka, Bangladesh

Md. Jahangir Alam
On-Farm Research Division
Bangladesh Agricultural Research Institute
Gaibandha, Bangladesh

Mohsina Anjum
Department of Soil Science & Agricultural Chemistry
University of Agricultural Sciences
Bangalore, India

Hirdayesh Anuragi
ICAR-Central Agroforestry Research Institute
Jhansi, UP, India

Jaya Arora
Laboratory of Biomolecular Technology
Department of Botany
M. L. Sukhadia University
Udaipur, Rajasthan, India

Muhamad Zohaib Aslam
Institute of Soil and Environmental Sciences
University of Agriculture
Faisalabad, Punjab, Pakistan

Sana Ashraf
College of Earth and Environmental Sciences
University of the Punjab
Lahore, Pakistan

Muhammad Ashar Ayub
Institute of Soil and Environmental Sciences
University of Agriculture Faisalabad
Punjab, Pakistan
Horticultural Sciences Department
Plant Root Biology Lab
Indian River Research and Education Center
Institute of Food and Agriculture Sciences
University of Florida
Fort Pierce, Florida, USA

Aditi Shreeya Bali
Dyal Singh College
Karnal, India

Samrat Banerjee
Department of Botany
UGC Centre for Advanced Studies
The University of Burdwan
West Bengal, India

Aaliya Batool
Department of Botany
University of Agriculture
Faisalabad, Punjab, Pakistan

Marian Brestic
Department of Botany and Plant Physiology
Faculty of Agrobiology, Food and Natural Resources
Czech University of Life Sciences Prague
Czechia
Department of Plant Physiology
Slovak University of Agriculture
Nitra, Slovak Republic

Brijesh Kumar Chaudhary
Institute of Agriculture Sciences
Banaras Hindu University
Varanasi, UP, India

Surendra Kumar Chaudhary
Institute of Agriculture Sciences
Banaras Hindu University
Varanasi, UP, India

Jyoti Chauhan
Institute of Agriculture Sciences
Banaras Hindu University
Varanasi, UP, India

Manoj Kumar Chitara
Department of Plant Pathology
College of Agriculture
GBPUAT
Pantnagar, Uttarakhand, India

Basant Kumar Dadrwal
Institute of Agriculture Sciences
Banaras Hindu University
Varanasi, UP, India

Mona F.A. Dawood
Department of Botany & Microbiology
Faculty of Science
Assiut University
Assiut, Egypt

Anupam Dikshit
Biological Product Laboratory
Department of Botany
University of Allahabad
Prayagraj, India

Ratnasekera Disna
Department of Agricultural Biology
Faculty of Agriculture
University of Ruhuna
Matara, Sri Lanka

Anamika Dubey
Metagenomics and Secretomics Research Laboratory
Department of Botany
Dr. Harisingh Gour University (A Central University)
Sagar, MP, India

Mujahid Farid
Department of Environmental Sciences
University of Gujrat
Hafiz Hayat Campus
Gujrat, Pakistan

Mohamed Gomaa
Department of Botany & Microbiology
Faculty of Science
Assiut University
Assiut, Egypt

Muhammad Bilal Hafeez
Department of Agronomy
University of Agriculture
Faisalabad, Punjab, Pakistan

Akash Hidangmayum
Institute of Agriculture Sciences
Banaras Hindu University
Varanasi, UP, India

Akbar Hossain
Bangladesh Wheat and Maize Research Institute
Dinajpur, Bangladesh

Usman Iftikhar
Department of Environmental Sciences and Engineering
Government College University Faisalabad,
Faisalabad, Pakistan

Muhammad Imran
Department of Entomology
Faculty of Agriculture
University of Poonch
Rawalakot, Pakistan

Muhammad Aamir Iqbal
Department of Agronomy
Faculty of Agriculture
University of Poonch
Rawalakot (AJK), Pakistan

Mohammad Sohidul Islam
Department of Agronomy
Hajee Mohammad Danesh Science and Technology
University
Basherhat, Bangladesh

Asad Jamil
Institute of Soil and Environmental Sciences
University of Agriculture
Faisalabad, Punjab, Pakistan

Hanuman Singh Jatav
S.K.N. Agriculture University
Jobner, Rajasthan, India

Surendra Singh Jatav
Department of Soil Science & Agricultural Chemistry
Institute of Agricultural Science
Banaras Hindu University
Varanasi, UP, India

Saket Jha
Biological Product Laboratory
Department of Botany
University of Allahabad
Prayagraj, UP, India

Abhishek Joshi
Laboratory of Biomolecular Technology
Department of Botany
M. L. Sukhadia University
Udaipur, Rajasthan, India

Bhanupriya Kanthaliya
Laboratory of Biomolecular Technology
Department of Botany
M. L. Sukhadia University
Udaipur, Rajasthan, India

Tanzeela Kokab
College of Earth and Environmental Sciences
University of the Punjab
Lahore, Pakistan

Ashwani Kumar
Metagenomics and Secretomics Research Laboratory
Department of Botany
Dr. Harisingh Gour University
Sagar, MP, India

Navneet Kumar
Institute of Agriculture Sciences
Banaras Hindu University
Varanasi, UP, India

Rajesh Kumar
PG Department of Botany
Mahatma Gandhi Governments and Arts College
Mahe, Puducherry, India

Utkarsh Kumar
ICAR-Vivekananda Parvatiya Krishi Anusandhan Sansthan (VPKAS)
Almora, Uttarakhand, India

Vivek Kumar
Institute of Agriculture Sciences
Banaras Hindu University
Varanasi, UP, India

S.P. Kushwaha
Chandra Shekhar Azad University of Agriculture & Technology
Kanpur, UP, India

Devidutta Lenka
Orissa University of Agriculture and Technology
Bhubaneswar, Odisha, India

Sagar Maitra
Department of Agronomy and Agroforestry
Centurion University of Technology and Management
Odisha, India

Supriya Meena
Laboratory of Biomolecular Technology
Department of Botany
M. L. Sukhadia University
Udaipur, Rajasthan, India

Tatiana Minkina
Academy of Biology and Biotechnology
Southern Federal University
Stachki, Rostov-on-Don, Russia

Preeti Mishra
Biological Product Laboratory
Department of Botany
University of Allahabad
Prayagraj, UP, India

Udit Nandan Mishra
Faculty of Agriculture
Sri Sri University
Cuttack, Odisha, India

Mehali Mitra
Department of Botany
UGC Centre for Advanced Studies
The University of Burdwan
West Bengal, India

Muhammad Mubeen
Department of Environmental Sciences
COMSATS University Islamabad
Vehari Campus, Pakistan

Naveed Ul Mushtaq
Department of Bioresources
School of Biological Sciences
University of Kashmir
Srinagar, Jammu and Kashmir, India

Rubina Naseer
Centre of Agricultural Biochemistry and Biotechnology
University of Agriculture
Faisalabad, Punjab, Pakistan

Misbah Naz
State Key Laboratory of Crop Genetics and Germplasm Enhancement
Nanjing Agricultural University
Nanjing, China

Sytar Okana
Department of Plant Physiology
Faculty of Agrobiology and Food Resources
Slovak University of Agriculture
Nitra, Slovakia

Manjulata Palei
Department of Botany
Centurion University of Technology and Management
Odisha, India

Anand Pandey
Department of Plant Sciences
Avvaiyar Government College for Women
University of Puducherry
Karaikal, India

Manoj Parihar
ICAR-Vivekananda Parvatiya Krishi Anusandhan Sansthan
(VPKAS)
Almora, Uttarakhand, India

Madhusmita Pradhan
PG Department of Botany
Utkal University
Bhubaneswar, Odisha, India

Afifa Qidwai
Biological Product Laboratory
Department of Botany
University of Allahabad
Prayagraj, UP, India

Julio E. Quinones III
Horticultural Sciences Department
Plant Root Biology Lab
Indian River Research and Education
Center
Institute of Food and Agriculture Sciences
University of Florida
Fort Pierce, Florida, USA

Junaid Rahim
Department of Entomology
Faculty of Agriculture
University of Poonch
Rawalakot, Pakistan

Md. Atikur Rahman
Species Research Centre
Bangladesh Agricultural Research Institute (BARI)
Bogra, Bangladesh

Vishnu D. Rajput
Academy of Biology and Biotechnology
Southern Federal University
Stachki, Rostov-on-Don, Russia

Amitava Rakshit
Department of Soil Science & Agricultural Chemistry
Institute of Agricultural Science
Banaras Hindu University
Varanasi, UP, India

Kiran Rana
Department of Agronomy
Institute of Agricultural Science
Banaras Hindu University
Varanasi, UP, India

Tahira Rasheed
Department of Botany
University of Agriculture
Faisalabad, Pakistan

Aadil Rasool
Department of Bioresources
School of Biological Sciences
University of Kashmir
Srinagar, Jammu and Kashmir, India

Muhammad Zia ur Rehman
Institute of Soil and Environmental Sciences
University of Agriculture
Faisalabad, Punjab, Pakistan

Reiaz Ul Rehman
Department of Bioresources
School of Biological Sciences
University of Kashmir
Srinagar, Jammu and Kashmir, India

Sujit Roy
Department of Botany
UGC Centre for Advanced Studies
The University of Burdwan
West Bengal, India

Aryadeep Roychoudhury
Post Graduate Department of Biotechnology
St. Xavier's College (Autonomous)
Kolkata, West Bengal, India

Ayman EL Sabagh
Department of Agronomy
Faculty of Agriculture
University of Kafrelsheikh
Egypt
Department of Field Crops
Faculty of Agriculture
Siirt University
Turkey

Ranjan Kumar Sahoo
Department of Biotechnology
Centurion University of Technology and Management
Odisha, India

Seerat Saleem
Department of Bioresources
School of Biological Sciences
University of Kashmir
Srinagar, Jammu and Kashmir, India

Santanu Samanta
Post Graduate Department of Biotechnology
St. Xavier's College (Autonomous)
Kolkata, West Bengal, India

Wasifa Hafiz Shah
Department of Bioresources
School of Biological Sciences
University of Kashmir
Srinagar, Jammu and Kashmir, India

Muhammad Bilal Shakoor
College of Earth and Environmental Sciences
University of the Punjab
Lahore, Pakistan

Kanval Shaukat
Department of Botany
University of Agriculture
Faisalabad, Pakistan
Department of Botany
University of Balochistan
Quetta, Pakistan

Ayesha Siddique
Institute of Soil and Environmental Sciences
University of Agriculture
Faisalabad, Punjab, Pakistan

Gagan Preet Singh Sidhu
Centre for Applied Biology in Environment Sciences
Kurukshetra University
Kurukshetra, India

Ankita Singh
Institute of Agriculture Sciences
Banaras Hindu University
Varanasi, UP, India

Ashish Kumar Singh
ICAR-Vivekananda Parvatiya Krishi Anusandhan Sansthan (VPKAS)
Almora, Uttarakhand, India

Nivedita Singh
Biological Product Laboratory
Department of Botany
University of Allahabad
Prayagraj, UP, India

Ravikant Singh
Department of Biotechnology
Swami Vivekanand University
Sagar, MP, India

Rajesh Kumar Singhal
ICAR-Indian Grassland and Fodder Research Institute
Jhansi, UP, India

Milan Skalicky
Department of Botany and Plant Physiology
Faculty of Agrobiology, Food and Natural Resources
Czech University of Life Sciences Prague
Czechia

Rekha Sodani
College of Agriculture
Agriculture University
Nagaur, Jodhpur, Rajasthan, India

Inayatullah Tahir
Department of Botany
University of Kashmir
Srinagar, Jammu and Kashmir, India

Md. Tahjib-Ul-Arif
Department of Biochemistry and Molecular Biology
Bangladesh Agricultural University
Mymensingh, Bangladesh

Arpita Tripathi
Biological Product Laboratory
Department of Botany
University of Allahabad
Prayagraj, UP, India

Sharad K. Tripathi
Biological Product Laboratory
Department of Botany
University of Allahabad
Prayagraj, UP, India

Wajid Umar
Institute of Environmental Science
Hungarian University of Agriculture and Life Sciences
Gödöllő, Hungary

Noreen Zahra
Department of Botany
University of Agriculture
Faisalabad, Pakistan

1

Heavy Metal Toxicity in Plants: Physiological and Molecular Adaptations

Santanu Samanta and Aryadeep Roychoudhury
Post Graduate Department of Biotechnology, St. Xavier's College (Autonomous), Kolkata, West Bengal, India

CONTENTS

1.1 Introduction

In the present times, heavy metal pollution is a serious environmental concern caused by several natural processes or anthropogenic activities. These are persistent and cannot be easily removed from the environment (Jalmi et al. 2018). Heavy metals are metallic elements having relatively high density and are toxic or poisonous even at low concentration (Gill 2014). They can displace beneficial metals from pigments or crucial enzymes by disrupting their vital functions. Metals render the land unsuitable for plant growth and development (Ghosh and Singh 2005). Heavy metals at high concentration can severely affect plant growth by altering physiological and biochemical processes, including photosynthesis and respiration, even leading to the death of the plants (Shahid et al. 2015). Some of these metals, such as manganese (Mn), cobalt (Co), nickel (Ni), zinc (Zn) and copper (Cu) are essential for crucial biological processes at low concentrations but display several toxic effects in plant parts beyond supraoptimal level. Other metals/metalloids such as cadmium (Cd), lead (Pb), arsenic (As) and mercury (Hg) are completely non-essential for plant growth and development (Dalcorso 2012; Kalaivanan and Ganeshamurthy 2016).

To counteract heavy metal stress, plants have adopted several physiological and molecular mechanisms, which are highly effective to control the uptake, accumulation, trafficking and detoxification of various metals. Some of these mechanisms are ubiquitous, i.e., part of homeostatic processes, whereas other mechanisms are activated only when a threshold level of metal toxicity is encountered. Additionally, increased levels of heavy metals are associated with elevated level of reactive oxygen species (ROS) (Rascio and Navari-Izzo 2011). These toxic radicals cause oxidative stress which alter the homeostasis between the prooxidant and antioxidant within the intracellular system of plants. This condition imposes multiple deteriorative disorders in plants such as imbalance in redox equilibrium, electrolyte leakage, oxidation of protein as well as

DOI: 10.1201/9781003155089-1

lipids, DNA breakage, and denaturation of cellular macromolecules which altogether are often correlated with the progression of programmed cell death (PCD) (Mourato et al. 2015).

1.2 Influence of Heavy Metals on Plants

Heavy metals can be categorized into two groups based on their physicochemical properties, (i) redox active metals which include Mn, Cu, Fe, and Cr and (ii) non-redox active metals which comprises Cd, Ni, Hg, Zn, and Aluminum (Al) (Schützendübel and Polle 2002; Yadav 2010). The redox active metals have the capacity to generate direct oxidative stress via Haber-Weiss and Fenton reactions, which has been demonstrated to stimulate formation of free radicals in plants (Schützendübel and Polle 2002; Halliwell 2006). On the contrary, non-redox active metals indirectly stimulate oxidative injury via depleting the glutathione (GSH) pool, enhancing the prooxidative enzymes like nicotinamide adenine dinucleotide phosphate (NADPH) oxidase, forming bond with sulfhydryl groups of proteins, or down regulating the activity of antioxidative enzymes (Solanki and Dhankher 2011). Some of these heavy metals are discussed to ascertain how they impose stress on plants, despite possessing different redox states.

1.2.1 Chromium

Chromium (Cr) has become a serious pollutant for the growth and development of plants due to its extensive use in industry (Panda and Choudhury 2005; Shanker et al. 2005). Chromium toxicity depends upon its valence state. Cr^{6+} is highly toxic and mobile, whereas Cr^{3+} is less toxic. In plants, carriers for sulfate or iron play a significant role in transportation and accumulation of Cr, which is not directly absorbed by plants due to lack of specific transport system (Becquer et al. 2003; Cervantes et al. 2001). Moreover, Cr^{6+} can be converted to its more toxic form, Cr^{3+} under reducing environment, which indirectly influences the soil pH, depending upon the prevailing condition of soil substrate (Peralta et al. 2001). The greater proportion of Cr occurs in the plant roots rather than in above ground tissues due to immobilization of Cr within the vacuole of the root cells (Kumar and Maiti 2013).

Cr has severe adverse effects on the seedling dry mass and inhibits the progression of leaves and stems during early growth stages (Zeid 2001). It was observed in an in vitro study in *Amaranthus viridis* that Cr decreased root growth by inhibiting cell division and disrupting the transport of calcium ion (Ca^{2+}) from plasma membrane to cytoplasm (Zou et al. 2006). In addition, Cr is capable of disrupting amylase activity, influencing the total soluble sugar levels and transport of carbohydrate reserve into embryonic tissues (Dey et al. 2009). Cr is also involved in altering the profile of other nutrients like Ca, Mn, Fe, Mg, phosphorus (P) and potassium (K) in both aerial and root parts of different plants (Samantaray et al. 1998; Kabir and Dietrich 2016).

1.2.2 Aluminum

Al ranks third in abundance among the earth's crust elements. Solubilization of this metal is enhanced especially in acidic soils with pH less than 5, in which its trivalent form (Al^{3+}) is prevalent (Delhaize and Ryan 1995). Although it has no known physiological functions in plants, a few reports demonstrated that it can stimulate plant growth at low concentrations (Kidd and Proctor 2000). The root apex region is the primary target of Al toxicity in plants (Bennet et al. 1985). Aluminum also causes reduction in respiration in plant roots and alters regulation of enzymes involved in sugar phosphorylation (Panda et al. 2009). Aluminum toxicity negatively affects the aboveground tissues of plants by hampering initial root damage, which significantly reduces nutrient uptake mechanism of roots (Sivaguru and Horst 1998).

Aluminum stress has severe deteriorating effect on chlorophyll content and rate of photosynthesis due to inhibition of electron transport in photosystem II (PSII). The impact of Al-toxicity in photosynthesis is indirect via disturbance in the chloroplast architecture (Zhang et al. 2001; Ali et al. 2008). The reduction in stomatal aperture is also reported to be caused by Al-toxicity (Vardar and Ünal 2007). Recently, it was reported that Al toxicity also perturbs the transportation of some macronutrients in plants like maize and sorghum (Bhalerao and Prabhu 2013). Panda et al. (2003) reported that prolonged exposure to Al-stress resulted in greatly increased lipid peroxidation in leaves of green gram (*Vigna radiata*).

1.2.3 Manganese

Mn acts as beneficial micronutrient that plays vital role in biochemical and molecular processes in plants including photosynthesis, respiration and activities of enzymes such as isocitrate dehydrogenase, nitrate reductase, and malic enzymes (Demirevska-Kepova et al. 2004). Furthermore, Mn is required for the biosynthesis of fatty acids and carotenoids, carbohydrate and nitrogen metabolism, as well as activation of phytohormones (Santos et al. 2017). Mn plays a pivotal role in water splitting mechanism of PSII, which provides the necessary electrons for photosynthesis and also provides protection of PSII from photodamages (Buchanan et al. 2000). The most stable and bioactive form of Mn in soil environment is Mn^{2+}, and this form is readily transported through root cells (Millaleo et al. 2010). In contrast to other such heavy metals like Al and Cu, Mn can easily translocate from roots to the aboveground tissues. This is the reason why symptoms of Mn toxicity are first visible in aerial parts of plants (Paschke et al. 2005).

1.2.4 Nickel

Ni is an essential element which is required by plants in very minute concentration (Ragsdale 1998; Nieminen et al. 2007). Ni acts as component for a number of enzymes, including glyoxalases I, ureases, peptide deformylases and a few superoxide dismutases (SODs) (Ermler et al. 1998; Küpper and Kroneck 2007). Ni is taken up by the plants mainly through root systems and some major factors like soil acidity and concentration of organic substances present in soil can affect its absorption (Seregin and Kozhevnikova 2006). Transportation of Ni from root to aboveground tissues is mediated by transpiration stream through xylem. More than half of Ni absorbed by the plants is retained within the root system, which may be due to the sequestration by the cation exchange sites present in the walls of xylem parenchyma cells

(Cataldo et al. 1978; Seregin and Kozhevnikova 2006). Excess Ni toxicity adversely affects the total dry mass and yield, and also hampers other physiological processes such as photosynthesis, water relations and mineral nutrition in plants (Gajewska and Skłodowska 2005; Hussain et al. 2013).

1.2.5 Copper

Cu plays several important functions in plants such as carbon assimilation and ATP biosynthesis. It participates in different physiological functions as well as acts as a constituent of cytochrome oxidase and plastocyanin, which are vital components for both photosynthesis and respiratory metabolic pathway (Landjeva et al. 2003). However, when its concentration rises above optimal level within the tissue, it displays toxic effects in plants. The availability of Cu greatly depends upon soil pH and its phytoavalability increases with declining pH. Cu exists in soil in different redox states, although plants uptake this metal in the form of Cu^{2+} (Alhamed and Shebany 2012). Several research studies in different plant species such as *Cucumis sativus*, *Syzygium aromaticum* and some *Eucalyptus* species suggest that Cu has a general tendency to accumulate in root tissues (Assareh et al. 2008; Alaoui-Sossé et al. 2004). Mediouni et al. (2006) reported that Cu has more pronounced effect in the induction of lipid peroxidation in young tissues. A negative correlation between root and shoot elongation with increasing concentration of Cu level was noted in spinach seedlings (Sharma et al. 2010). Excess Cu has severe cytotoxic role which causes leaf chlorosis and overall retardation of plant growth and metabolism (Lombardi and Sebastiani 2005).

1.2.6 Lead

Pb is an amphoteric trace metal, second most toxic heavy metal after As. It has no known physiological role and significance in plants. In general, Pb can be found as either free ions or as complex with inorganic or organic materials (Kumar and Prasad 2018). Unlike other heavy metals, Pb displays translocation restriction. In plants, Pb is accumulated mainly in the root system, and to a lesser degree in the shoot (Obiora et al. 2016). Pb toxicity causes severe damaging effects, leading to stunted growth with ultimate yield penalty of several crop plants. Several reports revealed its various toxic effects on plants, such as obstruction in germination process, impairment in nutrient uptake, perturbations in water status, substantial injury in internal macromolecules, etc. (Ma et al. 2016; Kuppusamy et al. 2016). Plant interaction with toxic Pb generally alters the photosynthetic pigment levels, thereby damaging photosynthetic activity (Mahdavian et al. 2017). It was established that Pb-toxicity influence considerably the level of other nutrient concentrations like Ca, Mg, K, Fe, Cu, and Zn (Zulfiqar et al. 2019). In addition, free radicals generated during Pb-stress causes damage of cytoskeleton, DNA strand breakage, microtubule depolymerization, and protein oxidation (Hesami et al. 2018).

1.2.7 Cadmium

Cd is non-essential element for plant growth, which is released into the environment from industrial effluents. It is recognized as extremely toxic pollutant due to its high solubility in water. It is not yet completely understood how Cd enters into the plant system. Thomine et al. (2000) found that *AtNramp3*, a metal transporter involved in iron uptake, is also capable of Cd^{2+} transport in *Arabidopsis* plant. Cd significantly interferes with the uptake and transport of several elements like P, K, Ca, Mg and also water by plants (Prasad 1995). Cd toxicity affects membrane permeability, ion content, inhibits water status within the tissue and also alters membrane functionality, including lipid peroxidation (Hasan et al. 2009; Gallego et al. 2012). Photosynthesis is hampered by Cd toxicity. Plants exposed to Cd displayed reduced photosynthetic rate due to distortion in chloroplast ultrastructure, resulting in inhibited activities of Calvin cycle in plants. Cd strongly binds with sulfhydryl groups of important enzymes which leads to inhibition of electron transport chain (Nazar et al. 2012; Wahid and Khaliq 2015).

1.2.8 Arsenic

Arsenic is a potentially toxic metalloid with prominent effect on plant growth and agronomic yield. It is considered to be biologically non-essential for all forms of life. Arsenic stress causes growth inhibition, physiological disorders, irreparable damage of important biomolecules and membrane deterioration in plants (Sharma 2012; Kumar et al. 2015). In nature, arsenic predominantly exists as two principal inorganic forms such as arsenate (As^{V}) and arsenite (As^{III}). Plants take up arsenate via high-affinity phosphate transporters, whereas nodulin 26-like intrinsic (NIP) aquaporin channels are responsible for uptake of arsenite in plants (Srivastava et al. 2016; Chen et al. 2017). Arsenic toxicity may be manifested by various physiological attributes like inhibition of root and shoot length, reduction of photosynthetic efficiency due to chlorophyll loss, and lowering of water potential within the cell (Li et al. 2016). Apart from these, arsenic induces changes in redox equilibrium, resulting in excess formation of ROS which leads to cellular oxidative damages (Samanta et al. 2020a; Shrivastava et al. 2015).

1.3 Physiological and Molecular Defense Mechanisms Employed by Plants against Heavy Metal Stress

Plants possess an intricate network of defense strategies to counteract heavy metal toxicity. During exposure to heavy metal stress, diverse cellular biomolecules are biosynthesized to encounter or neutralize metal toxicity; these include metal chelators like nicotinamide, organic acids, mugineic acids, and cellular exudates like flavonoids and phenolic compounds (Emamverdian et al. 2015; Ghori et al. 2019; Yang et al. 2005). When these strategies are not able to restrain metal poisoning, there occur changes in redox equilibrium, resulting in excess formation of ROS which leads to cellular oxidative damages (Nearing et al. 2014; Singh et al. 2016). Plant cells have developed antioxidant defense mechanisms which are composed of enzymatic antioxidants like SOD, catalase (CAT), glutathione-S-transferase (GST), guaiacol peroxidase (GPOX) and glutathione peroxidase (GPX), and non-enzymatic antioxidants like ascorbate (ASC), glutathione (GSH), carotenoids, flavonoids and tocopherols which act synergistically to combat overproduction of toxic free radicals during metal stress (Das and Roychoudhury 2014). Exploitation and up regulation of these mechanisms and biomolecules solely depend upon plant species and their level of metal tolerance. Some of these physiological and molecular defense mechanisms are widely utilized by various plant species against heavy metal stress as discussed in this chapter.

1.3.1 Amino Acids and Organic Acids

During exposure to metal toxicity, plants can generate a range of carboxylic acids like citrate, malate, pyruvate, oxalate, formate and amino acids, such as cysteine (Cys) and histidine (His) which act as potential ligands for heavy metals, thereby mediating tolerance and detoxification (Wang et al. 2011; Chen et al. 2009). The Zn- and Cd-citrate complexes are highly abundant in leaves, although malate is more prevalent. Citrate and His are the important ligands for Cu, Ni, and Zn in the xylem sap, transporting from root to shoot. A putative Zn-His complex has been identified in the roots of Zn-hyperaccumulator plant, *Thlaspi caerulescens* (Salt et al. 1999). In response to Ni-toxicity, a 36-fold up regulation in free His concentration has been observed in the xylem exudate of Ni-hyperaccumulator plant, *Alyssum lesbiacum* (Kramer et al. 1996). In an experiment in *Zea mays* it was found that the production of different organic acids varied with external concentration of Al ions, indicating that there is a strong correlation between metal exposure and organic acid production (Pintro et al. 1997). Additionally, Burke et al. (1990) reported that production of malic, citric or acotinic acid remained unchanged in Mn-tolerant variety, whereas such organic acids were slightly enhanced in the susceptible variety.

1.3.2 Phytochelatins

Many plants cope up with metal-induced toxicity by binding them via complexes with a class of peptides called phytochelatins (PCs). PCs are metal binding proteins, rich in Cys-polypeptides with general structure (γ-glutamyl-Cys)$_n$Gly, where the value of n can be from 2 to 11, depending upon the organisms (Cobbett 2000; Samanta et al. 2020b). PCs are rapidly induced in cells during exposure to heavy metal ions, such as Cd, Hg, Pb, Ni, Zn, Cu, and Ag, as well as anions like arsenate and selenite (Rauser 1995; Yang and Yang 2001). These metals strongly bind with the constitutively expressed enzyme γ-glutamyl cysteinyl dipeptidyl transpeptidase or phytochelatin synthase (PCS), which mediates the transpeptidation of γ-glutamyl cysteinyl dipeptides from GSH (Tiwari et al. 2013). GSH acts as a substrate for PCs, which is synthesized in two steps. The first step is catalyzed by γ-glutamyl-Cys synthetase, also known as γ-ECS, which is regulated by GSH via feedback mechanism and is dependent upon the availability of Cys, whereas glutathione synthase (GS) performs the second step (Cobbett 2008; Cobbett and Goldsbrough 2002). In plants, PCs act not only as a defensive agent against deleterious effects of metals but also as a part of other stressors, such as excess heat, UV-B, salt as well as herbicide (Stolt et al. 2003). PCs are mainly synthesized within the cytosol and actively form metal-PC complex and the complex is then accumulated within the vacuole as their final destination (Zagorchev et al. 2013).

The chain lengths of PCs may vary among plant species as well as with the type of metals present. Brunetti et al. (2011) reported that PC_4 was the most active form in tobacco seedlings, whereas PC_3 was of higher concentration in *Arabidopsis* upon exposure to Cd. It has been reported that longer chain PCs are most effective in legumes to detoxify Pb-induced toxicity than shorter PCs (Piechalak et al. 2002). An enhanced biosynthesis of PC production was coupled with escalated activity of antioxidative machinery in *Brassica chinensis* in order to detoxify Cd toxicity (Chen et al. 2008).

1.3.3 Metallothioneins

Metallothioneins (MTs) are naturally occurring, low molecular weight (6–7 kDa), Cys-rich peptides, having a strong binding affinity toward metals by tethering to sulfhydryl (–SH) groups of Cys residues (Cobbett and Goldsbrough 2002). They can be categorized into three different classes, according to their Cys residues and structure: Cys-Cys, Cys-X-Cys, and Cys-X-X-Cys motifs, in which X is any amino acid (Hassinen et al. 2011). Aromatic amino acids and His are not found in MTs. A number of important factors such as metals, cytotoxic agents and hormones enhanced the biosynthesis of MTs, by regulating at the transcription level. Several *MT* genes have been identified in plants. While PCs are enzymatically synthesized, *MTs* are genes encoded as a result of mRNA translation (Grennan 2011). Many studies have suggested that in addition to the chelation and metal homeostasis, they have diverse physiological functions in plants. In addition to their role in heavy metal detoxification, they also participate in a number of events, such as ROS homeostasis, and repair of plasma membrane and damaged DNA (Wong et al. 2004; Macovei et al. 2010; Mishra and Dubey 2006; Grennan 2011).

1.3.4 Transport Proteins and Metal Transporters

Using biochemical and molecular techniques, it has been well established that transport proteins play a pivotal role in metal homeostasis in terms of their mechanistic characteristics. There are various transport proteins present in plants like cation diffusion facilitator (CDF) family and natural resistance linked macrophage protein family (Nramp), which play an indispensable role in metal tolerance (Pottier et al. 2015; Hall and Williams 2003). Furthermore, *ZIP* gene family of transporters are also responsible for transport of several cations such as Zn^{2+}, Fe^{2+}, Mn^{2+}, Cd^{2+}, etc. (Guerinot 2000). It is evident that ZIP family of proteins was activated in *Arabidopsis* during exposure to Fe and Zn stress. Of these, ZIP1 and ZIP3 are expressed in roots, particularly under Zn stress (Milner et al. 2013). In *Arabidopsis*, AtZIP4 is expressed in roots and shoots and its expression was enhanced during Zn restriction, which suggests a role in Zn nutrition.

Nramps are highly conserved family of transporters, widely distributed in plants, animals, bacteria and fungi, which can transport a variety of metal ions (Williams et al. 2000). In plants, they were first identified in rice, where three Nramps

(OsNramp1, OsNramp2, and OsNramp3) were reported. Curie et al. (2000) reported that overexpression of AtNramp1 in transgenic *Arabidopsis thaliana* resulted in increased resistance of the plants toward Fe toxicity. The cation transporter family (CTR), another high-affinity transporter, plays a crucial role in Cu transportation in plants. This family of transporter localized in plasma membrane also play an important role in growth and pollen development in *Arabidopsis* (Colangelo and Guerinot 2006). Several other metal transporters are also engaged in the transport of metals in plants. ABC transporter (ATP-binding cassette) are diverse family of transporters, involved in a wide range of transport functions. Most ABC transporters harness energy from ATP hydrolysis and are involved in metal sequestration and compartmentalization within the vacuole (Theodoulou 2000; Do et al. 2018). A CPx-type ATPase transporter may also be involved in the transport of transition metals (Rao et al. 2018). Generally, this transporter is involved in efflux of Cd^{2+} and Zn^{2+} ions in xylem from neighboring tissues. Another transporter, LCT1 (low-affinity cation transporter) plays a potential role in the uptake of both Cd^{2+} and Ca^{2+} in plants.

1.4 Signal Transduction Pathways in Plants in Response to Heavy Metal Stress

Plants confront a wide range of heavy metal stress by boosting the various defense responses. This response displayed by the plants is the result of several signal transduction pathways functioning in the cell in order to transmit the extracellular stimuli into intracellular response. The crucial signaling components involved are ROS signaling, mitogen-activated protein kinase (MAPK) signaling and calcium signaling that are discussed in the following sections.

1.4.1 ROS Signaling Pathway

ROS generation is one of the most common responses of plants during exposure to metal stress. Such toxic free radicals then initiate a chain reaction and react with a large number of molecules that leads to irreversible damage within the tissues (Jaspers and Kangasjärvi 2010; Nafees et al. 2019). For many years, it was believed that ROS are harmful and toxic, but now it is widely accepted that they are important for plant defense. ROS are ideally suited to act as signaling molecules, because of their small size, so that they can diffuse over short distances (Suzuki et al. 2012). Because of the versatile nature of ROS, plants keep them under tight regulation, controlled by several non-enzymatic antioxidants, such as ASC, GSH, as well as enzymatic antioxidants like SOD, CAT, glutathione reductase (GR), ascorbate peroxidase, etc. (Mittler et al. 2011; Rodriguez-Serrano et al. 2006). Moreover, among the different ROS, H_2O_2 can easily diffuse through plasma membrane and directly function in cell-to-cell signaling. Several reports indicate that this molecule is involved in plant acclimatization which leads to activation of downstream signaling cascades (Choudhury et al. 2013). ROS can also modulate gene expression by controlling cellular redox state, although many aspects of ROS signaling still remain obscure.

1.4.2 MAPK Cascade

MAPK cascade plays a crucial role in signal transduction pathway, which is mainly composed of three modules, namely MAPK, MAPKK and MAPKKK. This cascade displays coordination and interaction with various other pathways in plants (Pitzschke et al. 2009). At the end of the cascade, MAPK phosphorylates another substrate in cells, which may be a transcription factor or other proteins. It has been reported that a Ca^{2+}-dependent protein kinase (CDPK) and phosphatidylinositol are necessary for activation of MAPK under Cd and Cu stress (Smékalová et al. 2014). The phosphatases are important regulators which deactivate the modules of MAPK cascade by dephosphorylation. Despite the presence of protein phosphatases, there are various other factors which can stimulate MAPK like phosphatidic acid, phosphatidylglycerol, phosphatidylcholine, phosphatidylethanolamine, and lipids which are generated by plasma membrane-bound phospholipase enzymes, in which phosphatidic acid is considered to be a crucial secondary messenger (Sinha et al. 2011; Chinnusamy et al. 2004). MAPK activation is transient during exposure to metal stress, ranging from 5 min to 1 h, although the detailed knowledge about MAPK cascade in plants is not known. Rao et al. (2011) studied the MAPK cascade in rice and found that OsMKK4/OSMPK3 could be the possible MAPK, activated under metal stress condition. Rentel et al. (2004) reported that H_2O_2 may induce MPK3 and MPK6 during metal toxicity in *Arabidopsis*. It is believed that several groups of transcription factors like MYC, MYB, and bZIP are recruited to downstream reaction and induce the MAPK signaling cascade under heavy metal stress (Zhang and Klessig 2001).

1.4.3 The Calcium-Calmodulin Pathway

The functions of calcium channels are generally altered in response to heavy metal stress which results in increment in calcium flux into the cell. Calcium acts as second messenger that stimulates calmodulin which ultimately regulates the uptake, transport and metabolism of heavy metals (Yang and Poovaiah 2003; Khan et al. 2014). The Ca^{2+}-calmodulin pathway was studied in detail under Pb, Ni, and Cd stress, which strongly indicates calmodulin being an important cellular target for various metals (Virdi et al. 2015; Das and Pandey 2010). Ouyang and Vogel (1998) observed that all four calcium-binding sites of calmodulin were occupied by Pb and enhance its activity by about 90%. In similar way, other metals like Zn or Mg either bind with low affinity to specific calcium binding sites on calmodulin or interact via auxiliary sites to modulate calmodulin functions. It is believed that most of the metals act as calcium analogs which induce calmodulin functions (Sánchez-Barrena et al. 2013).

1.5 Conclusion and Future Perspective

This chapter provides an overview on the mechanism by which different heavy metals cause damage to the plants and the different physiological and molecular strategies through which plants defend themselves against metal-induced toxicity.

Several reports are available that demonstrate plant strategies to protect themselves from heavy metal-induced toxicity. However, there seems to be a lack of substantial studies on molecular aspects of defensive molecules in different plants. It is important to understand the mechanism by which plants perceive environmental signals and further transmit these signals to cellular machinery to activate different adaptive responses. This understanding may help us in generating and selecting metal-tolerant plants which could be grown in highly contaminated agricultural fields. Our discussion also encompasses efforts to make plants more resistant to heavy metal uptake, which would minimize the entry of heavy metals into the food chain, thereby developing metal-resilient plants. Further research should be focused on molecular identification of signaling pathways since there is dearth of adequate knowledge regarding the precise physiological and biochemical response for each metal.

Acknowledgments

Financial assistance from Science and Engineering Research Board, Government of India through the grant [EMR/2016/004799] and Department of Higher Education, Science and Technology and Biotechnology, Government of West Bengal, through the grant [264(Sanc.)/ST/P/S&T/1G-80/2017] to Dr. Aryadeep Roychoudhury is gratefully acknowledged.

REFERENCES

Alaoui-Sossé, B., Genet, P., Vinit-Dunand, F., Toussaint, M.-L., Epron, D., Badot, P.-M. 2004. Effect of copper on growth in cucumber plants (*Cucumis sativus*) and its relationships with carbohydrate accumulation and changes in ion contents. *Plant Sci*, 166, 1213–1218.

Alhamed, M. F. A., Shebany, Y. M. 2012. Endophytic *Chaetomium globosum* enhances maize seedling copper stress tolerance. *Plant Biol*, doi:10.1111/j.1438-8677.2012.00608.x.

Ali, B., Hasan, S. A., Hayat, S., Hayat, Q., Yadav, S., Fariduddin, Q., Ahmad, A. 2008. A role for brassinosteroids in the amelioration of aluminium stress through antioxidant system in mung bean. *Environ Exp Bot*, 62, 153–159.

Assareh, M. H., Shariat, A., Ghamari-Zare, A. 2008. Seedling response of three *Eucalyptus* species to copper and zinc toxic concentrations. *Casp J Environ Sci*, 6, 97–103.

Becquer, T., Quantin, C., Sicot, M., Boudot, J. P. 2003. Chromium availability in ultramafic soils from New Caledonia. *Sci Total Environ*, 301, 251–261.

Bennet, R. J., Breen, C. M., Fey, M. V. 1985. Aluminum induced changes in the morphology of the quiescent center, proximal meristem and growth region of the root of *Zea mays*. *S Afr J Bot*, 51, 355–362.

Bhalerao, S. A., Prabhu, D. V. 2013. Aluminium toxicity in plants – a review. *J Applicable Chem*, 2, 447–474.

Brunetti, P., Zanella, L., Proia A., Paolis, A. D., Falasca, G., Altamura, M. M., di Toppi L. S., Costantino, P., Cardarelli, M. 2011. Cadmium tolerance and phytochelatin content of *Arabidopsis* seedlings over-expressing the phytochelatin synthase gene *AtPCS1*. *J Exp Bot*, 62, 5509–5519.

Buchanan, B., Grusen, W., Jones, R. 2000. Biochemistry and molecular biology of plants. American Society of Plant Physiologists Maryland, 1367.

Burke, D. G., Watkins, K., Scott, B. J. 1990. Manganese toxicity effects on visible symptoms, yield, manganese levels, and organic acid levels in tolerant and sensitive wheat cultivars. *Crop Sci*, 30, 275–280.

Cataldo, D. A., Garland, T. R., Wildung, R. E. 1978. Nickel in plants: I. Uptake kinetics using intact soybean seedlings. *Plant Physiol*, 62, 563–565.

Cervantes, C., Campos-García, J., Devars, S., Gutiérrez-Corona, F., Loza-Tavera, H., Torres-Guzmán J. C., Moreno-Sánchez, R. 2001. Interactions of chromium with microorganisms and plants. *FEMS Microbiol Rev*, 25, 335–347.

Chen, L., Guo, Y., Yang, L., Wang, Q. 2008. Synergistic defensive mechanism of phytochelatins and antioxidative enzymes in *Brassica chinensis* L. against Cd stress. *Chinese Sci Bull*, 53, 1503–1511.

Chen, W., Cui, P., Sun, H. 2009. Comparative effects of salt and alkali stresses on organic acid accumulation and ionic balance of seabuckthorn (*Hippophae rhamnoides* L.). *Ind Crops Prod*, 30, 351–358.

Chen, Y., Han, Y.-H., Cao, Y., Zhu, Y.-G., Rathinasabapathi, B., Ma, L. Q. 2017. Arsenic transport in rice and biological solutions to reduce arsenic risk from rice. *Front Plant Sci*, 8, 268.

Chinnusamy, V., Schumaker, K., Zhu, J.-K. 2004. Molecular genetic perspectives on cross-talk and specificity in abiotic stress signalling in plants. *J Exp Bot*, 55, 225–236.

Choudhury, S., Panda, P., Sahoo, L., Panda, S. K. 2013. Reactive oxygen species signaling in plants under abiotic stress. *Plant Signal Behav*, 8, e23681.

Cobbett, C. S. 2000. Phytochelatins and their roles in heavy metal detoxification. *Plant Physiol*, 123, 825–832.

Cobbett, C. S. 2008. Heavy metal detoxification in plants: phytochelatin biosynthesis and function. *IUBMB Life*, 51, 183–188.

Cobbett, C., Goldsbrough, P. 2002. Phytochelatins and Metallothineins: roles in heavy metal detoxification and homeostasis. *Annu Rev Plant Biol*, 53, 159–182.

Colangelo, E. P., Guerinot, M. L. 2006. Put the metal to the petal: metal uptake and transport throughout plants. *Curr Opin Plant Biol*, 9, 322–330.

Curie, C., Alonso, J. M., Le Jean, M., Ecker, J. R., Briat, J. F. 2000. Involvement of NRAMP1 from *Arabidopsis thaliana* in iron transport. *Biochem J*, 347, 749–755.

DalCorso, G. 2012. Heavy metal toxicity in plants. In: Furini, A. (Ed.), Plants and heavy metals, Springer: Dordrecht, 1–25.

Das, K., Roychoudhury, A. 2014. Reactive oxygen species (ROS) and response of antioxidants as ROS-scavengers during environmental stress in plants. *Front Plant Sci*, 2, 53.

Das, R., Pandey, G. 2010. Expressional analysis and role of calcium regulated kinases in abiotic stress signaling. *Curr Genomics*, 11, 2–13.

Delhaize, E., Ryan, P. R. 1995. Aluminum toxicity and tolerance in plants. *Plant Physiol*, 107, 315–321.

Demirevska-Kepova, K., Simova-Stoilova, L., Stoyanova, Z., Hölzer, R., Feller, U. 2004. Biochemical changes in barley plants after excessive supply of copper and manganese. *Environ Exp Bot*, 52, 253–266.

Dey, S. K., Jena, P. P., Kundu, S., 2009. Antioxidative efficiency of *Triticum aestivum* L. exposed to chromium stress. *J Environ Biol*, 30, 539–544.

Do, T. H. T., Martinoia, E., Lee, Y. 2018. Functions of ABC transporters in plant growth and development. *Curr Opin Plant Biol*, 41, 32–38.

Emamverdian, A., Ding, Y., Mokhberdoran, F., Xie, Y. 2015. Heavy metal stress and some mechanisms of plant defense response. *Sci World J*, 2015, 1–18.

Ermler, U., Grabarse, W., Shima, S., Goubeaud, M., Thauer, R. K. 1998. Active sites of transition-metal enzymes with a focus on nickel. *Curr Opin Struct Biol*, 8, 749–758.

Gajewska, E., Skłodowska, M. 2005. Antioxidative responses and proline level in leaves and roots of pea plants subjected to nickel stress. *Acta Physiol Plant*, 27, 329–340.

Gallego, S. M., Pena, L. B., Barcia, R. A., Azpilicueta, C. E., Iannone, M. F., Rosales, E. P., Zawoznik, M. S., Groppa, M. D., Benavides, M. P. 2012. Unravelling cadmium toxicity and tolerance in plants: insight into regulatory mechanisms. *Ecotoxicol Environ Saf*, 83, 33–46.

Ghori, N.-H., Ghori, T., Hayat, M. Q., Imadi, S. R., Gul, A., Altay, V., Ozturk, M. 2019. Heavy metal stress and responses in plants. *Int J Environ Sci Technol*, doi:10.1007/s13762-019-02215-8.

Ghosh, M., Singh, S. P. 2005. A review on phytoremediation of heavy metals and utilization of its byproducts. *As J Energy Env*, 6, 214–231.

Gill, M. 2014. Heavy metal stress in plants: a review. *Int J Adv Res*, 2, 1043–1055.

Grennan, A. K. 2011. Metallothioneins, a diverse protein family. *Plant Physiol*, 155, 1750–1751.

Guerinot, M. L. 2000. The ZIP family of metal transporters. *Biochim Biophys Acta*, 1465, 190–198.

Hall, J. L., Williams, L. E. 2003. Transition metal transporters in plants. *J Exp Bot*, 54, 2601–2613.

Halliwell, B. 2006. Reactive species and antioxidants. Redox biology is a fundamental theme of aerobic life. *Plant Physiol*, 141, 312–322.

Hasan, S. A., Fariduddin, Q., Ali, B., Hayat, S., Ahmad, A. 2009. Cadmium: toxicity and tolerance in plants. *J Environ Biol*, 30, 165–174.

Hassinen, V. H., Tervahauta, A. I., Schat, H., Kärenlampi, S. O. 2011. Plant metallothioneins – metal chelators with ROS scavenging activity? *Plant Biol*, 13, 225–232.

Hesami, R., Salimi, A., Ghaderian, S. M. 2018. Lead, zinc, and cadmium uptake, accumulation, and phytoremediation by plants growing around Tang-e Douzan lead–zinc mine. *Iran Environ Sci Pollut Res*, https://doi.org/10.1007/s11356-017-1156-y.

Hussain, M. B., Ali, S., Azam, A., Hina, S., Farooq, M. A., Ali, B., Bharwana, S. A., Gill, M. B. 2013. Morphological, physiological and biochemical responses of plants to nickel stress: a review. *Afr J Agric Res*, 8, 1596–1602.

Jalmi, S. K., Bhagat, P. K., Verma, D., Noryang, S., Tayyeba, S., Singh, K., Sharma, D., Sinha, A. K. 2018. Traversing the links between heavy metal stress and plant signaling. *Front Plant Sci*, 9, 12.

Jaspers P., Kangasjärvi, J. 2010. Reactive oxygen species in abiotic stress signaling. *Physiol Plant*, 138, 405–413.

Kabir, A. H., Dietrich, P. 2016. Biochemical and molecular changes in rice seedlings (*Oryza sativa* L.) to cope with chromium stress. *Plant Biol*, 18, 710–719.

Kalaivanan, D., Ganeshamurthy, A. N. 2016. Mechanisms of heavy metal toxicity in plants. In: Rao, N., Shivashankara, K., Laxman, R. (Eds.), Abiotic stress physiology of horticultural crops, Springer: New Delhi, 85–102.

Khan, M. N., Mohammad, F., Mobin, M., Saqib, M. A. 2014. Tolerance of plants to abiotic stress: a role of nitric oxide and calcium. In: Khan, M., Mobin, M., Mohammad, F., Corpas, F. (Eds.), Nitric oxide in plants: metabolism and role in stress physiology, Springer: Cham, 225–242.

Kidd P. S., Proctor, J. 2000. Effects of aluminium on the growth and mineral composition of *Betula pendula* Roth. *J Exp Bot*, 51, 1057–1066.

Kramer, U., Cotter-Howells, J. D., Charnock, J. M., Baker, A. J. M., Smith, J. A. C. 1996. Free histidine as a metal chelator in plants that accumulate nickel. *Nature*, 379, 635–638.

Kumar, A., Maiti, S. K. 2013. Availability of chromium, nickel and other associated heavy metals of ultramafic and serpentine soil/rock and in plants. *Int J Emerging Technol Adv Eng*, 3, 256–268.

Kumar, A., Prasad, M. N. V. 2018. Plant-lead interactions: transport, toxicity, tolerance, and detoxification mechanisms. *Ecotoxicol Environ Saf*, 166, 401–418.

Kumar, S., Dubey, R. S., Tripathi, R. D., Chakrabarty, D., Trivedi, P. K. 2015. Omics and biotechnology of arsenic stress and detoxification in plants: current updates and prospective. *Environ Int*, 74, 221–230.

Küpper, H., Kroneck, P. M. H. 2007. Nickel in the environment and its role in the metabolism of plants and cyanobacteria. In: Sigel, A., Sigel, H., Sigel, R. K. O. (Eds.), *Metal ions in life sciences*, John Wiley & Sons Ltd: New York, 2, 31–62.

Kuppusamy, S., Palanisami, T., Megharaj, M., Venkateswarlu, K., Naidu, R. 2016. In-situ remediation approaches for the management of contaminated sites: a comprehensive overview. In: de Voogt, P. (Ed.), *Reviews of environmental contamination and toxicology*, Springer International Publishing: Cham, 236, 1–115.

Landjeva, S., Merakchijska-Nikolova, M., Ganeva, G. 2003. Copper toxicity tolerance in *Aegilops* and *Haynaldia* seedlings. *Biol Plant*, 46, 479–480.

Li, N., Wang, J., Song, W.-Y. 2016. Arsenic uptake and translocation in plants. *Plant Cell Physiol*, 57, 4–13.

Lombardi, L., Sebastiani, L. 2005. Copper toxicity in *Prunus cerasifera*: growth and antioxidant enzymes responses of in vitro grown plants. *Plant Sci*, 168, 797–802.

Ma, Y., Egodawatta, P., McGree, J., Liu, A., Goonetilleke, A. 2016. Human health risk assessment of heavy metals in urban stormwater. *Sci Total Environ*, 557–558, 764–772.

Macovei, A., Ventura, L., Donà, M., Faé, M., Balestrazzi, A., Carbonera, D. 2010. Effects of heavy metal treatments on metallothionein expression profiles in white poplar (*Populus alba* L.) cell suspension cultures. *Analele Univ din Oradea-Fasc Biol*, 18, 274–279.

Mahdavian, K., Ghaderian, S. M., Torkzadeh-Mahani, M. 2017. Accumulation and phytoremediation of Pb, Zn, and Ag by plants growing on Koshk lead–zinc mining area. *Iran J Soils Sed*, 17, 1310–1320.

Mediouni, C., Benzarti, O., Tray, B., Ghorbel, M. H., Jemal, F. 2006. Cadmium and copper toxicity for tomato seedlings. *Agron Sustain Dev*, 26, 227–232.

Millaleo, R., Reyes-Diaz, M., Ivanov, A. G., Mora, M. L., Alberdi, M. 2010. Manganese as essential and toxic element for plants: transport, accumulation and resistance mechanisms. *J Soil Sci Plant Nutr*, 10, 470–481.

Milner, M. J., Seamon, J., Craft, E., Kochian, L. V. 2013. Transport properties of members of the ZIP family in plants and their role in Zn and Mn homeostasis. *J Exp Bot*, 64, 369–381.

Mishra, S., Dubey, R. S. 2006. Heavy metal uptake and detoxification mechanisms in plants. *Int J Agric Res*, 1, 122–141.

Mittler, R., Vanderauwera, S., Suzuki, N., Miller, G., Tognetti, V. B., Vandepoele, K., Gollery, M., Shulaev, V., Breusegem, F. V. 2011. ROS signaling: the new wave? *Trends Plant Sci*, 16, 300–309.

Mourato, M. P., Moreira, I. N., Leitão, I., Pinto, F. R., Sales, J. R., Martins, L. L. 2015. Effect of heavy metals in plants of the genus Brassica. *Int J Mol Sci*, 16, 17975–17998.

Nafees, M., Fahad, S., Shah, A. N., Bukhari, M. A., Maryam, Ahmed, I., Ahmad, S., Hussain, S. 2019. Reactive oxygen species signaling in plants. In: Hasanuzzaman, M., Hakeem, K. R., Nahar, K., Alharby, H. F. (Eds.), Plant abiotic stress tolerance: agronomic, molecular and biotechnological approaches, Springer: Cham, 259–272.

Nazar, R., Iqbal, N., Masood, A., Khan, M. I. R., Syeed, S., Khan, N. A. 2012. Cadmium toxicity in plants and role of mineral nutrients in its alleviation. *Am J Plant Sci*, 3, 1476–1489.

Nearing, M. M., Koch, I., Reimer, K. J. 2014. Arsenic speciation in edible mushrooms. *Environ Sci Technol*, 48, 14203–14210.

Nieminen, T. M., Ukonmaanaho, L., Rausch, N., Shotyk, W. 2007. Biogeochemistry of nickel and its release into the environment. *Met Ions Life Sci*, 2, 1–30.

Obiora, S. C., Chukwu, A., Toteu, S. F., Davies, T. C. 2016. Assessment of heavy metal contamination in soils around lead (Pb)-zinc (Zn) mining areas in Enyigba, southeastern Nigeria. *J Geol Soc*, 87, 453–462.

Ouyang, H., Vogel, H. J. 1998. Metal ion binding to calmodulin: NMR and fluorescence studies. *Biometals*, 11, 213–222.

Panda, S. K., Baluška, F., Matsumoto, H. 2009. Aluminum stress signaling in plants. *Plant Signal Behav*, 4, 592–597.

Panda, S. K., Choudhury, S. 2005. Chromium stress in plants. *Brazilian J Plant Physiol*, 17, 95–102.

Panda, S. K., Singha, L. B., Khan, M. H. 2003. Does aluminum phytotoxicity induce oxidative stress in green gram (*Vigna radiata*)? *Bulg J Plant Physiol*, 29, 77–86.

Paschke, M. W., Valdecantos, A., Redente, E. F. 2005. Manganese toxicity thresholds for restoration grass species. *Environ Pollut*, 135, 313–322.

Peralta, J. R., Gardea-Torresdey J. L., Tiemann, K. J., Gomez, E., Arteaga, S., Rascon, E., Parsons, J. G. 2001. Uptake and effects of five heavy metals on seed germination and plant growth in alfalfa (*Medicago sativa* L.). *Bull Environ Contam Toxicol*, 66, 727–734.

Piechalak, A., Tomaszewska, B., Baralkiewicz, D., Malecka, A. 2002. Accumulation and detoxification of lead ions in legumes. *Phytochemistry*, 60, 153–162.

Pintro, J., Barloy, J., Fallavier, P. 1997. Effects of low aluminum activity in nutrient solutions on the organic acid concentrations in maize plants. *J Plant Nutr*, 20, 601–611.

Pitzschke, A., Schikora, A., Hirt, H. 2009. MAPK cascade signalling networks in plant defence. *Curr Opin Plant Biol*, 12, 421–426.

Pottier, M., Oomen, R., Picco, C., Giraudat, J., Scholz-Starke, J., Richaud, P., Carpaneto, A., Thomine, S. 2015. Identification of mutations allowing natural resistance associated macrophage proteins (NRAMP) to discriminate against cadmium. *Plant J*, 83, 625–637.

Prasad, M. N. V. 1995. Cadmium toxicity and tolerance in vascular plants. *Environ Exp Bot*, 35, 525–545.

Ragsdale, S. W. 1998. Nickel biochemistry. *Curr Opin Chem Biol*, 2, 208–215.

Rao, K. P., Vani, G., Kumar, K., Wankhede, D. P., Misra, M., Gupta, M., Sinha, A. K. 2011. Arsenic stress activates MAP kinase in rice roots and leaves. *Arch Biochem Biophys*, 506, 73–82.

Rao, S., Zhou, Z., Miao, P., Bi, G., Hu, M., Wu, Y., Feng, F., Zhang, X., Zhou, J.-M. 2018. Roles of receptor-like cytoplasmic kinase VII members in pattern-triggered immune signaling. *Plant Physiol*, 177, 1679–1690.

Rascio, N., Navari-Izzo, F. 2011. Heavy metal hyperaccumulating plants: how and why do they do it? And what makes them so interesting? *Plant Sci*, 180, 169–181.

Rauser, W. E. 1995. Phytochelatins and related peptides. *Plant Physiol*, 109, 1141–1149.

Rentel, M. C., Lecourieux, D., Ouaked, F., Usher, S. L., Petersen, L., Okamoto, H., Knight, H., Peck, S. C., Grierson, C. S., Hirt, H. 2004. OXI1 kinase is necessary for oxidative burst-mediated signalling in *Arabidopsis*. *Nature*, 427, 858–861.

Rodriguez-Serrano, M., Romero-Puertas, M. C., Zabalza, A., Corpas, F. J., Gomez, M., del Río, L. A., Sandalio, L. M. 2006. Cadmium effect on oxidative metabolism of pea (*Pisum sativum* L.) roots. Imaging of reactive oxygen species and nitric oxide accumulation in vivo. *Plant Cell Environ*, 29, 1532–1544.

Salt, D. E., Prince, R. C., Baker, A. J. M., Raskin, I., Pickering, I. J. 1999. Zinc ligands in the metal hyperaccumulator *Thlaspi caerulescens* as determined using X-ray absorption spectroscopy. *Environ Sci Technol*, 33, 713–717.

Samanta, S., Banerjee, A., Roychoudhury, A. 2020a. Melatonin application differentially modulates the enzymes associated with antioxidative machinery and ascorbate-glutathione cycle during arsenate exposure in indica rice varieties. *Plant Biol*, https://doi.org/10.1111/plb.13181.

Samanta, S., Singh, A., Roychoudhury, A. 2020b. Involvement of sulfur in the regulation of abiotic stress tolerance in plants. In: Roychoudhury, A., Tripathi, D. K. (Eds.), Protective chemical agents in the amelioration of plant abiotic stress: biochemical and molecular perspectives, Wiley-Blackwell: Hoboken, NJ, 437–466.

Samantaray, S., Rout, G. R., Das, P. 1998. Role of chromium on plant growth and metabolism. *Acta Physiol Plant*, 20, 201–212.

Sánchez-Barrena, M., Martínez-Ripoll, M., Albert, A. 2013. Structural biology of a major signaling network that regulates plant abiotic stress: the CBL-CIPK mediated pathway. *Int J Mol Sci*, 14, 5734–5749.

Santos, E. F., Kondo, S., José, M., Paixão, A. P., Júnior, E. F., Lavres, J., Campos, M., Reis, A. R. d. 2017. Physiological highlights of manganese toxicity symptoms in soybean plants: Mn toxicity responses. *Plant Physiol Biochem*, 113, 6–19.

Schützendübel, A., Polle, A. 2002. Plant responses to abiotic stresses: heavy metal-induced oxidative stress and protection by mycorrhization. *J Exp Bot*, 53, 1351–1365.

Seregin, I. V., Kozhevnikova, A. D. 2006. Physiological role of nickel and its toxic effects on higher plants. *Russ J Plant Physiol*, 53, 257–277.

Shahid, M., Khalid, S., Abbas, G., Shahid, N., Nadeem, M., Sabir, M., Aslam, M., Dumat, C. 2015. Heavy metal stress and crop productivity. In: Hakeem, K. (Ed.), Crop production and global environmental issues. Springer, Cham, 1–25.

Shanker, A. K., Cervantes, C., Loza-Tavera, H., Avudainayagam, S. 2005. Chromium toxicity in plants. *Environ Int*, 31, 739–753.

Sharma, I. 2012. Arsenic induced oxidative stress in plants. *Biologia*, 67, 447–453.

Sharma, R. K., Devi, S., dan-Dhyani, P. P. 2010. Comparative assessment of the toxic effects of copper and cypermethrin using seeds of *Spinacia Oleracea* L. plants. *Trop Ecol*, 51, 375–387.

Shrivastava, A., Ghosh, D., Dash, A., Bose, S. 2015. Arsenic contamination in soil and sediment in India: sources, effects, and remediation. *Curr Pollut Rep*, 1, 35–46.

Singh, A. P., Dixit, G., Kumar, A., Mishra, S., Singh, P. K., Dwivedi, S., Trivedi, P. K., Chakrabarty, D., Mallick, S., Pandey, V., Dhankher, O. P., Tripathi, R. D. 2016. Nitric oxide alleviated arsenic toxicity by modulation of antioxidants and thiol metabolism in rice (*Oryza sativa* L.). *Front Plant Sci*, 6, 1272.

Sinha, A. K., Jaggi, M., Raghuram, B., Tuteja, N. 2011. Mitogen-activated protein kinase signaling in plants under abiotic stress. *Plant Signal Behav*, 6, 196–203.

Sivaguru, M., Horst, W. J. 1998. The distal part of the transition zone is the most aluminium sensitive apical root zone of maize. *Plant Physiol*, 116, 155–163.

Smékalová, V., Doskočilová, A., Komis, G., Šamaj, J. 2014. Crosstalk between secondary messengers, hormones and MAPK modules during abiotic stress signalling in plants. *Biotechnol Adv*, 32, 2–11.

Solanki, R., Dhankhar, R. 2011. Biochemical changes and adaptive strategies of plants under heavy metal stress. *Biologia*, 66, 195–204.

Srivastava, S., Upadhyay, M. K., Tripathi, R. D., Dhankher, O. P. 2016. Arsenic toxicity, metabolism and toxicity in plants. *Int J Plant Environ*, 2, 17–28.

Stolt, J. P., Sneller, F. E. C., Bryngelsson, T., Lundborg, T., Schat, H. 2003. Phytochelatin and cadmium accumulation in wheat. *Environ Exp Bot*, 49, 21–28.

Suzuki, N., Koussevitzky, S., Mittler, R., Miller, G. 2012. ROS and redox signalling in the response of plants to abiotic stress. *Plant Cell Environ*, 35, 259–270.

Theodoulou, F. L. 2000. Plant ABC transporters. *Biochim Biophys Acta*, 1465, 79–103.

Thomine, S., Wang, R., Ward, J. M., Crawford, N. M., Schroeder, J. I. 2000. Cadmium and iron transport by members of a plant metal transporter family in *Arabidopsis* with homology to *Nramp* genes. *Proc Natl Acad Sci*, 97, 4991–4996.

Tiwari, M., Sharma, D., Dwivedi, S., Singh, M., Tripathi, R. D., Trivedi, P. K. 2013. Expression in *Arabidopsis* and cellular localization reveal involvement of rice NRAMP, OsNRAMP1, in arsenic transport and tolerance. *Plant Cell Environ*, 37, 140–152.

Vardar, F., Ünal, M. 2007. Aluminum toxicity and resistance in higher plants. *Adv Mol Biol*, 1, 1–12.

Virdi, A. S., Singh, S., Singh, P. 2015. Abiotic stress responses in plants: roles of calmodulin-regulated proteins. *Front Plant Sci*, 6, 809.

Wahid, A., Khaliq, S. 2015. Architectural and biochemical changes in embryonic tissues of maize under cadmium toxicity. *Plant Biol*, 17, 1005–1012.

Wang, H., Wong, M. H., Lan, C., Qin, Y., Shu, W., Qiu, R., Ye, Z. 2011. Organic acids in two arsenic hyperaccumulators and a non-hyperaccumulator of *Pteris* exposed to elevated arsenic concentrations. *Intern J Environ Anal Chem*, 91, 241–254.

Williams, L. E., Pittman, J. K., Hall, J. L. 2000. Emerging mechanisms for heavy metal transport in plants. *Biochim Biophys Acta Biomembr*, 1465, 104–126.

Wong, H. L., Sakamoto, T., Kawasaki, T., Umemura, K., Shimamoto, K. 2004. Down-regulation of metallothionein, a reactive oxygen scavenger, by the small GTPase OsRacl in rice. *Plant Physiol*, 135, 1447–1456.

Yadav, S. K. 2010. Heavy metals toxicity in plants: an overview on the role of glutathione and phytochelatins in heavy metal stress tolerance of plants. *S Afr J Bot*, 76, 167–179.

Yang, T., Poovaiah, B. W. 2003. Calcium/calmodulin-mediated signal network in plants. *Trend Plant Sci*, 8, 505–512.

Yang, X., Feng, Y., He, Z., Stoffella, P. J. 2005. Molecular mechanisms of heavy metal hyperaccumulation and phytoremediation. *J Trace Elem Med Biol*, 18, 339–353.

Yang, X. E., Yang, M. J. 2001. Some mechanisms of zinc and cadmium detoxification in a zinc and cadmium hyperaccumulating plant species (*Thalaspi*). In: Horst, W. J. et al. (Eds.), *Plant nutrition-food security and sustainability of agro-ecosystems through basic and applied research*, Kluwer Academic Publishers: Dordrecht, The Netherlands, 444–445.

Zagorchev, L., Seal, C., Kranner, I., Odjakova, M. 2013. A central role for thiols in plant tolerance to abiotic stress. *Int J Mol Sci*, 14, 7405–7432.

Zeid, I. M. 2001. Responses of *Phaseolus vulgaris* to chromium and cobalt treatments. *Biol Plant*, 44, 111–115.

Zhang, S., Klessig, D. F. 2001. MAPK cascades in plant defense signaling. *Trends Plant Sci*, 6, 520–527.

Zhang, W., Ryan, P., Tyerman, S. 2001. Malate-permeable channels and cation channels activated by aluminum in the apical cells of wheat root roots. *Plant Physiol*, 125, 1459–1472.

Zou, J., Wang, M., Jiang, W., Liu, D. 2006. Chromium accumulation and its effects on other mineral elements in *Amaranthus viridis* L. *Acta Biol Crac Ser Bot*, 48, 7–12.

Zulfiqar, U., Farooq, M., Hussain, S., Maqsood, M., Hussain, M., Ishfaq, M., Ahmad, M., Anjum, M. Z. 2019. Lead toxicity in plants: impacts and remediation. *J Environ Manage*, 250, 109557.

2

Heavy Metal Toxicity in Plants: Impact, Signaling, and Remediation Strategies

Preeti Mishra, Afifa Qidwai, Nivedita Singh, and Anupam Dikshit
Biological Product Laboratory, Department of Botany, University of Allahabad, Prayagraj, UP, India

S.C. Agrawal
Phycology Laboratory, Department of Botany, University of Allahabad, Prayagraj, UP, India

CONTENTS

2.1 Introduction

Heavy metals belong to the prestigious category of utterly economically useful resources to mankind both pre and post the industrial revolution. However, population explosion has led to an exponential rise in the process of industrialization of our flora, subsequently leading to unprecedented challenging issues of expeditious depletion of our limited resources and annihilating rates of triphibian pollution (Igiri *et al.* 2018; Siddiquee *et al.* 2015). Heavy metal-induced environmental pollution has become lethal for the fauna thriving in our biosphere (Deepa and Suresha 2014; Hrynkiewicz and Baum 2014; Okolo *et al.* 2016; Siddiquee *et al.* 2015; Su 2014). Additionally, not only is there an evident reduction in the yield of crops but also the quality of food seems to be approaching a nadir. This can majorly be attributed to excessive usage of fertilizers and pesticides which have caused heavy metal contamination of soils (Su 2014). Moreover, contrary to their organic contemporaries, heavy metal contaminants are not degradable and remain in the environment indefinitely, leading to their high concentration and bearing detrimental effects to the plant metabolism (Ferraz *et al.* 2012). Heavy metal (HM) contamination can be attributed to major sources including

DOI: 10.1201/9781003155089-2

industrial solid waste, agricultural, natural, inland affluent, atmospheric sources, etc. Other man-made activities, such as mining, electroplating, metallurgical smelting actions, and the extensive usage of agricultural pesticides and fertilizers have also resulted in widespread contamination across the globe (Zhang *et al.* 2011). Pesticides can be either inorganic or organic, with the antecedent being the vast majority, and the minority comprising of pure minerals or having traces of other heavy metals like zinc (Zn), copper (Cu), mercury (Hg), arsenic (As), etc. (Arao *et al.* 2010). It is interesting to note that microorganisms carry out metabolic processes which require chemical contaminants as the sources of energy. The solution to this budding issue seems to exist in nature, through a process termed as parametrization. It is defined as a spontaneous process where microbiological process is employed to atrophy the hazardous contaminants into their lesser or non-toxic counterparts, subsequently ameliorating and removing contaminants from our surrounding media, according the EPA 2010 (Environmental Protection Agency (EPA) 2010). But a roadblock to this approach seems to be the inhibition of the microbes owing to the presence of inorganic nutrients in the soil in excess (Ahirwar *et al.* 2016). There is an undeniable demand for treatment technologies that can remove heavy metal ions from the amphibian media. Various microbes have been vouched as economical and efficient alternatives for the eradication of heavy metals from the amphibian media (Ahirwar *et al.* 2016). Such microbes have been biochemically rumbled and their ability to combat heavy metals like Zn and Cu is presently unknown. However, with recent advancements in the field of development and modeling of innovative bioremediation techniques that are low, eco-friendly, and effective in environment restoration, there seems to be an absence of a single wheel-like-invention in bioremediation techniques, primarily because of the nature and/or type of pollutant. However autochthonous (indigenous) microorganisms show some promising signs in resolving the issues pertaining to biodegradation and remediation of pollutants as long as the environmental conditions support their growth and metabolism (Azubuike *et al.* 2016). Moreover, pollutant removal is dependent on the nature of the pollutant, that consist of agrochemicals, chlorinated compounds, dyes, greenhouse gases, heavy metals, hydrocarbons, nuclear waste, plastics, and sewage. Bioremediation techniques are categorized into in-situ or ex-situ depending on the application site. The selection of a bioremediation technique depends these parameters – nature of pollutant, pollution depth and degree, environment type, cost and location, and environmental policies (Smith *et al.* 2015). The performance criteria consisting of the concentration of oxygen and nutrients, temperature, pH, and abiotic factors, indicate the success of a bioremediation are considered majorly prior to its implementation. Microbial remediation is a breakthrough and seemingly promising techniques for resolving the budding issues of heavy metal pollution in the environment. In the ongoing pages, we delve further into the role of microbes as tools for bioremediation of heavy metals.

2.2 Heavy Metals

In biology, heavy metals (HM) refer to metals with potential human and biosphere toxicity (Tchounwou *et al.* 2012). Geological and anthropogenic activities increase heavy metal concentration in the soil which is hazardous to plants and animals (Chibuike and Obiora 2014). The defining attributes of heavy metals are density, atomic number, and chemical properties (Igiri *et al.* 2018; Siddiquee *et al.* 2015). Heavy Metal pollution that is deadly to living organisms relies on absorbed dose and heavy metal bioavailability (Deepa and Suresha 2014; Hrynkiewicz and Baum 2014; Okolo *et al.* 2016; Rasmussen *et al.* 2000; Siddiquee *et al.* 2015; Su 2014). Mechanisms, such as the breaking of fatal enzymatic functions, reactions as redox catalysts while producing reactive oxygen species (ROS), decimating the regulation of ion, and hampering protein and DNA formation together make up heavy metal toxicity (Gauthier *et al.* 2014). Moreover, heavy metals alter the physiological and biochemical properties of microbes (Ahirwar *et al.* 2016). Heavy metals travel as gases and particulate over long distances leading to their brisk accumulation in triphibian media (Adriano *et al.* 2005). Heavy Metal accumulation in agricultural soils is caused by agrochemicals, urban sewage sludge, incineration, and disposal of waste and vehicle exhausts. The plants absorb heavy metals from the polluted soils and pass it into humans via the food chain (Zhuang *et al.* 2013, 2014). Heavy metal absorption through underground and aboveground surfaces affects the plant health (Patra *et al.* 2004), leading to cytoplasmic enzyme inhibition and cell structure decimation caused by oxidative stress (Jadia and Fulekar 2009). Oxidative stress relates to ROS and cytotoxic compounds like methylglyoxal (MG) formation and perturbation the equilibrium of ionic homeostasis within the plant cells (Hossain *et al.* 2012; Sytar *et al.* 2013). Glutathione depletion, binding to sulfhydryl groups of proteins are indirect ways in which some heavy metals out oxidative stress (Valko *et al.* 2005), inhibiting antioxidative enzymes, or inducing ROS producing enzymes like NADPH (nicotinamide adenine dinucleotide phosphate) oxidases (Bielen *et al.* 2013). Moreover, reduction or even complete cessation of all metabolic activities is experienced by pants exposed to heavy metals. Though plants possess several defense strategies to avoid or tolerate HM intoxication but beyond certain limits these mechanisms fail and survival of plant is jeopardized (Clemens and Ma 2016). Thus, removal of the accumulated heavy metals for normal functioning of plant is important. Physical processes used to clean heavy metal pollutants such as soil washing, excavation, in situ fixation are expensive and inefficient (Schnoor 1997). However, Phytoremediation (the use of growing plants reduces the concentration of HM in the soil) and use of rhizospheric microbes are emerging as alternatives ensuring high efficiency and better performance. Particularly, rhizospheric microbes protect the plant from HM stress and prevent their accumulation from soil. Microbes also metabolic capabilities supported by molecular machinery that allow performance in high concentration of HM. According to Pires *et al.* (2017) in the contaminated sites Firmicutes, Proteobacteria, and Actinobacteria from the bacterial population and most represented genera belong to Bacillus, Pseudomonas, and Arthrobacter. Rhizobia found in the rhizosphere are also very important plant growth-promoting (PGP) microbes. Moreover, HMT rhizobial strains effectively carry out symbiotic nitrogen fixation whereas Legume–rhizobia symbiosis detoxifies improves the quality of contaminated soils (Checcucci *et al.* 2017). In fungi, Ascomycota and Basidiomycota are the most reported from HM contaminated soils (Narendrula-Kotha and Nkongolo 2017), whereas poor nutrient HM contaminated soil is primarily colonized by arbuscular mycorrhizal (AM) fungi (Khan *et al.* 2000). Various intracellular functions of AM fungi and other rhizosphere microbes are results of metal ion binding of the external environment on the cell surface or their transportation into the cell (Ehrlich 1997). Metal speciation, toxicity, mobility, dissolution, and deterioration can be nudged by it (Gadd 2010). These microbes transform vast types of metals. Their interaction in rhizosphere is rigorous and relies on physico-chemical character of soil, concentration and category of metal species, and the metabolism and variety of microbes (Khan 2005; Giller *et al.* 2009; Gadd 2010; Kong and Glick 2017). Human activities have also led to direct or indirect heavy metal pollution in open water and aquatic ecosystems (McEldowney *et al.* 1993). There are two distinctions of water pollution, point sources and non-point sources (McEldowney *et al.* 1993). The point source of pollution takes place due to direct exposure of a single, identifiable hazardous substance in a water body such as the Exxon Valdez oil spills in 1989 that ran aground in the Prince William Sound, causing expelling of 11 million gallons of crude oil into the Alaska environment (Siddiquee *et al.* 2011). On the other hand, non-point source pollution is difficult to stop as it is caused by multiple diffusion sources which makes finding a specific solution tedious.

2.3 Bioremediation

The technique that can potentially decimate or subjugate contaminants through natural bioactivity in the ecosystem is termed as bioremediation (Siddiquee *et al.* 2015). According to Qazilbash (2004), it is the biodegradation of organic and inorganic waste to non-toxic substances. This is promoted through the addition of an electron acceptor, nutrient, and other factors and is generally cost-effective uses, inexpensive technology, resulting in a high public acceptance and is mostly carried out on site (Su 2014). It's application frequently employs environmental configuration to speed up the process of microbial growth and degradation. Effective bioremediation is caused by the enzymatic attack of microorganisms on the pollutants and its conversion into non-toxic products (Su 2014). Reactions that are a part of the microorganisms' metabolic processes convert the contaminants (Siddiquee *et al.* 2015). Various principles of bioremediation include biofilters, bioventing, biosorption, composting, bioaugmentation, bioreactor, land farming, and biostimulation (Qazilbash 2004). Khan *et al.* (1997) elaborated that complex factors like control and optimization of bioremediation processes consist of the presence of a microbial population capable of degrading pollutants, contaminant availability to the microbes, and environmental factors like soil type, temperature, pH, the presence of oxygen, or other electron acceptors, and nutrients. Certain microbes show promising results when it comes to pollutant removal or its conversion into benign states and utilization in metabolic processes (Qazilbash 2004).

2.4 Heavy Metal Signaling

Heavy metal stress signal transduction commences only after the receptors/ion channels perceive the stress signal(s) succeeded by the non-protein messengers, such as cyclic nucleotides, calcium, and hydrogen ions (Figure 2.1). Gene expression of various TFs and synthesis of metal-detoxifying peptides takes place due to the relaying of the stress signal caused by various kinases and phosphates (Rao *et al.* 2011; Islam *et al.* 2015; Kumar and Trivedi 2016). Activation of distinct signaling pathways in plants like calcium-dependent signaling, mitogen-activated protein kinase signaling, ROS signaling, and hormone signaling due to heavy metal(s) result in the enhancement of the expression of TFs and/or stress-responsive genes (Dubey *et al.* 2014; Kumar and Trivedi 2016). Plants that can sense, decode, and convey the modifications in cytosolic Ca^{2+} concentration for the stress response have diverse Ca^{2+} sensors like calmodulins (CaMs), CaM-like proteins, calcineurin B-like proteins (CBLs), and Ca^{2+}dependent protein kinases (CDPKs) (Conde *et al.* 2011; Steinhorst and Kudla 2014). According to researchers fluctuating levels of phytohormones alter plant response to metal stress. Role of phytohormones ABA, SA, JA, and ethylene have also been elucidated in PGPR inoculated plants under stressed conditions (Tiwari *et al.* 2016, 2017b). The induction of these genes that are central to heavy metal stress signaling, when the plant-associated microbes are present as well indicate the sophisticated cross-talk of plants, microbes, and heavy metals in stress response and tolerance. Thus, an understanding of the complex metal stress signaling pathways and the existing crosstalk among the networks of plant–microbe–metal interaction is quintessential to elucidate the stress-responsive networks in plants.

2.5 Strategies Involved in Heavy Metal Remediation

2.5.1 Bioremediation Capacity of Microorganisms on Heavy Metals

Bacteria, fungi, and algae clean up heavy metal-contaminated environments using metal-resistant strains in single, consortium, and immobilized forms as remediation yields effective results, while the immobilized form leads to higher chemisorption sites that adsorb heavy metals (Kim *et al.* 2015).

2.5.2 Bacterial Remediation Capacity of Heavy Metal

Pretreatment and experimental conditions govern biosorption in each microbial cell. Microbial cells shall attune to the alteration of physical, chemical, and bioreactor configuration to enhance biosorption (Ayangbenro and Babalola 2017). Bacteria possess ubiquity, size, and ability to grow under controlled conditions and resilience to environmental conditions (Srivastava *et al.* 2015) and high surface-to-volume ratios and potential active chemisorption sites (teichoic acid) on the cell wall (Mosa *et al.* 2016) making them great biosorbents.

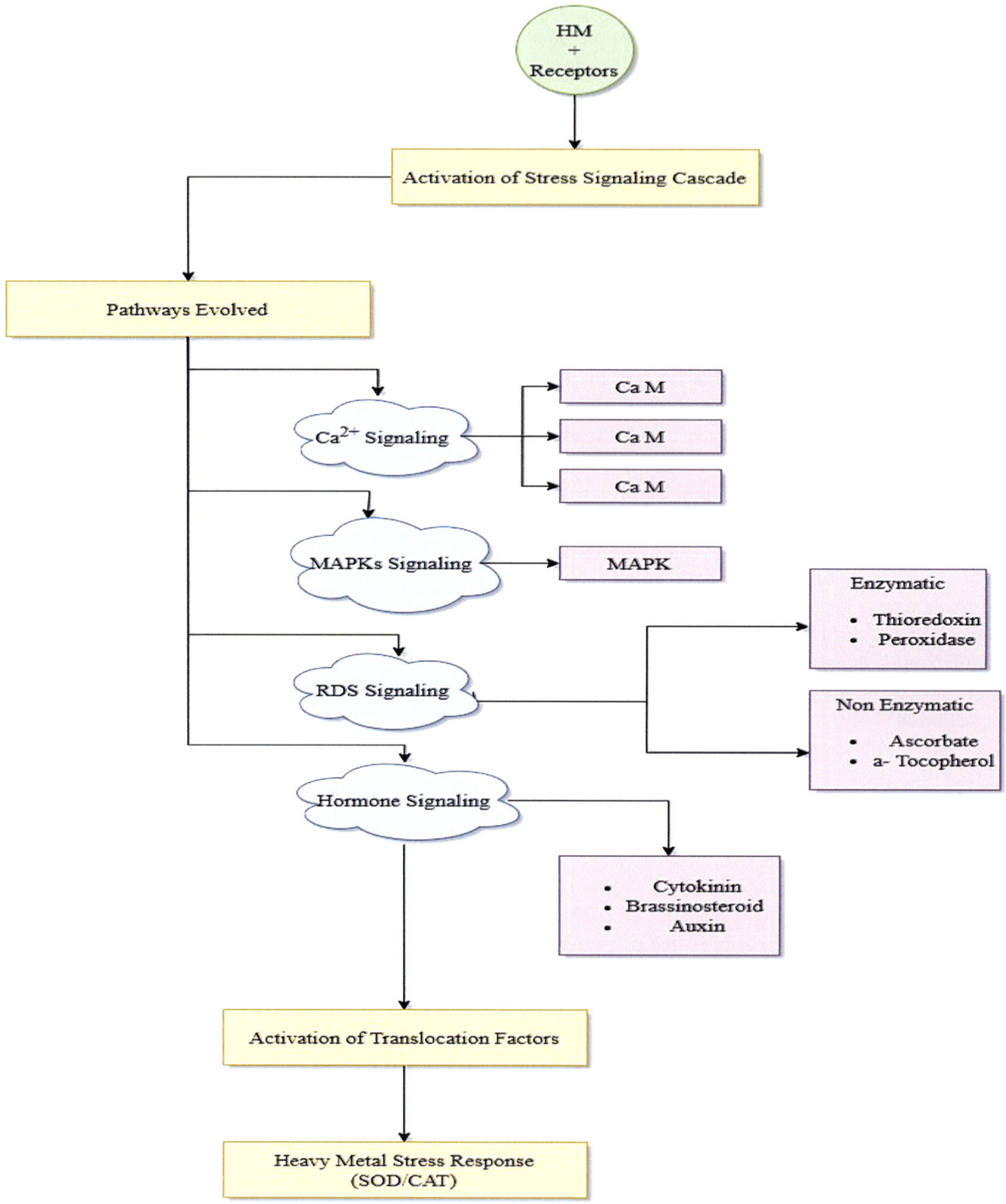

FIGURE 2.1 A schematic representation of heavy metal stress signaling cascade in plants and the existing cross-talk among the networks of plant–microbe–metal interaction. These signaling pathways include MAPKs, calcium, ROS, and hormone signaling molecules that mediate signal transduction to enhance the expression of stress-responsive genes.

Moreover, in mixed cultures they are more stable and survive more (Sannasi *et al.* 2006) and hence metabolically superior for the biosorption and preferred for field applications (Kader *et al.* 2007). De *et al.* (2008) reported a 78% reduction of chromium (Cr) employing bacterial syndicate of *Acinetobacter* sp. and *Arthrobacter* sp. at 16 mg/L metal ion concentration. *Micrococcus luteus* helped remove a large amount of Pb from a synthetic medium. In ideal environments, the elimination ability was 1965 mg/g (Puyen *et al.* 2012). Abioye and coworkers (Abioye *et al.* 2018) investigated the biosorption of Pb, Cr, and Cd in tannery effluent using *Bacillus subtilis, B. megaterium, Aspergillus niger*, and *Penicillium* sp. *B. megaterium* recorded the highest Pb reduction (2.13–0.03 mg/L), followed by *B. subtilis* (2.13–0.04 mg/L). *A. niger* portrayed the highest ability to reduce the concentration of Cr (1.38–0.08 mg/L) followed by *Penicillium* sp. (1.38–0.13 mg/L), while *B. subtilis* showed the best ability to decrease the concentration of Cd (0.4–0.03 mg/L) succeeded by *B. megaterium* (0.04–0.06 mg/L) post 20 days. Kim and coauthors (Kim *et al.* 2015) made a batch system using zeolite-immobilized *Desulfovibrio desulfuricans* for the removal of $Cr6^{+}$, Cu, and Ni with removal efficiencies of 99.8%, 98.2%, and 90.1%, respectively. Abbas *et al.* (2014)

reported proficient eradication of Cr, Zn, cadmium (Cd), lead (Pb), Cu, and cobalt by bacterial consortia at approximately 75% to 85% in fewer than 2 h of contact duration.

2.5.3 Fungi Remediation Capacity of Heavy Metal

Fungi are widely used as biosorbents (Fu *et al.* 2012) and active and lifeless fungal cells are quintessential in the adhesion of inorganic chemicals (Tiwari *et al.* 2013). Srivastava and Thakur 2006 reported that *Aspergillus* sp. usage resulted in 85% of Cr removal at pH 6 in a bioreactor system from the synthetic medium, compared to 65% removal from the tannery effluent, due to organic pollutants that hinder the growth of the organism. *Coprinopsis atramentaria* was studied for its ability to bioaccumulate 76% of Cd^{2+} at a concentration of 1 mg/L of Cd^{2+}, and 94.7% of Pb^{2+}, at a concentration of 800 mg/L of Pb^{2+} hence making it an effective accumulator of heavy metal ions for mycoremediation (Lakkireddy and Kues 2017). Park and his coauthors (Park *et al.* 2005) reported that dead fungal biomass of *A. niger, Rhizopus oryzae, Saccharomyces cerevisiae*, and *Penicillium chrysogenum* could be used to convert toxic Cr (VI) to less toxic or nontoxic Cr (III). Luna *et al.* (2016) also observed that *Candida sphaerica* produce biosurfactants with removal efficiencies of 95%, 90%, and 79% for Fe, Zn, and Pb, respectively. Biosurfactants have gained interest in recent years owing to their low toxicity, biodegradability nature, and diversity. Mulligan *et al.* (2001) assessed the viability of using surfactin, rhamnolipid, and sophorolipid for the removal of heavy metals (Cu and Zn). A single wash with 0.5% rhamnolipid removed 65% of Cu and 18% of Zn, whereas 4% sophorolipid removed 25% of the Cu and 60% of Zn. Several strains of yeast such as *Hansenula polymorpha, S. cerevisiae, Yarrowia lipolytica, Rhodotorula pilimanae, Pichia guilliermondii*, and *Rhodotorula mucilage* have been used to bioconvert Cr (VI) to Cr (III) (Chatterjee *et al.* 2012).

2.5.4 Heavy Metal Removal Using Biofilm

The use of biofilm as an effective bioremediation tool and a biological stabilization agent is well documented. Biofilms possess high tolerance against toxic inorganic elements even at high concentrations that are noxious. A study on *Rhodotorula mucilaginosa concluded* that metal removal efficiency was from 4.79% to 10.25% for planktonic cells and from 91.71% to 95.39% for biofilm cells (Goher *et al.* 2016). Biosorbent or exopolymeric substances containing surfactant or emulsifier property possessing molecules carry out bioremediation (El-Masry *et al.* 2004).

2.5.5 Algae Remediation Capacity of Heavy Metal

Algae are autotrophic and hence require low nutrients and produce enormous biomass compared to other microbial biosorbents. These biosorbents have also been used for heavy metal removal with a high sorption capacity (Abbas *et al.* 2014). Algae biomass is used for bioremediation of heavy metal-polluted effluent via adsorption or by integration into cells. Phycoremediation is the use of various types of algae and cyanobacteria for the remediation of heavy metals by either removal or degradation of toxicants (Chabukdhara *et al.* 2017). Algae have various chemical moieties on their surface, such as hydroxyl, carboxyl, phosphate, and amide, which act as metal-binding sites (Abbas *et al.* 2014). Hussian and Napiorkowska-Krzebietke *et al.* (Abbas *et al.* 2014) used dead cells of Chlorella vulgaris to remove Cd^{2+}, Cu^{2+}, and Pb^{2+} ions from aqueous solutions under various conditions of pH, biosorbent dosage, and contact time. These results suggest that the biomass of *C. vulgaris* is an extremely efficient biosorbent for the removal of Cd^{2+}, Cu^{2+}, and Pb^{2+} at 95.5%, 97.7%, and 99.4%, respectively, from a mixed solution of a 50 mg/dm^3 of each metal ion could be considered as the agents in the bioremediation process.

2.5.6 Heavy Metal Remediation Using Microbes

HMT-PGP microbes enhance bioavailability and trigger detoxification by regulation plant growth and physico-chemical soil property and through acidification, chelation, complexation, precipitation, and redox reactions. Acidic pH conditions favor bioavailability and adsorption of HM in rhizosphere (Merdy *et al.* 2009). Moreover, organic acids released by HMT-PGP microbes decrease soil pH and sequester soluble metal ions (Turnau and Kottke 2005). Experimental evidence suggests that a wide array of bacteria and fungi produce organic acids as natural chelating agents of HM (Seneviratne *et al.* 2017). Gluconic, oxalic, acetic, and malic acids are mainly reported for HM solubilization by soil microbes (Ullah *et al.* 2015; Gube 2016). In a study, Fomina *et al.* (2005) showed that over secretion of organic acids (oxalic and citric) by HM tolerant Beauveria caledonica solubilized Cd, Cu, Pb, and Zn metals. The oxalate crystals produced by mycorrhizal fungi are also known to immobilize and detoxify HM (Gadd *et al.* 2014). Their filamentous hyphal structure deeply penetrates in to the deeper soil aggregates and chelates or adsorbs HM. A study by Kaewdoung *et al.* (2016) involving scanning electron microscopy equipped with energy dispersive X-ray microanalysis (SEM-EDXA) and X-ray powder diffraction (XRPD) revealed that oxalate crystals produced by wood-rotting fungi *Fomitopsis* cf. meliae and *Ganoderma* aff. Steyaertanum contributed in metal tolerance by transforming the metals into less toxic forms (zinc sulfate into zinc oxalate dihydrate, copper sulfate into copper oxalate hydrate, cadmium sulfate into cadmium oxalate trihydrate, and lead nitrate into lead oxalate). Root exudates contain organic acids, amino acids, and phytochelatins (PC) which perform as intracellular binding compounds for HM. Release of protons (HC) and enzymes with root exudates helps in acidification and electron transfer in the rhizosphere which leads to enhanced metal bioavailability (Ma *et al.* 2016). Changes in concentrations of exudate compounds in the presence of particular HM can also help in developing biomarkers.

Recently, based on gas chromatography-mass spectrometry (GC-MS) and metabolomics methods, Luo *et al.* (2017) showed that Pb accumulating and Sedum alfredii can significantly change the types of root exudates, and 15 compounds were identified and assumed to be potential biomarkers of Pb contamination. Microbially mediated redox reactions also have profound effect on transformation of HM to less or non-toxic forms (Amstaetter *et al.* 2010). Outer membrane c-type cytochromes (OM c-Cyts), trans-outer membrane porin–cytochrome protein complex (Pcc), or MtrABC extracellular electron conduit play key role in microbial metal reduction processes (Shi *et al.* 2016). Such systems are well investigated in Shewanella and Geobacter species. Furthermore, HM may also be oxidized by specific enzymes. For example, multicopper oxidases such as CueO or CuiD and/or CopR are essentially required in Cu efflux. Whereas ChrA of chromate reductase perform reduction of Cr^{+6} to Cr^{+3}. For Hg, the protein MerA reduces Hg^{2+} to lesser toxic Hg^0. These proteins are up-regulated under toxic HM stress (Figure 2.2). There are several instances where HMT bacteria provide substantial aid in detoxification of HM in plants. A study of Chatterjee *et al.* (2009) on Cr-tolerant bacteria *Cellulosimicrobium cellulans* showed transformation of toxic Cr6C to non-toxic Cr3C and also its enhanced uptake in the shoot and root of green chili. Majumder *et al.* (2013) reported biotransformation of toxic As3C to less toxic As5C by Asoxidizing bacteria *Bacillus* sp. and *Geobacillus* sp. isolated from As-contaminated soils. Bioaccumulation is also largely responsible for HM uptake and further detoxification by HMT-PGP microbes.

There are two combined processes which are responsible for bioaccumulation of HM. Passive uptake or "biosorption" is metabolism independent accumulation of metals by living or inactive nonliving biomass or biological materials, whereas "active uptake" occurs only in living cells, requires metabolism and energy for the transport of metals (Gutierrez-Corona *et al.* 2016). Biosorption may involve one or a combination of different processes including complexation, coordination, chelation, ion exchange, microprecipitation, and entrapment (Pokethitiyook and Poolpak 2016). Cell walls and associated functional groups like –SH, –OH, and –COOH, and other biomolecules have affinity for HM that helps in the biosorption process. Metal binding also involves chelators and metal-binding peptides, such as PC (glutathione-derived peptides) and metallothioneins (MT). PC and MT are produced by rhizospheric bacteria and fungi as well as by plants in response to HM stress and may result in the deposition of HM in microbial or plant cells (Miransari 2011). MT are cysteine-rich metal peptides with high affinity for Cd, Cu, and Hg metals (Ahemad 2014).

In a study, Murthy *et al.* (2011) found an increase in the MT biosynthesis in Bacillus cereus when it was exposed to increased Pb concentrations. Similarly, Sharma *et al.* (2017) also showed the role of MT assisted periplasmic Pb sequestration by HMT Providencia vermicola strain SJ2A. Detoxification of HM via MT biosynthesis is also very well studied in HMT fungi. However, expression of the MT-related genes and their production in the presence HM has gained

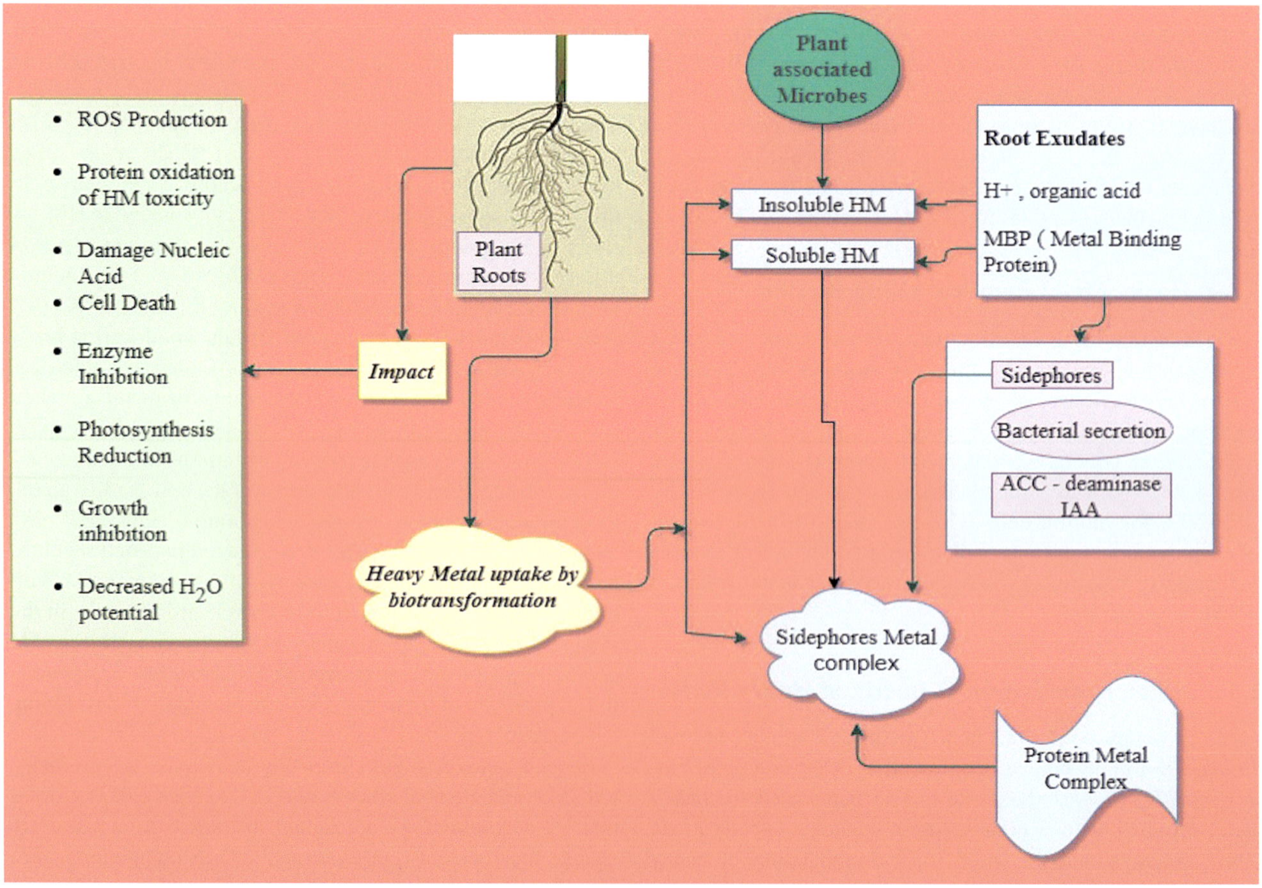

FIGURE 2.2 Mechanism of microbial HM bioremediation, physicochemical properties by root exudates, and bacterial secretion to enhance metal bioavailability and biotransformation that leads to rapid detoxification and impact on plant physiology.

more attention in members of mycorrhizal fungi (Lanfranco *et al.* 2002; Lanfranco 2007; Hložková *et al.* 2016). After entering into the cell final step of HM detoxification involves their sequestration or compartmentalization into different subcellular organelles. In mycorrhizal fungi, mainly the vacuolar compartmentation of HM is noticed. Vacuolar compartmentalization of Zn, Cu, and Cd was observed in extraradical mycelium of Glomus intraradices renamed as Rhizophagus irregularis (González-Guerrero *et al.* 2008). Similarly, Yao *et al.* (2014) also showed vacuolar accumulation of Cd in Cd-exposed extraradical mycelium of R. irregularis in symbiosis with clover. Microbial communities in the rhizosphere also excrete extracellular polymeric substances (EPS), such as polysaccharides, glycoprotein, lipopolysaccharide, and soluble peptide which possess substantial quantity of anion functional groups and help to remove or recover metals from the rhizosphere through biosorption (Ayangbenro and Babalola 2017). EPS production by certain PGP microbes induces biofilm formation in response to the exposure of toxic HM. Biofilm formation enhances tolerance of microbial cells by forming a protective sheath as well as transforming toxic metal ions into non-toxic forms after adsorption. EPS produced by rhizobia and other PGP microbes with multiple numbers of anionic groups are reported to sequester several types of HM (Gupta and Diwan 2017).

2.6 Bioremediation Mechanisms of Heavy Metal-Contaminated Environment

The different bioremediation mechanisms are explained in Table 2.1.

2.7 Mechanisms of Rhizobacteria Influencing Heavy Metal Accumulation

2.7.1 Rhizobacteria Secretion

Phytoremediation can be assisted by rhizobacteria as indirect mechanisms like preventing phytopathogens from inhibiting plant growth and development while direct mechanisms include nitrogen fixation; synthesis of siderophores which can solubilize and sequester iron from the soil; production of phytohormones such as auxins and cytokinins, that enhance plant growth; and solubilization of minerals such as phosphorus (Kloepper *et al.* 1989; Glick 1995; Glick *et al.* 1999; Patten and Glick 1996). Rhizobacteria produce metal-chelating agents called siderophores that help in acquisition of several heavy metals (Leong 1986). These organic substances scavenge Fe^{3+} and enhance the bioavailability of soil bound iron (Kanazawa *et al.* 1994). Moreover, the production of siderophores by PGP

TABLE 2.1

Bioremediation Mechanisms of Various Heavy Metals

Biological Organism	Bioremediation	Heavy Metal	Reference
Bacteria	*Acinetobacter* sp. *Bacillus subtilis* *Staphylococcus*	Cr	Bhattacharya *et al.* (2014) Kim *et al.* (2015) Kumar *et al.* (2011)
	Methylobacterium organophilum	Pb	Bharagava and Mishra (2018)
	Vibriofluvialis	Co	Jafari *et al.* (2015)
	Klebsiellap neumoniae	Hg	Jafari *et al.* (2015)
	Bacillus firmus *Pseudomonas* sp.	Zn	Salehizadeh and Shojaosadati (2003) Kumaran *et al.* (2011)
Fungi	*Aspergillus* sp., *Saccharomyces cerevisiae*	Cr	Congeevaram *et al.* (2007) Benazir *et al.* (2010)
	Candidapara psilosis	Hg	Muneer *et al.* (2013)
	Aspergillus versicolor, *Aspergillus* sp., *Aspergillus niger*	Ni	Taştan *et al.* (2010)
Algae	*Spirogyra* sp. *Spirulina* sp.	Cr	Mane and Bhosle (2012)
	Nostoc sp.	Pb	Kumaran *et al.* (2011)
	Spirogyra sp.	Cu	Mane and Bhosle (2012)
	Nostoc sp.	Cd	Kumaran *et al.* (2011)
Plant	*Pteris vittata*	As, Cr	Kalve *et al.* (2011)
	Phytolacca americana	Cd	Peng *et al.* (2008)
	Haumaniastrum robertii	Co	Marques *et al.* (2009)
	Eleocharis acicularis	Cu	Sakakibara *et al.* (2011)
	Achillea millefolium	Hg	Wang *et al.* (2012)w
	Schima superb	Mn	Yang *et al.* (2008)
	Psychotria douarrei	Ni	Cunningham and Ow (1996)
	Medicago sativa	Pb	Koptsik (2014)
	Eleocharis acicularis	Zn	Sakakibara *et al.* (2011)

bacteria helps plants obtain sufficient iron in iron deficient plants grown in contaminated soil (Burd *et al.* 2000; Wallace *et al.* 1992). Microbial siderophores are used as iron chelating agents that iron availability in the plant rhizosphere (Bar-Ness *et al.* 1992; Loper and Henkels 1999). The affinity of the siderophore for iron ultimately decides the rhizosphere population structure on these governing factors, concentration of various types of siderophore, kinetics of exchange, and availability of Fe-complexes to microbes as well as plants (Loper and Henkels 1999). A number of PGPR, which stimulate root growth of different plant species including Indian mustard (Burd *et al.* 1998; Belimov *et al.* 2001), contain the enzyme ACC deaminase, which hydrolyses and decreases the amount of ACC, an ethylene precursor of the plant hormone ethylene, in plants and, as a result, to decrease ethylene biosynthesis by plants (Glick *et al.* 1994, 1998; Hall et al. 1996). The model which represents how a PGPR bound to either a seed or plant root lowers the ethylene concentration and thereby prevents ethylene inhibition of root elongation was previously proposed by Glick *et al.* (1998). In some of the plants, ACC is exuded from roots or seeds and then taken up by the bacterium and cleaved by ACC deaminase to ammonia and α-ketobutyrate (Glick *et al.* 1998). The bacteria utilize the ammonia evolved from ACC as a nitrogen source and thereby decrease ACC within the plant (Penrose and Glick 2001) with the concomitant reduction of plant ethylene and promoting root elongation (Burd *et al.* 1998; Mayak *et al.* 1999; Grichko and Glick 2001; Belimov *et al.* 2002). To maintain the gradient between internal and external ACC levels, the plant must exude increasing amounts of ACC. The lowering of ACC levels within the plant results in a reduction in the amount of plant ethylene and a decreased extent of ethylene inhibition of plant seedling root elongation. This model may also be invoked to explain how PGP bacteria lower the concentration of stress ethylene in plants. Evidence for this model includes the fact that the ability of a bacterium to promote root elongation is positively correlated with both the ACC deaminase activity of the bacterium and the ACC content (measured by high-pressure liquid chromatography) of the plant tissues. In addition, depending on the conditions, plant root growth may also be stimulated by IAA produced by PGPR bound to the seeds or roots (Patten and Glick 2002). As a matter of fact, low levels of IAA produced by rhizobacteria promote primary root elongation, whereas high levels of IAA stimulate lateral and adventitious root formation (Glick *et al.* 1995) but inhibit primary root growth (Xie *et al.* 1996). Thus, PGP bacteria can facilitate plant growth by altering the hormonal balance within the affected plant (Glick *et al.* 1999). Similarly, although an ethylene pulse is important in breaking seed dormancy, too much ethylene can inhibit plant seed germination (Bewley and Black 1985; Mayer and Poljakoff-Mayber 1989; Smalle and van der Straeten 1997). As just described above, a significant portion of the damage to plants from infection with fungal phytopathogens may occur as a direct result of the response of the plant to the increased level of stress ethylene (van Loon 1984). In the presence of fungal pathogens, not only does exogenous ethylene increase the severity of a fungal infection but also inhibitors of ethylene synthesis can significantly decrease the severity of infection. Since the enzyme ACC deaminase, when present in PGP bacteria, can act to modulate the level of ethylene in a plant, lower the stress placed on plants by the presence of heavy metals and therefore ameliorate some of the apparent toxicity of heavy metals to plants.

2.7.2 High Surface Area-to-Volume Ratio

Soil rhizobacteria, having high surface area-to-volume ratio due to small size and providing large contact area, has the potential to be microbial chelates associated with phytoremediation (Anderson *et al.* 1993; Kärenlampi *et al.* 2000; Sitaula *et al.* 1999). Indian mustard plants germinated on Se-containing media from axenic seeds coated with bacteria produced more root hairs and accumulated more Se than plants grown from axenic seeds (de Souza *et al.* 1999a). Since bacteria does not help HM accumulation, increased root size is not solely responsible. Antibiotic experiments performed with bulrush and rabbit foot grass supplied with chromate and arsenate showed no contrast between Cr and As accumulation in antibiotic-supplied and untreated plants (de Souza *et al.* 1999b).

2.7.3 Transform Toxic Heavy Metals

Bacteria transform toxic heavy metals to forms that are more readily taken up into roots. For example, bacteria could enhance Se accumulation in plants by reducing selenate to organic Se, and organo-selenium forms like SeMet are known to be taken up at faster rates into roots than inorganic forms (Zayed *et al.* 1998). Huang *et al.* (2005) further depict the relative changes as the percentages of the speciation concentration difference between bulked soil and rhizosphere to the concentration of bulked soil. Results showed that the relative changes of organic bound Cu, Zn, and Pb were, respectively, +5%, +23%, +3% in the infected rhizosphere, and 0.8%, –3%, –2% in the non-infected rhizosphere. Thus, significant amounts of Cu, Zn, and Pb were bounded by organic matter in the infected rhizosphere. Soil rhizobacteria alters their chemical properties, such as pH, organic matter content, redox state, etc., and helps in the leaching process. The bioavailability of heavy metals in soils is a function of its solubility (Ernst 1996) with pH and organic matter content being the main controlling factors (Gray *et al.* 1998). For example, a strain of *Pseudomonas maltophilia* was shown to reduce the mobile and toxic Cr^{6+} to nontoxic and immobile Cr^{3+}, and also to minimize environmental mobility of other toxic ions such as Hg^{2+}, Pb^{2+}, and Cd^{2+} (Blake *et al.* 1993; Park *et al.* 1999).

2.7.4 Inhibition of Plant Pathogens

PGPR mechanisms for pathogen suppression include competition for nutrients and space (Elad and Baker 1985; Elad and Chet 1987), antibiosis by producing antibiotics viz., pyrrolnitrin, pyocyanine, 2,4-diacetyl phloroglucinol (Pierson and Thomashow 1992), and production of siderophores (fluorescent yellow-green pigment), viz., pseudobactin that limits the availability of iron needed for pathogen growth (Kloepper *et al.* 1980; Lemanceau *et al.* 1992). Production of lytic enzymes such as chitinases and β-1,3-glucanases degrades chitin and glucan present in the cell wall of fungi (Fridlender *et al.* 1993;

Lim *et al.* 1991; Potgieter and Alexander 1996; Velazhahan *et al.* 1999), HCN production (Défago *et al.* 1990) and degradation of toxin produced by pathogens (Borowitz *et al.* 1992; Duffy and Défago 1997).

2.7.5 Stimulation of Transport Protein

Transition metals like manganese, Zn, and iron and their uptake and sequestration govern bacteria survival and proliferation. For instance, cells strictly regulate intracellular Zn levels, as its high concentration is toxic to cellular functions and has evolved multiple protein types involved in binding and transport of Zn (Claverys 2001). Bacteria also stimulates the sulfate transport protein, located in the root plasma membrane, which also transports selenate (Leggett and Epstein 1956). Inorganic Hg uptake in higher plants is linked to the passive uptake of lipophilic chloride complexes in phytoplankton (Mason *et al.* 1996).

2.8 Factors Affecting Microbial Remediation of Heavy Metals

Heavy metals are either stimulatory or inhibitory to microorganisms depending on the total metal ion concentrations, chemical forms of the metals, and related factors such as redox potential. Environmental factors such as temperature, pH, low molecular weight organic acids, and humic acids alter the transformation, transportation, valance state of heavy metals, and the bioavailability of heavy metals (Rasmussen *et al.* 2000). Formation of free ionic species at acidic pH levels, increasing protons availability to saturate metal-binding sites result in more toxicity as at higher hydrogen ion concentrations, the adsorbent surface is more positively charged, and reduces the attraction between adsorbent and metal cations. Also, an increase in temperature improves the rate of adsorbate diffusion across the external boundary layer and heavy metal solubility and bioavailability (Bandowe *et al.* 2014). Bioremediation is accelerated when temperature is increased within a suitable range, thereby enhancing microbial metabolism and enzyme activity. The microbes-metal complex stability depends on the sorption sites, microbial cell wall configuration, and ionization of chemical moieties on the cell wall. The degradation process outcome relies on the substrate and the range of environmental factors.

Conclusion and Future Prospects

Tackling on a priority basis the potentially catastrophic issue of heavy metal contamination using cutting edge and economical techniques is the need of the hour. Plant-microbe synergy is an optimistic technique that might help restore triphibian contaminated media. Although HMT-PGP microorganisms improve soil quality, enhance plant growth, and detoxify the heavy metals, more research is needed to develop suitable bioformulations using HMT-PGP microbes for remediation and utilization of contaminated soils. Heavy metal elimination in plant tissues hampers remediation (Ma *et al.* 2011), but when soil is contaminated with different heavy metals, HMT-PGP microbes having additives (nutrients) is efficacious. There is an increase in the mobilization and uptake of As and Hg in *B. juncea* and *L. albus* grown in contaminated soil when thiosulfate with used with HMT-PGP microbes' microcosm-scale phytoextraction experiments, as shown by Franchi *et al.* 2017. For heavy metal phytoextraction using genetically engineered microbes (GEM) adjusted for local conditions and targeting the use of non-food crops (ones used for commercial purposes) brings about efficient removal of heavy metals and avoids its addition in the food chain (Das *et al.* 2016; Gupta and Singh 2017). Even local microbial biostimulation by nutrient addition (Fulekar *et al.* 2012) and pathway engineering techniques for microbial modification(Mosa *et al.* 2016) carry out detoxification. A recent study by Migahed *et al.* (2017) proved that mixture of HMT bacterial biomass and fungal spores successfully removed Cr and Fe ions from industrial effluents. Entomopathogenic fungi also removes HM from contaminated soils (Gola *et al.* 2016). For HM remediation through GEM and transgenic plants, overexpression of metal-binding proteins, chelators, metal transforming, and detoxifying enzymes are quintessential. Genetic engineering induced "microbial biosensors" with enhanced potential of rapid detection of accurate measurement of degree of contamination at sites is also a propitious technology (Dixit *et al.* 2015). In conclusion, it has been shown that studies of plant microbiome from contaminated soils accelerates prevalent phytoremediation technology for remediation of HM (Thijs *et al.* 2017).

REFERENCES

Abbas, S. H., Ismail, I. M., Mostafa, T. M., & Sulaymon, A. H. (2014). Biosorption of heavy metals: A review. *Journal of Chemical Science and Technology*, *3*(4), 74–102.

Abioye, O. P., Oyewole, O. A., Oyeleke, S. B., Adeyemi, M. O., & Orukotan, A. A. (2018). Biosorption of lead, chromium and cadmium in tannery effluent using indigenous microorganisms. *Brazilian Journal of Biological Sciences*, *5*(9), 25–32. https://doi.org/10.21472/bjbs.050903.

Adriano, D. C., Bolan, N. S., Vangronsveld, J., & Wenzel, W. W. (2005). Heavy metals. In: D. Hillel (Ed.), *Encyclopedia of Soils in the Environment*. Amsterdam: Elsevier, 175–182. doi: 10.1016/B0-12-348530-4/00196-X.

Ahemad, M. (2014). Remediation of metalliferous soils through the heavy metal resistant plant growth promoting bacteria: paradigms and prospects. *Arab. J. Chem.* doi: 10.1016/j.arabjc.2014.11.020.

Ahirwar, N. K., Gupta, G., Singh, R., & Singh, V. (2016). Isolation, identification and characterization of heavy metal resistant bacteria from industrial affected soil in central India. *International Journal of Pure & Applied Bioscience*, *4*(6), 88–93. https://doi. org/10.18782/2320-7051.2424.

Amstaetter, K., Borch, T., Larese-Casanova, P., & Kappler, A. (2010). Redox transformation of arsenic by Fe (II)-activated goethite (a-FeOOH). *Environmental Science & Technology*, *44*, 102–108. doi: 10.1021/es901274s.

Anderson, T. A., Guthrie, E. A., & Walton, B. T. (1993). Bioremediation in the rhizosphere: Plant roots and associated microbes clean contaminated soil. *Environmental Science & Technology*, *27*(13), 2630–2636. doi: 10.1021/es00049a001.

Ayangbenro, A. S., & Babalola, O. O. (2017). A new strategy for heavy metal polluted environments: A review of microbial biosorbents. *International Journal of Environmental Research and Public Health*, *14*(1), 94. https://doi.org/10.3390/ijerph14010094.

Azubuike, C. C., Chikere, C. B., & Okpokwasili, G. C. (2016). Bioremediation techniques classification based on site of application: Principles, advantages, limitations and prospects. *World Journal of Microbiology & Biotechnology*, *32*(11), 180. https://doi.org/10.1007/s11274-016-2137-x.

Bandowe, B. A. M., Bigalke, M., Boamah, L., Nyarko, E., Saalia, F. K., & Wilcke, W. (2014). Polycyclic aromatic compounds (PAHs and oxygenated PAHs) and trace metals in fish species from Ghana (West Africa): Bioaccumulation and health risk assessment. *Environment International*, *65*, 135–146. https://doi.org/10.1016/j.envint.2013.12.018.

Bar-Ness, E., Hadar, Y., Chen, Y., Shanzer, A., & Libman, J. (1992). Iron uptake by plant from microbial siderophores. *Plant Physiology*, *99*(4), 1329–1335.

Belimov, A. A., Safronova, V. I., Sergeyeva, T. A., Egorova, T. N., Matveyeva, V. A., Tsyganov, V. E., ... & Stepanok, V. V. (2001). Characterization of plant growth promoting rhizobacteria isolated from polluted soils and containing 1-aminocyclopropane-1-carboxylate deaminase. *Canadian Journal of Microbiology*, *47*(7), 642–652. doi: 10.1139/cjm-47-7-642.

Benazir, J. F., Suganthi, R., Rajvel, D., Pooja, M. P., & Mathithumilan, B. (2010). Bioremediation of chromium in tannery effluent by microbial consortia. *African Journal of Biotechnology*, *9*(21), 3140–3143.

Bewley, J. D., & Black, M. (1985). *Dormancy and the Control of Germination, Seeds: Physiology of Development and Germination*. New York: Plenum Press, 175–235.

Bharagava, R. N., & Mishra, S. (2018). Hexavalent chromium reduction potential of Cellulosimicrobium sp. isolated from common effluent treatment plant of tannery industries. *Ecotoxicology and Environmental Safety*, *147*, 102–109. https://doi.org/10.1016/j.ecoenv.2017.08.040.

Bhattacharya, A., Gupta, A., Kaur, A., & Malik, D. (2014). Efficacy of Acinetobacter sp. B9 for simultaneous removal of phenol and hexavalent chromium from co-contaminated system. *Applied Microbiology and Biotechnology*, *98*(23), 9829–9841. https://doi.org/10.1007/s00253-014-5910-5.

Blake, R. C., Choate, D. M., Bardhan, S., Revis, N., Barton, L. L., & Zocco, T.G. (1993). Chemical transformation of toxic metals by a Pseudomonas strain from a toxic waste site. *Environmental Toxicology and Chemistry*, *12*(5), 1365–1376.

Borowitz, J. J., Stankie-Dicz, M., Lewicka, T., & Zukowska, Z. (1992). Inhibition of fungal cellulase, pectinase and xylanase activity of plant growth promoting fluorescent pseudomonads. *Bull. OILB/SROP*, *15*(4), 103–106.

Burd, G. I., Dixon, D. G., & Glick, B. R. (1998). A plant growth-promoting bacterium that decreases nickel toxicity in seedlings. *Applied and Environmental Microbiology*, *64*(3), 3663–3668.

Burd, G. I., Dixon, D. G., Glick, B. R. (2000). Plant growth promoting bacteria that decrease heavy metal toxicity in plants. *Canadian Journal of Microbiology*, *46*(3), 237–245. doi: 10.1139/cjm-46-3-237.

Chabukdhara, M., Gupta, S. K., & Gogoi, M. (2017). Phycoremediation of heavy metals coupled with generation of bioenergy. In: *Algal Biofuels*. Springer, Cham pp. 163–188.

Chatterjee, S., Sau, G. B., & Mukherjee, S. K. (2009). Plant growth promotion by a hexavalent chromium reducing bacterial strain, Cellulosimicrobium cellulans KUCr3. *World Journal of Microbiology & Biotechnology*, *25*, 1829–1836. doi: 10.1007/s11274-009-0084-5.

Claverys, J. P. (2001). A new family of high-affinity ABC manganese and zinc permeases. *Research in Microbiology*, *152*(3–4), 231–243. doi: 10.1016/S0923-2508(01)01195-0.

Congeevaram, S., Dhanarani, S., Park, J., Dexilin, M., & Thamaraiselvi, K. (2007). Biosorption of chromium and nickel by heavy metal resistant fungal and bacterial isolates. *Journal of Hazardous Materials*, *146*(1–2), 270–277. https://doi.org/10.1016/j.jhazmat.2006.12.017.

Cunningham, S. D., & Ow, D. W. (1996). Promises and prospects of phytoremediation. *Plant Physiology*, *110*, 715. doi: 10.1104/pp.110.3.715.

Das, S., Dash, H. R., & Chakraborty, J. (2016). Genetic basis and importance of metal resistant genes in bacteria for bioremediation of contaminated environments with toxic metal pollutants. *Applied Microbiology and Biotechnology*, *100*, 2967–2984. doi: 10.1007/s00253-016-7364-4.

Deepa, C. N., & Suresha, S. (2014). Biosorption of lead (II) from aqueous solution and industrial effluent by using leaves of Araucaria cookii: Application of response surface methodology. *Journal of Environmental Science, Toxicology and Food Technology (IOSR-JESTFT)*, *8*(7), 67–79. https://doi.org/10.9790/2402-08716779.

Défago, G., Berling, C. H., Burger, U., Hass, D., Kahr, G., Keel, C., Voisard, C., Wirthner, P., & Wuthrich, B. (1990). Suppression of black root rot of tobacco and other root diseases by strains of *Pseudomonas fluorescens*: Potential applications and mechanisms. In: D. Hornby (Ed.), *Biological Control of Soilborne Plant Pathogens*. Wallingford, Oxon, UK: CAB International, 93–108.

De Souza, M. P., Chu, D., Zhao, M., Zayed, A. M., Ruzin, S. E., Schichnes, D., & Terry, N. (1999a). Rhizosphere bacteria enhance selenium accumulation and volatilization by Indian mustard. *Plant Physiology*, *119*(2), 565–573. doi: 10.1104/pp.119.2.565.

De Souza, M. P., Huang, C. P., Chee, N., & Terry, N. (1999b). Rhizosphere bacteria enhance the accumulation of selenium and mercury in wetland plants. *Planta*, *209*(2), 259–263. doi: 10.1007/s004250050630

Dixit, R., Malaviya, D., Pandiyan, K., Singh, U. B., Sahu, A., Shukla, R., ... & Paul, D. (2015). Bioremediation of heavy metals from soil and aquatic environment: an overview of principles and criteria of fundamental processes. *Sustainability*, *7*(2), 2189–2212. doi: 10.3390/su7022189

Duffy, B. K., & Défago, G. (1997). Zinc improves biocontrol of *Fusarium* crown and root rot of tomato by *Pseudomonas fluorescens* and represses the production of pathogen metabolites inhibitory to bacterial antibiotic biosynthesis. *Phytopathology*, *87*(12), 1250–1257.

Elad, Y., & Baker, R. (1985). The role of competition for iron and carbon in suppression of chlamydospore germination of *Fusarium oxysporum*. *Phytopathology*, *75*, 190–195.

Elad, Y., & Chet, I. (1987). Possible role of competition for nutrition in biocontrol of Pythium damping-off by bacteria. *Phytopathology, 77*, 190–195.

El-Masry, M. H., El-Bestawy, E., & Nawal, I. (2004). Bioremediation of vegetable oil and grease from polluted wastewater using a sand biofilm system. *World Journal of Microbiology & Biotechnology, 20*(6), 551–557. https://doi.org/10.1023/B:WIBI.0000043162.17813.17.

Environmental Protection Agency (EPA). (2010). *Assessment and remediation of contaminated sediments (ARCS) program, final summary report* (EPA-905-S- 94-001). EPA Chicago.

Ernst, W. H. O. (1996). Bioavailability of heavy metals and decontamination of soils by plants. *Applied Geochemistry, 11*(1–2), 163–167. doi: 10.1016/0883-2927(95)00040-2.

Ferraz, P., Fidalgo, F., Almeida, A., & Teixeira, J. (2012). Phytostabilization of nickel by the zinc and cadmium hyperaccumulator Solanum nigrum L. Are metallothioneins involved? *Plant Physiology and Biochemistry, 57*, 254–260. https://doi.org/10.1016/j.plaphy.2012.05.025.

Fomina, M., Hillier, S., Charnock, J. M., Melville, K., Alexander, I. J., & Gadd, G. M. (2005). Role of oxalic acid overexcretion in transformations of toxic metal minerals by Beauveria caledonica. *Applied and Environmental Microbiology, 71*, 371–381. doi: 10.1128/aem.71.1.371-381.2005.

Franchi, E., Rolli, E., Marasco, R., Agazzi, G., Borin, S., Cosmina, P., ... & Petruzzelli, G. (2017). Phytoremediation of a multi contaminated soil: mercury and arsenic phytoextraction assisted by mobilizing agent and plant growth promoting bacteria. *Journal of soils and sediments, 17*(5), 1224–1236. doi: 10.1007/s11368-015-1346-5.

Fridlender, M., Inbar, J., & Chet, I. (1993). Biological control of soilborne plant pathogens by a β-1,3-glucanase-producing *Pseudomonas cepacia. Soil Biology and Biochemistry, 25*(9), 1211–1221. doi: 10.1016/0038-0717(93)90217-Y.

Fu, Y. Q., Li, S., Zhu, H. Y., Jiang, R., & Yin, L. F. (2012). Biosorption of copper (II) from aqueous solution by mycelial pellets of Rhizopus oryzae. *African Journal of Biotechnology, 11*(6), 1403–1411.

Fulekar, M. H., Sharma, J., & Tendulkar, A. (2012). Bioremediation of heavy metals using biostimulation in laboratory bioreactor. *Environmental Monitoring and Assessment, 184*, 7299–7307. doi: 10.1007/s10661-011-2499-3.

Gadd, G. M., Bahri-Esfahani, J., Li, Q., Rhee, Y. J., Wei, Z., Fomina, M., & Liang, X. (2014). Oxalate production by fungi: significance in geomycology, biodeterioration and bioremediation. *Fungal Biology Reviews, 28*(2–3), 36–55. doi: 10.1016/j. fbr.2014.05.001.

Gauthier, P. T., Norwood, W. P., Prepas, E. E., & Pyle, G. G. (2014). Metal–PAH mixtures in the aquatic environment: A review of co-toxic mechanisms leading to more-than-additive outcomes. *Aquatic Toxicology, 154*, 253–269. https://doi.org/10.1016/j.aquatox.2014.05.026.

Glick, B. R., Jacobson, C. B., Schwarze, M. M. K., & Pasternak, J. J. (1994). 1-aminocyclopropane-1-carboxylic acid deaminase mutants of the plant growth promoting rhizobacterium *Pseudomonas putida* GR 12-2 do not stimulate canola root elongation. *Canadian Journal of Microbiology, 40*(2), 911–915.

Glick, B. R., Karaturovic, D. M., & Newell, P. C. (1995). A novel procedure for rapid isolation of plant growth promoting pseudomonads. *Canadian Journal of Microbiology, 41*(6), 533–536.

Glick, B. R., Patten, C. L., Holguin, G., & Penrose, D. M. (1999). *Biochemical and Genetic Mechanisms Used by Plant Growth-Promoting Bacteria*. London: Imperial College Press.

Glick, B. R., Penrose, D. M., & Li, J. P. (1998). A model for the lowering of plant ethylene concentrations by plant growth-promoting bacteria. *Journal of Theoretical Biology, 190*(1), 63–68. doi: 10.1006/jtbi.1997.0532.

Goher, M. E., AM, A. E. M., Abdel-Satar, A. M., Ali, M. H., Hussian, A. E., & Napiorkowska-Krzebietke, A. (2016). Biosorption of some toxic metals from aqueous solution using non-living algal cells of Chlorella vulgaris. *Journal of Elementology, 21*(3).

Gola, D., Dey, P., Bhattacharya, A., Mishra, A., Malik, A., Namburath, M., & Ahammad, S. Z. (2016). Multiple heavy metal removal using an entomopathogenic fungi Beauveria bassiana. *Bioresource technology*, 218, 388–396. doi: 10.1016/j. biortech.2016.06.096.

González-Guerrero, M., Melville, L. H., Ferrol, N., Lott, J. N., Azcon-Aguilar, C., & Peterson, R. L. (2008). Ultrastructural localization of heavy metals in the extraradical mycelium and spores of the arbuscular mycorrhizal fungus Glomus intraradices. *Canadian Journal of Microbiology, 54*, 103–110. doi: 10.1139/w07-119.

Gray, C. W., McLaren, R. G., Roberts, A. H. C., & Condron, L. M. (1998). Sorption and desorption of cadmium from some New Zealand soils: effect of pH and contact time. *Australian Journal of Soil Research, 36*(2), 199–216. doi: 10.1071/S97085.

Grichko, V. P., & Glick, B. R. (2001). Amelioration of flooding stress by ACC deaminase-containing plant growth-promoting bacteria. *Plant Physiology and Biochemistry, 39*(1), 11–17. doi: 10.1016/S0981-9428(00)01212-2.

Gube, M. (2016). Fungal molecular response to heavy metal stress. In: D. Hoffmeister (Ed.), *Biochemistry and Molecular Biology*. Cham: Springer International Publishing, 47–68.

Gupta, P., & Diwan, B. (2017). Bacterial exopolysaccharide mediated heavy metal removal: a review on biosynthesis, mechanism and remediation strategies. *Biotechnology Reports, 13*, 58–71. doi: 10.1016/j.btre.2016.12.006.

Gupta, S., & Singh, D. (2017). Role of genetically modified microorganisms in heavy metal bioremediation. In: R. Kumar, A. K. Sharma, & S. S. Ahluwalia (Eds.), *Advances in Environmental Biotechnology*. Singapore: Springer, 197–214.

Gutierrez-Corona, J. F., Romo-Rodriguez, P., Santos-Escobar, F., Espino-Saldana, A. E., & Hernandez-Escoto, H. (2016). Microbial interactions with chromium: basic biological processes and applications in environmental biotechnology. *World Journal of Microbiology & Biotechnology, 32*, 191. doi: 10.1007/s11274-016-2150-0.

Hall, J. A., Peirson, D., Ghosh, S., & Glick, B. R. (1996). Root elongation in various agronomic crops by the plant growth promoting rhizobacterium *Pseudomonas putida* GR12-2. *Israel Journal of Plant Sciences, 44*(2), 37–42.

Hložková, K., Matěnová, M., Žáčková, P., Strnad, H., Hršelová, H., Hroudová, M. (2016). Characterization of three distinct metallothionein genes of the Ag-hyperaccumulating ectomycorrhizal fungus Amanita strobiliformis. *Fungal Biology*, *120*, 358–369. doi: 10.1016/j.funbio.2015.11.007.

Hossain, M. A., Piyatida, P., da Silva, J. A. T., & Fujita, M. (2012). Molecular mechanism of heavy metal toxicity and tolerance in plants: central role of glutathione in detoxification of reactive oxygen species and methylglyoxal and in heavy metal chelation. *Journal of Botany*, *2012*, 872875. doi: 10.1155/2012/872875

Hrynkiewicz, K., & Baum, C. (2014). Application of microorganisms in bioremediation of environment from heavy metals. In: *Environmental Deterioration and Human Health*. Springer.

Huang, Y., Tao, S., & Chen, Y. J. (2005). The role of arbuscular mycorrhiza on change of heavy metal speciation in rhizosphere of maize in wastewater irrigated agriculture soil. *J. Environ. Sci.*, *17*(2), 276–280 (in Chinese).

Igiri, B. E., Okoduwa, S. I., Idoko, G. O., Akabuogu, E. P., Adeyi, A. O., & Ejiogu, I. K. (2018). Toxicity and bioremediation of heavy metals contaminated ecosystem from tannery wastewater: A review. *Journal of Toxicology*, *2018*, 1–16. https://doi.org/10.1155/2018/2568038.

Jafari, S. A., Cheraghi, S., Mirbakhsh, M., Mirza, R., & Maryamabadi, A. (2015). Employing response surface methodology for optimization of mercury bioremediation by Vibrio parahaemolyticus PG02 in coastal sediments of Bushehr, Iran. *CLEAN – Soil, Air, Water*, *43*(1), 118–126. https://doi.org/10.1002/clen.201300616.

Kader, J., Sannasi, P., Othman, O., Ismail, B. S., & Salmijah, S. (2007). Removal of Cr (VI) from aqueous solutions by growing and non-growing populations of environmental bacterial consortia. *Global Journal of Environmental Research*, *1*(1), 12–17.

Kaewdoung, B., Sutjaritvorakul, T., Gadd, G. M., Whalley, A. J. S., & Sihanonth, P. (2016). Heavy metal tolerance and biotransformation of toxic metal compounds by new isolates of wood-rotting fungi from Thailand. *Geomicrobiology Journal*, *33*, 283–288. doi: 10.1080/01490451.2015.1048394.

Kalve, S., Sarangi, B. K., Pandey, R. A., & Chakrabarti, T. (2011). Arsenic and chromium hyperaccumulation by an ecotype of Pteris vittata–prospective for phytoextraction from contaminated water and soil. *Current Science India*, *100*, 888–894.

Kanazawa, K., Higuchi, K., Nishizawa, N. K., Fushiya, S., Chino, M., & Mori, S. (1994). Nicotianamine aminotransferase activities are correlated to the phytosiderophore secretion under Fe-deficient conditions, in Gramineae. *Journal of Experimental Botany*, *45*(12), 1903–1906. doi: 10.1093/jxb/45.12.1903.

Kärenlampi, S., Schat, H., Vangronsveld, J., Verkleij, J. A. C., van der Lelie, D., Mergeay, M., & Tervahauta, A. I. (2000). Genetic engineering in the improvement of plants for phytoremediation of metal polluted soils. *Environmental Pollution*, *107*(2), 225–231. doi: 10.1016/S0269-7491(99)00141-4.

Khan, A. G., Bari, A., Chaudhry, T. M., & Qazilbash, A. A. (1997). Phytoremediation-a strategy to decontaminate heavy metal polluted soils and to conserve the biodiversity of Pakistan soils. In: S. A. Mufti, C. A. Woods, & S. A. Hasan (Eds.), *Biodiversity of Pakistan Museum of Natural History*. Islamabad and Florida Museum of Natural History.

Kim, I. H., Choi, J. H., Joo, J. O., Kim, Y. K., Choi, J. W., & Oh, B. K. (2015). Development of a microbe-zeolite carrier for the effective elimination of heavy metals from seawater. *Journal of Microbiology, Biotechnology*, *25*(9), 1542–1546. https://doi.org/10.4014/jmb.1504.04067.

Kloepper, J. W., Lifshitz, R., & Zablotowicz, R. M., 1989. Free-living bacterial inocula for enhancing crop productivity. *Trends in Biotechnology*, *7*(2), 39–44. doi: 10.1016/0167-7799(89)90057-7.

Koptsik, G. (2014). Problems and prospects concerning the phytoremediation of heavy metal polluted soils: A review. *Eurasian Soil Science*, *47*, 923–939. doi: 10.1134/S1064229314090075.

Kumar, R., Bhatia, D., Singh, R., Rani, S., & Bishnoi, N. R. (2011). Sorption of heavy metals from electroplating effluent using immobilized biomass Trichoderma viride in a continuous packed-bed column. *International Biodeterioration & Biodegradation*, *65*(8), 1133–1139. https://doi.org/10.1016/j.ibiod.2011.09.003.

Kumaran, N. S., Sundaramanicam, A., & Bragadeeswaran, S. (2011). Adsorption studies on heavy metals by isolated cyano bacterial strain (Nostoc sp.) from Uppanar estuarine water, southeast coast of India. *Journal of Applied Sciences Research*, *7*(11), 1609–1615.

Lakkireddy, K., & Kues, U. (2017). Bulk isolation of basidiospores from wild mushrooms by electrostatic attraction with low risk of microbial contaminations. *AMB Express*, *7*(1), 28. https://doi.org/10.1186/s13568-017-0326-0.

Lanfranco, L. (2007). The fine-tuning of heavy metals in mycorrhizal fungi. *New Phytology*, *174*, 3–6. doi: 10.1111/j.1469-8137.2007.02029.x.

Lanfranco, L., Bolchi, A., Ros, E. C., Ottonello, S., & Bonfante, P. (2002). Differential expression of a metallothionein gene during the presymbiotic versus the symbiotic phase of an arbuscular mycorrhizal fungus. *Plant Physiology*, *130*, 58–67. doi: 10.1104/pp.003525.

Leggett, J. E., & Epstein, E. (1956). Kinetics of sulfate adsorption by barley roots. *Plant Physiology*, *31*, 222–226.

Lemanceau, P., Bakker, P. A. H. M., Dekogel, W. J., Alabouvette, C., & Schippers, B. (1992). Effect of pseudobactin 358 produced by *Pseudomonas putida* WSC358 on suppression of *Fusarium* wilt of carnations by non pathogenic *Fusarium oxysporum*. *Applied and Environmental Microbiology*, *58*(3), 2978–2980.

Lim, H., Kim, Y., & Kim, S. (1991). *Pseudomonas stutzeri* YLP-1 genetic transformation and antifungal mechanism against *Fusarium solani*, an agent of plant root rot. *Applied and Environmental Microbiology*, *57*(2), 510–516.

Loper, J. E., & Henkels, M. D. (1999). Utilization of heterologous siderophore enhances levels of iron available to *Pseudomonas putida* in rhizosphere. *Applied and Environmental Microbiology*, *65*(12), 5357–5363.

Luna, J. M., Rufino, R. D., & Sarubbo, L. A. (2016). Biosurfactant from Candida sphaerica UCP0995 exhibiting heavy metal remediation properties. *Process Safety and Environmental Protection*, *102*, 558–566. https://doi.org/10.1016/j.psep.2016.05.010.

Luo, Q., Wang, S., Sun, L.-N., & Wang, H. (2017). Metabolic profiling of root exudates from two ecotypes of Sedum alfredii treated with Pb based on GC-MS. *Scientific Reports*, *7*, 39878. doi: 10.1038/srep39878.

Ma, Y., Prasad, M. N. V., Rajkumar, M., & Freitas, H. (2011). Plant growth promoting rhizobacteria and endophytes accelerate phytoremediation of metalliferous soils. *Biotechnology Advances*, *29*, 248–258. doi: 10.1016/j.biotechadv.2010.12.001.

Majumder, A., Bhattacharyya, K., Bhattacharyya, S., & Kole, S. C. (2013). Arsenictolerant, arsenite-oxidising bacterial strains in the contaminated soils of West Bengal, India. *Science of the Total Environment*, *46*, 1006–1014. doi: 10.1016/j.scitotenv.2013.06.068.

Mane, P. C., & Bhosle, A. B. (2012). Bioremoval of some metals by living *Algae Spirogyra sp.* and *Spirullina sp.* from aqueous solution. 571–576.

Marques, A. P., Rangel, A. O., & Castro, P. M. (2009). Remediation of heavy metal contaminated soils: Phytoremediation as a potentially promising clean-up technology. *Crit. Rev. Env. Sci. Technol.*, *39*, 622–654. doi: 10.1080/10643380701798272.

Mason, R. P., Reinfelder, J. R., & Morel, F. M. M. (1996). Uptake, toxicity, and trophic transfer of mercury in a coastal diatom. *Environmental Science & Technology*, *30*(6), 1835–1845. doi: 10.1021/es950373d.

Mayak, S., Tirosh, T., & Glick, B. R. (1999). Effect of wild-type and mutant plant growth promoting rhizobacteria on the rooting of mung bean cuttings. *Journal of Plant Growth Regulation*, *18*(2), 49–53. doi: 10.1007/PL00007047

Mayer, A. M., & Poljakoff-Mayber, A. (1989). *The Germination of Seeds*. Oxford: Pergamon Press.

Merdy, P., Gharbi, L. T., & Lucas, Y. (2009). Pb, Cu and Cr interactions with soil: sorption experiments and modelling. *Colloids and Surfaces A: Physicochemical and Engineering Aspects*, *347*, 192–199. doi: 10.1016/j.colsurfa.2009.04.004.

Migahed, F., Abdelrazak, A., & Fawzy, G. (2017). Batch and continuous removal of heavy metals from industrial effluents using microbial consortia. *International Journal of Environmental Science & Technology*, *14*, 1169–1180. doi: 10.1007/s13762-016-1229-3.

Miransari, M. (2011). Hyperaccumulators, arbuscular mycorrhizal fungi and stress of heavy metals. *Biotechnology Advances*, *29*, 645–653. doi: 10.1016/j.biotechadv.2011.04.006.

Mosa, K. A., Saadoun, I., Kumar, K., Helmy, M., & Dhankher, O. P. (2016). Potential biotechnological strategies for the cleanup of heavy metals and metalloids. *Frontiers in Plant Science*, *7*, 303. https://doi.org/10.3389/fpls.2016.00303.

Mulligan, C. N., Yong, R. N., & Gibbs, B. F. (2001). Remediation technologies for metal-contaminated soils and groundwater: An evaluation. *Engineering Geology*, *60*(1–4), 193–207. https://doi.org/10.1016/S0013-7952(00)00101-0.

Muneer, B., Iqbal, M. J., Shakoori, F. R., & Shakoori, A. R. (2013). Tolerance and biosorption of mercury by microbial consortia: Potential use in bioremediation of wastewater. *Pakistan Journal of Zoology*, *45*(1), 247–254.

Murthy, S., Bali, G., & Sarangi, S. (2011). Effect of lead on metallothionein concentration in lead resistant bacteria Bacillus cereus isolated from industrial effluent. *African Journal of. Biotechnology*, *10*, 15966–15972. doi: 10.5897/AJB11.1645.

Okolo, N. V., Olowolafe, E. A., Akawu, I., & Okoduwa, S. I. R. (2016). Effects of industrial effluents on soil resources in Challawa industrial area, Kano, Nigeria. *Journal of Global Ecology and Environment*, *5*, 1–10.

Park, C. H., Keyhan, M., & Matin, A. (1999). Purification and characterization of chromate reductase in *Pseudomonas putida. Abstract General Meeting American Society for Microbiology*, *99*(4), 536–548.

Park, D., Yun, Y. S., Jo, J. H., & Park, J. M. (2005). Mechanism of hexavalent chromium removal by dead fungal biomass of Aspergillus niger. *Water Research*, *39*(4), 533–540. https://doi.org/10.1016/j.watres.2004.11.002.

Patten, C. I., & Glick, B. R. (1996). Bacterial biosynthesis of indole-3-acetic acid. *Canadian Journal of Microbiology*, *42*(3), 207–220.

Peng, K., Luo, C., You, W., Lian, C., Li, X., & Shen, Z. (2008). Manganese uptake and interactions with cadmium in the hyperaccumulator – Phytolacca Americana L. *Journal of Hazardous Materials*, *154*, 674–681. doi: 10.1016/j.jhazmat.2007.10.080.

Penrose, D. M., & Glick, B. R. (2001). Levels of 1-aminocyclopropane-1-carboxylic acid (ACC) in exudates and extracts of canola seeds treated with plant growth-promoting bacteria. *Canadian Journal of Microbiology*, *47*(4), 368–372. doi: 10.1139/cjm-47-4-368

Pierson, L. S., & Thomashow, L. S. (1992). Cloning and heterologous expression of the phenazine biosynthetic locus from *Pseudomonas aureofaciens. Molecular Plant-Microbe Interactions*, *5*(4), 330–339.

Pokethitiyook, P., & Poolpak, T. (2016). Biosorption of heavy metal from aqueous solutions. In: A. A. Ansari, S. S. Gill, R. Gill, G. R. Lanza, & L. Newman (Eds.), *Phytoremediation: Management of Environmental Contaminants*, Vol. 3. Cham: Springer International Publishing, 113–114.

Potgieter, H., & Alexander, M. (1996). Susceptibility and resistance of several fungi to microbial lysis. *Journal of Bacteriology*, *91*(4), 1526–1532.

Puyen, Z. M., Villagrasa, E., Maldonado, J., Diestra, E., Esteve, I., & Sole, A. (2012). Biosorption of lead and copper by heavy-metal tolerant Micrococcus luteus DE2008. *Bioresource Technology*, *126*, 233–237. https://doi.org/10.1016/j.biortech.2012.09.036.

Qazilbash, A. A. (2004). *Isolation and characterization of heavy metal tolerant biota from industrially polluted soils and their role in bioremediation* [Doctoral dissertation]. Quaid-i-Azam University Islamabad.

Rasmussen, L. D., Sorensen, S. J., Turner, R. R., & Barkay, T. (2000). Application of a mer-lux biosensor for estimating bioavailable mercury in soil. *Soil Biology & Biochemistry*, *32*(5), 639–646. https://doi.org/10.1016/S0038-0717(99)00190-X.

Sakakibara, M., Ohmori, Y., Ha, N. T. H., Sano, S., & Sera, K. (2011). Phytoremediation of heavy metal-contaminated water and sediment by Eleocharis acicularis. *Clean Soil Air Water*, *39*, 735–741. doi: 10.1002/clen.201000488.

Salehizadeh, H., & Shojaosadati, S. A. (2003). Removal of metal ions from aqueous solution by polysaccharide produced from Bacillus firmus. *Water Research*, *37*(17), 4231–4235. https://doi.org/10.1016/S0043-1354(03)00418-4.

Sannasi, P., Kader, J., Othman, O., & Salmijah, S. (2006). Single and multi-metal removal by an environmental mixed bacterial isolate. In: *Modern Multidisciplinary Applied Microbiology: Exploiting Microbes and their Interactions*, 136–141.

Schnoor, J. (1997). *Phytoremediation. Technology Evaluation Report TE-98-01*. Pittsburgh, PA: Groundwater Remediation Technologies Analysis Center.

Seneviratne, M., Seneviratne, G., Madawala, H., & Vithanage, M. (2017). Role of rhizospheric microbes in heavy metal uptake by plants. In: J. S. Singh & G. Seneviratne (Eds.), *Agro-Environmental Sustainability: Managing Environmental Pollution*, Vol. 2. Cham: Springer International Publishing, 147–163.

Sharma, J., Shamim, K., Dubey, S. K., & Meena, R. M. (2017). Metallothionein assisted periplasmic lead sequestration as lead sulfite by Providencia vermicola strain SJ2A. *Science of the Total Environment*, *579*, 359–365. doi: 10.1016/j.scitotenv.2016.11.089.

Shi, L., Dong, H., Reguera, G., Beyenal, H., Lu, A., Liu, J., *et al.* (2016). Extracellular electron transfer mechanisms between microorganisms and minerals. *Nature Reviews Microbiology*, *14*, 651–662. doi: 10.1038/nrmicro.2016.93.

Siddiquee, S., Rovina, K., Azad, S. A., Naher, L., Suryani, S., & Chaikaew, P. (2015). Heavy metal contaminants removal from wastewater using the potential filamentous fungi biomass: A review. *Journal of Microbial and Biochemical Technology*, *7*(6), 384–395. https://doi.org/10.4172/1948-5948.1000243.

Sitaula, B. K., Almas, A., Bakken, L. R., & Singh, B. R. (1999). Assessment of heavy metals associated with bacteria in soil. *Soil Biology and Biochemistry*, *31*(2), 315–316. doi: 10.1016/S0038-0717(98)00104-7.

Smalle, J., & van der Straeten, J. D. (1997). Ethylene and vegetative development. *Physiologia Plantarum*, *100*(3), 593–605. doi: 10.1034/j.1399-3054.1997.1000322.x.

Smith, E., Thavamani, P., Ramadass, K., Naidu, R., Srivastava, P., & Megharaj, M. (2015). Remediation trials for hydrocarbon-contaminated soils in arid environments: Evaluation of bioslurry and biopiling techniques. *International Biodeterioration & Biodegradation*, *101*, 56–65. https://doi.org/10.1016/j.ibiod.2015.03.029.

Srivastava, S., Agrawal, S. B., & Mondal, M. K. (2015). A review on progress of heavy metal removal using adsorbents of microbial and plant origin. *Environmental Science and Pollution Research*, *22*(20), 15386–15415. https://doi.org/10.1007/s11356-015-5278-9.

Srivastava, S., & Thakur, I. S. (2006). Isolation and process parameter optimization of Aspergillus sp. for removal of chromium from tannery effluent. *Bioresource Technology*, *97*(10), 1167–1173. https://doi.org/10.1016/j.biortech.2005.05.012.

Su, C. (2014). A review on heavy metal contamination in the soil worldwide: Situation, impact and remediation techniques. *Environmental Skeptics and Critics*, *3*(2), 24.

Sytar, O., Kumar, A., Latowski, D., Kuczynska, P., Strzałka, K., & Prasad, M. N. V. (2013). Heavy metal-induced oxidative damage, defense reactions, and detoxification mechanisms in plants. *Acta Physiologiae Plantarum*, *35*, 985–999. doi: 10.1007/s11738-012-1169-6

Taştan, B. E., Ertuğrul, S., & Dönmez, G. (2010). Effective bioremoval of reactive dye and heavy metals by Aspergillus versicolor. *Bioresource Technology*, *101*(3), 870–876. https://doi.org/10.1016/j.biortech.2009.08.099.

Tchounwou, P. B., Yedjou, C. G., Patlolla, A. K., & Sutton, D. J. (2012). Heavy metals toxicity and the environment. *EXS*, *101*, 133–164. doi: 10.1007/978-3-7643-8340-4_6.

Thijs, S., Sillen, W., Weyens, N., & Vangronsveld, J. (2017). Phytoremediation: state-of-the-art and a key role for the plant microbiome in future trends and research prospects. *International Journal of Phytoremediation*, *19*, 23–38. doi: 10.1080/15226514.2016.1216076.

Tiwari, S., Singh, S. N., & Garg, S. K. (2013). Microbially enhanced phytoextraction of heavy-metal fly-ash amended soil. *Communications in Soil Science and Plant Analysis*, *44*(21), 3161–3176. https://doi.org/10.1080/00103624.2013.832287.

Turnau, K., & Kottke, I. (2005). Fungal activity as determined by microscale methods with special emphasis on interactions with heavy metals. In: J. Dighton & J. F. White (Eds.), *The Fungal Community*. Boca Raton: CRC Press, 287–305.

Ullah, A., Heng, S., Munis, M. F. H., Fahad, S., & Yang, X. (2015). Phytoremediation of heavy metals assisted by plant growth promoting (PGP) bacteria: A review. *Environmental and Experimental Botany*, *117*, 28–40. doi: 10.1016/j.envexpbot.2015.05.001.

Van Loon, L. C., 1984. Regulation of pathogenesis and symptom expression in diseased plants by ethylene. In: Y. Fuchs & E. Chalutz (Eds.), *Ethylene: Biochemical, Physiological and Applied Aspects*. The Hague, The Netherlands: Martinus Nijhoff/Dr. W. Junk, 171–180.

Velazhahan, R., Samiyappan, R., & Vidhyasekaran, P. (1999). Relationship between antagonistic activities of *Pseudomonas fluorescens* isolates against *Rhizoctonia solani* and their production of lytic enzyme. *Journal of Plant Diseases and Protection*, *106*(3), 244–250.

Wang, J., Feng, X., Anderson, C. W., Xing, Y., & Shang, L. (2012). Remediation of mercury contaminated sites – A review. *Journal of Hazardous Materials*, *221*, 1–18. doi: 10.1016/j.jhazmat.2012.04.035.

Wallace, A., Wallace, G. A., & Cha, J. W. (1992). Some modifications in trace elements toxicities and deficiencies in plants resulting from interactions with other elements and chelating agents. The special case of iron. *Journal of Plant Nutrition*, *15*(2), 1589–1598.

Xie, H., Pasternak, J. J., & Glick, B. R. (1996). Isolation and characterization of mutants of the plant growth-promoting rhizobacterium *Pseudomonas putida* GR12-2 that overproduce indoleacetic acid. *Current Microbiology*, *32*(2), 67–71. doi: 10.1007/s002849900012.

Yang, S. X., Deng, H., & Li, M. S. (2008). Manganese uptake and accumulation in a woody hyperaccumulator, Schima superba. *Plant, Soil and Environment 54*, 441–446. doi: 10.17221/401-PSE.

Yao, Q., Yang, R., Long, L., & Zhu, H. (2014). Phosphate application enhances the resistance of arbuscular mycorrhizae in clover plants to cadmium via polyphosphate accumulation in fungal hyphae. *Environmental and Experimental Botany 108*, 63–70. doi: 10.1016/j.envexpbot.2013.11.007.

Zayed, A. M., Lytle, C. M., & Terry, N. (1998). Accumulation and volatilization of different chemical species of selenium by plants. *Planta*, *206*(2), 284–292. doi: 10.1007/s004250050402.

Zhang, W. J., Jiang, F. B., & Ou, J. F. (2011). Global pesticide consumption and pollution: With China as a focus. *Proceedings of the International Academy of Ecology and Environmental Sciences*, *1*(2), 125–144.

Zhuang, P., Li, Z. A., McBride, M. B., & Zou, B. (2013). Health risk assessment for consumption of fish originating from ponds near Dabaoshan mine, South China. *Environmental Science and Pollution Research*, *20*, 5844–5854. doi: 10.1007/s11356-013-1606-0.

Zhuang, P., Lu, H., Li, Z., Zou, B., & McBride, M. B. (2014). Multiple exposure and effects assessment of heavy metals in the population near mining area in South China. *PLoS ONE*, *9*, e94484. doi: 10.1371/journal.pone.0094484.

3

Responses and Adaptation of Photosynthesis and Respiration under Heavy Metal Stress

Ranjan Kumar Sahoo
Department of Biotechnology, Centurion University of Technology and Management, Odisha, India

Madhusmita Pradhan
PG Department of Botany, Utkal University, Bhubaneswar, Odisha, India

Manjulata Palei
Department of Botany, Centurion University of Technology and Management, Odisha, India

Sagar Maitra
Department of Agronomy and Agroforestry, Centurion University of Technology and Management, Odisha, India

CONTENTS

3.1 Introduction

Plants typically remobilize nutrients from mature, senescent leaves to new and reproductive leaves. Heavy metal exposure is one of the main causes of leaf senescence. Heavy metal is a concept that is used to cover a large variety of chemical elements with specific weights greater than 5 g/cm^3 (Sall et al. 2020). Some metals are important nutrients necessary for plant growth in plant nutrition, but others can be potentially toxic if they are overly consumed by the plant and can be transformed into altered biochemical and physiological activities (Bhat et al. 2019). Photosynthesis, cellular elongation, respiration, and many other physiological processes may be inhibited. The toxicity of heavy metals is the direct generation of reative oxygen species (ROS) by oxidation and Fenton reaction, or by blocking important biomolecular function groups, or by substituting other essential metal ions for essential metal ions (Sall et al. 2020). Transition metals (Cu, Fe, Mn, etc.) in their orbitals have an unpaired electron that allows them to accept or donate a single electron. It is possible to transfer this electron to ground-state oxygen O_2 and thus produce ROS. The inactivation of antioxidant enzymes responsible for the scavenging of ROS is another mechanism for targeting heavy metals. In general, toxicity of heavy metals is induced when they attain a critical concentration value and especially when they are in their accessible form (Nagajyoti et al. 2010). In plants, the most common effect of heavy metals is their attack on the photosynthetic apparatus. These properties are common to all heavy metals and are not unique to a specific metal, making the measurement of photosynthetic activities a successful screening tool for heavy metal stress detection (Appenroth 2010). The influence on respiration rates of elevated metal concentrations depends on the degree of metal stress. Compared with control conditions, mild metal stress increases dark respiration rates, while O_2 intake or CO_2 release decreases under extreme metal stress, indicating metabolic damage. If natural control conditions are compared to slightly elevated metal concentrations, metal deficiency can

DOI: 10.1201/9781003155089-3

lead to increased respiration rates when micronutrient efficiency is decreased in metal-tolerant organisms. Excess Cu or Mn concentration reduces respiration rates (Błażewicz et al. 2013). Heavy metals are likely to directly impact the cytoplasm and disrupt the structures of the mitochondria. In terrestrial plants, where treatment with cadmium (Cd) or zinc (Zn) results in increased respiration, the exclusion or sequestration of metals will effectively protect sensitive enzymes at relatively high external concentrations from excess metal ions. Increased energy demand for active heavy metal exclusion or sequestration can be fulfilled by increased net breathing rates. In addition, decreased photophosphorylation in heavy metal-sensitive chloroplasts can boost mitochondrial energy supply requirements (Shackira and Puthur 2019).

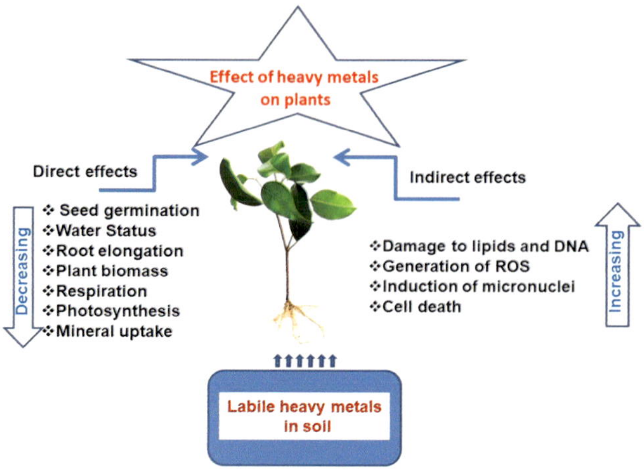

FIGURE 3.1 Direct and indirect effects of heavy metals on plants.

3.2 Heavy Metal Stress on Photosynthesis

An area of concern across the globe is heavy metal pollution of the atmosphere from various sources, including manufacturing and agriculture. Different anthropogenic practices, such as increased environmental emissions from industry, agriculture, and mining, have led to unwanted toxic accumulations of heavy metals, while heavy metals are natural components of soils at trace levels. Phytoremediation is a metal-contaminated soil cleanup technology. Different types of trees include some model phytoremediators; cottonwood, poplar, and willows are examples. Trees are suitable for heavy metal remediation because, due to their large biomass, they can tolerate higher levels of contaminants (Paz-Alberto and Sigua 2013). Owing to their size and vast root systems, they have the ability to accumulate significant quantities of toxins in their systems. In addition, trees are able to stabilize an area, prevent erosion, and minimize the distribution of pollutants. With minimal risk, they can be easily harvested and removed from the field, enabling pollutants to be removed effectively (Paz-Alberto and Sigua 2013).

For various metabolic activities in plants, some heavy metals, such as copper (Cu), Zn, cobalt (Co), and iron (Fe) are important in trace amounts. However, any kind of metal excess (essential or non-essential) adversely affects the metabolism of plants (Asati et al. 2016). Metals exert their toxic action in plants mainly by destroying chloroplasts and disrupting photosynthesis. Photosynthesis inhibition is a result of metal ion interaction with photosynthetic enzymes and chloroplast membranes (Aggarwal et al. 2011). Photosynthesis is indirectly decreased in higher plants by the accumulation of heavy metal in leaves, which affects the functioning of stomata and thus affects overall photosynthesis and transpiration rates (Figure 3.1). The reduction of photosynthetic pigments by heavy metals indirectly affects photosynthesis, so the use of non-destructive methods and ease of calculation enables photosynthetic pigments to be regularly used for regulatory purposes to assess stress (Shah et al. 2017).

The exposure of plants to heavy metals depends on an interrelated network of physiological and molecular processes, such as (i) the absorption and accumulation of metals by binding extracellular exudates and components of the cell wall; (ii) the efflux of heavy metals from cytoplasm into extracentric compartments, including vacuoles; (iii) the complexation of heavy metal ions within the cells of different substances (Cho et al. 2003).

In various heavy metals, various poplar genotypes have been used to assess their responses at physiological and molecular levels. Sebastiani et al. (2004) investigated the effects of heavy metal (Zn, Cu, chromium [Cr], and cadmium [Cd]) enriched organic waste in two poplar clones (*Populus deltoids* x maximowiczii-clone Eridano and *P.* x euramericana-clone I-214) on biomass portioning and heavy metal accumulation in plant organs. The physiological and molecular reactions of Cd stress in two *Populus nigra* L. were contrasted by Gaudet et al. (2011). Genotypes from contrasting backgrounds. The capacity of hybrid poplar for boron (B) phytomanagement was investigated by Robinson et al. (2007) using a lysimeter experiment and a field trial on B contaminated wood waste. *In P. nigra* and Salix alba, Zacchini et al. (2011) studied the effects of accumulation and tolerance of Cd. The basic mechanisms involved in heavy metal accumulation and detoxification have yet to be well established, despite numerous studies. In improving the potential of Salicaceae plants for phytoremediation, the characterisation of these mechanisms may be important.

3.3 Mechanism of Heavy Metal Toxicity towards Photosynthesis: Indirect and Direct Effects

Heavy metals directly or indirectly impact both light and dark photosynthesis reactions. The degree to which photosynthesis can be impaired by the toxicity of these metals depends on the way these events are studied (in vitro or in vivo) and the age of the plants. The direct effects of heavy metals on light reactions are on the evolution of O_2 photosynthesis, Nicotinamide adenine dinucleotide phosphate (NADP) decrease, and photophosphorylation. It indirectly impacts the process of synthesis and degradation of chlorophyll, changes in the ratio of Chl a/b, rivalry with other essential metals, etc. Some information concerning the potential photosynthesis pathways of heavy metal toxicity is given below.

3.3.1 Indirect Effects

Photosynthesis is indirectly decreased in higher plants by heavy metal accumulation in leaves that influences the functioning of stomata. Cd decreased net photosynthesis and transpiration in excised *Acer sacchanium* L leaves (Schlegel et al. 2006). Its primary effect is therefore proposed to be on stomatal function. Other heavy metals like Cd, Ni, and Pb also cause stomatal closure in epidermal peels on detached leaves of *Helienthus annus* L. floating on a solution containing either of these elements (Bazzaz et al. 2006). This simultaneously led to inhibition of photosynthesis and transpiration as these are stomata-related functions. This was also revealed by Ni in Zea mays L. In leaves of metal treated plants the net CO_2 fixation rate per unit chlorophyll increased, but this activity decreased when expressed on a leaf area basis (Heraut-Bron et al. 2011). Moreover Zn, Cd also decreased the water use efficiency (amount of CO_2 photosynthetically fixed per mol of water transpired) (Szopiński et al. 2019). Reduction of photosynthetic pigments by the heavy metals also indirectly influences photosynthesis (Figure 3.1).Photosynthetic pigments are readily observable and widely used for regulatory purposes to assess stress. Heavy metals decrease algae chlorophyll content, chlorophyll a/b ratio, pheophytin levels, but in most cases, they increase protochlorophyll levels and carotenoid/chlorophyll ratios (Piotrowska-Niczyporuk et al. 2012). Under the excessive concentrations of Hg, Cu, Cr, Ni, Cd, and Zn, etc., photosynthetic pigments were found to decrease (Chandra and Kang 2016). Heavy metals were found to increase or decrease the chlorophyll ratio (Yang et al.2020), depending on the metals used for experiments. A substantial decrease in the ratio of Chl a and Chl b was found in Scenedesmus treated with Cr relative to control ratios (Yang et al. 2020). The simplest explanation consistent with those findings is that metals inhibit the reductive steps in the biosynthetic path of photosynthetic pigments, in view of the high redox potential of many heavy metals. In addition, it is well known that heavy metals inhibit the main enzyme, protochlorophyllide reductase, which is involved in reducing protochlorophyll to chlorophyll (Yang et al. 2020). In addition to the above effects, heavy metals also induce a deficiency of other essential elements that inhibit metal-ion-dependent reactions, such as water photolysis, whereas the deficient essential metal can also be replaced by the abundantly toxic metal-ion reaction in metalloproteins. Patsikka et al. (2002) demonstrated in vivo that excess Cu predisposes photosystems II to photoinhibition in bean plants; the explanation behind photoinhibition is not just because of Cu toxicity, but because it competes with Fe at higher Cu concentrations and Fe deficiency causes leaf chlorophyll to decrease. For the evaluation of their potential mechanisms of action on photosynthesis, the specific information on the concentration of polluting heavy metals on the leaf and in the respective cell compartments is of significant importance.

3.3.2 Direct Effects

Through binding to the different sensitive sites of the photosynthetic apparatus, heavy metals directly influence the photosynthetic machinery. Heavy metals in chloroplast interfere with the thylakoid membrane architecture, which in turn alters some light reaction processes, especially those associated with PSII. Excess Cu has a direct impact on the fine structure of chloroplast, resulting in grana stacks and stroma lamellae degradation and a rise in the amount and size of plastoglobules and intrathylakoidal inclusions (Rottet et al.2015). These disruptions were associated with processes of enhanced lipid peroxidation by Sandmann and Boger (1980). Different authors have shown that the toxic effect of metal on phototropic species appears to be closely correlated with increased levels of lipid peroxidation and protein carbonylation, as well as the development of antioxidant protection systems (Devi and Prasad 2005; Tripathi et al. 2006). In photosynthesis, proteins and lipids located embedded in thylakoid membranes are directly involved. The inhibition of electron flow may be primarily responsible for metal-induced oxidative damage to these essential proteins and lipids (Figure 3.1). The oxidative damage is merely a function of metal toxicity, and that direct metal contact with either the donor or the PSII acceptor site contributes to the disruption of electron transport (Jaishankar et al. 2014). In photosynthetic behaviour, defence mechanisms for organisms against ROS also play an important role. Knauert and Knauer (2008) concluded that variations in species-specific sensitivities, measured as photosynthetic activity, were not due to differences in the algae's cellular ROS material, but possibly due to different ROS protection mechanisms specific to the species.

3.4 Target Sites of Heavy Metal Interaction with Photosynthesis

3.4.1 Photosynthetic Pigments

Metal stress inhibits chlorophyll biosynthesis (Küpper et al. 2002). Two main enzymes of the chlorophyll biosynthetic pathway are found to be responsive to metals: d-aminolevulinic acid (ALA)-dehydratase and protochlorophyllide reductase (Vangrosveld and Clijsters 1994). d-ALA-dehydratase, a non-membrane-bound chloroplastic enzyme, catalyses the conversion of d-ALA into porphobilinogen during chlorophyll synthesis. This enzyme is found to be sensitive to different metals (Jan et al.2015). In the green alga Chlorella vulgaris grown in media containing sublethal mercury concentrations, chlorophyll biosynthesis was strongly inhibited and protochlorophyll accumulated (Lenti et al. 2002). A more metal sensitive alga, Euglena gracilis showed a decrease in chlorophyll and a marked enhancement in protochlorophyllide levels in the presence of sublethal concentrations of Zn, Cd, and Hg (Aggarwal et al. 2011). In these reports it was speculated that metals inhibit protochlorophyllide reductase, which is responsible for the final reductive step of chlorophyll biosynthesis.

3.4.2 Photosynthetic Enzymes

In the photosynthetic electron transport chain, at least two possible metal sensitive sites can be derived from in vitro experiments: water splitting enzymes on the oxidizing side of PSII and NADPH oxidoreductase on the reduction side of PSII. The decrease in photosynthetic potential under metal stress in Euglena gracilis can be due to the inhibition of both the water splitting enzyme and NADPH-oxidoreductase (González-Moreno et al. 2006). The in vitro studies of Zn, Cd, and Hg treated plants clearly indicate that certain carbon fixation pathway enzymes, in particular energy-dependent enzymes, are also influenced by metals. The main enzymes of the Calvin cycle, ribulose-1, 5-bisphosphate carboxylase/oxygenase (Küpper et al. 2002), and phosphofructokinase are influenced by the majority of metals. Heavy metal interaction with functional SH groups is usually suggested as the mechanism of inhibition of heavy metals.

3.4.3 Photosystems

The photosystem (PS) is the most sensitive site for metal ions in the photosynthetic electron transport chain (Roach and Krieger-Liszkay 2014). All the metals examined are possible PSII inhibitors, while it has been stated that PSI is less sensitive. In algae and plants, different authors have documented the inhibitory effect of metals on the reduction side of the PSI electron transport mechanism. PSII is a large protein complex that consists of around 20 subunits of protein and is located embedded in the chloroplasts' thylakoid membrane. It catalyses the reduction of plastoquinone by light-driven electrons from water oxidized to molecular oxygen (Müh et al. 2012). Inhibition of oxygen evolution followed by quenching of variable fluorescence is the most obvious result of the toxic action of metal (Roach and Krieger-Liszkay 2014). It is found that both the donor and the acceptor sides of PSII are inhibited by metal ions, but the most susceptible site of the metal inhibitor is located on the oxidation (donor) side of PSII (Roach and Krieger-Liszkay 2014), where there is a reversible inhibition of Tyrz (the D1 protein redox-active tyrosine residue). The most vulnerable component is cytochrome b559 (cyt b559) in its high potential form, in addition to the oxygen-evolving complex. Yruela et al. (1996) found that Cu decreased the photoreduced cytb559 level and slowed down its photoreduction rate, while Jegerschold et al. (1995) observed low-potential form conversion at high concentrations of Cu. Burda et al. (2002) indicate that tyrosine Z, high and low potential cytb559, and chlorophylls are the primary target sites in photosystems II for Cu, and that these sites are the source of Cu-induced fluorescence quenching and inhibition of oxygen evolution in PSII. In addition to these heavy metal interactions with photosynthesis, the photosynthesis process also plays an important role in recovering photosynthetic species from previous heavy metal exposure. Tripathi et al. (2004) demonstrated this in Scenedesmus sp. After the move of algal cells to the basal medium, the photosynthesis was quickly restored. This is due to the photosynthesis used by the test algae as acclimatory responses as they prepare a context for the recovery of other parameters, including growth, energy generation, photosynthesis formation, and the usual catabolism formed to achieve normal conditions.

3.5 Respiration under the Influence of Heavy Metals

Less attention has been paid to the influence of transition metals and Al on plant catabolism than to their effect on other metabolic traits. Applied and ecological studies focus on the role of metal exclusion from symplast or metal sequestration in plant survival vacuoles on toxic metal concentrated substrates (Yadav 2010). Effects on photosynthesis and more recently, complex gene expression are the subject of physiological research on metabolic effects of excess heavy metal levels (Tomsett and Thurman 1988). The lack of thorough insight into heavy metal interactions with metabolic processes and other ions impedes causal understanding of mechanisms of toxicity and tolerance. In the case of heavy metal impact on plant respiration, this statement is especially valid. Higher doses of heavy metals partially impact energy fixation enzymes and carbon assimilation pathways rather than enzymes involved in dissimilation processes (Van Assche and Clijsters 1990). As a result, the harmful effects of catabolism are also less noticeable. Plant respiration, however, can be modified by higher heavy metal concentrations. The locations at which respiratory procedures take place are isolated by barriers, namely the cation-exchange ability of the cell wall and the plasma membrane, from potentially toxic metals in the atmosphere. By metal chelation or precipitation by SH-rich molecules, organic anions, or phosphate, glycolytic enzymes may be shielded from toxic metals in the cytoplasm. Before metals have immediate access to the functional elements of the citrate cycle, electron transport chain and ATPsynthase, an additional metal transfer through the outer and partly also the inner mitochondrial membrane is needed. Thus these barriers shield sensitive macromolecules from unsafe contact with metal concentrations that are too high. In microorganisms, an alternate approach has been found –the expression or activation of unique, less metal-sensitive isoenzymes. For example, some enzymes of the Krebs cycle are more Cu-resistant in tolerant strains than in sensitive strains in Mycobacterium tuberculosis avium (Wolschendorf et al. 2011) and in Saccharomyces cerevisiae (Adamo et al. 2012). However, no example of constitutive or induced expression of enzymes with greater metal tolerance in metal resistant genotypes is documented in higher plants compared to sensitive genotypes. The effects of Cu on the function of malate dehydrogenase in the roots of tolerant and intolerant clones of Agrostis stolonifera did not indicate such differential respiratory enzyme tolerance (Cox and Hutchinction 2006). Metal concentrations resulting in a 50 percent inhibition of the function of isocitrate dehydrogenase (ICDH: 0.16–0.47 mM Zn, 0.4–0.42 mM Cu and malate dehydrogenase (MDH:1.4–1.46 mM Zn, 0.38–0.39 mM Cu) were independent of the origin of metal tolerant or intolerant.

3.6 Heavy Metal Influence on Respiratory Gas Exchange

The influence on respiration rates of elevated metal concentrations depends on the degree of metal stress. Compared with control conditions, mild metal stress increases dark respiration rates, while O_2 intake or CO_2 release decreases under extreme metal stress, indicating metabolic damage. If "natural control conditions are compared to slightly elevated metal concentrations, metal deficiency can result in increased respiration rates when micronutrient efficiency in metal-tolerant species (e.g. different armeria ecotypes) is decreased." (Lindberg and Greger 2002). Excess Cu or Mn concentration decreases respiration rates (Adrees et al. 2015). Similarly, Zn^{2+}, Cd^{2+}, and Hg^{2+} treatments result in lower rates of oxygen consumption in all cases in the unicellular alga Euglena and in other aquatic plants (Adrees et al. 2015). Heavy metals are likely to directly impact the cytoplasm and disrupt the structures of the mitochondria. In terrestrial plants, where treatment with Cd or Zn results in increased respiration, the exclusion or sequestration of metals will effectively protect sensitive enzymes at relatively high external concentrations from excess metal ions. Increased energy demand for active heavy metal exclusion or sequestration can be fulfilled by increased net breathing rates. In addition, decreased photophosphorylation in heavy metal-sensitive chloroplasts can boost mitochondrial energy supply requirements (Lindberg and Greger 2002). Higher ATP requirements can lead to decreased electron flow through the alternative respiratory chain pathway. Furthermore, consumption of oxygen and oxidation of NADH can result from peroxidase activities (Kang et al. 2013). In conclusion, there is no stoichiometric relationship between net exchange of O_2 or CO_2 and consumption of tissue energy. Therefore, the metabolic expense of heavy metal stress cannot be specifically derived from its effect on the exchange of gases.

3.7 Conclusion

Human activities are strongly influenced by air quality. After the inauguration of a paved road on its south side, the Bentael Nature Reserve-Lebanon–is facing a huge threat, but not even useful research has been done in this field. Human activities are strongly influenced by air quality. After the inauguration of a paved road on its south side, the Bentael Nature Reserve-Lebanon–is facing a huge threat, but not even useful research has been done in this field. A significant source of heavy metal pollution is the transportation of cars. Aluminium is a catalyst; Cd is mainly released by diesel engines. Furthermore, lubricants contain lead and Cd. The measurements of "Urginea maritime" photosynthetic pigments demonstrated its battle against stress. One of the most commonly used techniques for calculating the degree of many abiotic stresses and following the evolution of leaf senescence over time is photosynthetic pigment content. It has been shown that by acclimating to environmental conditions, the plant is battling stress, regenerating chlorophyll after its decrease, along with a constant level of pheophytin and a higher level of carotenoids. Useful measures should be taken as a way of avoiding airborne emissions. The introduction of weather stations that provide climate pollution detectors, the physico-chemical properties of air pollutants, and their time in the atmosphere have an effect on the fauna and flora surrounding them. An urgent need is the development of biomonitoring programmes. With low prices, the use of phytoremediation of soils from heavy metal deposition is of great interest. The organization of Green Belts across cities is the most significant procedure that can be carried out.

REFERENCES

Adamo, G.M., Brocca, S., Passolunghi, S., Salvato, B., Lotti, M. 2012. Laboratory evolution of copper tolerant yeast strains. *Microb Cell Fact.*11, 1.https://doi.org/10.1186/1475-2859-11-1

Adrees, M., Ali, S., Rizwan, M., Ibrahim, M., Abbas, F., Farid, M., Zia-ur-Rehman, M., Irshad, H.M.K., Bharwana, S. 2015. The effect of excess copper on growth and physiology of important food crops: a review. *Environ Sci Pollut Res Int.* 22, 8148–8162. https://doi.org/10.1007/s11356-015-4496-5

Aggarwal, A., Sharma, I., Tripathi, B., Munjal, A., Baunthiyal, M., Sharma, V. 2011. Metal Toxicity and Photosynthesis. In book: *Photosynthesis: Overviews on Recent Progress and Future Perspective*, 16 , IK International Publishing House, New Delhi, India. pp. 2011.

Appenroth, K.J. 2010. Definition of "Heavy Metals" and their role in biological systems. pp. 19–29. https://doi.org/10.1007/978-3-642-02436-8_2

Asati, A., Pichhode, M., Nikhil, K. 2016. Effect of heavy metals on plants: An overview. *IJAIEM.* 5, 56–66. https://doi.org/10.13140/RG.2.2.27583.87204

Bazzaz, F.A., Rolfe, G., Carlson, R. 2006. Effect of Cd on photosynthesis and transpiration of excised leaves of corn and sunflower. *Physiol Plant.* 32, 373–376. https://doi.org/10.1111/j.1399-3054.1974.tb03154.x

Bhat, S.A., Hassan, T., Majid, S. 2019. Heavy metal toxicity and their harmful effects on living organisms – a review. *IJMSDR.*3, 106–122.

Błażewicz, A., Klatka, M., Astel, A., Partyka, M., & Kocjan, R. (2013). Differences in trace metal concentrations (Co, Cu, Fe, Mn, Zn, Cd, And Ni) in whole blood, plasma, and urine of obese and nonobese children. *Biol Trace Elem Res.* *155*(2), 190–200. https://doi.org/10.1007/s12011-013-9783-8

Burda, K., Kruk, J.S., Schmid, G.H. 2003. Stimulation of oxygen evolution in photosystem II by copper (II) ions. *Z Naturforsch.* 57c, 853–857. https://doi.org/10.1042/bj20021265

Chandra, R., Kang, H. 2016. Mixed heavy metal stress on photosynthesis, transpiration rate, and chlorophyll content in poplar hybrids, *Forest Sci Technol.* 12, 55–61. https://doi.org/10.1080/21580103.2015.1044024

Cho, M., Chardonnens, A.N., Dietz, K.J. 2003. Differential heavy metal tolerance of Arabidopsis halleri and Arabidopsis thaliana: A leaf slice test. *New Phytol.* 158, 287–293. https://doi.org/10.1046/j.1469-8137.2003.00746.x

Cox, R., Hutchinson, T. 2006. The response of root acid phosphatase activity to heavy metal stress in tolerant and non-tolerant clones of two grass species. *New Phytol.* 86, 359–364. https://doi.org/10.1111/j.1469-8137.1980.tb01676.x

Devi, S.R., Prasad, M.N.V. 2005. Antioxidant capacity of Brassica juncea plants exposed to elevated levels of copper. *Russ J Plant Physiol.* 52, 205–208. https://doi.org/10.1007/s11183-005-0031-8

Gaudet, M., Pietrini, F., Beritognolo, I., Iori, V., Zacchini, M., Massacci, A., Mugnozza, G.S., Sabatti, M. 2011. *Intraspecific variation of physiological and molecular response to cadmium stress in Populus nigra L. Tree Physiol.* 31, 1309–1318. https://doi.org/10.1093/treephys/tpr088

González-Moreno, S., Gómez-Barrera, J., Perales, H., Moreno-Sanchez, Rafael. 2006. Multiple effects of salinity on photosynthesis of protist *Euglena gracilis*. *Physiol Plant.* 101, 777–786. https://doi.org/10.1111/j.1399-3054.1997.tb01063.x

Heraut-Bron, V., Robin, C., Varlet-Grancher, C., Afif, D. 2011. Light quality (red:far-red ratio): Does it affect photosynthetic activity, net CO2 assimilation, and morphology of young white clover leaves?. *Can J Bot.* 77, 1425–1431. https://doi.org/10.1139/b99-099

Jaishankar, M., Tseten, T., Anbalagan, N., Mathew, B.B., Beeregowda, K.N. 2014. Toxicity mechanism and health effects of some heavy metals. *Interdiscip Toxicol.* 7, 60–72.

Jan, A.T., Azam, M., Siddiqui, K., Ali, A., Choi, I., Haq, Q.M. 2015. Heavy metals and human health: mechanistic insight into toxicity and counter defense system of antioxidants. *Int J Mol Sci.*16, 29592–29630. https://doi.org/10.3390/ijms161226183

Jegerschold, C., Arellano, J.B., Schroder, W.P., van Kan, P.J.M., Baron, M., Styring, S. 1995. Copper (II) inhibition of electron transfer through photosystem II studied by EPR spectroscopy. *Biochemistry* 34, 12747–12754.

Kang, T.S., Korber, D.R., Tanaka, T. 2013. Influence of oxygen on NADH recycling and oxidative stress resistance systems in *Lactobacillus panis* PM1. *AMB Express.* 3, 10. https://doi.org/10.1186/2191-0855-3-10

Knauert, S., Knauer, K. 2008. The role of reactive oxygen species in copper toxicity to two fresh water green algae. *J Phycol.* 44, 311–319.

Küpper, H., Setlik, I., Spiller, M., Küpper, F., Prasil, O. 2002. Heavy metal-induced inhibition of photosynthesis: Targets of in vivo heavy metal chlorophyll formation. *J Phycol.* 38, 429–441. https://doi.org/10.1046/j.1529-8817.2002.t01-1-01148.x

Lenti, K., Fodor, F., Böddi, Bela. 2002. Mercury Inhibits the Activity of the NADPH: Protochlorophyllide oxidoreductase (POR). *Photosynthetica.* 40, 145–151. https://doi.org/10.1023/A:1020143602973

Lindberg, S., Greger, M. 2002. Plant genotypic differences under metal deficient and enriched conditions. https://doi.org/10.1007/978-94-017-2660-3_14

Müh, F., Glöckner, C., Hellmich, J., Zouni, A. 2012. Light-induced quinone reduction in photosystem II. *Biochim Biophys Acta.* 1817, 44–65. https://doi.org/10.1016/j.bbabio.2011.05.021

Nagajyoti, P.C., Lee, K.D., Sreekanth, T.V.M. 2010. Heavy metals, occurrence and toxicity for plants: A review. *Environ Chem Lett.* 8, 199–216. https://doi.org/10.1007/s10311-010-0297-8

Patsikka, E., Kairavuo, M., Sersen, F., Aro, E., Tyystjarvi, E. 2002. Excess Copper Predisposes Photosystem II to Photoinhibition in vivo by outcompeting Iron and causing decrease in leaf Chlorophyll. *Plant Physiol.* 129, 1359–1367.

Paz-Alberto, A.M., Sigua, G.C. 2013. Phytoremediation: A green technology to remove environmental pollutants. *Am J Clim Change.* 2, 71–86. https://doi.org/10.4236/ajcc.2013.21008

Piotrowska-Niczyporuk, A., Bajguz, A., Zambrzycka, E., Godlewska-Żyłkiewicz, B. 2012. Phytohormones as regulators of heavy metal biosorption and toxicity in green alga Chlorella vulgaris (Chlorophyceae). *Plant Physiol Biochem.* 52, 52–65. https://doi.org/10.1016/j.plaphy.2011.11.009

Roach, T., Krieger-Liszkay, A. 2014. Regulation of photosynthetic electron transport and photoinhibition. *Curr Protein Pept Sci.* 15(4), 351–362. https://doi.org/10.2174/1389203715666140327105143

Robinson, B., Green, S., Chancerel, B., Mills, T., Clothier, B. 2007. Poplar for the phytomanagement of boron contaminated sites. *Environ pol.* (Barking, Essex : 1987). 150, 225–233. https://doi.org/10.1016/j.envpol.2007.01.017

Rottet, S., Besagni, C., Kessler, F. 2015. The role of plastoglobules in thylakoid lipid remodeling during plant development. *Biochim Biophys Acta.* 1847. https://doi.org/10.1016/j.bbabio.2015.02.002

Sall, M.L., Diaw, A.K.D., Gningue-Sall, D., Efremova, Aaron, S., Aaron, J.J. 2020. Toxic heavy metals: impact on the environment and human health, and treatment with conducting organic polymers, a review. *Environ Sci Pollut Res Int.* 27, 29927–29942. https://doi.org/10.1007/s11356-020-09354-3

Sandmann, G., Boger, P. 1980. Copper mediated lipid peroxidation process in photosynthetic membranes. *Plant Phyisiol.* 63, 797–800. https://doi.org/10.1104/pp.66.5.797

Schlegel, H., Godbold, D., Hüttermann, A. 2006. Whole plant aspects of heavy metal induced changes in CO_2, uptake and water relations of spruce (*Picea abies*) seedlings. *Physiol Plant.* 69, 265–270. https://doi.org/10.1111/j.1399-3054.1987.tb04285.x

Sebastiani, L., Scebba, F., Tognetti, R. 2004. Heavy metal accumulation and growth responses in poplar clones Eridano (*Populus deltoides x maximowiczii*) and I-214 (*P. euramericana*) exposed to industrial waste. *Environ. Exp Bot.* 52, 7988. https://doi.org/10.1016/j.envexpbot.2004.01.003

Shackira, A. M., Puthur, J.T. 2019. Cd^{2+} influences metabolism and elemental distribution in roots of *Acanthus ilicifolius* L., *Int J Phytoremediation*, 21(9), 866–877. https://doi.org/10.1080/15226514.2019.1577356

Shah, S.H., Houborg, R., McCabe, M.F. 2017. Response of chlorophyll, carotenoid and SPAD-502 measurement to salinity and nutrient stress in wheat (*Triticum aestivum* L.) *Agronomy.* 7, 61. https://doi.org/10.3390/agronomy7030061

Szopiński, M., Sitko, K., Gieroń, Z., Rusinowski, S., Corso, M., Hermans, C., Verbruggen, N., Małkowski, E. 2019. Toxic effects of Cd and Zn on the photosynthetic apparatus of the *Arabidopsis halleri* and *Arabidopsis arenosa* Pseudo-Metallophytes. *Front Plant Sci.* 10, 748. https://doi.org/10.3389/fpls.2019.00748

Tomsett, A.B., Thurman, D.A. 1988. Molecular biology of metal tolerances of plants. *Plant Cell Environ.* 11, 383–394. https://doi.org/10.1111/j.1365-3040.1988.tb01362.x

Tripathi, B.N., Mehta, S.K., Guar, J.P. 2004. Recovery of uptake and assimilation of nitrate in *Scenedesmus* sp. Previously exposed to elevated levels of Cu2+ and Zn2+. *J Plant Physiol.* 161, 543–547. https://doi.org/10.1078/0176-1617-01238

Tripathi, B.N., Mehta, S.K., Amar, A., Guar, J.P. 2006. Oxidative stress in Scenedesmus sp. during short- and long–term exposure to Cu^{2+} and Zn^{2+}. *Chemosphere.* 62, 538–44.

Tripathi, R.D., Srivastava, S., Mishra, S., Singh, N., Tuli, R., Gupta, D.K., Maathuis, F.J.M. 2006. Arsenic hazards: Strategies for tolerance and remediation by plants. *Trends Biotech.* 25, 158–165. https://doi.org/10.1007/978-3-642-21408-0_9

Van Assche, F., Clijsters, H. 1990. Effect of metals on enzyme activity in plants. *Plant Cell Environ.* 13, 195–206. https://doi.org/10.1111/j.1365-3040.1990.tb01304.x

Vangronsveld, J., Clijsters, H. 1994. Toxic effects of metals. In: ME Farago (ed.), Plant and the chemical elements: Biochemistry, Uptake, Tolerance and Toxicity. VCH Verlagsgesellschaft, Weinheim, pp. 150–177.

Wolschendorf, F., Ackart, D., Shrestha, T.B., Hascall-Dove, L., Nolan, S., Lamichhane, G., Wang, Y., Bossmann, S.H., Basaraba, R.J., Niederweis, M. 2011. Copper resistance is essential for virulence of Mycobacterium tuberculosis. *Proc Natl Acad Sci U S A.* 108, 1621–1626. https://doi.org/10.1073/pnas.1009261108

Yadav, S.K. 2010. Heavy metals toxicity in plants: An overview on the role of glutathione and phytochelatins in heavy metal stress tolerance of plants. *S AFR J BOT.* 76, 167–179.

Yang, Y., Zhang, L., Huang, X., Zhou, Y., Quan, Q., Li, Y., Zhu, X. 2020. Response of photosynthesis to different concentrations of heavy metals in Davidia involucrata. *PLoS ONE.* 15(3), e0228563. https://doi.org/10.1371/journal.pone.0228563

Yruela, I., Pueyo, J.J., Alonso, P.J., Picorel, R. 1996. Photoinhibition of photosystems II from higher plants: effect of copper inhibition. *J Biol Chem.* 271, 27408–27415. https://doi.org/10.1074/jbc.271.44.27408

Zacchini, M., Iori, V., Mugnozza, G., Pietrini, F., Massacci, A. 2011. Cadmium accumulation and tolerance in Populus nigra and Salix alba. *Biol Plant.* 55, 383–386. https://doi.org/10.1007/s10535-011-0060-4

4

Transcription Factors Involved in Plant Responses to Heavy Metal Stress Adaptation

Wasifa Hafiz Shah, Aadil Rasool, Seerat Saleem, and Naveed Ul Mushtaq
Department of Bioresources, University of Kashmir, Srinagar, Jammu and Kashmir, India

Inayatullah Tahir
Department of Botany, University of Kashmir, Srinagar, Jammu and Kashmir, India

Reiaz Ul Rehman
Department of Bioresources, University of Kashmir, Srinagar, Jammu and Kashmir, India

CONTENTS

4.1 Introduction

Plants are sessile, which depend on stimuli perception to react to an unfavorable natural condition. They are ceaselessly exposed to numerous biotic and abiotic stresses. Abiotic stress is the negative impact of non-living factors on the living organisms in a specific environment. Among the abiotic stress influencing plants, heavy metal (HM) stress is considered one of the most deleterious because of its exponential increase in the ecosystem due to anthropogenic activities. Industrialization, ever-expanding urbanization, and pollutants from industries are directly responsible for the augmentation of HM in the environment which in turn, results to be phytotoxic by strongly affecting plant physiology. Aggregation of heavy metals is an issue of principal significance for biological, natural, and health reasons because of the ramifications instigated by human influence on characteristic biogeochemical cycles (Ali et al. 2013; Emamverdian et al. 2015). HM alludes to any metallic component with a moderately high thickness that is harmful even in low quantities. By and large, metals, and metalloids with an atomic density of more than 4 g·cm^{-3} may be categorized into heavy metals (Hawkes 1997). HMs are non-biodegradable, inorganic compounds with a nuclear mass of more than 20 units, with cytotoxic, genotoxic, and mutagenic impacts on plants and animals.

DOI: 10.1201/9781003155089-4

There are two sorts of metals found in soils, referred to as essential and non-essential micronutrients required for typical plant development (Zhou et al. 2014). Most HMs does not assume a crucial role in plants however, metals such as manganese (Mn), zinc (Zn), nickel (Ni), copper (Cu), molybdenum (Mo), and cobalt (Co) are indispensable for vital biological processes in average concentrations (Shahid et al. 2015; Tiwari and Lata 2018). Common heavy metals like cadmium (Cd), silver (Ag), chromium (Cr), platinum (Pt), lead (Pb), arsenic (Ar), iron (Fe), Zn, Co, and Ni present beyond suboptimal concentrations diminish the crop productivity (Pierart et al. 2015). They negatively change the texture, pH, and chemistry of the soil. Heavy metals cause direct and indirect reduction of plant growth by adversely affecting various physiological and molecular attributes (Hassan et al. 2017). HMs spoil natural ways of life, soil, water systems. They can accumulate both underground and over soil surfaces (Wuana and Okieimen 2011; Patra et al. 2004). Although plants require essential elements in minuscule amounts for their development, they play a crucial role in proteins and enzyme structure. The concentration of both these essential and non-essential metals is of prime importance, as their presence in abundance can prompt decreased development in plants. HM stress poses a severe danger to human well-being (Khan et al. 2015). As heavy metals can be transported through the food chain into animals and humans, their presence causes a significant threat to human health (Nagajyoti et al. 2010). HMs reduce biomass accumulation, inhibit growth and photosynthesis, alter water equilibrium and nutrient absorption, and cause senescence. Toxic levels of HMs hamper ordinary plant functioning and pose an obstruction to metabolic cycles in an assortment of ways. HMs form bonds with sulfhydryl groups of proteins and render them functionless (Hall 2002; Hossain et al. 2012). They block functional groups of important cellular molecules, disrupt membranes, and pigments (Farid et al. 2013; Hossain et al. 2012). Thus, they bring restraint to essential functions in plants, such as photosynthesis and respiration (Hossain et al. 2012; Kim et al. 2014). When the HM stress is prolonged, they prompt the plant to produce reactive oxygen species (ROS) like superoxide, hydroxyl, peroxyl, and alkoxyl species. In some instances, they may produce hypochlorous acid, singlet oxygen, hydrogen peroxide (Tamás et al. 2017). ROS can be generated in various ways under heavy metal stress like directly through the Haber-Weiss/Fenton reaction, by stimulating NADPH oxidases (NOXs), or by displacement of essential cations through enzyme inhibition (Shahid et al. 2014).

Plants have developed numerous defense mechanisms to withstand HM toxicity. Plants reduce heavy metal uptake and sequester metals into vacuoles and activate various antioxidants (Shahid et al. 2015). Whenever confronted with the stressful condition caused by the high concentrations of HMs, plants adopt an avoidance strategy to preclude the onset of stress by immobilization of metals by mycorrhizal association, metal sequestration, or complexation by exuding organic compounds from the root. Thus restricting metal uptake from soil or excluding it, preventing metal entry into plant roots (Patra et al. 2004; Dalvi and Bhalerao 2013). If the stress persists, HMs enter inside plant tissues, which activates detoxification mechanisms like metal sequestration and compartmentalization in various intracellular compartments like vacuoles (Patra et al. 2004). Also, the accumulation of osmolytes and osmoprotectants like proline and intracellular chelation of metal ions by releasing several substances, like, organic acids, polysaccharides, phytochelatins, and metallothioneins takes place (Dalvi and Bhalerao 2013; John et al. 2009). After that, the activation of antioxidant defense mechanisms is stimulated. Various omics approaches such as transcriptomics, proteomics, and metabolomics are employed to know the necessary regulatory machinery involved in response to HM tolerance. This review gives a comprehensive account of how plants activate their transcription machinery to withstand HM toxicity. This chapter majorly emphasizes on impact and mechanism of action of HM, and transcription factors (TFs) triggered by heavy metal stress, and the effect of HM on downstream plant signaling defense responses.

4.2 Source of Heavy Metals

HMs are vital only in trace amount while their excess amount is deleterious. Though there is noprecise definition for heavy metal but literally they are described as elements with higher atomic weight and density than water (Bánfalvi 2011). These metals don't decay easily and remain in soil for longer periods. Their prolonged stay reduces the agricultural capacity of soil and leads to food insecurity. Various different sources of heavy metals are discussed below.

4.2.1 Natural Source

The natural weathering processes of parent rocks are the main source of heavy metals and produce HMs at trace levels (<1000 mg kg^{-1}) which are rarely toxic (Kabata-Pendias and Pendias 2001; Pierzynski 2000). The acceleration of the same process by anthropogenic activity results in accumulation of these metals above the defined safe values, high enough to pose threat to living beings (D'Amore et al. 2005). Different parent rocks are the source of different heavy metals to soils like igneous rocks, like olivine, hornblende and augite add a significant amount of Ni, Mn, Cu, Co, and Zn. In soil, shale crust, a sedimentary rock augments the concentrations of Cr, Mn, Co, Ni, Cu, Zn, Cd, Hg, and Pb. Sandstone and limestone are considered as an important source of diverse HM. Volcanic eruptions and forest fires release HMs into the environment in larger amounts which accumulate in the soil. Fossils also add different HMs to the soil which mainly include Cr, Mn, Co, Ni, Cu, Zn, Cd, Sn, Hg, and Pb to the soil.

4.2.2 Anthropogenic Source

Various anthropogenic activities lead to the accumulation of heavy metal in the soil like rapidly expanding industrial areas, previous use of site, mining, metal waste, paints, fertilizers, manures, sewage waste, domestic effluents, pesticides, fungicides, irrigation using wastewater, use of contaminated fill for leveling the site, coal combustion, spillage of petrochemicals, bonfires, dumping of waste (Khan et al. 2008; Zhang et al. 2010). Main anthropogenic sources of heavy metals are discussed below.

4.2.2.1 *Industrial Source*

Due to processing of metal ores in industries at high temperature, HMs enter the atmosphere in particulate and vapor forms. HMs like As, Cd, Cu, Pb, Sn, and Zn in vapor form combine with water and form aerosols, which are dispersed by wind or precipitated with rainfall resulting in contamination of soil. Likewise, Mining, erosion of mine wastes, metal corrosion, and leaching of HMs also contaminate soil and groundwater system. Industries involving plastics, textiles, microelectronics, wood, and paper processing contribute HMs to the soil.

4.2.2.2 *Fertilizers*

Some soils lack the potential heavy metals which play an essential role in plant growth and are manually added to the soil. HMs like Co, Cu, Mn, Fe, Ni, Mo, and Zn are supplied to the soil either as soil fertilizer or foliar spray. Agricultural soils are treated with large quantities of fertilizers on regular basis to provide adequate nutrients for proper growth of crop plants. Most of these fertilizers contain cadmium and lead as impurities, which significantly add their content in the soil. Certain phosphate fertilizer adds Cd, F, Hg, and Pb to the soil inadvertently (Jones et al. 1981).

4.2.2.3 *Pesticides*

Broad variety of pesticides is used in agriculture which contains significant amounts of metal to the soil. Various fungicides are copper based like Bordeaux mixture (copper sulfate) and copper oxy-chloride add a considerable amount of copper to the soil (Jones et al. 1981). Lead arsenate and different arsenic compounds used as pesticides for fruit orchards and banana crops add As to the soil.

4.2.2.4 *Manures*

Manures obtained from livestock, compost and sewage waste added to the soil considerably accumulate HMs like As, Cd, Cr, Cu, Pb, Hg, Ni, Mo, and Zn in it. Cu, Zn, and As are usually added to the diets of cattle, pigs, and poultry as growth regulators, and the manures obtained from them can potentially add Cu and Zn to the soil if applied repeatedly (Sumner 2000). Bio-solids obtained from the sewage waste, produced by wastewater treatment processes added to the soil for agricultural purposes have the potential to amass the heavy metals in it (USEPA, 2000). Likewise, the compost materials like garden waste, sawdust, and straw dumped in the soil also add HMs to it. Metal concentrations in manures depend on nature and intensity of the industrial activity and on the type of process followed in bio-solids treatment (Mattigod and Page 1983).

4.2.3 Dynamics of Heavy Metals in Soil

Soil dynamics of HMs depend on the type of soil and its properties like pH, moisture content, temperature, composition, and nutrient content (Gu et al. 2016; Hu et al. 2017a). The most abundantly accumulated HMs in the soil include Pb, Cr, As, Zn, Cd Cu, and Hg. Mostly the HMs are adsorbed by the inorganic, organic, and colloidal components, which tend to moderate the HM mobility in soil. This results in the potential availability of metals to the living beings especially plants. These metals have the capability of decreasing crop production by accumulating in the plant parts and hampering their vital processes. HMs are mostly found adsorbed on the colloidal particles like humus, soluble salts (calcium carbonate), alumino-silicate clays, and hydrous oxides of metals like Fe, Mn, and Al. On humus, metals bound to a range of sites, potentially at the sites where charged ions are present. The anions generate a slight coulombic attraction toward the positively charged heavy metals. Like Nitrogen and sulfur shows bonding site for Cu and Hg, respectively (Skyllberg et al. 2006; Tipping 2002). HM binds to humus either through carboxylic or hydroxyl groups. Humic and fulvic acids are the main ingredients of humus which are responsible for mineral adsorption. For adsorption on Fe or Mn hydrous oxides, HMs compete with humic and fulvic acids for their preferred adsorption sites. HMs prefer bonding with single or multiple groups on a single humic molecule (Christl et al. 2005). The mobility of HMs in the soil is governed by their chemical form and speciation. In soil, heavy metals are adsorbed immediately with fast reaction in few minutes or hours and then followed by slow adsorption reactions taking days or years (Shiowatana et al. 2001). Distribution of HMs is believed to be controlled by reactions of heavy metals in soils such as (i) mineral precipitation and dissolution, (ii) ion exchange, adsorption and desorption, (iii) aqueous complexation, (iv) biological immobilization and mobilization, and (v) plant uptake (Levy et al. 1992).

4.3 Uptake and Transport of Heavy Metals

Plants have developed exceptionally explicit and extremely proficient systems to acquire micronutrients from the soil. Plant roots induce pH changes and produce various chelating agents to positively change the soil chemistry for solubilization and uptake of micronutrients even though present at shallow levels. Plants have advanced mechanisms likewise to move and store micronutrients. Plants adjust the cell's molecular and biochemical machinery when unfavorable concentrations of HMs are detected. This reaction evokes a significant signal transduction network in plant cells visible by orchestrating metal transporter proteins and metal restricting proteins (Singh et al. 2016). As the transport of essential nutrients, plants take up HMs into cytoplasmic spaces by proton pumps, transporters, and channel proteins. The transport is dependent on various factors like the pH of the soil, overall plant health, root framework, temperature, soil chemistry, and association of different microorganisms. HM present in either soluble or easily solubilized form is readily absorbed by plants (Chibuike and Obiora 2014). Plants produce various organic acids on exposure to HMs. These acids repress metal take-up by forming a complex with the rhizosphere (Larsen et al. 1998).

In non-stress conditions, plants employ different carrier proteins to maintain a threshold of essential metal ions; however, they are responsible for the build-up of ions inside the plant in stress conditions. Various transporters like the zinc-regulated, iron-regulated transporter-like proteins (ZIP), natural

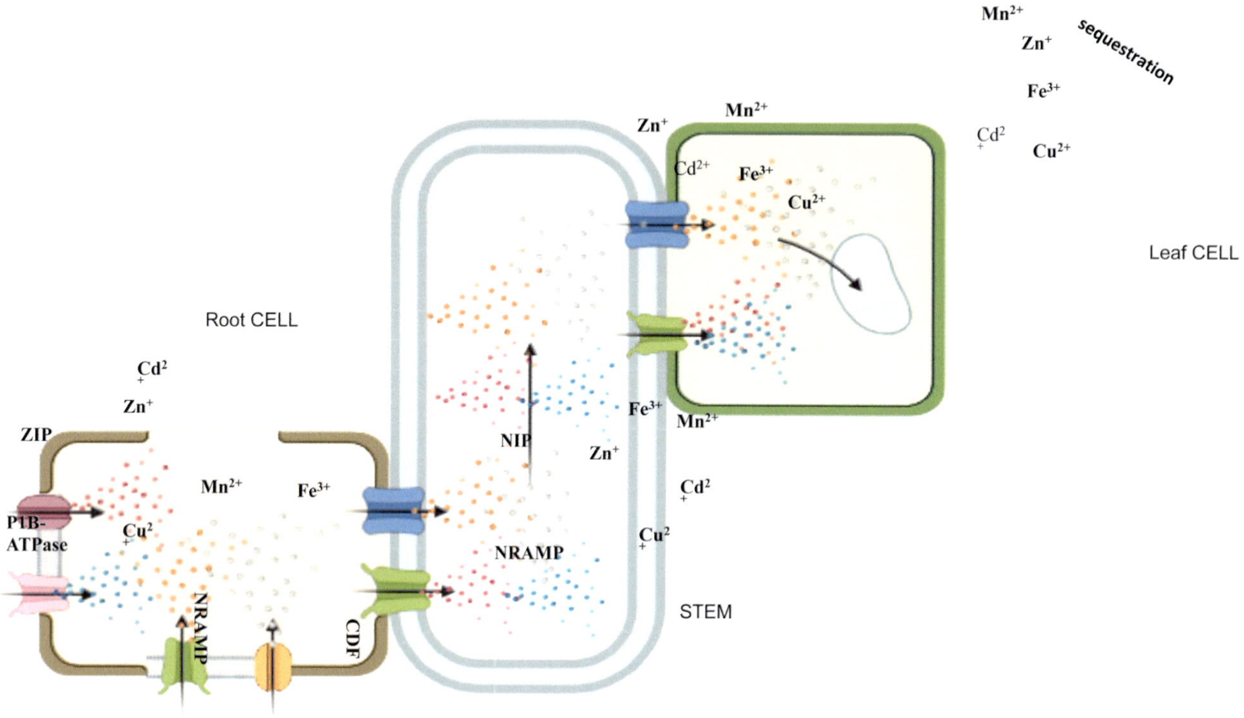

FIGURE 4.1 Transport and translocation of various heavy metals through various different transporters.

resistance-associated macrophage protein (NRAMP), P1B-ATPase, and cation diffusion facilitator (CDF) facilitate the transport of HMs (Maestri et al. 2010) (Figure 4.1). The ZIP family is involved in cellular uptake of Zn, its intracellular trafficking, and detoxification in plants. ZIP family transporters also transport other divalent metal cations such as Cadmium (Cd^{2+}), Iron (Fe^{2+}), Manganese (Mn^{2+}), Cobalt (Co^{2+}), Nickel (Ni^{2+}), and Copper Cu^{2+} (Krishna et al. 2020; Pedas and Husted 2009). The ZIP transporters are responsible for homeostasis, uptake, and usage under Zn stress conditions (Palmgren et al. 2008). Under stress conditions, the ZIP transporters located in different cell organelles are effectively engaged in Zn homeostasis. They are responsible for adapting and acclimatizing plants to low and high Zn soils (Tiong et al. 2015). The molecular characterization of the ZIP family carriers is inadequate and necessitates to be further studied (Kavitha et al. 2015). Similarly, the NRAMP family is known to transport divalent metals like Fe^{2+} and Mn^{2+} across the plasma membrane. The first NRAMP protein (NRAMP1) was found in mice's phagosomal membrane, protecting against intracellular parasites (Vidal et al. 1993). In contrast to NRAMP1, NRAMP2, also known as DMT1, is a transporter of divalent metal ions like Fe, Mn, zinc, cadmium, and lead (Garrick et al. 2006). These toxic metal ions, such as cadmium, enter plant cells by such carriers under stress conditions. Another class of metal ion transporters includes the P1B-ATPases, which couple ATP hydrolysis to transport metal cation. These proteins have been reported to transport Cu, Zn, Co. Besides transport, they perform various critical physiological tasks like ion homeostasis, micronutrient nutrition, delivery of essential metals to target proteins, and toxic metal detoxification (Williams and Mills 2005). Similarly, the CDF is another family of transporters involved in metal ion transport in plants. CDF members are also known as metal tolerance proteins (MTPs) (Migocka et al. 2015b). MTP proteins from different plant species have been divided into seven groups according to the phylogenetic analysis results and annotation performed on Arabidopsis MTPs (Gustin et al. 2011). They can transport different metals, including Zn, Cd, Co, Ni, or Fe, into plant cells' vacuoles (Migocka et al. 2015a). Once inside, various proteins mediate the transport of ions from the root to the shoot. After the roots take the metal ions, they push toward the xylem (Migeon et al. 2010). Chelation is very important for the further transport of metal ions. Various organic acids like citrates and malate's act as chelators of these metal ions (Manara 2012). The primary element controlling the movement of metal-containing sap is root pressure, xylem loading in the roots, interactions with cell walls during acropetal transport and selective removal from the xylem sap (Lasat 2002; Miyadate et al. 2011; Mari et al. 2006). The metal ions tend to accumulate primarily in photosynthetically active leaves. Metal ions like Zn, Ni, and Mn are readily transported via the xylem to the shoot. Mn is practically immobile in phloem, while Zn is more gradually reallocated via the phloem. The phloem's simplistic transport relies upon source/sink pattern (Turgeon and Wolf 2009; van Bel 2003). The transport through phloem depends on various factors, including the nature of metal ions like Zn can be relocated through the phloem to developing plant parts (Page et al. 2006; Riesen and Feller 2005). Likewise, essential metal ions like Mn and Fe are not efficiently refocused from adult organs to roots and young leaves. Metal ions can be redistributed from the shoots in various ways. They can be reallocated from the senescing leaves via the phloem to growing vegetative parts and maturing fruits or transferred to the phloem before the xylem sap reaches mesophyll cells (Page et al. 2012)). Although some translocators of HMs are known, yet we are far away from knowing the organization of translocator proteins engaged with the take-up into the roots and the complex machinery associated with translocation of HMs.

4.4 Competition of HMs with Essential Nutrients

Plants are delicate both to the inadequacy and to the abundance of HMs. As for plants, HMs like Cd, Hg, As are emphatically toxic to the metabolic machinery. Damage of agricultural soil by HMs has become a primary ecological problem. Heavy metals have shown toxic effects on plants and animals, and their degrees of toxicities vary from species to species and from metal to metal. HMs interfere with the take up and metabolism of essential micronutrients and therefore have serious consequences on plant health (Thys et al. 1991; Hernández et al. 1998). Plants fail to germinate and have serious developmental issues when exposed to even low levels of HMs like Cd (Hernández et al. 1998; Li et al. 2005). HMs cause genotoxicity by associating hereditary material by binding with the bases of DNA (Hossain and Huq 2002).

The plant development seriously reduces under high convergences of HMs (Chaves et al. 2011; Manivasagaperumal et al. 2011). The development and metabolic cycles of crops are influenced, having genuine ramifications on yield (John et al. 2009; Sinha et al. 2005). Primary plant processes like photosynthesis are adversely affected by oxidative stress brought about by HMs (Le Guédard et al. 2012). HMs disturb the organized coordination between fundamental components by changing them into futile or poisonous forms (Gill et al. 2012). HMs seriously influence other physiological processes like reduction in leaf area, protein assembly and synthesis, cell division, respiration, and root and shoot development (Soares et al. 2001). Moreover, HMs affect plant structure and physiology. Plants experience non-appearance of palisade structure, decreased leaf thickness, underlying changes in cell membrane, chloroplast, and mitochondria (Bini et al. 2012). The noxiousness of HMs on cell structure and development depends on the harmful capability of HMs (Saraf and Samant 2013). HM like Fe, when present in higher concentrations produce free radicals which may harm cell membrane, protein, and DNA and modify cell structure (Arora et al. 2002; de Dorlodot et al. 2005). Additionally, Cu, which is a basic component that assumes a significant part in different physiological capacities and plant development, when present in higher concentrations, gets harmful and influences plant development and meddles with ordinary physiological capacities (Bouazizi et al. 2010; Hänsch and Mendel 2009; Upadhyay and Panda 2009). Increased Cu concentrations cause leakage in the plasma membrane from the cells and hinder root development (Bouazizi et al. 2010). Likewise, plants have negative effects on plant structure and physiological processes. Plants experience the absence of palisade structure, reduced leaf thickness, structural changes in cell membrane, chloroplast, and mitochondria (Bini et al. 2012). The toxicity on cellular structure and function of organisms varies according to toxic potential of heavy metal (Saraf and Samant 2013). HM like Fe, when present in higher concentrations produce free radicals which may damage cell membrane, protein, and DNA and alter cellular structure (Arora et al. 2002; de Dorlodot et al. 2005). Similarly, Cu, which is an essential element that plays an important role in various physiological functions and plant growth, when present in higher concentration, becomes toxic and affects plant growth and interferes with normal physiological functions (Bouazizi et al. 2010; Hänsch and Mendel 2009; Upadhyay and Panda 2009). Excess Cu may cause root growth inhibition and damage the plasma membrane resulting in ion leakage from the cells (Bouazizi et al. 2010).

4.5 Mechanism of Plants to Alleviate HM Stress

The sessile nature of plants snatches the opportunity from them to escape from HMs in the soil. HMs make their way into the plants through specific transporters and stimulates a wide range of physiological and metabolic responses. Remarkably, the concentration of these trace metals in soil is a critical factor for determining whether the metals possess inhibitory or stimulating effects on plant growth (Emamverdian et al. 2018). Like zinc and selenium are proved to have growth stimulating and protective effects on plants, however, their high concentration can be detrimental (Rasool et al. 2020; ul Hassan et al. 2017). HMs destroy the nucleic acids by either directly binding to them and/or by cleaving them. HMs disrupt the structural conformation of the cell membrane by binding to its proteins and lipids (Aslam et al. 2014). HMs are responsible for oxidative as well as osmotic stress. HMs are known to hamper the vital metabolic processes by excessive generation of ROS which damage the nucleic acid, proteins, and lipids (Pandey et al. 2009). HM toxicity in plants results in reduced seed germination, seedling growth, and photosynthesis. HMs lead to loss of turgor, destabilize the osmotic balance, and lead to nutrient imbalance, chlorosis, necrosis, and early senescence (Gamalero et al. 2009; Rucińska-Sobkowiak 2016). In order to adapt and harbor in these environments, plants respond accordingly either by avoiding mechanisms or by tolerance. Plant avoids the HM entry by excreting cellular/root exudates, which inhibit the entry of the HMs into the cell. However, if they make their way into the cell, transporters and complex signal transduction and its components play a pivot role at biochemical and molecular levels for tolerance mechanism. The signal transduction includes sensing of stress, downstream signal relay, and initiation of specific measures to alleviate HM toxicity. Under the extreme HM toxicity, the plants' routine defense mechanism may not suffice the job, so plant activates a specialized mechanism for metal detoxification. These specialized mechanisms include metal chelation, transport, sequestration, detoxification inside the vacuole. Extensive synthesis of stress-related compounds like osmolytes, phytochelatins, metallothioneins, proteins, antioxidants, phytohormones, and different signaling molecules (Hossain et al. 2012).

4.6 Role of Transporter Proteins in Alleviating HM Stress

When plant roots perceive HM toxicity, a general response is elicited and a plant modulates its biochemical and molecular machinery accordingly. The response evoked ultimately regulates the transporter and metal-binding proteins, which helps the plant in alleviating the HM toxicity within the plant (Singh et al. 2016). Several transporters play a considerable role in heavy metal homeostasis which are present on plasma membrane or tonoplast membrane and majorly includes zinc–iron protein/permease (ZIP), NRAMPSs, copper transporters (CTRs), CPx-type ATPases/heavy metal ATPases (HMA), multidrug and toxic compound extrusion (MATE) transporters, ABC transporters, CDF family /metal transport proteins (MTPs) (Jalmi et al. 2018). ZIP family transport different divalent cations like Cd, Fe, Mn, and Zn (Guerinot 2000). In response to uptake of Zn and Fe through roots, ZIP proteins are formed. In Arabidopsis, ZIP1 and ZIP3 are induced in roots and ZIP4 are induced in both roots and shoots during Zn stress (Manara 2012). Likewise, NRAMPS are metal transporters expressed in plasma membrane and tonoplast and have the capability to translocate variety of HMs (Nevo and Nelson 2006). In *A. thaliana*, overexpression of AtNramp1 resulted in increased iron toxicity resistance (Curie et al. 2000). In Arabidopsis, overexpression of AtNRAMP3 increases Cd sensitivity and reduced accumulation of manganese (Ghori et al. 2019). Similarly for Cu transport across the plants, the CTRs family located on plasma membrane plays a very vital role due to their high affinity(Puig and Thiele 2002). In addition to Cu transport, they are also involved in growth and pollen development. CPx-type ATPases/ HMA are ATP-driven heavy metal pumps involved in the transport of soft metal cations across biological membranes. They are also responsible for the efflux of metal ions from the cell. In Arabidopsis, overexpression of AtHMA4 increases cadmium and zinc translocation from root to shoot and confers tolerance (Verret et al. 2004). In addition to the efflux of metals, certain transporters are capable of sequestering the metals into the vacuoles. ABC transporters of vacuoles are known for efficient metal chelation and transport of metal ions from cytosol into the vacuole. Among ABC transporters, half-molecule ABC transporter (HMT1) is known to transport phytochelatin-cadmium complexes into the vacuoles (Song et al. 2014). Two subfamilies of ABC transporters viz., multidrug resistance-associated protein (MRP) and the pleiotropic drug resistance (PDR) chelate, sequester and transport the metal ions into the vacuole (Ortiz et al. 1995). CDF/MTP transporter family also transports HM from cytosol to vacuole, apoplast, and endoplasmic reticulum (Kawachi et al. 2008). In addition to the transporters, vacuoles possess an ability to directly sequester the HM known as vacuole sequestration capacity (VSC). This capability of vacuoles plays an important role in metal allocation and is regulated in changing environments by the interaction of membrane-localized transporters and ion chelators (Peng and Gong 2014).

4.7 Role of Signal Transduction and Its Components in Alleviating HM Toxicity

The primary indication of HM toxicity is as similar as osmotic and oxidative stress in plants along with reduced photosynthesis and nutrient imbalance (Rucińska-Sobkowiak 2016). Different signaling networks like mitogen-activated protein kinase (MAPK) cascade, the Ca-calmodulin pathway, phytohormonal response, and ROS mediated signaling are immediately activated in response to HM toxicity. These cascades altogether lead to the regulation of different functions at physiological like accumulation of osmolytes and secondary metabolites and molecular levels like induction of transcriptional and proteomic pattern (Zarattini and Forlani 2017). MAPK cascade has a great significance and comprises of three important kinases, namely MAPK kinase kinase (MAPKKK), the MAPK kinase (MAPKK), and the MAPK. These kinases are activated by the process of phosphorylation, where the last phosphorylated kinase further phosphorylates the substrate in the cell. These phosphorylated substrates like the nuclear TFs and different proteins regulate the stress specific responsive genes. It is reported that in *Medicago sativa*, four MAPK isoforms are activated under Cu/Cd stress (Jonak et al. 2004). This cascade has been elucidated in rice and it was predicted that during heavy metal stress OsMKK4/OSMPK3 are stimulated (Rao et al. 2011). Some reports suggest that certain TFs activate the MAPK cascade under HM toxicity (Roelofs et al. 2008). While others suggest MAPK cascade either activates or represses the TF like WRKY and Zinc finger TF which further regulates the gene expression to alleviate metal stress (Pitzschke et al. 2009). In Arabidopsis, MPK3 and MPK6 are induced by $CdCl_2$ and $CuSO_4$ (Sethi et al. 2014). In response to cadmium and copper stress, rice plants showed increased transcripts of OsMPK3, OsMPK7, and OsMPK20 in leaves and roots (Yeh et al. 2007; Rao et al. 2011). Exogenous treatment of rice roots with glutathione reduced Fe-induced damage and activates MAPK (Tsai and Huang 2006). Al stress in wheat stimulates 48-kDa MAPK, which plays an important role in Al-resistant in wheat (Mossor-Pietraszewska 2001). Another important signaling cascade is calcium–calmodulin pathway. Under HM stress, the function of calcium channels gets altered and results in increased calcium influx. Calcium being the universal secondary messenger mediates the stress response and regulates the stress specific genes for tolerance. Several studies showed that exogenously applied Ca to the heavy metal stressed plant modulate their physiological machinery accordingly. Exogenous application of Ca results in increased activity of antioxidative enzymes like superoxide dismutase, ascorbate peroxidase, glutathione peroxidase (Ahmad et al. 2015). Plants possess different calcium-sensing proteins such as calmodulins (CaMs), CaM like proteins (CMLs), calcineurin B-like proteins (CBLs), and Ca^{2+}-dependent protein kinases (CDPKs). These proteins bind to Ca^{2+} and elicit different downstream signaling pathways. Under HM toxicity, Ca activates its important target protein viz., calmodulin which is actually responsible for stress alleviation. Calmodulin is the actual target for heavy metals, where they bind to the calcium binding sites like four calcium-binding sites of calmodulin are occupied by Pb and excites its activity over 90% (Ouyang and Vogel 1998). Likewise, K, Na, Mg, and Zn ions bind to the auxiliary site of calmodulin and have varying effects on its functions (Snedden and Fromm 2001). Another important component of signaling in plants involves phytohormones. Phytohormones have the capability of modifying different physiological reactions in order to combat HM stress (Fahad et al. 2015). They are required in low concentrations to control the cell permeability, secondary metabolite production, enzyme activity, plant growth, and development (Wani et al. 2016). Development and growth of plant directly depend on its photosynthetic ability and their correlation under HM is studied by Gururani et al. (2015). It was observed that the energy trapping capacity of PSII is increased by auxins under HM toxicity (Ouzounidou and Ilias 2005). Likewise, brassinosteroids improve chlorophyll content and stomatal conductivity in plants under HM toxicity (Ali et al. 2008). Exogenous application of brassinosteroids alleviated the HM toxicity by increasing the photosynthetic efficiency (Sytar et al. 2019). The crosstalk and convergence of these above mentioned signaling pathways ends up regulating TFs which in-turn regulate HM stress specific genes mostly related to metal transporters and biosynthesis of chelating compounds which ultimately helps plant to adapt in HM stress.

4.8 Role of Transcription Factors in Alleviating HM Stress

Regulation of stress-responsive genes is crucial for determining the chance of plant growth and development under HM toxic conditions. HM toxicity triggers different genes and proteins to elicit a response for stress tolerance and adaptation. The genes induced by the elicitor are grouped into two categories viz., regulatory gene and the functional genes. The TFs belong to the regulatory group and are responsible for the regulation of stress-responsive genes, either independently or synergistically. TF is known to control the target gene expression by binding to their promoter at cis-acting element by their DNA-binding domain (Nakashima et al. 2009). They also bound to the target genes by protein interacting domain responsible for their oligomerization of TFs with other regulators, which is termed as regulon (Nakashima et al. 2009). Transcriptomic analysis of different heavy metal stressed plants reveals that HM toxicity regulates different transcriptional processes by their corresponding TFs (LeDuc et al. 2006). Different classes of TFs are responsible for conferring tolerance against variable heavy metals, which mainly include, MYB (MYeloBlastosis protein), WRKY, bZIP (basic leucine Zipper), ERF (ethylene-responsive factor), ARF, DREB/CBF bHLH (Farinati et al. 2010). A summary of TFs involved in heavy metal tolerance and adaptation is given in Table 4.1.

TABLE 4.1

List of TFs Involved in Heavy Metal Tolerance and Adaptation

TF Family	Plant	TF	Metal Tolerance	Reference
WRKY	*Arabidopsis thaliana*	WRKY22 WRKY25 WRKY29	Copper, Cadmium	Opdenakker et al. (2012)
		WRKY6	Arsenic	Castrillo et al. (2013)
	Arabidopsis Spp.	WRKY45	Zinc Iron	van de Mortel et al. (2006)
Zinc-finger C2H2	*Oryza sativa*	ART1 ASR5	Aluminium	Yamaji et al. (2009), Takahashi et al. (2012)
	Arabidopsis thaliana	OXS2	Cadmium	Blanvillain et al. (2011)
		STOP1	Aluminium	Iuchi et al. (2007)
		ACE1	Copper	Xu et al. (2009)
	Nicotiana tabacum	OSISAP1	Copper, cadmium, Manganese, Calcium Zinc Lithium	Mukhopadhyay et al. (2004)
AP2/ERF	*Pinus Virginiana*	CaPF1	Cadmium Copper Zinc	Tang et al. (2005)
	Physcomitrella patens	ACP1	Cadmium Copper	Cho et al. (2007)
bHLH	*Arabidopsis thaliana*	bHLH38 bHLH39	Cadmium Iron	Wu et al. (2012)
		PYE	Iron	Long et al. (2010)
	Arabidopsis Spp.	bHLH100	Zinc Iron	van de Mortel et al. (2006)
bZIP	*Glycine max*	bZIP62	Cadmium	Chmielowska-Bąk et al. (2014)
	Brassica juncea	BjCdR15	Cadmium	Farinati et al. (2010)

4.9 MYB Transcription Factors

MYeloBlastosis (MYB) protein TF having a diverse function in eukaryotes, plays a pivot role in abiotic stress tolerance (Dubos et al. 2010; Ambawat et al. 2013; Li et al. 2015). MYB domains are conserved having 51–52 amino acid residue and on the basis of MYB domain repeats, MYB in plants are categorized into four subfamilies viz.,1R-MYB/MYB-related, R2R3-MYB, R1R2R3-MYB, and 4R-MYB (Du et al. 2013; Dubos et al. 2010). JrVHAG1 *Juglans regia*, a candidate gene confers CdCl2 stress tolerance by regulating the response through ABA and MYB transcription regulation networks (Xu et al. 2018). In Arabidopsis, Cd and Zn toxicity induce MYB4, MYB28, MYB43, MYB48, MYB72, and MYB124 (van de Mortel et al. 2008). OsMYB45 plays an important role of in Cd toxicity by showing increased Cd sensitivity with increased H_2O_2 content and decreased catalase activity (Hu et al. 2017b). OsARM1 (*ARSENITE-RESPONSIVE MYB1*) regulates As-related transporters by binding to the conserved MYB binding sites to the promoters of *OsLsi1*, *OsLsi2*, and *OsLsi6*genes (Wang et al. 2017). In *Raphanus sativus*, it was characterized that RsMYB1 plays a pivotal role in heavy metal stress by overexpressing the genes related to metal detoxification and antioxidant activity (Ai et al. 2018). Also in *Boehmeria nivea, BnMYB2* transcription was observed to up-regulated significantly under Cd stress and its overexpression in *Arabidopsis* enhanced Cd tolerance and accumulation (Zhu et al. 2020).

4.10 WRKY Transcription Factors

WRKY family is a unique TF family of higher plants having an important role in abiotic stress tolerance (Zhang and Wang 2005). WRKY TFs comprise of two structures viz., N-terminal DNA binding domain and the C-terminal zinc-finger structure (Phukan et al. 2016). DNA binding domain is based on heptapeptide, mostly WRKYGQK and also WRKYGKK, WRKYGMK, WSKYGQK, WKRYGQK, WVKYGQK, and WKKYGQK (Huang et al. 2012; Xie et al. 2005). While the zinc-finger structures mainly include C2H2 type and C2HC type (Eulgem et al. 2000). The zinc finger motifs are also found in the form of CX29HXH and CX7CX24HXC (Huang et al. 2012). WRKY TFs precisely bind to W-box in the promoters of stress-responsive genes. Short-term exposure to Cu modulates the expression of C2H2-type and ZAT12 Zinc finger TFs in plants (Opdenakker et al. 2012). On Cu and Cd metal exposure, WRKY TFs showed increased expression. It was observed that in response to Cu, WRKY22, WRKY25, and WRKY29 were overexpressed in roots and only WRKY25 and WRKY29 showed overexpression under Cd stress (Opdenakker et al. 2012). Likewise under Cd stress, overexpression of WRKY53 was observed in *T. caerulescens* (Wei et al. 2008). In Arabidopsis, Cd inducesWRKY13which modulatesPDR8 gene expression responsible for inhibiting Cd accumulation and Cd extrusion conferring Cd tolerance (Sheng et al. 2019).

4.11 bZIP Transcription Factors

Basic leucine zipper (bZIP) TFs plays a vital role in various biological processes in plants including defense against heavy metal stress. The bZIP binding domain contains a conserved region and a leucine zipper region (Hurst 1995). The conserved region comprises of N-x7-R/K motif of 18 amino acid responsible for DNA identification residues while leucine zipper region contains numerous leucine repeats responsible for DNA binding (Liu and Chu 2015; Hu et al. 2016; Glover and Harrison 1995). Based on their conserved motifs, these TFs are divided into several subfamilies. An important subfamily of bZIP TFs having abscisic acid (ABA)-responsive elements (AREB) in the promoter region, named ABA-responsive element-binding proteins or AREB binding factors (ABFs) have been recognized to play an important role in stress signaling (Yoshida et al. 2010; Fujita et al. 2013). bZIP TFs are known to be induced on Cd exposure (Ramos et al. 2007). In *Brassica juncea*, bZIP TF BjCdR15 regulates Cd transport and accumulation (Farinati et al. 2010). Likewise in transgenic tobacco, overexpression of bZIP1 from *Tamarix hispida* increased Cd tolerance (Wang et al. 2010). Overexpression of bZIP gene, BnbZIP3 from ramie (*Boehmeria nivea*) regulates heavy metal stress tolerance by improving root growth (Huang et al. 2016).

4.12 AP2/ERF Transcription Factor

APETALA 2/ethylene-responsive element binding factor (AP2/ERF) superfamily is one of the largest and classified into four main groups (Singh et al. 2002). These APETALA 2 (AP2), ethylene-responsive factor (ERF), related to ABI3 and VP1 (RAV) and dehydration-responsive element-binding proteins (DREBs) (Rashid et al. 2012). Cadmium stress is known to induce ERF1 and ERF5 in *A. thaliana* and *A. halleri* (Weber et al. 2006; Herbette et al. 2006). DREB TFs are also reported to be up-regulated upon heavy metal treatment likeDREB1A and DREB1B were overexpressed in rice under Cd stress (Ogawa et al. 2009). In transgenic tobacco, overexpression of LbDREB of *Limonium bicolor* increased the proline and soluble protein content and also elevates the K^+ to Na^+ ratio to confer tolerance under copper stress (Ban et al. 2011). Overexpression of the EsDREB2B (*Eremosparton songoricum*) conferred heavy metal tolerance in transgenic tobacco (Li et al. 2014).

4.13 NAC Transcription Factors

The plant-specific NAC (NAM, ATAF1,2, and CUC2) proteins constitute a major TF family renowned for their important implications in plant development. Observational studies have demonstrated a highly prevalence across the plant kingdom with remarkable diversity even after having the highly conserved DNA binding domain. Due to their role in regulating stress, they received great attention. Overexpression of AtSAP13 (STRESS-ACTIVATED PROTEIN KINASE) in transgenic plants confers significant tolerance toward heavy metals (Zn, Cd, and AsIII) and NAC TFs were found to interact with the AtSAP13 promoter (Shen et al. 2017; Xin et al. 2011). Overexpression of AemNAC2 genes from *Aegilops markgrafii* in wheat significantly enhances Cd tolerance (Du et al. 2020). Expression of TaHMA2 was also suppressed in transgenic wheat which demonstrates that overexpression of AemNAC2 inhibited both Cd absorption and transport in transgenic wheat (Du et al. 2020). NAC TF genes (Han682, Han2027, and Han2724) expressed differently in different tissues of the Cr stressed sunflower plant were up-regulated, especially in roots (Yuce et al. 2019).

Conclusion

Globally around 80% population of rural areas depend on agriculture for their livelihood, which contributes 30% to the overall economic growth (FAO 2012). Assumptions are in middle of the twenty-first century, population will be about 10 billion and food security is the major concern (Naika et al. 2013). To worsen the scenario, anthropogenic activities are degrading the important environmental components like soil, water, and air. Heavy metal toxicity is one such factor which is affecting the global crop productivity endlessly. Therefore, demand of adaptation strategies in HM stressed plants is need of the hour. Predominantly, expression of stress-responsive genes, particularly the TFs which regulate the signaling or other regulatory pathways, should be of prime focus.

REFERENCES

Ahmad A, Hadi F, Ali N (2015) Effective phytoextraction of cadmium (Cd) with increasing concentration of total phenolics and free proline in Cannabis sativa (L) plant under various treatments of fertilizers, plant growth regulators and sodium salt. International journal of phytoremediation 17 (1):56–65.

Ai TN, Naing AH, Yun B-W, Lim SH, Kim CK (2018) Overexpression of RsMYB1 enhances anthocyanin accumulation and heavy metal stress tolerance in transgenic petunia. Frontiers in plant science 9:1388.

Ali B, Hasan SA, Hayat S, Hayat Q, Yadav S, Fariduddin Q, Ahmad A (2008) A role for brassinosteroids in the amelioration of aluminium stress through antioxidant system in mung bean (Vigna radiata L. Wilczek). Environmental and experimental botany 62 (2):153–159.

Ali H, Khan E, Sajad MA (2013) Phytoremediation of heavy metals – concepts and applications. Chemosphere 91 (7):869–881.

Ambawat S, Sharma P, Yadav NR, Yadav RC (2013) MYB transcription factor genes as regulators for plant responses: an overview. Physiology and molecular biology of plants 19 (3):307–321.

Arora A, Sairam RK, Srivastava GC (2002) Oxidative stress and antioxidative system in plants. Current science 82 (10):1227–1238.

Aslam R, Ansari MYK, Choudhary S, Bhat TM, Jahan N (2014) Genotoxic effects of heavy metal cadmium on growth, biochemical, cyto-physiological parameters and detection of DNA polymorphism by RAPD in Capsicum annuum L. – An important spice crop of India. Saudi journal of biological sciences 21 (5):465–472.

Ban Q, Liu G, Wang Y (2011) A DREB gene from Limonium bicolor mediates molecular and physiological responses to copper stress in transgenic tobacco. Journal of plant physiology 168 (5):449–458.

Bánfalvi G (2011) Heavy metals, trace elements and their cellular effects. In: Cellular effects of heavy metals. Springer, pp. 3–28.

Bini C, Wahsha M, Fontana S, Maleci L (2012) Effects of heavy metals on morphological characteristics of *Taraxacum officinale* Web growing on mine soils in NE Italy. Journal of geochemical exploration 123:101–108.

Blanvillain R, Wei S, Wei P, Kim JH, Ow DW (2011) Stress tolerance to stress escape in plants: role of the OXS2 zinc-finger transcription factor family. The EMBO journal 30 (18):3812–3822.

Bouazizi H, Jouili H, Geitmann A, El Ferjani E (2010). Copper toxicity in expanding leaves of *Phaseolus vulgaris* L.: antioxidant enzyme response and nutrient element uptake. Ecotoxicology and environmental safety 73 (6):1304–1308.

Castrillo G, Sánchez-Bermejo E, de Lorenzo L, Crevillén P, Fraile-Escanciano A, Mohan TC, Mouriz A, Catarecha P, Sobrino-Plata J, Olsson S (2013) WRKY6 transcription factor restricts arsenate uptake and transposon activation in Arabidopsis. The plant cell 25 (8):2944–2957.

Chaves LHG, Estrela MA, de Souza RS (2011) Effect on plant growth and heavy metal accumulation by sunflower. Journal of phytology 3 (12):04–09

Chibuike GU, Obiora SC (2014) Heavy metal polluted soils: effect on plants and bioremediation methods. Applied and environmental soil science. http://dx.doi.org/10.1155/2014/752708, Article ID 752708.

Chmielowska-Bąk J, Gzyl J, Rucińska-Sobkowiak R, Arasimowicz-Jelonek M, Deckert J (2014) The new insights into cadmium sensing. Frontiers in plant science 5:245.

Cho SH, Hoang QT, Phee JW, Kim YY, Shin HY, Shin JS (2007) Modified suppression subtractive hybridization identifies an AP2-containing protein involved in metal responses in Physcomitrella patens. Molecules & cells (Springer Science & Business Media BV) 23 (1).

Christl I, Metzger A, Heidmann I, Kretzschmar R (2005) Effect of humic and fulvic acid concentrations and ionic strength on copper and lead binding. Environmental science & technology 39 (14):5319–5326.

Curie C, Alonso JM, Jean MLE, Ecker JR, Briat J-F (2000) Involvement of NRAMP1 from Arabidopsis thaliana in iron transport. Biochemical journal 347 (3):749–755.

D'Amore JJ, Al-Abed SR, Scheckel KG, Ryan JA (2005) Methods for speciation of metals in soils: a review. Journal of environmental quality 34 (5):1707–1745.

Dalvi AA, Bhalerao SA (2013) Response of plants towards heavy metal toxicity: an overview of avoidance, tolerance and uptake mechanism. Annals of plant sciences 2 (9):362–368.

De Dorlodot S, Lutts S, Bertin P (2005). Effects of ferrous iron toxicity on the growth and mineral composition of an interspecific rice. Journal of plant nutrition 28 (1):1–20.

Du HAI, Wang Y-B, Xie YI, Liang ZHE, Jiang S-J, Zhang S-S, Huang Y-B, Tang Y-X (2013) Genome-wide identification and evolutionary and expression analyses of MYB-related genes in land plants. DNA research 20 (5):437–448.

Du X, He F, Zhu B, Ren M, Tang H (2020) NAC transcription factors from Aegilops markgrafii reduce cadmium concentration in transgenic wheat. Plant soil 449:39–50.

Dubos C, Stracke R, Grotewold E, Weisshaar B, Martin C, Lepiniec L (2010) MYB transcription factors in Arabidopsis. Trends in plant science 15 (10):573–581.

Emamverdian A, Ding Y, Mokhberdoran F, Xie Y (2015) Heavy metal stress and some mechanisms of plant defense response. The scientific world journal 2015.

Emamverdian A, Ding Y, Mokhberdoran F, Xie Y (2018) Antioxidant response of bamboo (Indocalamus latifolius) as affected by heavy metal stress. Journal of elementology 23 (1):341–352.

Eulgem T, Rushton PJ, Robatzek S, Somssich IE (2000) The WRKY superfamily of plant transcription factors. Trends in plant science 5 (5):199–206.

Fahad S, Hussain S, Bano A, Saud S, Hassan S, Shan D, Khan FA, Khan F, Chen Y, Wu C (2015) Potential role of phytohormones and plant growth-promoting rhizobacteria in abiotic stresses: consequences for changing environment. Environmental science and pollution research 22 (7):4907–4921.

FAO. FAOSTAT database Rome: Food and Agriculture Organizatio. http://www.fao.org/faostat/en/#data/QCAccessed 04 May 2020.

Farid M, Shakoor MB, Ehsan S, Ali S, Zubair M, Hanif MS (2013) Morphological, physiological and biochemical responses of different plant species to Cd stress. International journal of chemical and biochemical sciences 3 (2013):53–60.

Farinati S, DalCorso G, Varotto S, Furini A (2010) The Brassica juncea BjCdR15, an ortholog of Arabidopsis TGA3, is a regulator of cadmium uptake, transport and accumulation in shoots and confers cadmium tolerance in transgenic plants. New phytologist 185 (4):964–978.

Fujita Y, Yoshida T, Yamaguchi-Shinozaki K (2013) Pivotal role of the AREB/ABF-SnRK2 pathway in ABRE-mediated transcription in response to osmotic stress in plants. Physiologia plantarum 147 (1):15–27.

Gamalero E, Lingua G, Berta G, Glick BR (2009) Beneficial role of plant growth promoting bacteria and arbuscular mycorrhizal fungi on plant responses to heavy metal stress. Canadian journal of microbiology 55 (5):501–514.

Garrick MD, Singleton ST, Vargas F, Kuo HC, Zhao L, Knöpfel M, Davidson T, Costa M, Paradkar P, Roth JA (2006) DMT1: which metals does it transport? Biological research 39 (1):79–85.

Ghori NH, Ghori T, Hayat MQ, Imadi SR, Gul A, Altay V, Ozturk M (2019) Heavy metal stress and responses in plants. International journal of environmental science and technology 16 (3):1807–1828.

Gill SS, Khan NA, Tuteja N (2012) Cadmium at high dose perturbs growth, photosynthesis and nitrogen metabolism while at low dose it up regulates sulfur assimilation and antioxidant machinery in garden cress (Lepidium sativum L.). Plant science 182:112–120.

Glover JNM, Harrison SC (1995) Crystal structure of the heterodimeric bZIP transcription factor c-Fos–c-Jun bound to DNA. Nature 373 (6511):257–261.

Gu Y-G, Lin Q, Gao Y-P (2016) Metals in exposed-lawn soils from 18 urban parks and its human health implications in southern China's largest city, Guangzhou. journal of cleaner production 115:122–129.

Guerinot ML (2000) The ZIP family of metal transporters. Biochimica et Biophysica Acta (BBA)-Biomembranes 1465 (1–2):190–198.

Gururani MA, Mohanta TK, Bae H (2015) Current understanding of the interplay between phytohormones and photosynthesis under environmental stress. International journal of molecular sciences 16 (8):19055–19085.

Gustin JL, Zanis MJ, Salt DE (2011) Structure and evolution of the plant cation diffusion facilitator family of ion transporters. BMC evolutionary biology 11 (1):76.

Hänsch R, Mendel RR (2009). Physiological functions of mineral micronutrients (cu, Zn, Mn, Fe, Ni, Mo, B, cl). Current opinion in plant biology 12 (3):259–266.

Hall Já (2002) Cellular mechanisms for heavy metal detoxification and tolerance. Journal of experimental botany 53 (366):1–11.

Hassan TU, Bano A, Naz I (2017) Alleviation of heavy metals toxicity by the application of plant growth promoting rhizobacteria and effects on wheat grown in saline sodic field. International journal of phytoremediation 19 (6):522–529.

Hawkes SJ (1997) What is a "heavy metal"? Journal of chemical education 74 (11):1374.

Herbette S, Taconnat L, Hugouvieux V, Piette L, Magniette ML, Cuine S, Auroy P, Richaud P, Forestier C, Bourguignon J (2006) Genome-wide transcriptome profiling of the early cadmium response of Arabidopsis roots and shoots. Biochimie 88 (11):1751–1765.

Hernández LE, Lozano-Rodrıguez E, Gárate A, Carpena-Ruiz R (1998) Influence of cadmium on the uptake, tissue accumulation and subcellular distribution of manganese in pea seedlings. Plant science 132 (2):139–151.

Hossain MA, Piyatida P, da Silva JAT, Fujita M (2012) Molecular mechanism of heavy metal toxicity and tolerance in plants: central role of glutathione in detoxification of reactive oxygen species and methylglyoxal and in heavy metal chelation. Journal of botany 2012.

Hossain Z, Huq F (2002) Studies on the interaction between Cd2+ ions and nucleobases and nucleotides. Journal of inorganic biochemistry 90 (3–4):97–105.

Hu B, Jia X, Hu J, Xu D, Xia F, Li Y (2017a) Assessment of heavy metal pollution and health risks in the soil-plant-human system in the Yangtze River Delta, China. International journal of environmental research and public health 14 (9):1042.

Hu S, Yu Y, Chen Q, Mu G, Shen Z, Zheng L (2017b) OsMYB45 plays an important role in rice resistance to cadmium stress. Plant science 264:1–8.

Hu W, Yang H, Yan Y, Wei Y, Tie W, Ding Z, Zuo J, Peng M, Li K (2016) Genome-wide characterization and analysis of bZIP transcription factor gene family related to abiotic stress in cassava. Scientific reports 6:22783.

Huang C, Zhou J, Jie Y, Xing H, Zhong Y, She W, Wei G, Yu W, Ma Y (2016) A ramie (Boehmeria nivea) bZIP transcription factor BnbZIP3 positively regulates drought, salinity and heavy metal tolerance. Molecular breeding 36 (8):120.

Huang S, Gao Y, Liu J, Peng X, Niu X, Fei Z, Cao S, Liu Y (2012) Genome-wide analysis of WRKY transcription factors in Solanum lycopersicum. Molecular genetics and genomics 287 (6):495–513.

Hurst HC (1995) Transcription factors 1: bZIP proteins. Protein profile 2 (2):101–168.

Iuchi S, Koyama H, Iuchi A, Kobayashi Y, Kitabayashi S, Kobayashi Y, Ikka T, Hirayama T, Shinozaki K, Kobayashi M (2007) Zinc finger protein STOP1 is critical for proton tolerance in Arabidopsis and coregulates a key gene in aluminum tolerance. Proceedings of the national academy of sciences 104 (23):9900–9905.

Jalmi SK, Bhagat PK, Verma D, Noryang S, Tayyeba S, Singh K, Sharma D, Sinha AK (2018) Traversing the links between heavy metal stress and plant signaling. Frontiers in plant science 9:12.

John R, Ahmad P, Gadgil K, Sharma S (2009) Heavy metal toxicity: Effect on plant growth, biochemical parameters and metal accumulation by *Brassica juncea* L. International journal of plant production 3 (3):65–76.

Jonak C, Nakagami H, Hirt H (2004) Heavy metal stress. Activation of distinct mitogen-activated protein kinase pathways by copper and cadmium. Plant physiology 136 (2):3276–3283.

Jones LHP, Jarvis SC, Green DJ, Hayes MHB (1981) The fate of heavy metals. In: The chemistry of soil processes. John Wiley & sons, New York, NY. pp. 593–620.

Kabata-Pendias A, Pendias H (2001) Trace elements in soils and plants. CRC Press Inc. Boca Raton, FL.

Kavitha PG, Kuruvilla S, Mathew MK (2015) Functional characterization of a transition metal ion transporter, OsZIP6 from rice (Oryza sativa L.). Plant physiology and biochemistry 97:165–174.

Kawachi M, Kobae Y, Mimura T, Maeshima M (2008) Deletion of a histidine-rich loop of AtMTP1, a vacuolar Zn2+/H+ antiporter of Arabidopsis thaliana, stimulates the transport activity. Journal of biological chemistry 283 (13):8374–8383.

Khan A, Khan S, Khan MA, Qamar Z, Waqas M (2015) The uptake and bioaccumulation of heavy metals by food plants, their effects on plants nutrients, and associated health risk: a review. Environmental science and pollution research 22 (18):13772–13799.

Khan S, Cao Q, Zheng YM, Huang YZ, Zhu YG (2008) Health risks of heavy metals in contaminated soils and food crops irrigated with wastewater in Beijing, China. Environmental pollution 152 (3):686–692.

Kim Y-H, Khan AL, Kim D-H, Lee S-Y, Kim K-M, Waqas M, Jung H-Y, Shin J-H, Kim J-G, Lee I-J (2014) Silicon mitigates heavy metal stress by regulating P-type heavy metal ATPases, Oryza sativa low silicon genes, and endogenous phytohormones. BMC plant biology 14 (1):1–13.

Krishna TPA, Maharajan T, Roch GV, Ignacimuthu S, Ceasar SA (2020) Structure, function, regulation and phylogenetic relationship of ZIP family transporters of plants. Frontiers in plant science 11.

Larsen PB, Degenhardt J, Tai C-Y, Stenzler LM, Howell SH, Kochian LV (1998) Aluminum-resistant Arabidopsis mutants that exhibit altered patterns of aluminum accumulation and organic acid release from roots. Plant physiology 117 (1):9–17.

Lasat MM (2002) Phytoextraction of toxic metals: a review of biological mechanisms. Journal of environmental quality 31 (1):109–120.

Le Guédard M, Faure O, Bessoule J-J (2012) Early changes in the fatty acid composition of photosynthetic membrane lipids from Populus nigra grown on a metallurgical landfill. Chemosphere 88 (6):693–698.

LeDuc DL, AbdelSamie M, Móntes-Bayon M, Wu CP, Reisinger SJ, Terry N (2006) Overexpressing both ATP sulfurylase and selenocysteine methyltransferase enhances selenium phytoremediation traits in Indian mustard. Environmental pollution 144 (1):70–76.

Levy DB, Barbarick KA, Siemer EG, Sommers LE (1992) Distribution and partitioning of trace metals in contaminated soils near Leadville, Colorado. Journal of environmental quality 21 (2):185–195.

Li C, Ng CKY, Fan L-M (2015) MYB transcription factors, active players in abiotic stress signaling. Environmental and experimental botany 114:80–91.

Li W, Khan MA, Yamaguchi S, Kamiya Y (2005) Effects of heavy metals on seed germination and early seedling growth of Arabidopsis thaliana. Plant growth regulation 46 (1):45–50.

Li X, Zhang D, Li H, Wang Y, Zhang Y, Wood AJ (2014) EsDREB2B, a novel truncated DREB2-type transcription factor in the desert legume Eremosparton songoricum, enhances tolerance to multiple abiotic stresses in yeast and transgenic tobacco. BMC plant biology 14 (1):1–16.

Liu X, Chu Z (2015) Genome-wide evolutionary characterization and analysis of bZIP transcription factors and their expression profiles in response to multiple abiotic stresses in Brachypodium distachyon. BMC genomics 16 (1):1–15.

Long TA, Tsukagoshi H, Busch W, Lahner B, Salt DE, Benfey PN (2010) The bHLH transcription factor POPEYE regulates response to iron deficiency in Arabidopsis roots. The plant cell 22 (7):2219–2236.

Maestri E, Marmiroli M, Visioli G, Marmiroli N (2010) Metal tolerance and hyperaccumulation: costs and trade-offs between traits and environment. Environmental and experimental botany 68 (1):1–13.

Manara A (2012) Plant responses to heavy metal toxicity. In: Plants and heavy metals. Springer, pp. 27–53.

Manivasagaperumal R, Vijayarengan P, Balamurugan S, Thiyagarajan G (2011) Effect of copper on growth, dry matter yield and nutrient content of Vigna radiata (L.) Wilczek. Journal of phytology 3 (3):53–62.

Mari S, Gendre D, Pianelli K, Ouerdane L, Lobinski R, Briat J-F, Lebrun M, Czernic P (2006) Root-to-shoot long-distance circulation of nicotianamine and nicotianamine–nickel chelates in the metal hyperaccumulator Thlaspi caerulescens. Journal of experimental botany 57 (15):4111–4122.

Mattigod S, Page A (1983) Assessment of metal pollution in soils. In: Applied environmental geochemistry (ed. Thornton, I), Chap. 12. Academic Press.

Migeon A, Blaudez D, Wilkins O, Montanini B, Campbell MM, Richaud P, Thomine S, Chalot M (2010) Genome-wide analysis of plant metal transporters, with an emphasis on poplar. Cellular and molecular life sciences 67 (22):3763–3784.

Migocka M, Kosieradzka A, Papierniak A, Maciaszczyk-Dziubinska E, Posyniak E, Garbiec A, Filleur S (2015a) Retracted: Two metal-tolerance proteins, MTP1 and MTP4, are involved in Zn homeostasis and Cd sequestration in cucumber cells. Oxford University Press UK.

Migocka M, Papierniak A, Kosieradzka A, Posyniak E, Maciaszczyk-Dziubinska E, Biskup R, Garbiec A, Marchewka T (2015b) Cucumber metal tolerance protein Cs MTP 9 is a plasma membrane H+-coupled antiporter involved in the Mn2+ and Cd2+ efflux from root cells. The plant journal 84 (6):1045–1058.

Miyadate H, Adachi S, Hiraizumi A, Tezuka K, Nakazawa N, Kawamoto T, Katou K, Kodama I, Sakurai K, Takahashi H (2011) OsHMA3, a P1B-type of ATPase affects root-to-shoot cadmium translocation in rice by mediating efflux into vacuoles. New phytologist 189 (1):190–199.

Mossor-Pietraszewska T (2001) Effect of aluminium on plant growth and metabolism. Acta biochimica polonica 48 (3):673–686.

Mukhopadhyay A, Vij S, Tyagi AK (2004) Overexpression of a zinc-finger protein gene from rice confers tolerance to cold, dehydration, and salt stress in transgenic tobacco. Proceedings of the national academy of sciences 101 (16):6309–6314.

Nagajyoti PC, Lee KD, Sreekanth TVM (2010) Heavy metals, occurrence and toxicity for plants: a review. Environmental chemistry letters 8 (3):199–216.

Naika M, Shameer K, Mathew OK, Gowda R, Sowdhamini R (2013) STIFDB2: an updated version of plant stress-responsive transcription factor database with additional stress signals, stress-responsive transcription factor binding sites and stress-responsive genes in Arabidopsis and rice. Plant and cell physiology 54 (2):e8–e8.

Nakashima K, Ito Y, Yamaguchi-Shinozaki K (2009) Transcriptional regulatory networks in response to abiotic stresses in Arabidopsis and grasses. Plant physiology 149 (1):88–95.

Nevo Y, Nelson N (2006) The NRAMP family of metal-ion transporters. Biochimica et biophysica acta (BBA) – Molecular cell research 1763 (7):609–620.

Ogawa I, Nakanishi H, Mori S, Nishizawa NK (2009) Time course analysis of gene regulation under cadmium stress in rice. Plant and soil 325 (1–2):97.

Opdenakker K, Remans T, Vangronsveld J, Cuypers A (2012) Mitogen-activated protein (MAP) kinases in plant metal stress: regulation and responses in comparison to other biotic and abiotic stresses. International journal of molecular sciences 13 (6):7828–7853.

Ortiz DF, Ruscitti T, McCue KF, Ow DW (1995) Transport of metal-binding peptides by HMT1, a fission yeast ABC-type vacuolar membrane protein. Journal of biological chemistry 270 (9):4721–4728.

Ouyang H, Vogel HJ (1998) Metal ion binding to calmodulin: NMR and fluorescence studies. Biometals 11 (3):213–222.

Ouzounidou G, Ilias I (2005) Hormone-induced protection of sunflower photosynthetic apparatus against copper toxicity. Biologia plantarum 49 (2):223.

Page V, Blösch RM, Feller U (2012) Regulation of shoot growth, root development and manganese allocation in wheat (Triticum aestivum) genotypes by light intensity. Plant growth regulation 67 (3):209–215.

Page V, Weisskopf L, Feller U (2006) Heavy metals in white lupin: uptake, root-to-shoot transfer and redistribution within the plant. New phytologist 171 (2):329–341.

Palmgren MG, Clemens S, Williams LE, Krämer U, Borg S, Schjørring JK, Sanders D (2008) Zinc biofortification of cereals: problems and solutions. Trends in plant science 13 (9):464–473.

Pandey N, Pathak GC, Pandey DK, Pandey R (2009) Heavy metals, Co, Ni, Cu, Zn and Cd, produce oxidative damage and evoke differential antioxidant responses in spinach. Brazilian journal of plant physiology 21 (2):103–111.

Patra M, Bhowmik N, Bandopadhyay B, Sharma A (2004) Comparison of mercury, lead and arsenic with respect to genotoxic effects on plant systems and the development of genetic tolerance. Environmental and experimental botany 52 (3):199–223.

Pedas P, Husted S (2009) Zinc transport mediated by barley ZIP proteins are induced by low pH. Plant signaling & behavior 4 (9):842–845.

Peng J, Gong J (2014) Vacuolar sequestration capacity and long-distance metal transport in plants. Frontiers in plant science 5:19.

Phukan UJ, Jeena GS, Shukla RK (2016) WRKY transcription factors: molecular regulation and stress responses in plants. Frontiers in plant science 7:760.

Pierart A, Shahid M, Séjalon-Delmas N, Dumat C (2015) Antimony bioavailability: knowledge and research perspectives for sustainable agricultures. Journal of hazardous materials 289:219–234.

Pierzynski GM (2000) Methods of phosphorus analysis for soils, sediments, residuals, and waters. North Carolina State University Raleigh.

Pitzschke A, Djamei A, Bitton F, Hirt H (2009) A major role of the MEKK1–MKK1/2–MPK4 pathway in ROS signalling. Molecular plant 2 (1):120–137.

Puig S, Thiele DJ (2002) Molecular mechanisms of copper uptake and distribution. Current opinion in chemical biology 6 (2):171–180.

Ramos J, Clemente MR, Naya L, Loscos J, Pérez-Rontomé C, Sato S, Tabata S, Becana M (2007) Phytochelatin synthases of the model legume Lotus japonicus. A small multigene family with differential response to cadmium and alternatively spliced variants. Plant physiology 143 (3):1110–1118.

Rao KP, Vani G, Kumar K, Wankhede DP, Misra M, Gupta M, Sinha AK (2011) Arsenic stress activates MAP kinase in rice roots and leaves. Archives of biochemistry and biophysics 506 (1):73–82.

Rashid M, Guangyuan H, Guangxiao Y, Hussain J, Xu Y (2012) AP2/ERF transcription factor in rice: genome-wide canvas and syntenic relationships between monocots and eudicots. Evolutionary bioinformatics 8:EBO–S9369.

Rasool A, Shah WH, Tahir I, Alharby HF, Hakeem KR, Rehman R (2020) Exogenous application of selenium (Se) mitigates NaCl stress in proso and foxtail millets by improving their growth, physiology and biochemical parameters. Acta physiologiae plantarum 42 (7):1–13.

Riesen O, Feller U (2005) Redistribution of nickel, cobalt, manganese, zinc, and cadmium via the phloem in young and maturing wheat. Journal of plant nutrition 28 (3):421–430.

Roelofs D, Aarts MGM, Schat H, Van Straalen NM (2008) Functional ecological genomics to demonstrate general and specific responses to abiotic stress. Functional ecology 22 (1):8–18.

Rucińska-Sobkowiak R (2016) Water relations in plants subjected to heavy metal stresses. Acta physiologiae plantarum 38 (11):257.

Saraf A, Samant A (2013). Evaluation of some minerals and trace elements in *Achyranthes aspera* Linn. International journal of pharma sciences 3 (3):229–233.

Sethi V, Raghuram B, Sinha AK, Chattopadhyay S (2014) A mitogen-activated protein kinase cascade module, MKK3-MPK6 and MYC2, is involved in blue light-mediated seedling development in Arabidopsis. The plant cell 26 (8):3343–3357.

Shahid M, Khalid S, Abbas G, Shahid N, Nadeem M, Sabir M, Aslam M, Dumat C (2015) Heavy metal stress and crop productivity. In: Crop production and global environmental issues. Springer, Cham, pp. 1–25.

Shahid M, Pourrut B, Dumat C, Nadeem M, Aslam M, Pinelli E (2014) Heavy-metal-induced reactive oxygen species: phytotoxicity and physicochemical changes in plants. Reviews of environmental contamination and toxicology, 232:1–44.

Shen J, Lv B, Luo L, He J, Mao C, Xi D, Ming F (2017) The NAC-type transcription factor OsNAC2 regulates ABA-dependent genes and abiotic stress tolerance in rice. Scientific reports 7:40641.

Sheng Y, Yan X, Huang Y, Han Y, Zhang C, Ren Y, Fan T, Xiao F, Liu Y, Cao S (2019) The WRKY transcription factor, WRKY13, activates PDR8 expression to positively regulate cadmium tolerance in Arabidopsis. Plant, cell & environment 42 (3):891–903.

Shiowatana J, McLaren RG, Chanmekha N, Samphao A (2001) Fractionation of arsenic in soil by a continuous-flow sequential extraction method. Journal of environmental quality 30 (6):1940–1949.

Singh KB, Foley RC, Oñate-Sánchez L (2002) Transcription factors in plant defense and stress responses. Current opinion in plant biology 5 (5):430–436.

Singh S, Parihar P, Singh R, Singh VP, Prasad SM (2016) Heavy metal tolerance in plants: role of transcriptomics, proteomics, metabolomics, and ionomics. Frontiers in plant science 6:1143.

Sinha S, Pandey K, Gupta AK, Bhatt K (2005) Accumulation of metals in vegetables and crops grown in the area irrigated with river water. Bulletin of environmental contamination and toxicology 74 (1):210–218.

Skyllberg U, Bloom PR, Qian J, Lin C-M, Bleam WF (2006) Complexation of mercury (II) in soil organic matter: EXAFS evidence for linear two-coordination with reduced sulfur groups. Environmental science & technology 40 (13):4174–4180.

Snedden WA, Fromm H (2001) Calmodulin as a versatile calcium signal transducer in plants. New phytologist 151 (1):35–66.

Soares CRFS, Grazziotti PH, Siqueira JO, Carvalho JGd, Moreira FMS (2001) Zinc toxicity on growth and nutrition of Eucalyptus maculata and Eucalyptus urophylla in nutrient solution. Pesquisa agropecuária brasileira 36 (2):339–348.

Song W-Y, Park J, Eisenach C, Maeshima M, Lee Y, Martinoia E (2014) ABC transporters and heavy metals. In: Plant ABC transporters. Springer, Cham, pp. 1–17.

Sumner ME (2000) Beneficial use of effluents, wastes, and biosolids. Communications in soil science and plant analysis 31 (11–14):1701–1715.

Sytar O, Kumari P, Yadav S, Brestic M, Rastogi A (2019) Phytohormone priming: regulator for heavy metal stress in plants. Journal of plant growth regulation 38 (2):739–752.

Takahashi H, Kawakatsu T, Wakasa Y, Hayashi S, Takaiwa F (2012) A rice transmembrane bZIP transcription factor, OsbZIP39, regulates the endoplasmic reticulum stress response. Plant and cell physiology 53 (1):144–153.

Tamás L, Mistrík I, Zelinová V (2017) Heavy metal-induced reactive oxygen species and cell death in barley root tip. Environmental and experimental botany 140:34–40.

Tang W, Charles TM, Newton RJ (2005) Overexpression of the pepper transcription factor CaPF1 in transgenic Virginia pine (Pinus virginiana Mill.) confers multiple stress tolerance and enhances organ growth. Plant molecular biology 59 (4):603–617.

Thys C, Vanthomme P, Schrevens E, De Proft M (1991) Interactions of cd with zn, cu, mn and fe for lettuce (Lactuca-sativa l) in hydroponic culture. Plant cell and environment 14 (7):713–717.

Tiong J, McDonald G, Genc Y, Shirley N, Langridge P, Huang CY (2015) Increased expression of six ZIP family genes by zinc (Zn) deficiency is associated with enhanced uptake and root-to-shoot translocation of Zn in barley (Hordeum vulgare). New phytologist 207 (4):1097–1109.

Tipping E (2002) Cation binding by humic substances, vol 12. Cambridge University Press, UK.

Tiwari S, Lata C (2018) Heavy metal stress, signaling, and tolerance due to plant-associated microbes: an overview. Frontiers in plant science 9:452.

Tsai TM, Huang HJ (2006) Effects of iron excess on cell viability and mitogen-activated protein kinase activation in rice roots. Physiologia plantarum 127 (4):583–592.

Turgeon R, Wolf S (2009) Phloem transport: cellular pathways and molecular trafficking. Annual review of plant biology 60:207–221.

USEPA (2020), US Environmental protection agency. A plan English guide to the EPA part 503 Biosolids Rule. http://www.epa.gov/owm/mtb/biosolids/503pe-/pdf-aug21. 2001.

Ul Hassan Z, Ali S, Rizwan M, Hussain A, Akbar Z, Rasool N, Abbas F (2017) Role of zinc in alleviating heavy metal stress. In: Essential plant nutrients. Springer, Cham, pp. 351–366.

Upadhyay RK, Panda SK (2009) Copper-induced growth inhibition, oxidative stress and ultrastructural alterations in freshly grown water lettuce (*Pistia stratiotes* L.). Comptes rendus biologies, 332 (7):623–632.

van Bel AJE (2003) The phloem, a miracle of ingenuity. Plant, cell & environment 26 (1):125–149.

van de Mortel JE, Schat H, Moerland PD, Van Themaat EVL, Van Der Ent S, Blankestijn H, Ghandilyan A, Tsiatsiani S, Aarts MGM (2008) Expression differences for genes involved in lignin, glutathione and sulphate metabolism in response to cadmium in Arabidopsis thaliana and the related Zn/Cd-hyperaccumulator Thlaspi caerulescens. Plant, cell & environment 31 (3):301–324.

van de Mortel JE, Villanueva LA, Schat H, Kwekkeboom J, Coughlan S, Moerland PD, van Themaat EVL, Koornneef M, Aarts MGM (2006) Large expression differences in genes for iron and zinc homeostasis, stress response, and lignin biosynthesis distinguish roots of Arabidopsis thaliana and the related metal hyperaccumulator Thlaspi caerulescens. Plant physiology 142 (3):1127–1147.

Verret F, Gravot A, Auroy P, Leonhardt N, David P, Nussaume L, Vavasseur A, Richaud P (2004) Overexpression of AtHMA4 enhances root-to-shoot translocation of zinc and cadmium and plant metal tolerance. FEBS letters 576 (3):306–312.

Vidal SM, Malo D, Vogan K, Skamene E, Gros P (1993) Natural resistance to infection with intracellular parasites: isolation of a candidate for Bcg. Cell 73 (3):469–485.

Wang X, Wang C, Sheng H, Wang Y, Zeng J, Kang H, Fan X, Sha L, Zhang H, Zhou Y (2017) Transcriptome-wide identification and expression analyses of ABC transporters in dwarf polish wheat under metal stresses. Biologia plantarum 61 (2):293–304.

Wang Y, Gao C, Liang Y, Wang C, Yang C, Liu G (2010) A novel bZIP gene from Tamarix hispida mediates physiological responses to salt stress in tobacco plants. Journal of plant physiology 167 (3):222–230.

Wani SH, Kumar V, Shriram V, Sah SK (2016) Phytohormones and their metabolic engineering for abiotic stress tolerance in crop plants. The crop journal 4 (3):162–176.

Weber M, Trampczynska A, Clemens S (2006) Comparative transcriptome analysis of toxic metal responses in Arabidopsis thaliana and the Cd2+-hypertolerant facultative metallophyte Arabidopsis halleri. Plant, cell & environment 29 (5):950–963.

Wei W, Zhang Y, Han L, Guan Z, Chai T (2008) A novel WRKY transcriptional factor from Thlaspi caerulescens negatively regulates the osmotic stress tolerance of transgenic tobacco. Plant cell reports 27 (4):795–803.

Williams LE, Mills RF (2005) P1B-ATPases–an ancient family of transition metal pumps with diverse functions in plants. Trends in plant science 10 (10):491–502.

Wu H, Chen C, Du J, Liu H, Cui Y, Zhang Y, He Y, Wang Y, Chu C, Feng Z (2012) Co-overexpression FIT with AtbHLH38 or AtbHLH39 in Arabidopsis-enhanced cadmium tolerance via increased cadmium sequestration in roots and improved iron homeostasis of shoots. Plant physiology 158 (2):790–800.

Wuana RA, Okieimen FE (2011) Heavy metals in contaminated soils: a review of sources, chemistry, risks and best available strategies for remediation. ISRN ecology 2011.

Xie Z, Zhang Z-L, Zou X, Huang J, Ruas P, Thompson D, Shen QJ (2005) Annotations and functional analyses of the rice WRKY gene superfamily reveal positive and negative regulators of abscisic acid signaling in aleurone cells. Plant physiology 137 (1):176–189.

Xin M, Wang Y, Yao Y, Song N, Hu Z, Qin D, Xie C, Peng H, Ni Z, Sun Q (2011) Identification and characterization of wheat long non-protein coding RNAs responsive to powdery mildew infection and heat stress by using microarray analysis and SBS sequencing. BMC plant biology 11 (1):61.

Xu J, Tian Y-S, Peng R-H, Xiong A-S, Zhu B, Jin X-F, Gao J-J, Hou X-L, Yao Q-H (2009) Yeast copper-dependent transcription factor ACE1 enhanced copper stress tolerance in Arabidopsis. Bmb reports 42 (11):752–757.

Xu Z, Ge Y, Zhang W, Zhao Y, Yang G (2018) The walnut JrVHAG1 gene is involved in cadmium stress response through ABA-signal pathway and MYB transcription regulation. BMC plant biology 18 (1):1–13.

Yamaji N, Huang CF, Nagao S, Yano M, Sato Y, Nagamura Y, Ma JF (2009) A zinc finger transcription factor ART1 regulates multiple genes implicated in aluminum tolerance in rice. The plant cell 21 (10):3339–3349.

Yeh C-M, Chien P-S, Huang H-J (2007) Distinct signalling pathways for induction of MAP kinase activities by cadmium and copper in rice roots. Journal of experimental botany 58 (3):659–671.

Yoshida T, Fujita Y, Sayama H, Kidokoro S, Maruyama K, Mizoi J, Shinozaki K, Yamaguchi-Shinozaki K (2010) AREB1, AREB2, and ABF3 are master transcription factors that cooperatively regulate ABRE-dependent ABA signaling involved in drought stress tolerance and require ABA for full activation. The plant journal 61 (4):672–685.

Yuce M, Taspinar MS, Aydin M, Agar G (2019) Response of NAC transcription factor genes against chromium stress in sunflower (Helianthus annuus L.). Plant Cell, Tissue and organ culture (PCTOC) 136 (3):479–487.

Zarattini M, Forlani G (2017) Toward unveiling the mechanisms for transcriptional regulation of proline biosynthesis in the plant cell response to biotic and abiotic stress conditions. Frontiers in plant science 8:927.

Zhang M-K, Liu Z-Y, Wang H (2010) Use of single extraction methods to predict bioavailability of heavy metals in polluted soils to rice. Communications in soil science and plant analysis 41 (7):820–831.

Zhang Y, Wang L (2005) The WRKY transcription factor superfamily: its origin in eukaryotes and expansion in plants. BMC evolutionary biology 5 (1):1–12.

Zhou B, Yao W, Wang S, Wang X, Jiang T (2014) The metallothionein gene, TaMT3, from Tamarix androssowii confers Cd2+ tolerance in tobacco. International journal of molecular sciences 15 (6):10398–10409.

Zhu S, Shi W, Jie Y, Zhou Q, Song C (2020) A MYB transcription factor, BnMYB2, cloned from ramie (Boehmeria nivea) is involved in cadmium tolerance and accumulation. PloS one 15 (5):e0233375.

5

Role of Transporters during Heavy Metals Toxicity in Plants

Misbah Naz
*State Key Laboratory of Crop Genetics and Germplasm Enhancement,
Nanjing Agricultural University, Nanjing, China*

Mohammad Sohidul Islam
Department of Agronomy, Hajee Mohammad Danesh Science and Technology University, Basherhat, Bangladesh

Muhammad Aamir Iqbal
Department of Agronomy, Faculty of Agriculture, University of Poonch, Rawalakot (AJK), Pakistan

Sytar Okana
*Department of Plant Physiology, Faculty of Agrobiology and Food Resources,
Slovak University of Agriculture, Nitra, Slovakia*

Ratnasekera Disna
Department of Agricultural Biology, Faculty of Agriculture, University of Ruhuna, Matara, Sri Lanka

Akbar Hossain
Bangladesh Wheat and Maize Research Institute, Dinajpur, Bangladesh

Muhammad Mubeen
Department of Environmental Sciences, COMSATS University Islamabad, Vehari Campus, Pakistan

Junaid Rahim and Muhammad Imran
Department of Entomology, Faculty of Agricultural, University of Poonch, Rawalakot, Pakistan

Md. Tahjib-Ul-Arif
Department of Biochemistry and Molecular Biology, Bangladesh Agricultural University, Mymensingh, Bangladesh

Sharif Ahmed
International Rice Research Institute, Bangladesh Office, Dhaka, Bangladesh

Anamika Dubey and Ashwani Kumar
*Metagenomics and Secretomics Research Laboratory, Department of Botany,
Dr. Harisingh Gour University (A Central University), Sagar, MP, India*

Milan Skalicky and Marian Brestic
*Department of Botany and Plant Physiology, Faculty of Agrobiology, Food and Natural Resources,
Czech University of Life Sciences Prague, Czechia
Department of Plant Physiology, Slovak University of Agriculture, Nitra, Slovak Republic*

Ayman EL Sabagh
*Department of Agronomy, Faculty of Agriculture, University of Kafrelsheikh, Egypt
Department of Field Crops, Faculty of Agriculture, Siirt University, Turkey*

DOI: 10.1201/9781003155089-5

CONTENTS

Abbreviations

Al, Aluminum
Cd, Cadmium
CDF, Cation diffusion facilitator
CEC, Cation exchange capacity
Cr, Chromium
GDH, Glutamate dehydrogenase
GOGAT, Glutamate synthase
GS, Glutamine synthetase
Hg, Mercury
HMs, Heavy metals
MAPK, Mitogen-activated protein kinase
MTs, Metallothioneins
Ni, Nickel
NR, Nitrate reductase
NRAMP, Natural resistance-associated macrophage protein
Pb, Lead
ROS, Reactive oxygen species

5.1 Introduction

Heavy metals (HMs) are metallic elements having comparatively higher densities and atomic weights. The major sources of HMs contamination in the agroecosystem are corrosion of metals, soil erosion and subsequent leaching and suspended sediments, industrial wastes and extensive use of HM-contaminated chemical fertilizers (Turkyilmaz et al. 2019). The most well-known HM-pollutants are Zn^{2+}, Fe^{2+}, Cu^{2+}, Mn^{2+} and Co that are toxic to plants whereas some other metals like Cd^{2+}, Hg^{2+} and Pb^{2+} are not yet known as highly toxic to plants. Even at lower concentrations, HMs tend to alter plant growth owing to their effect on physiological, metabolic processes and enzymatic activities in plants (Singh and Kalamdhad 2011; Sytar et al. 2013). Most importantly, higher concentrations of HMs in cells trigger the production of excessive reactive oxygen species (ROS) that cause gradual cell death, leading to a substantial reduction in growth and yield of crops (Nasim and Dhir 2010; Hasanuzzaman et al. 2020; Saleem et al. 2020). The availability of mineral nutrients in the soil hinders many fundamental cellular processes and downregulates the overall plant growth performances (Nihorimbere et al. 2011). The presence of excessive amounts of HMs in the rhizosphere can reduce the uptake and translocation of essential nutrients by the plants (Seshadri et al. 2015). Moreover, exposure to certain HMs can alter uptake and assimilation of nitrogen (Chandrakar et al. 2018) and can create an artificial deficiency of any particular nutrient (Schjoerring et al. 2019). Usually, nutrient deficiency symptoms in plants include stunted growth, senescence and yellowing of leaves, plant tissue death, etc., caused by the reduction of chlorophylls and other pigments needed for photosynthesis (Singh and Kalamdhad 2011). Reduced crop yield and quality are the common consequences of nutrient deficiency. It has been inferred that HMs toxicity has emerged as a serious threat to the environment and food security globally.

Many transporter proteins are embedded in the cellular and vacuolar membranes as well as in the membranes of different subcellular organelles that are mainly involved in HMs transportation from one organelle to the other and further eliminate it out from the plants (Hedrich et al. 2015). In each cellular compartment, there exists a set of peculiar transporters that perform versatile functions related to transportation and homeostasis of HMs (Forbes and Gros 2001). Among those, HM ATPases, natural resistance-associated macrophage protein (NRAMP) family, cation diffusion facilitator (CDF) family, Zn transporters (ZIP) family and the cation antiporter stook are significant attention in recent years (Nevo and Nelson 2006; Goswami et al. 2012; Cotrim et al. 2019). It is important to understand the interactive functions of HM transporters for the breeding of HM-tolerant crops. This study aims to critically synthesize and analyze the crucial features of HMs transporters in plants about their structures, functions and regulatory roles in diversified organisms. It is important to describe the transporter's role in HMs uptake, transportation, accumulation and reduction by plant root.

5.2 Heavy Metal-Induced Phytotoxicity

Certain trace metals (Cu^{2+}, Mn^{2+}, Fe^{2+} and Zn^{2+}) at an optimal cellular concentration play important roles as co-factors of many enzymes while at excessive levels cause toxicity (Sudre 2009). Toxic metals like Cd^{2+}, Co^{2+} and Ni^{2+} lead to the suppression of plant growth by disrupting transport activities and cellular homeostasis. The NRAMP protein family members are crucial owing to their function of transporting various ions (such as Mn^{2+}, Zn^{2+}, Cu^{2+}, Fe^{2+}, Cd^{2+}, Ni^{2+} and Co^{2+}) while acting as metal ion transporters (Nevo and Nelson 2004, 2006).

Excessive Cu^{2+} may exert detrimental effects on crop growth and production (Ravet and Pilon 2013, Tahjib-Ul-Arif et al. 2020). Therefore, plants need to control Cu^{2+} homeostasis strictly to maintain an equilibrium condition at the tissue

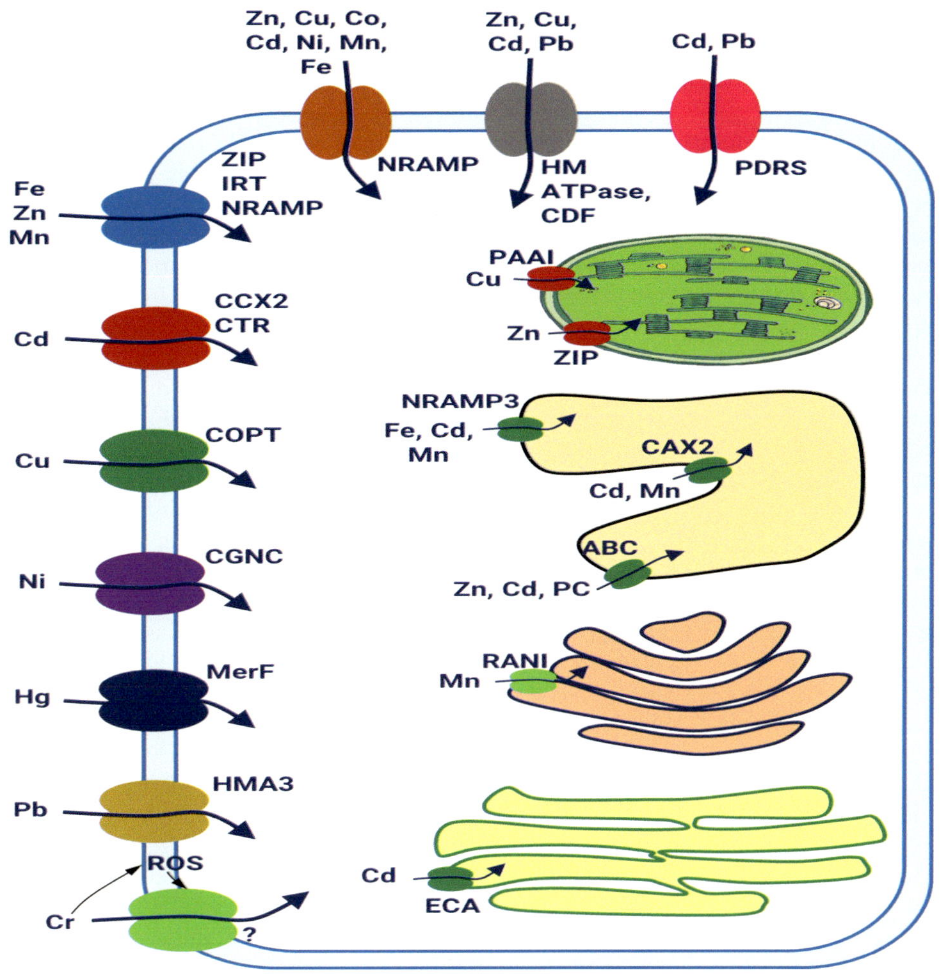

FIGURE 5.1 Cellular mechanisms involved in transport and uptake of HMs with metal transporters (Modified from Soliman et al. 2019; Williams et al. 2000; Guerinot 2000; Jain et al. 2018). **Abbreviations:** ABC – ATPtions; Cd – cadmium; CDF – cation diffusion facilitator; Co – cobalt; Cr – chromium; Cu – copper; Fe – iron; Mn – manganese; Ni – nickel; Hg on to the HMs – heavy metals; Pb – lead; Zn – zinc ROS – reactive oxygen species; NRAMP – natural resistance-associated macrophage protein; PDRS – pleiotropic drug resistance transporters; ZIP – ZRT-IRT-related protein.

level (Figure 5.1). Contrarily, high-affinity nitrate transporters belonging to the NRT2 family up-regulate under low to medium Cu^{2+} levels, but this up-regulation remains ineffective in restoring N absorption by roots to original pre-toxicity levels (Puig 2014). Under Cu^{2+} toxicity, plants display diversified responses caused by genes modulating which tend to regulate nitrate uptake using low-affinity nitrate transporters as well as proton pumps.

A higher concentration of Al^{3+} is toxic for many crop plants (Krstic et al. 2012). It has been regarded as a non-essential element for plant growth; rather its excessive concentration leads to toxic effects on plants (Gupta et al. 2013; Krstic et al. 2012). Al^{3+} toxicity is dominant in acid soils, so excess Al^{3+} inhibits the uptake of nutrients from acidic soils. The cessation in root elongation has been reported as the primary symptom of Al^{3+} phytotoxicity, which may happen within 1 h after the exposure to Al^{3+} stress (Gupta et al. 2013; Krstic et al. 2012). Al^{3+} significantly affects the cell wall, plasma membrane, cytoskeleton and signal transduction pathways and triggers oxidative stress which hampers cytosolic calcium homeostasis, magnesium uptake, and auxin polar transport consequently leads to inhibition of cell elongation and cell division (Lal 2010). Al^{3+} also inhibits NO_3^- transport across the membrane (Zhao and Shen 2018). Thus, the inhibition of NO_3^- uptake may happen due to the binding of NO_3^- transporters by the intracellular Al^{3+} (Kronzucker et al. 1999). However, the activity of nitrate reductase (NR) is inhibited by Al^{3+}exposure owing to a sharp decline in the accumulation of internal NO_3^- content (Wang et al. 2013). The NR activity is stimulated by Al^{3+} in lower concentration caused by enhanced uptake of NO_3^- (Jiang et al. 2018). The effects of Al^{3+} on the activities of close-in action enzymes in plant Jiang, metabolisms such as glutamine synthetase (GS), glutamate synthase (GOGAT) and glutamate dehydrogenase (GDH), and potential aims for improving N use efficiency in the plant are still uncertain (Zhang et al. 2018). Plants detoxify Al^{3+} in two ways viz., by excluding Al^{3+} from the root tips (exclusion mechanism), and by sequestering excessiveAl^{3+} that enters the plant system (internal tolerance mechanism) (Lal, 2010).

Nickel (Ni^{2+}) in the plant is usually absorbed in ionic form, but being the non-essential element is not needed for plant growth and development. Ni^{2+} has effects on root and soil because its mobility within the plant is usually higher than the other HMs (Gill 2014). Exposure to higher Ni^{2+}concentrations significantly inhibited plant growth (Maheshwari and Dubey 2007).

Lead (Pb) is found in the inorganic form in the environment. Among known soil pollutants, Pb is regarded as one of the most toxic metal pollutants (Dutta et al. 2018), which gets released as a byproduct of numerous industrial processes and exists in all environmental compartments (soil, water and living organisms). Plants get exposed to Pb through its transportation from soil solution as well as in the form of aerosols from the air. The low mobility of Pb in the soil supports its accumulation mostly in the topsoil (Ahmadipour et al. 2014). The Pb uptake by plants is coordinated by cation exchange capacity (CEC) and pH of growth media along with the size of Pb particles and other agro-botanical characteristics. Excessive Pb causes several toxicity syndromes in plants such as chlorosis, stunted growth and blackening of the root system (Sharma and Dubey 2005 The Pb toxicity leads to the inhibition of photosynthesis, unbalancing of minerals and water uptake, alteration of hormones synthesis and membrane structure, and their permeability levels (Sharma et al. 2016). The biological function of Pb is not yet is known but it can cause morphological, physiological and biochemical dysfunctions in plants (Fahr et al. 2013).

The plants use diversified morphological, physiological and biochemical adaptations in response to Pb toxicity, which assists in its detoxification and imparts tolerance against Pb stress (Jiang et al. 2020). Under Pb stress, plants have developed numerous tolerance mechanisms especially related to changes in root systems for lower Pb absorption (Pourrut et al. 2011). Gill et al. (2014) postulated that plant roots rapidly respond in the response of Pb either (i) by callose synthesis, which creates a barrier against Pb entrance (ii) by Pb uptake in higher concentration and subsequent sequestration in the vacuole along with modifications in root growth (iii) by Pb translocation to the leaves.

5.3 The Role of the Transporter in the Regulation of Heavy/Toxic Ions in Cellular Environments

Plants tend to maintain optimal nutrients content under adverse soil conditions by altering root architecture, activation of transport systems in roots, and developing associations with soil microbes. Plants are required to initiate nutrients uptake from the solution of soil and transporting the acquired nutrients through the xylem system to leaves. Besides free diffusion, mechanisms of nutrients transport have been classified into three groups including facilitated diffusion, group translocation and active transport (Yadav 2010; Viehweger 2014). There are several nutrients transporters to uptake the elements from the soil by roots (Wang et al. 2006), such as high-affinity transporter nitrate transporter (NRTs) (Fan et al. 2017; Naz et al. 2019), nitrate transporter *DsNRT3.1* (also known as the nitrate assimilation-related protein), ammonium transporters (AMTs) to uptake and transport ammonium root to shoot (Lima et al. 2010), Mn transporter *NRAMP* (Lanquar et al. 2010), yellow stripe-like (*YSL*), ZIP (zinc-regulated transporter/iron-regulated transporter [*ZRT/IRT1*]-related protein), cation exchanger (CAX) (Socha and Guerinot 2014), calcium cation exchangers (CCX), *CDF/MTP*, P-type ATPase, vacuolar Fe transporter (VIT), etc. Among Zn transporters present in roots, ZIP1, ZIP2, ZIP9 and IRT3 are the most active and prominent (Moreira et al. 2018), while ZIP4 plays a vital role in transporting Zn across the membranes of the chloroplast (Bin et al. 2018). These proteins transporter varies in their binding potential and specificities to Zn transportation. Potassium channels and secondary potassium transporters can mediate potassium transporter membrane transport of K^+. Moreover, three potassium transporters have been identified including K^+-permeases (KT/HAK/KUP), the K^+-Trk/HKT, and cation proton antiporters (CPAs). For the transportation of sulfate, S-metabolite transporters have been recognized as the most crucial *S*-transporters (Takahashi 2019). Furthermore, numerous transporters are reported to participate in the transport of inorganic phosphate (Pi). For instance, a protein transporter called Pht1 mediates Pi acquisition from the soil solution and likely to be involved in Pi transport to plant tissues (Li et al. 2019). Different factors such as topology, type of functions, expression sites of the genes, etc., assist to determine the role of these metal transporter families. Fe, Zn, Mn uptake by ZIP, IRT and NRAMP transporter while the CCX2 for cadmium, As COTP1, is in charge for Cu^{2+} uptake, Zinc transporters ZRT1 and ZRT2 expression sequence correspondence to IRT1 and could resourcefully transport zinc. CGNP is responsible for nickel uptake and transfer from the soil. MerF is in charge of Hg in bacteria while in the higher plant not know thus hypothetically possibility MerF may be responsible also in the plant. HMA3 transporter for Pb, while in Arabidopsis have been discord HMA4 (be appropriate to the Zn/Cd//Pb/Co group). For Cr, many mechanisms are working under stress like ROS defense mechanism, antioxidant enzymes (SOD: superoxide dismutase, CAT: catalase, APX: ascorbate peroxidase, GPX: glutathione peroxidase, GSH: glutathione reduce, and GSSG: glutathione oxidase). However, PCs and MTs are chelating properties that chelate the toxic ions by breaking down complex compounds (Jain et al. 2018).

For example, 18 YSL genes have been found to responsible for the transport of Fe in rice. It is interesting to note that this transporter gets up-regulated Fe-limited conditions deficiency and resided at the plasma membrane of the root epidermis. The vacuolar Fe transporter 1 member (VIT1) of the *NRAMPs* gene family resolve the uptake of different divalent cations. In *Arabidopsis*, *AtNramp1*, *AtNramp3 and AtNramp4* can complement the Fe uptake mutant of yeast, revealing that these proteins can mediate Fe transport in the (Table 5.1) different nutrients transporter on Arabidopsis, AMT1;1, AMT1;3 ammonium uptake capacity under lack viability of nitrogen, AtNRT2.1 promotes an effect on root growth, AtNramp2 Mn-influx transporter and to remobilization of Mn from old to young leaves. AtMRS2-11 magnesium transport into chloroplasts/plastids. AtSultr, S-nutrition is up-regulated by inadequate S-nutrition. KUP7 K^+ uptake in Arabidopsis roots and also can participate in K^+ transport into xylem sap, TRH1 is essential for root hair/lateral root elongation. BOR1 is required for endocytosis. BOR2 root elongation under low B supply. We

TABLE 5.1
Different Nutrients Transporters in Crop Plants and Their Role in Uptake Transportation and Assimilations

Transporters	Metals/Nutrients	Plant Species	Functions	References
Different nutrient transporter in various crops				
OsNRT2.1	NO_3^-	Rice	Overexpressing OsNRT2.1 stimulates root growth upon NO_3^- treatment under active polar auxin transport.	Naz et al. (2019); Fan et al. (2017)
OsNRT2.1, OsNRT2.2 & OsNRT2.3a	NO_3^-	Rice	Uptake under the low and high concentration ranges.	Yan et al. (2011)
OsNRT2.3b	NO_3^-	Rice	Nitrate transporter OsNRT2.3b can improve NUE activity under plant adaption to varied N supply forms.	Fan et al. (2016)
OsNRT2.4	NO_3^-	Rice	OsNRT2.4 responded to changes in auxin supply, unlike all the other NRT2 genes.	Feng et al. (2011)
AMT1;1 and AMT1;3	NH_4^+	Arabidopsis	AMT1; 1 and AMT1; 3 boost uptake of ammonium under nitrogen-limited conditions.	Loqué et al. (2006)
AtAMT1;4	NH_4^+	Arabidopsis	AtAMT1; 4 mediates uptake of ammonium across pollen's plasma membrane and thus contributes to nitrogen accumulation.	Yuan et al. (2009)
AMT1;3	NH_4^+	Arabidopsis	Ammonium multiplies nitrate effect for boosting root development by stimulating lateral root branching.	Lima et al. (2010)
PttAMT1.2	NH_4^+	Populustrichocarpa	Ammonium could be a major N source delivered from the fungus toward the plant in symbiosis.	Selle et al. (2005)
NRAMP1	Mn	Arabidopsis	NRAMP1 is a physiological manganese transporter in Arabidopsis.	Cailliatte et al. (2010)
AtNramp2	Mn	Arabidopsis	The vascular bundle-localized Mn-influx transporter OsNRAMP3 participated in Mn distribution and contributes to Mn remobilization from old to young leaves.	Yang et al. (2013)
OsNramp3	Mn	Rice	OsNramp3 localized at the nodes reacts to carious Mn concentrations by differing responses to changes in Mn levels.	Shao et al. (2017)
AtNRAMP3 and AtNRAMP4	Mn	Arabidopsis	It plays a vital role in Mn translocation from the vacuole to chloroplasts of mesophyll cells.	Lanquar et al. (2010)
HvIRT1	Mn	Barley (*Hordeum vulgare*)	HvIRT1 is responsible for Mn^{2+} uptake in roots of barley and contributes to genotypic differences in Mn^{2+} uptake kinetics.	Pedas et al. (2008)
Nramp5	Mn and Cd	Rice	It is a main Mn and Cd transporter from soil solution to root cells.	Sasaki et al. (2012)
NRAMP 2	Mg	Arabidopsis	NRAMP2 performs Mn remobilization in Golgi apparatus for boosting root growth.	Gao et al. (2017)
AtHMA3	Cd/Pb	Arabidopsis	AtHMA3, a plant P1B-ATPase is Cd/Pb transporter in yeast.	Gravotet al. (2004)
OsCCX2	Cd	Rice	It is a vital transporter of Cd which leads to accumulation in grain through Ca translocation pathway.	Zhou et al. (2019)
OsLCT1	Cd	Rice	Transporter gene for phloem Cd transport in plants.	Uraguchi et al. (2011)
SaNramp6	Cd	Arabidopsis/cloned *Sedum alfredii*	SaNramp6 may play a significant role in improving Cd accumulation, transgenic plants for phytoremediation.	Chen et al. (2017)
(YSL) transporter	Fe		Facilitates mechanisms for long-distance Fe transport.	Bashir et al. (2013)
TaVIT2	Fe	Wheat	Triggers transportation of vacuolar Fe in the endosperm.	Connorton et al. (2017)
IRT2	Fe	Arabidopsis	IRT2 transporter under iron-limited conditions is responsible for iron and zinc uptake from the soil.	Vert et al. (2001)
IRT1	Fe	The plant-soil interface	Radial transportation of iron via vascular tissues involving efflux transporters.	Barberon et al. (2014)
IRT1	Fe	Arabidopsis	It is also called peanut Fe transporter especially under Fe-limited conditions.	Ding et al. (2010)
OsNRAMP1	Fe	Rice	The iron transporter OsNRAMP1 in Cd uptake and transport in rice.	Takahashi et al. (2011)
Fe transporter 1 (VIT1)	Fe	Crops	Crystal structure of plant vacuolar iron transporter VIT.	Kato et al. (2019)
OsFRDL1	Fe	Rice	Essential for efficient translocation of iron in rice.	Yokosho et al. (2009)

(Continued)

TABLE 5.1 *(Continued)*

Different Nutrients Transporters in Crop Plants and Their Role in Uptake Transportation and Assimilations

Transporters	Metals/Nutrients	Plant Species	Functions	References
Different nutrient transporter in various crops				
IRT1	Fe	Arabidopsis	IRT1 expression under iron deficiency in Arabidopsis roots.	Eide et al. (1996)
TgVit1,	Fe	Arabidopsis	TgVit1 is responsible for the blue coloration as a vacuolar Fe transporter in tulip petals.	Momonoi et al. (2009)
AhNRAMP1	Fe	Peanut	The AhNRAMP1 is possibly involved in iron acquisition in peanut plants.	Xiong et al., 2012)
LeIRT1	Fe	Tomato	LeIRT1 expression, localization and regulation which is iron transporter in tomato roots.	Schikora et al. (2006)
OsZIP4	Zn	Rice	It is a Zn transporter which translocates Zn within rice plants.	Ishimaru et al. (2005)
OsZIP5	Zn	Rice	It facilitates Zn distribution within the rice.	Lee et al. (2010)
OsZIP1, OsZIP3 and OsZIP4	Zn	Rice	OsZIP1, OsZIP3 and OsZIP4 in the roots and that of OsZIP4 in the shoots of both genotypes, indicating that all these genes may participate in rice zinc nutrition.	Chen et al. (2008)
OsZIP3	Zn	Rice	Zn distribution in the developing rice tissues.	Sasaki et al. (2015)
OsZIP1, OsZIP3, OsZIP4, OsZIP5, OsZIP8,	Zn	Rice	The genes which participate in the different growth stages of rice.	Bashir et al. (2012)
OsZIP7	Zn	Rice	OsZIP7 is a good candidate for developing Zn bio-fortified rice.	Ricachenevsky et al. (2018)
OsZIP1	Zn	Rice	OsZIP1 function in epigenetic modification under Cd stress.	Liu et al. (2019)
OsMGT1	Mg	Rice	OsMGT1 supports Mg uptake through roots. The up-regulation of OsMGT1 confers Al tolerance in rice via enhancing Mg concentration in the cell.	Chen et al. (2012)
AtMRS2-11	Mg	Arabidopsis	The AtMRS2-11 gene playing a role in magnesium transport into chloroplasts/plastids.	Drummond et al. (2006)
OsMGT1	Mg	Rice	OsMGT1 is regulating transport activity of OsHKT1under salt stress; a main transporter for the removal of Na^+ from the xylem sap at the root mature zone.	Chen et al. (2017)
OsMGT1	Mg	Rice	OsMGT1 plays an important role in rice Mg import and is required for the resistance to Mg deficiency.	Zhang et al. (2019)
PtrMGT5	Mg	*Poncirus trifoliata*	PtrMGT5 supports Mg homeostasis in citrus plants by getting involved in the absorption and transport of Mg.	Liu et al. (2019)
CorA/MRS2/ MGT-type;ZmMGT10	Mg	Arabidopsis.	Mg transporter of maize which can enhance the tolerance to Mg deficient conditions by improving Mg^{2+} uptake in the transgenic plants of Arabidopsis.	Li et al. (2017)
ZmMGT12	Mg	Maize	ZmMGT12 possessed Mg^{2+} transportability and characterized by a high protein sequence identity with Arabidopsis AtMGT6 and rice OsMRS2-6.	Li et al. (2018)
AtMGT1 and AtMGT10	Mg	Arabidopsis	AtMGT1 and AtMGT10) are Mg^{2+} transporters which can complement bacterial or yeast mutant lacking Mg^{2+} transport capacity.	Chen et al. (2009)
OsPht1; 4	P	Rice	OsPht1;4 is phosphate transporter which involved in rice phosphate homeostasis.	Ye et al. (2015)
OsPT1	P	Rice	It accumulates in plant shoot in double concentration compared to wild-type plants.	Seo et al. (2008)
OsPT8	P	Rice	It triggers Pi homeostasis in rice and promotes plant growth.	Jia et al. (2011)
OsPT1	P	Rice	OsPT1 is involved in the OsPHO2-regulated Pi pathway.	Sun et al. (2012)
OsPT4	P	Rice	It facilitates As (V) uptake and transportation along with producing low As-accumulating rice.	Ye et al. (2015)

(Continued)

TABLE 5.1 ***(Continued)***

Different Nutrients Transporters in Crop Plants and Their Role in Uptake Transportation and Assimilations

Transporters	Metals/Nutrients	Plant Species	Functions	References
OsNRT2. 3b	P	Rice	OsNRT2.3b may upregulate expression levels of OsPT2, OsPT8 and OsPHR2 in roots and of OsPT1 and OsPHR2 in shoots of rice. OsNRT2.3b overexpression can improve P uptake and accumulation in rice.	Feng et al. (2017)
TaPT2 Pi-	P	Pi-deprived Wheat	It regulates P level in specified tissues in both monocots and dicots.	Tittarelli et al. (2007)
OsNLA1	P	Rice	Roles of OsNLA1 in maintaining Pi homeostasis in rice.	Yang et al. (2017)
OsSultr1;1	S	Rice	OsSultr1;1 is a rice high-affinity sulfate transporter that participates in maintaining sulfur demand under arsenic (As) stress.	Kumar and Khare (2019)
OsLsi1 and OsLsi2	S	Rice	OsLsi1 and OsLsi2 are potent transporters of As in rice. Greater S supply immobilized the As in roots and lowers its translocation to shoot.	Dixit et al. (2016)
AtSultr2;2	S	Arabidopsis	Roots play a significant role to transport sulfate via the phloem.	Davidian and Kopriva (2010)
AtSultr2;1	S	Arabidopsis	It is synthesized in roots during sulfate deficiency which increases mRNA level.	Buchner et al. (2004)
AtSultr1;3	S	Arabidopsis	It restricts sulfate movement to other tissues from cotyledons.	Hawkesford (2003)
AtSultr	S	Arabidopsis	It is a set of four transporters synthesized under inadequate S-nutrition which assists S uptake and translocation.	Tombuloglu et al. (2017)
AtHAK5, AKT1 K^+	K^+	Arabidopsis	AtHAK5 is supporting a high-affinity K^+ uptake system in Arabidopsis roots.	Gierth et al. (2005); Pyo et al. (2010)
TRH1 and AKT1	K^+	*Arabidopsis thaliana*	These are functional in the root hair and transport potassium in varying concentrations.	Desbrosses et al. (2003)
OsHAK5	K^+	Rice	It boosts K acquisition by roots and facilitates its transport to shoots from roots under K-deficient conditions.	Yang et al. (2014)
KUP7	K^+	*Arabidopsis thaliana*	It is involved in K^+ uptake with a possible role in K^+ transportation into xylem sap.	Han et al. (2016)
TRH1	K^+	*Arabidopsis thaliana*	It mediates K^+ uptake and translocation along with promotes root hair elongation.	Rigas et al. (2001)
BOR1	B	*Arabidopsis thaliana*	It has a non-essential role in endocytosis but plays a crucial function in multi-vesicular bodies for subsequent degradation in vacuoles.	Takano et al. (2005)
BOR1	B	*Arabidopsis thaliana*	Its functions are regulated by B availability as it tends to hamper excessive accumulation of B in shoots.	Kasai et al. (2011)
BOR1	B	Rice	It promotes efficient uptake of B and subsequent loading to the xylem.	Nakagawa et al. (2007)
AtBOR1	B	*Arabidopsis thaliana*	AtBOR1 encodes an efflux boron transporter. It is shown different expression throughout the reproductive development of *Vitis vinifera* L.	Pérez-Castro et al. (2012)
BnaC4.BOR1;1c	B	*Brassica napus*	BnaC4.BOR1;1c is an AtBOR1-homologous B transporter gene expressing in both shoot nodes and roots.	Zhang et al. (2017)
OsBOR4	B	Rice	OsBOR4 is a boron efflux transporter responsible for normal pollen germination and/or tube elongation.	Tanaka et al. (2013)
BOR2	B	*Arabidopsis thaliana*	BOR2, a B transporter that supports root elongation and cross-linking of RG-II under low B supply.	Takada et al. (2014)
Lsi1 and Lsi2	Si	Rice	Lsi1 is required for the transport of Si through the cells of the exodermis and endodermis.	Ma and Yamaji (2008)
Lsi1, Lsi2 and Lsi6	Si	Rice	It mediates the uptake of Si from the soil solution.	
HvLsi1	Si	Barley	It serves as a Si influx transporter which performs radial transport through the layers of epidermal and cortical membranes in basal roots.	Chiba et al. (2008)
TaLsi1	Si	Wheat	TaLsi1 is indeed a functional Si transporter as its expression in Arabidopsis leads to increased Si uptake.	Montpetit et al. (2012)

discuss some transporter for each metal that transports and uptake the nutrient in the medium and soil and how is important for plant and crop growth.

This transporter is responsible for transporting nutrients, helping in plant growth and development. There are some transporters responsible for dual nutrient transport and are directly correlated to each other. For example, Nramp5 is responsible to transport Mn and Cd in rice crop, AtHMA3 in Arabidopsis crop, and Cd/Pb SaNramp6 may play a significant role in improving Cd accumulation in transgenic plants for phytoremediation.

5.4 Frequent Mechanism Working as Network inside the Cell

Numerous mechanisms are involved in the HMs transfer across cell membranes in plant roots (Table 5.2). Some of the proteins are highly selective and tend to interfere with the movement of essential metals from the extracellular environment. For instance, DCT1 (divalent cation transporter1; also known as NRAMP2 or DMT1) serves as vital transporter of Fe^{2+} and other divalent metals, however, it has also been known to mediate uptake of toxic metals (Cd^{2+}, Pb^{2+}, etc.) (Ballatori 2002). The underlying mechanism is the potential of HMs to interact with the binding sites reserved for essential metals through which these gain access to various cellular compartments leading to disorganizing of normal biochemical and physiological functions. Another major mechanism by which metals traverse cell membranes and produce cell injury is by forming complexes whose overall structures mimic those of endogenous molecules (Ballatori 2002).

It is known that chromate and molybdate can mimic sulfate to support normal cellular functions (Bridges and Zalups 2005). The arsenate and vanadate for transport and metabolism can compete with phosphate in biological systems (Finnegan and Chen 2012). From the first step at root exudation, and some microorganisms are surrounding the rhizosphere which can tolerate the movement of metal ions from the soil. The various networks are involved many transporters to uptake and accumulate HMs. Under HMs, stress conditions increasing the synthesis of different phytohormones, shown a greater level of oxidative ROS and some ions such as Ca^{2+} ions, K^{+} ions, Ca–Calmodulin system. The mitogen-activated protein kinase (MAPK) system is also up-regulating the genes related to metal toxicity. Emamverdian et al. (2015) reported that the high level of Ca^{2+} is identified in the plants under HMs stress effects and able to assist in the metabolism process in changed stress environments. There are some chemicals in the plant chloroplast that cannot able to dissolve in the cell, and that time plants produce chelation and metallothioneins (MTs) that able to bind those chemicals with metals and isolate them to stable forms. These plant toxic metal-chelating proteins such as MTs which contain two classes 1 MTs and 2 MTs (Robinson et al. 1993).

Later, the protein functions as a defensive element against free radicals' toxicity for plants while essential nutrients help to down-regulate the toxic metals by MTs genes. The heavy and toxic metals affect the uptake of essential nutrients by roots, while the toxic effects of HMs are reduced by essential nutrients. After the transporters, the plant cytoplasm transported these metals to other cell parts to eliminate them from cells. The transporters like PAA1 and ZIP in the plastid, NRAMP 3, ABC and CAX2 in vacuoles, RAN1 in the Golgi apparatuses organelle and ECA transporter proteins in the endoplasmic reticulum are involved to translocate the heavy metal ion (HMI) in plants (Jain et al. 2018).

5.5 Conclusions

This chapter provides comprehensive knowledge about the starring role of metal transporters on plant growth and development under HMs stress. It was highlighted that how metals transporters' mechanisms are involved in response to HMs stress in plants and their particular role in metallic toxicity in crops or soils. The different nutrients transporter in crops and plants that are responsible for uptake transport and

TABLE 5.2

Mechanisms Involvement in Response to HM Toxicity in Plants

Mechanism	Expects	Functions	Organelle	References
CDF	Metals chelate		Cytoplasm	Chen et al. (2009)
CA-CalmodulinMPK signaling	Increased proteins synthesis under metal toxicity	Signal transduction response for HMs.	Vacuole	Maksymiec (2007)
Plasma membrane (PM) ATPase	Signal transduction in response to metals	Metal ions entrance into roots from the soil solution.	Plasma membrane	Janicka-Russak et al. (2008)
Metal ion uptake from the soil	Signal transduction in response to metals	Surrounding root.	Outside root cell	Callahan et al. (2006)
Chelating compound like PCs MTs	HMs chelation in the cytosol	Metals transportation to shoots from the root.	Cytosol	Jain et al. (2018)
SOD, CAT	Oxidative stress mechanism	Metal ions distribution to organelles cells.	Plasma membrane	Sharma et al. (2016)
Receptors	The presence of roots and some microorganisms near the root layer	Receptors allow metal ions to move from the soil.	Outside root cell	Skórzyńska-Polit (1998); Maksymiec (2007)

assimilations were described together with their functions. In addition to the microbial effects, interactions between essential metals and HMs may impart effects on plant N-fixation potential. In conclusion, the HMs not only can affect plants and soil microorganisms but also the functions of different metal transporters which are essential in molecular studies for understanding all interactions between essential nutrients, HMs and transporter in the system plant crop and soil microorganisms.

5.6 Future Perspectives and Strategies

a. There is a dire need to identify and develop mechanisms to regulate the transporter genes involved in the uptake of heavy and toxic metals from the soil, to reduce metal toxicity for plants.
b. There is a need to identify the nutrient transporters link to each other such as up-regulation and down-regulation genes during toxic effects of HMs.
c. There exists the need to conduct further molecular and mechanism enzymatic studies in vegetables for being the most sensitive to HMs stress.
d. Studies must be executed to introduce CRISPER-case9 technology to reduce all kinds of toxic effects of HMs on plants and crops under soil contamination and environmental pollution.
e. Development of the microbial organism also holds potential as these maintain close interaction with plant roots and thus have the potential to assimilate HMs without causing additional toxic effects for plants.

5.7 Conflict of Interest

Authors declare no conflict of interest.

REFERENCES

Ahmadipour F, Bahramifar N, Ghasempouri, SM (2014) Fractionation and mobility of cadmium and lead in soils of Amol area in Iran, using the modified BCR sequential extraction method. Chemi Speci Bioava 26:31–36.

Ballatori N (2002) Transport of Toxic Metals by Molecular Mimicry. Environ Health Perspec110 Suppl 5(Suppl 5):689–694.

Barberon M, Dubeaux G, Kolb C, Isono E, Zelazny E, Vert G (2014) Polarization of iron-regulated transporter 1 (IRT1) to the plant-soil interface plays crucial role in metal homeostasis. ProcNational AcadSci 111(22). https://doi.org/10.1073/pnas.1402262111.

Bashir MK, Steven S, Pandit R (2012) The determinants of rural household food security in the Punjab, Pakistan: an econometric analysis. Working paper 1203, School of Agricultural and Resource Economics, University of Western Australia, Crawley, Australia. http://ageconsearch.umn.edu/bitstream/122526/2/working%20paper%201203.pdf.

Bashir K, Takahashi R, Nakanishi H, Nishizawa NK (2013) The road to micronutrient biofortification of rice: progress and prospects. Front Plant Sci. 4:1–7.

Bin B, Seo J, Kim S T (2018) Function, structure, and transport aspects of ZIP and ZnT zinc transporters in immune cells. J Immuno Res. https://doi.org/10.1155/2018/9365747.

Bridges CC, Zalups RK (2005) Molecular and ionic mimicry and the transport of toxic metals. Toxicol Appl Pharmacol 204(3):274–308.

Buchner P, Takahashi H, Hawkesford MJ (2004) Plant sulphate transporters: co-ordination of uptake, intra cellular and long-distance transport. J Exp Bot 55(404):1765–1773.

Cailliatte R, Schikora A, Briat JF, Mari S, Curie C (2010) High-affinity manganese uptake by the metal transporter NRAMP1 is essential for Arabidopsis growth in low manganese conditions. Plant Cell 22:904–917.

Callahan DL, Baker AJ, Kolev SD, Wedd AG (2006) Metal ion ligands in hyper accumulating plants. JBIC J Biol Inorg Chem 11(1):2–12.

Chandrakar V, Dubey A, Keshavkant S (2018) Modulation of arsenic-induced oxidative stress and protein metabolism by diphenyleneiodonium, 24-epibrassinolide and proline in *Glycine max* L. Acta Bot Croat 77:51–61.

Chen C, Xia S, Deng R, Liu C, Shi G (2017) AhIRT1 and AhNRAMP1 metal transporter expression correlates with Cd uptake in peanuts under iron deficiency 2017. https://doi.org/10.1371/journal.pone.0185144.

Chen J, Li LG, Liu ZH, Yuan YJ, Guo LL, Mao DD, Tian LF, Chen LB, Luan S, Li DP (2009) Magnesium transporter AtMGT9 is essential for pollen development in Arabidopsis. Cell Res 19:887–898.

Chen M, Shen X, Li D, Ma L, Dong J, Wang T(2009) Identification and characterization of MtMTP1, a Zn transporter of CDF family, in the Medicago truncatula. Plant Physiol Biochem 47(11–12):1089–1094.

Chen WR, Feng Y, Chao YE (2008) Genomic analysis and expression pattern of OsZIP1, OsZIP3, and OsZIP4 in two rice (*Oryza sativa* L.) genotypes with different zinc efficiency. Russian J Plant Physiol 55(3):400–409.

Chen ZC, Yamaji N, Horie T, Che J, Li J, An G, Ma JF (2017) A magnesium transporter OsMGT1 plays acritical role in salt tolerance in rice. Plant Physiol 174:1837–1849.

Chen ZC, Yamaji N, Motoyama R, Nagamura Y, Ma JF (2012) Up-regulation of a magnesium transporter gene OsMGT1is required for conferring aluminum tolerance in rice. Plant Physiol 159:1624–1633.

Chiba Y, Mitani N, Yamaji N, Ma JF (2008) HvLsi1 is a silicon influx transporter in barley. The Plant J 57(5):810–818.

Connorton JM, Jones ER, Rodriguez-Ramiro I, Fairweather-Tait S, Uauy C, Balk J (2017) Vacuolar iron transporter TaVIT2 transports Fe and Mn and is effective for biofortification. Plant Physiol174:2434–2444.

Cotrim CA, Jarrott RJ, Martin JL, Drew D (2019) A structural overview of the zinc transporters in the cation diffusion facilitator family. Acta Crystallogr D Struct Biol 75(4), 357–367.

Davidian JC, Kopriva S (2010) Regulation of sulphate uptake and assimilation – the same or not the same? Mol Plant Adv 3:314–325.

Desbrosses G, Josefsson C, Rigas S, Hatzopoulos P, Dolan L (2003) AKT1 and TRH1are required during root hair elongation in Arabidopsis. J Exp Bot 54(383): 781–788.

Ding H, Duan L, Li J, Yan H, Zhao M, Zhang F, Li WX (2010) Cloning and functional analysis of the peanut iron transporter AhIRT1 during iron deficiency stress and intercropping with maize. J Plant Physiol. 167:996–1002.
Dixit G, Singh AP, Kumar A, Mishra S, Dwivedi S, Kumar S, Trivedi PK, Pandey V, Tripathi RD (2016) Reduced arsenic accumulation in rice (*Oryza sativa* L.) shoot involves sulfur mediated improved thiol metabolism, antioxidant system and altered arsenic transporters. Plant Physiol Biochem 99:86–96.
Drummond RSM, Tutone A, Li Y, Gardner RC (2006) A putative magnesium transporter AtMRS2-11 is localized to the plant chloroplast envelope membrane system. Plant Sci 170:78–89.
Dutta A, Dutta Gupta S, Gupta A, Sarkar J, Roy S, Mukherjee A, Sar P (2018) Exploration of deep terrestrial subsurface microbiome in Late Cretaceous Deccan traps and underlying Archean basement, India. Sci Rep 8:17459.
Eide D, Broderius M, Fett J, Guerinot ML (1996) A novel iron-regulated metal transporter from plants identified by functional expression in yeast. Proc Natl Acad Sci USA 93:5624–5628.
Emamverdian A, Ding Y, Mokhberdoran F, Xie Y (2015) Heavy metal stress and some mechanisms of plant defense response. Sci World J 2015, e756120.
Fahr M, Laplaze L, Bendaou N, Hocher V, El Mzibri M, Bogusz D, Smouni A (2013) Effect of lead on root growth. Front Plant Sci. https://doi.org/10.3389/fpls.2013.00175.
Fan X, Naz M, Fan X, Xuan W, Miller AJ, Xu G (2017) Plant nitrate transporters: from gene function to application. J Exp Bot 68: 2463–2475.
Fan X, Tang Z, Tan Y, Zhang Y, Luo B, Yang M, Lian X, Shen Q, Miller AJ, Xu G (2016) Over expression of a pH sensitive nitrate transporter in rice increases crop yields. Proc Natl Acad Sci USA. 113:7118–7123.
Feng H, Li B, Zhi Y, Chen J, Li R, Xia X, Xu G, Fan X (2017) Overexpression of the nitrate transporter, OsNRT2.3b, improves rice phosphorus uptake and translocation. Plant Cell Rep 36:1287–1296.
Feng H, Yan M, Fan X, Li B, Shen Q, Miller AJ, Xu G (2011) Spatial expression and regulation of rice high-affinity nitrate transporters by nitrogen and carbon status. J Exp Bot 62:2319–2332.
Finnegan PM, Chen W (2012) Arsenic toxicity: the effects on plant metabolism. Front Physiol 3:182. https://doi.org/10.3389/fphys.2012.00182.
Forbes JR, Gros P (2001) Divalent-metal transport by NRAMP proteins at the interface of host-pathogen interactions. Trends Microbiol 9(8):397–403.
Gao H, Xie W, Yang C, Xu J, Li J, Wang H, Chen X, Huang C-F (2017) NRAMP2, a trans-Golgi network-localized manganese transporter, is required for Arabidopsis root growth under manganese deficiency. New Phytologist 217(1):179–193.
Gierth M, Mäser P, Schroeder JI (2005) The potassium transporter AtHAK5 functions in K+ deprivation-induced high-affinity K+ uptake and AKT1 K+ channel contribution to K+ uptake kinetics in Arabidopsis roots. Plant Physiol 137:1105–1114.
Gill LW, Ring P, Higgins NMP, Johnston PM (2014) Accumulation of heavy metals in a constructed wetland treating road runoff. Ecol Eng 70:133–139.
Gill M (2014) Heavy metal stress in plants: a review. Int J Adv Res 2:1043–1055.
Goswami D, Kaur J, Surade S, Grell E, Michel H (2012) Heterologous production and functional and thermodynamic characterization of cation diffusion facilitator (CDF) transporters of mesophilic and hyperthermophilic origin. Biol Chem 393, 617–629.
Gravot A, Lieutaud A, Verret F, Auroy P, Vavasseur A, Richaud P (2004) AtHMA3, a plant P1B-ATPase, functions as a Cd/Pb transporter in yeast. FEBS Lett 561:22–28.
Guerinot ML (2000) The ZIP family of metal transporters. Biochim Biophys Acta 1465(1–2):190–198.
Gupta N, Gaurav SS, Kumar A (2013) Molecular basis of aluminum toxicity in plants: a review. Am J Plant Sci 4:21–37.
Han M, Wu W, Wu WH, Wang Y (2016) Potassium transporter KUP7 is involved in K+ acquisition and translocation in Arabidopsis root under K+-limited conditions. Mol Plant 9:437–446.
Hasanuzzaman M, Bhuyan MHMB, Zulfiqar F, Raza A, Mohsin SM, Mahmud JA, Fujita M, Fotopoulos V (2020) Reactive oxygen species and antioxidant defense in plants under abiotic stress: revisiting the crucial role of a universal defense regulator. Antioxi (Basel) 9(8):681. https://doi.org/10.3390/antiox9080681.
Hawkesford MJ (2003) Transporter gene families in plants: the sulphate transporter gene family-redundancy or specialization? Physiol Planta 117:155–165.
Hedrich R, Sauer N, Neuhaus HE (2015) Sugar transport across the plant vacuolar membrane: nature and regulation of carrier proteins. Curr Opin Plant Biol 25, 63–70.
Ishimaru Y, Suzuki M, Kobayashi T, Takahashi M, Nakanishi H, Mori S, Nishizawa NK (2005) OsZIP4, a novel zinc-regulated zinc transporter in rice. J Exp Bot 56(422):3207–3214.
Jain S, Muneer S, Guerriero G, Liu S, Vishwakarma K, Chauhan DK, Dubey NK, Tripathi DK, Sharma S (2018) Tracing the role of plant proteins in the response to metal toxicity: a comprehensive review. Plant Sign Behav 13(9):e1507401.
Janicka-Russak M, Kabała K, Burzyński M, Kłobus G (2008) Response of plasma membrane H+-ATPase to heavy metal stress in Cucumissativu s roots. J Exp Bot 59(13):3721–3728.
Jia H, Ren H, Gu M, Zhao J, Sun S, Zhang X, Chen J, Wu P, Xu G (2011) The phosphate transporter gene OsPht1;8 is involved in phosphate homeostasis in rice. Plant Physiol 156:1164–1175.
Jiang HH, Cai LM, Wen HH, Hu GC, Chen LG, Luo J (2020) An integrated approach to quantifying ecological and human health risks from different sources of soil heavy metals. Sci Total Environ 701:134466.
Jiang C, Liu L, Li X, Han R, Yongxi YW (2018) Insights into aluminum-tolerance pathways in Stylosanthes as revealed by RNA-Seq analysis. Sci Rep. https://doi.org/10.1038/s41598-018-24536-3.
Kasai K, Takano J, Miwa K, Toyoda A, Fujiwara T (2011) High boron-induced ubiquitination regulates vacuolar sorting of the BOR1 borate transporter in *Arabidopsis thaliana*. J Biol Chem 286:6175–6183.
Kato A, Kumazaki K, Wada M, Taniguchi R, Nakane T, Yamashita K, Hirata K, Ishitani R, Ito, K, Nishizawa T, Nureki O (2019) Crystal structure of plant vacuolar iron transporter VIT1. Nat Plant 5(3):308–315.

Kronzucker HJ, Glass AD, Yaeesh Siddiqi M (1999) Inhibition of nitrate uptake by ammonium in Barley. Analysis of Component Fluxes. Plant Physio 120:283–291.

Krstic D, Djalovic I, Nikezic D, Bjelic D (2012) Aluminium in acid soils: chemistry, toxicity and impact on maize plants food production – approaches, challenges and tasks. InTechOpen. https://doi.org/10.5772/33077.

Kumar S, Khare RP (2019) Arsenic-responsive high-affinity rice sulphate transporter, OsSultr1;1, provides abiotic stress tolerance under limiting sulphur condition. J Hazard Mat 373. https://doi.org/10.1016/j.jhazmat.2019.04.011.

Lal R (2010) Managing soils and ecosystems for mitigating anthropogenic carbon emissions and advancing global food security. Bio Sci 60(9):708–721.

Lanquar V, Ramos MS, Lelièvre F, Barbier-Brygoo H, Krieger-Liszkay A, Krämer U, Thomine S (2010) Export of vacuolar manganese by AtNRAMP3 and AtNRAMP4 is required for optimal photosynthesis and growth under manganese deficiency. Plant Physiol 152:1986–1999.

Lee S Jeong HJ, Kim SA, Lee J, Guerinot ML, An G (2010) OsZIP5 is a plasma membrane zinc transporter in rice. Plant Mol Biol 73:507–517. https://doi.org/10.1007/s11103-010-9637-0.

Li CX, Li YY, Li YJ, Fu GZ (2018) Cultivation techniques and nutrient management strategies to improve productivity of rainfed maize in semi-arid regions. Agril Water Manage 210:149–157.

Li H, Wang N, Ding J, Liu C, Du H, Huang K, Cao M, Lu Y, Gao S, Zhang S (2017) The maize CorA/MRS2/MGT-type Mg transporter, ZmMGT10, responses to magnesium deficiency and confers low magnesium tolerance in transgenic Arabidopsis. Plant Mol Biol 95:269–278. https://doi.org/10.1007/s11103-017-0645-1.

Li X, He H, Zhang X, Yan X, Six J, Cai Z, Barthel M, Zhang J, Necpalova M, Ma Q, Li Z (2019) Distinct responses of soil fungal and bacterial nitrate immobilization to land conversion from forest to agriculture. Soil Biol Biochem 134:81–89.

Lima JE, Kojima S, Takahashi H, von Wiren N (2010) Ammonium triggers lateral root branching in Arabidopsis in an ammonium transporter1; 3-dependent manner. Plant Cell 22:3621–3633.

Liu X, Guo L, Luo L, Liu Y, Peng S (2019) Identification of the magnesium transport (MGT) family in Poncirustrifoliata and functional characterization of PtrMGT5 in magnesium deficiency stress. Plant Mol Biol 101:551–560. https://doi.org/10.1007/s11103-019-00924-9.

Loqué D, Yuan L, Kojima S, Gojon A, Wirth J, Gazzarrini S, Ishiyama K, Takahashi H, von Wirén N (2006) Additive contribution of AMT1;1 and AMT1;3 to high-affinity ammonium uptake across the plasma membrane of nitrogen-deficient Arabidopsis roots. Plant J 48:522–534.

Ma JF, Yamaji N (2008) Functions and transport of silicon in plants. Cell Mol Life Sci 65:3049–3057.

Maheshwari R, Dubey RS (2007) Nickel toxicity inhibits ribonuclease and protease activities in rice seedlings: protective effects of proline. Plant Growth Regul 51:231–243. https://doi.org/10.1007/s10725-006-9163-x.

Maksymiec W (2007) Signaling responses in plants to heavy metal stress. Acta Physiol Plant. 29(3):177. https://doi.org/10.1007/s11738-007-0036-3.

Momonoi K, Yoshida K, Mano S, Takahashi H, Nakamori C, Shoji K, Nitta A, Nishimura M (2009) A vacuolar iron transporter in Tulip TgVit1 is responsible for blue coloration in petal cells through iron accumulation. Plant J 59(3):437–47. https://doi.org/10.1111/j.1365-313X.2009.03879.x.

Montpetit J, Vivancos J, Mitani-Ueno N, Yamaji N, Remus-Borel W, Belzile F, Ma JF, Belanger RR (2012) Cloning, functional characterization and heterologous expression of TaLsi1, a wheat silicon transporter gene. Plant Mol Biol 79:35–46. https://doi.org/10.1007/s11103-012-9892-3.

Moreira A, Cardoso MLA, Reis A (2018) The molecular genetics of zinc uptake and utilization efficiency in crop plants. In: Plant Micronutrient Use Efficiency: Molecular and Genomic Perspectives in Crop Plants, pp. 87–108. https://doi.org/10.1016/B978-0-12-812104-7.00006-X.

Nakagawa Y, Hanaoka H, Kobayashi M, Miyoshi K, Miwa K, Fujiwara T (2007) Cell-type specificity of the expression of Os BOR1, a rice efflux boron transporter gene, is regulated in response to boron availability for efficient boron uptake and xylem loading. Plant Cell 19:2624–2635. https://doi.org/10.1105/tpc.106.049015.

Nasim S, Dhir (2010) Heavy metals alter the potency of medicinal plants. In: Whitacre D (ed). Reviews of Environmental Contamination and Toxicology (Continuation of Residue Reviews), vol. 203. Springer, New York. https://doi.org/10.1007/978-1-4419-1352-4_5.

Naz M, Luo B, Guo X, Li B, Chen J, Fan X (2019) Overexpression of nitrate transporter OsNRT2.1 enhances nitrate-dependent root elongation. Genes 10(4):290. https://doi.org/10.3390/genes10040290.

Nevo Y, Nelson N (2004) The mutation F227I increases the coupling of metal ion transport in DCT1. J Biol Chem 279(51):53056–53061.

Nevo Y, Nelson N (2006) The NRAMP family of metal-ion transporters. Biochimica et Biophysica Acta Mole Cell Res 1763:609–620.

Nihorimbere V, Ongena M, Smargiassi M, Thonart P (2011) Beneficial effect of the rhizosphere microbial community for plant growth and health. Biotechnol Agron Soc 15:327–337.

Pérez-Castro R, Kasai K, Gainza-Cortés F, Ruiz-Lara S, Casaretto JA, Peña-Cortés H, Tapia J, Fujiwara T, González E (2012) VvBOR1, the grapevine ortholog ofAtBOR1, encodes an efflux boron transporter that is differentially expressed throughout reproductive development of *Vitis vinifera* L. Plant Cell Physiol 53(2):485–494. https://doi.org/10.1093/pcp/pcs001.

Pedas P, Ytting CK, Fuglsang AT, Jahn TP, Schjoerring JK, Husted S (2008) Manganese efficiency in barley: identification and characterization of the metal ion transporter HvIRT1. Plant Physiol 148:455–466.

Pourrut B, Shahid M, Dumat C, Winterton P, Pinelli E (2011) Lead uptake, toxicity, and detoxification in plants. Rev Environ Conta Toxico 213:113–136.

Puig S (2014) Function and regulation of the plant COPT family of high-affinity copper transport proteins. Adv Bot 2014:1–9. https://doi.org/10.1155/2014/476917.

Pyo YJ, Gierth M, Schroeder JI, Cho MH (2010) High-affinity K+ transport in Arabidopsis: AtHAK5 and AKT1 are vital for seedling establishment and postgermination growth under low-potassium conditions. Plant Physiol 153:863–875.

Ravet R, Pilon M (2013) Copper and iron homeostasis in plants: the challenges of oxidative stress. Antioxid Redox Signal 19(9):919–932. https://doi.org/10.1089/ars.2012.5084.

Ricachenevsky FK, Punshon T, Lee S, Oliveira BHN, Trenz TS, Maraschin FDS, Hindt MN, Danku J, Salt DE, Fett JP, Guerinot ML (2018) Elemental profiling of rice FOX lines leads to characterization of a new Zn plasma membrane transporter, OsZIP7. Front Plant Sci 9:865.

Rigas S, Debrosses G, Haralampidis K, Vicente Agullo F, Feldmann K, Grabov A, Dolan L, Hatzopoulos P (2001) Trh1 encodes a potassium transporter required for tip growth in Arabidopsis root hairs. The Plant Cell 13:139–151.

Robinson NJ, Tommey AM, Kuske C, Jackson PJ (1993) Plant metallothioneins. Biochem J 295(1):1. https://doi.org/10.1042/bj2950001.

Saleem MH, Fahad S, Khan SU et al (2020) Copper-induced oxidative stress, initiation of antioxidants and phytoremediation potential of flax (*Linum usitatissimum L.*) seedlings grown under the mixing of two different soils of China. Environ Sci Pollut Res 27, 5211–5221. https://doi.org/10.1007/s11356-019-07264-7.

Sasaki A, Yamaji N, Mitani Ueno N, Kashino M, Ma JF (2015) A node localized transporter OsZIP3 is responsible for the preferential distribution of Zn to developing tissues in rice. Plant J 84(2). https://doi.org/10.1111/tpj.13005.

Sasaki A, Yamaji N, Yokosho K, Ma JF (2012) Nramp5 is a major transporter responsible for manganese and cadmium uptake in rice. Plant Cell 24(5):2155–67. https://doi.org/10.1105/tpc.112.096925.

Schikora A, Thimm O, Linke B, Buckhout TJ, Müller M, Schmidt W (2006) Expression, localization, and regulation of the iron transporter LeIRT1 in tomato roots. Plant Soil 284(1):101–108. https://doi.org/10.1007/s11104-006-0046-8.

Schjoerring JK, Cakmak I, White PJ (2019) Plant nutrition and soil fertility: synergies for acquiring global green growth and sustainable development. Plant Soil 434:1–6. https://doi.org/10.1007/s11104-018-03898-7.

Selle A, Willmann M, Grunze N, Geßler A, Weiß M, Nehls U (2005) The high affinity poplar ammonium importer PttAMT1.2 and its role in ectomycorrhizal symbiosis. New Phytol 168(3):697–706. https://doi.org/10.1111/j.1469-8137.2005.01535.x.

Seo HM, Jung Y, Song S, Kim Y, Kwon T, Kim DH, Jeung SJ, Yi YB, Yi G, Nam MH, Nan J (2008) Increased expression of OsPT1, a high-affinity phosphate transporter, enhances phosphate acquisition in rice. Biotechnol Lett 30:1833–1838. https://doi.org/10.1007/s10529-008-9757-7.

Seshadri B, Bolan N, Naidu R (2015) Rhizosphere-induced heavy metal (loid) transformation in relation to bioavailability and remediation. J Soil Sci Plant Nutr 15:524–548. http://dx.doi.org/10.4067/S0718-95162015005000043.

Shao JF, Yamaji N, Shen RF, Ma JF (2017) The key to Mn homeostasis in plants: regulation of Mn transporters. Trends Plant Sci 22: 215–224.

Sharma P, Dubey RS (2005) Lead Toxicity in Plants. Brazilian J Plant Physiol 17(1):45–52. https://doi.org/10.1590/S1677-04202005000100004.

Sharma P, Kumar A, Bhardwaj R (2016) Plant steroidal hormone epibrassinolide regulate-heavy metal stress tolerance in *Oryza sativa* L. by modulating antioxidant defense expression. Environ Exp Bot 122:1–9.

Singh J, Kalamdhad AS (2011) Effects of heavy metals on soil, plants, human health and aquatic life. Int J Res Chem Environ 1:15–21.

Skórzyńska-Polit E, Tukendorf A, Selstam E, Baszyński T (1998) Calcium modifies Cd effect on runner bean plants. Environ Exp Bot 40(3):275–286. https://doi.org/10.1016/S0098-8472(98)00045-8.

Socha AL, Guerinot ML (2014) Mn-euvering manganese: the role of transporter gene family members in manganese uptake and mobilization in plants. Front Plant Sci 5:106. https://doi.org/10.3389/fpls.2014.00106.

Soliman M, Potlakayala S, Millar D, Weeden H, Bogush D, Deguchi M, Rudrabhatla S (2019) Comparing a review of heavy metal uptake and their toxicity on plant and human health. Int J Plant Anim Environ Sci 9(3):182–189.

Sudre D (2009) Identification of membrane proteins involved in heavy metal detoxification and transport in *Arabidopsis thaliana* (Doctoral dissertation). University of Zurich, Faculty of Science.

Sun SB, Gu M, Cao Y, Huang XP, Zhang X, Ai PH, Zhao JN, Fan XR, Xu GH (2012) A constitutive expressed phosphate transporter, OsPht1;1, modulates phosphate uptake and translocation in phosphate replete rice. Plant Physiol 159:1571–1581.

Sytar O, Cai Z, Brestic M, Kumar A, Prasad MNV, Taran N, Smetanska I (2013) Foliar applied nickel on buckwheat (*Fagopyrum esculentum*) induced phenolic compounds as potential antioxidants. Clean Soil Air Water 41:1129–1137. https://doi.org/10.1002/clen.201200512.

Tahjib-Ul-Arif M, Al Mamun Sohag A, Mostofa MG, Polash MAS, Mahamud ASU, Afrin S, Hossain A, Hossain MA, Murata Y, Phan Tran LS (2020) Comparative effects of ascobin and glutathione on copper homeostasis and oxidative stress metabolism in mitigation of copper toxicity in rice. Plant Biol. https://doi.org/10.1111/PLB.13222.

Takada S, Miwa K, Omori H, Fujiwara T, Naito S, Takanao J (2014) Improved tolerance to boron deficiency by enhanced expression of the boron transporter BOR2. Soil Sci Plant Nutr 60:341–348.

Takahashi H (2019) Sulfate transport systems in plants: functional diversity and molecular mechanisms underlying regulatory coordination. J Experi Bot 70 (16): 4075–4087. https://doi.org/10.1093/jxb/erz132.

Takahashi T, Nakashima S, Masuda T, Yoneda S, Hwang GW, Naganuma A (2011) Overexpression of CLN1, CLN2, or ERG13 increases resistance to adriamycin in *Saccharomyces cerevisiae*. J Toxicol Sci 36(6):855–857.

Takano J, Miwa K, Yuan L, von Wirén N, Fujiwara T (2005) Endocytosis and degradation of BOR1, a boron transporter of *Arabidopsis thaliana*, regulated by boron availability. Proc Natl Acad Sci USA 102:12276–12281.

Tanaka N, Uraguchi S, Saito A, Kajikawa M, Kasai K, Sato Y, Nagamura Y, Fujiwara T (2013) Roles of pollen-specific boron efflux transporter, OsBOR4, in the rice fertilization process. Plant Cell Physiol 54:2011–2019. https://doi.org/10.1093/pcp/pct136.

Tittarelli A, Milla L, Vargas F, Morales A, Neupert C, Meisel LA, Salvo-G H, Peñaloza E, Muñoz G, Corcuera LJ, Silva H (2007) Isolation and comparative analysis of the wheat TaPT2 promoter: identification in silico of new putative regulatory motifs conserved between monocots and dicots. J Exp Bot 58(10):2573–2582. https://doi.org/10.1093/jxb/erm123.

Tombuloglu H, Filiz E, Aydın M, Koc I (2017) Genome-wide identification and expression analysis of sulphate transporter (SULTR) genes under sulfur deficiency in *Brachypodium distachyon*. J Plant Biochem Biotechnol 26:263–273. https://doi.org/10.1007/s13562-016-0388-0.

Turkyilmaz A, Sevik H, Isinkaralar K, Cetin M (2019) Use of tree rings as a bioindicator to observe atmospheric heavy metal deposition. Environ Sci Pollu Res 26:5122–5130. https://doi.org/10.1007/s11356-018-3962-2.

Uraguchi S, Kamiya T, Sakamoto T, Kasaia K, Satoc Y, Nagamurac Y, Yoshidab A, Kyozukab J, Isikawa S, Fujiwara T (2011) Low-affinity cation transporter (OsLCT1) regulates cadmium transport into rice grains. Proc Nat Aca Sci 108:20959–20964.

Vert G, Briat JF, Curie C (2001) Arabidopsis IRT2 gene encodes a root-periphery iron transporter. Plant J 26:181–189.

Viehweger K (2014) How plants cope with heavy metals. Bot Stud 55:35. https://doi.org/10.1186/1999-3110-55-35.

Wang AS, Angle JS, Rufus LC, Delorme TA, Reeves RD (2006) Soil pH effects on uptake of Cd and Zn by *Thlaspi caerulencens*. Plant Soil 281:325–337.

Wang L, Zheng J, Luo YZ, Xu T, Zhang QX, Fan YL (2013) Construction of a genome-wide RNAi mutant library in rice. Plant Biotechnol J 11:997–1005.

Williams LE, Pittman JK, Hall JL (2000) Emerging mechanisms for heavy metal transport in plants. Biochim Biophys Acta-Biomembranes 1465(1–2):104–126. https://doi.org/10.1016/S0005-2736(00)00133-4.

Xiong H, Kobayashi T, Kakei Y, Senoura T, Nakazono M, Takahashi H, Nakanishi H, Shen H, Duan P, Guo X, Nishizawa NK, Zuo Y (2012) AhNRAMP1 iron transporter is involved in iron acquisition in peanut. J Exp Bot 63(12):4437–4446.

Yadav SK (2010) Heavy metals toxicity in plants: an overview on the role of glutathione and phytochelatins in heavy metal stress tolerance of plants. S Afr J Bot 76:167–179. https://doi.org/10.1016/j.sajb.2009.10.007.

Yan M, Fan X, Feng H, Miller AJ, Shen Q, Xu G (2011) Rice OsNAR2.1 interacts with OsNRT2.1, OsNRT2.2 and OsNRT2.3a nitrate transporters to provide uptake over high and low concentration ranges. Plant Cell and Environ 34(8):1360–1372. https://doi.org/10.1111/j.1365-3040.2011.02335.x.

Yang C, Li Z, Shi Z, He K, Tian A, Wu J, Zhang Y, Li Z (2014) Regulation of cell survival by the HIP-55 signaling network. Mol Biosyst 10(6):1393–1399.

Yang J, Wang L, Mao C, Lin H (2017) Characterization of the rice NLA family reveals a key role for OsNLA1 in phosphate homeostasis. Rice (10):1–6.

Yang W, Gao M, Yin X, He ZH (2013) Control of rice embryo development, shoot apical meristem maintenance, and grain yield by a novel cytochrome P450. Mol Plant 6:1945–1960.

Ye Y, Yuan J, Chang X, Yang M, Zhang L, Lu K, Lian X (2015) The phosphate transporter gene OsPht1;4 is involved in phosphate homeostasis in rice. PLoS ONE 10: e0126186.

Yokosho K, Yamaji N, Ueno D, Mitani N, Ma JF (2009) OsFRDL1 is a citrate transporter required for efficient translocation of iron in rice. Plant Physiol 149:297–305.

Yuan W, Xie J, Long C, Erdjument-Bromage H, Ding X, Zheng Y, Tempst P, Chen S, Zhu B, Reinberg D (2009) Heterogeneous nuclear ribonucleoprotein L is a subunit of human KMT3a/Set2 complex required for H3 Lys-36 trimethylation activity in vivo. J Biol Chem 284(23):15701–15707.

Zhang L, Peng Y, Li J, Tian X, Chen Z (2019) OsMGT1 confers resistance to magnesium deficiency by enhancing the import of Mg in rice. Int J Mol Sci 20:207.

Zhang Q, Chen H, He M, Zhao Z, Cai H, Ding G, Shi L, Xu F (2017) The boron transporter BnaC4.BOR1;1c is critical for inflorescence development and fertility under boron limitation in *Brassica napus*. Plant Cell Environ 40(9): 1819–1833. https://doi.org/10.1111/pce.12987.

Zhang W, Li Y, Zhu B, Zheng X, Liu C, Tang J, Su F, Zhang C, Ju X, Deng J (2018) A process-oriented hydro-biogeochemical model enabling simulation of gaseous carbon and nitrogen emissions and hydrologic nitrogen losses from a subtropical catchment. Sci Total Environ 616:305–317.

Zhao ZQ, Shen RF (2018) Aluminum-nitrogen interactions in the soil-plant system. Front Plant Sci 9:807. https://doi.org/10.3389/fpls.2018.00807.

Zhou JQ, Jiang YR, Ming XQ, Wang JR, Tang WB, Sun L (2019) Introgressing the allelic variation of a major locus in reducing the grain cadmium accumulation in Indica rice hybrids. Mol Breed 39:84.

6

Role for Genetically Modified Plants in Protection against Heavy Metal Toxicity

Aaliya Batool and Noreen Zahra
Department of Botany, University of Agriculture, Faisalabad, Pakistan

Rubina Naseer
Centre of Agricultural Biochemistry and Biotechnology, University of Agriculture, Faisalabad, Pakistan

Kanval Shaukat
Department of Botany, University of Agriculture, Faisalabad, Pakistan
Department of Botany, University of Balochistan, Quetta, Pakistan

Tahira Rasheed
Department of Botany, University of Agriculture, Faisalabad, Pakistan

Muhammad Bilal Hafeez
Department of Agronomy, University of Agriculture, Faisalabad, Pakistan

CONTENTS

DOI: 10.1201/9781003155089-6

6.1 Introduction

Plants frequently face biotic and abiotic stress conditions. Human activities like sewage disposal, smelting and the use of fertilizers are the major reasons for heavy metal (HMs) accumulations i.e. copper (Cu), zinc (Zn), cobalt (Co), nickel (Ni), iron (Fe), manganese (Mn), cadmium (Cd), argon (Ar) and mercury (Hg) in soil (Aydinalp and Marinova 2009). Anthropogenic activities are the main reason due to which HMs leached into nearby water resources or accumulate over the soil surface (Gupta and Ali 2002; Ali et al. 2009; Hakeem et al. 2015; Ozturk et al. 2015a, 2015b; Basheer 2018a). These HMs are considered as non-biodegradable as they cannot be decomposed from environment naturally. Plants uptake these HMs from soil through root system by the process of endocytosis, diffusion or with the help of metal transporters (Ali et al. 2017b, 2018; Alharbi et al. 2018; Basheer 2018b; Burakova et al. 2018). However, some HMs like Cu, Zn and Ni are required by plants in minute quantities as they act as cofactor whereas other HMs like Pb and Cd are required with optimal limits, and eventually becomes toxic and dangerous for plants and environment as well as human health if accumulates in food chain (Raskin and Ensley 2000; Sharma and Ali 2011; Ali et al. 2016, 2017a). The toxicity of HMs is dependent on species type, its accumulation and tolerance level; as some species may become toxic at faster rate. Plants exposed to HMs stress exhibit the following symptoms like reduced growth, leaf chlorosis, browning of roots, decline and then finally death (Ozturk et al. 2008, 2015b). Plants exposure to HMs stress may results in the inactivation and denaturation of several significant enzymes and proteins, restricting substitution reaction of metallic ion that disrupts membranes integrity which in turn alters basic metabolic reactions like respiration, homeostasis and photosynthesis (Hossain et al. 2012). Further, it results in the production of reactive oxygen species (ROS) that include superoxide radical (O^{-2}), hydrogen peroxide (H_2O_2) and hydroxyl radical (OH). Plants possess several physiological and molecular procedures to cope with HMs stress including procedures of complex genomic and biochemical level.

Plants detoxify these HMs via activation of specialized mechanisms in order to survive under stress conditions; these mechanisms include HMs chelation, transportation, sequestration or detoxification into vacuole. The activation of any of this mechanism requires the stimulation of various stress alleviating agents such as hormones, stress-related proteins, signaling molecules and antioxidants (Tena et al. 2001; Jonak et al. 2002). In plants' metal defense mechanism, GSH (reduced glutathione) is considered as a main redox buffer for homeostasis of ions and detoxification of HMs as it provides protection against ROS via ascorbate-glutathione cycle (Foyer and Noctor 2005; Helbig et al. 2008). The production of more anthocyanin provides antioxidants that helps to scavenge ROS for plants' survival under stress conditions (Cheng et al. 2013; Naing et al. 2017). HMs accumulation and tolerance can be enhanced by the incorporation of functional genes into plants with genetic engineering (Helbig et al. 2008). It has been reported that glutathione and phytochelatin synthase (PCS) gene plays a significant role in reduction of HMs stress and scavenging of ROS (Shao et al. 2008; Ai et al. 2018). One of the highly effective, economical and environment friendly way of HMs removal is known as phytoremediation. This process results in the extraction or stabilization of HMs and their metabolites into tissues (Raskin and Ensley 2000). Considerable efforts have been made for the identification of genes involved in phytoremediation (Chaudhary et al. 2016). The overexpression of *IbMYB1* gene in transgenic potato and snapdragon Delila (Del) gene in transgenic tobacco resulted in increased production of anthocyanin and showed improved tolerance to abiotic stress. The role of *GSH*, *PCS* and *MYB* in tolerance of HMs stress has been documented in walnut and maize (Li et al. 2017; Xu et al. 2018). The overexpression of *RsMYB1* in transgenic plants having more anthocyanin helps to tolerate HMs stress (Ai et al. 2017). Some other genes such as HM ATPases; comprising *HMA2*, *HMA3* and *HMA4* are also involved in phytoremediation. The expression of certain genes also involved in HMs uptake, transportation, sequestration and detoxification resulting in enhanced plant yield (Chaudhary et al. 2016). The goal of this chapter is to discuss the new strategies of HMs tolerance in transgenic plants via utilizing genetic engineering protocols together with the study of important genes that plays crucial role in HMs tolerance.

6.2 Role of Genes to Improve Plant Growth under Stress Conditions

As plants are sessile, they cannot move from one place to another, so they have to face environmental fluctuations. The plant adopts several mechanisms to cope with these situations and minimize possible damage. These adaptations help plants to survive under critical situations or specific growth habits (Allan and Fluhr 2001). A study by Jaspers and Kangasjärvi (2010) reported that signaling cascades are induced under stress conditions that result in the activation of ionic channels, ROS production, kinase cascades and several hormone accumulation. Particular subsets of defense genes are activated in response to these stress signals responsible for the whole defense system. According to Nematshahi et al. (2012) for the period of early growth stage, chromium (Cr) harshly minimizes dry matter production of seedling as well as hinders leaves and stems development; through inhibiting the processes of cell division and elongation resulting in shortage of roots length. Therefore, absorption of nutrients and water is strictly limited, reducing shoot growth (Shanker et al. 2005). Similarly, a study by Zou et al. (2006) reported decreased growth of root under Cr stress in *Amaranthus viridis* L. It indicated that Cr inhibits cell division and creates a Cs^{2+} imbalance in cells, thus reducing root growth. Two techniques such as genome-wide transcriptome profiling and metabolome analysis were used to study root growth of *Oryza sativa* under Cr (VI) stress. It was shown by microarray analysis that 1,138 were up-regulated and 1,610 genes were down-regulated out of 1,279 total genes. The expression pattern of these genes was responsible for stress response, defense, growth and development of plants (Dubey et al. 2010). In addition, a study by Liu et al. (2014) stated the inhibitory effect of aluminum (Al) on plant growth specifically in acidic soils having lowest pH values ranging between 5 and less than 5 with a dominant phytotoxic form of Al^{3+}. Further, a recent study by Bouzroud et al. (2020) reported that root development and density are promoted by antisense down-regulation of SIARF4 under a stressful environment. It was indicated that plants displayed better growth and tolerance by upregulation of Cu/ZnSOD and mdhar genes in ARF4 as plants.

6.3 Omics Techniques for Heavy Metal Tolerance

The new "omics" techniques like metabolomics, genomics, transcriptomics and proteomics help in the identification of genes behind stress tolerance in plants. These techniques are involved in monitoring of factors that affects the growth and development of plants. Under stress conditions, plants revealed a change in gene expression that may result in alteration of the transcriptome, proteome and metabolome composition (Pérez-Alfocea et al. 2011) (Table 6.1).

6.3.1 Metabolomics

Arbona et al. (2013) reported the term metabolomics that deals with identifying and quantifying metabolites having lower-molecular weight that are necessary for developmental stages of an organism. It was indicated that various metabolites play a crucial role in tolerance of HMs stress. Several studies suggest that amino acids and their derived compounds are involved in metal ions chelation that may result in tolerance of HMs. It was found that particular amino acids such as proline and histidine carry out chelation of metal ions in cells along with sap of xylem (Rai 2002; Sharma and Dietz 2006). Similarly, Siripornadulsil et al. (2002) demonstrated that an enhanced level of proline minimizes free radical production, and by increasing the level of glutathione (GSH), it also sustains the reducing environment. Kerkeb and Krämer (2003) discussed Ni uptake simultaneously in *B. juncea*. Alterations in the content of GSH play a significant role in tolerance of metal stress (Sharma and Dietz 2006). Similarly, another study by Stephan and Scholz (1993) indicated the role of GSH and amino acid derivative known as NA (amino carboxylate) produced by three S-adenosyl-L-methionine condensations in metal ion chelation.

A recent study by Osmolovskaya et al. (2018) reported that several organic acids are secreted under HMs, such as oxalic acid, malic acid and citric acid, responsible for sequestrations of HMs into vacuoles. This study demonstrates the role of these organic acids in A^{3+} detoxification through complex chelating formation. Similarly, several studies suggest the same role of organic acids to Al^{3+} induced stress in various plants such as tomato, spinach and amaranth. Another study by Pietrini et al. (2003) reported the crucial role of GSH considered as water-soluble tripeptide having low molecular weight in tolerance of HMs stress. Likewise, Talukdar and Talukdar (2014) reported that GSH reacts with ROS produced as a result of HMs stress. In addition, according to Metwally et al. (2005), GSH is responsible for phytochelatins (PCs) productions that are involved in HMs sequestration. Several authors indicated that GSH level was increased in the case of Cr, Al and Cd stress in *Pisum sativum*.

TABLE 6.1
Various Omics Techniques for HM Stress Tolerance

Omics Techniques for HMs Stress Tolerance	Name of Genes/TFs and Techniques	Studied Plants	Plant Responses	References
Genomics	*CAT3*	*Nicotiana tobaccum*	Helps in Cd tolerance, better seedling growth and longer roots	Gichner et al. (2004)
	AtPCS1	*N. tobaccum*	Helps in HMs detoxification specifically Cd stress	Pomponi et al. (2006)
	Zn/SOD and APX	*Festuca arundinacea*	Increase Cu, As and Cd stress tolerance by elevating oxidative stress	Lee et al. (2007)
	MDHAR/DHAR	*N. tobaccum*	DHAR increases tolerance to Al by maintaining ascorbate level	Yin et al. (2010)
	MT1	*N. tobaccum*	Increase accumulation and tolerance of Hg stress	Ruiz et al. (2011)
Transcriptomics	*WRKY6 (WRKY family)*	*Arabidopsis thaliana*	Plant shows As^{V} tolerance and transposon gene silencing	Castrillo et al. (2013)
	ZIP39 (bZIP family)	*Oryza sativa*	Regulates ER stress response	Takashi et al. (2012b)
	bHLH38 (bHLH family)	*Arabisopsis thaliana*	Involved in Cd tolerance by enhancing Cd sequestration in root and also improves Fe homeostasis	Wu et al. (2012)
	bHLH100 (bHLH family)	Arabidopsis sp.	Responsible for Fe and Zn stress response and homeostasis	Gao et al. (2013)
	ERF1 and ERF2 (AP2/ ERF family)	Arabidopsis	Responsible for Cd stress tolerance	Xie et al. (2019)
	OSISAP1 (Zinc-finger protein)	*N. tobaccum*	Overexpression may Pb to tolerance against various abiotic stresses including HMs stress	Mukhopadhyay et al. (2004)
Proteomics	*2DE, MALDITOF-MS, LC-ESIQTOF-MS*	*A. thaliana*	Changes in these proteins in roots help plants to cope with Cd stress by modulating S assimilation	Roth et al. (2006)
	2-DE, Maldi-TOFTOF-MS	*Suaeda salsa*	Changes in these proteins help to tolerated Hg stress	Liu et al. (2013)

6.3.2 Genomics

According to Hossain et al. (2012), plants have a well-developed gene expression system and signaling in response to HMs stress. Molecular genetics plays an essential role in the consideration of HMs detoxification in plants. In a study by Ortega-Villasante et al. (2007), it was recognized through Rt-PCR that alteration in the expression level of certain genes such as *GR1, GR2, GS, hGS, GPX, ECS* and *PCS* was noticed in response to exposure durability of Cd and mercury (Hg). Similarly, Cd stress when exposed to the root part of rice results in enhanced *GST* and *APX* gene expression (Lee et al. 2010). Likewise, a study demonstrates the direct involvement of GST in Cd^{2+} sequestration through GSH-Cd complex synthesis (Adamis et al. 2004). Further, it was revealed that expression patterns of genes for GSH, PCs and metallothioneins (MTs) play a crucial role in the sequestration and detoxification of HMs. A new technique, namely CRISPR/CAS acts as a tool for genome editing and can be used to make plants resistant to HMs. A recent study reported the role CRISPR/CAS9 in developing new lines of indica rice that hold lower levels of Cd in grains through knockdown of metal transporter gene (*OsNramp5*) without affecting yield potential (Tang et al. 2017). It was stated that *CAT3* was responsible for tolerance of Cd stress, it also results in enhanced seedling growth and longer roots in *Nicotiana tobaccum* (Gichner et al. 2004). Similarly, another study reported the role of *ZnSOD/APX* in improved arsenic (As), Cd and Cu tolerance via elevating oxidative stress in *Festuca arundinacea* (Lee et al. 2007). Another study by Yin et al. (2010) indicated that *DHAR* is involved in tolerance of Al stress in *N. tobaccum*. Likewise, Ruiz et al. (2011) reported Hg accumulation and tolerance in *N. tobaccum* due to *MT1*.

6.3.3 Transcriptomics

It is well known that as the fate of plant growth and HMs tolerance is determined by gene expression, so it is necessary to observe the regulation of gene expression at a molecular level. Many genes and few proteins are activated in response to stress conditions through linking signal pathways (Umezawa et al. 2006; Valliyodan and Nguyen 2006; Manavalan et al. 2009; Tran et al. 2010). Some other studies also reported about these regulatory genes that several transcription factors (TFs) are encoded by these genes. These TFs can activate several genes that respond to stress conditions individually or in cooperation and hence, form a network of genes. Though, metabolic compounds like sugars, alcohols and amines are encoded by genes of functional group that are involved in tolerance of HMs. So, TFs are considered to be master regulators of gene expression (Wray et al., 2003; Nakashima et al. 2009).

Similarly, according to some other studies various families of transcription factors like MYB, MYC, WRKY, DREB1/CBF AREB/ABF, etc., are responsible for stress tolerance in plants (Singh et al. 2002; Shiu et al. 2005; Shameer et al. 2009). Liang et al. (2013) revealed that *FER* is a regulatory gene responsible for Fe uptake in tomato under Fe deficiency. A similar role is played by its functional analog, namely FER-like deficiency-induced transcription factor (FIT) in *A. thaliana* (Yuan et al. 2005). Further, a study by Ramos et al. (2007) carried out the transcriptomic analysis of *A. thaliana* and *B. juncea*, and the results showed that basic region leucine zipper (bZIP) and Zn finger TFs are induced during Cd stress exposure. Likewise, another study reported the same role of TFs in HMs stress tolerance. It was suggested that two TFs, namely ERF1 and ERF2 that belongs to superfamily of AP2/ERF are induced upon exposure to Cd stress in *Arabidopsis* (Xie et al. 2019).

6.3.4 Proteomics

Plants cope with stress conditions by altering gene expression that alters transcriptome, metabolome and proteome composition in plants. The study of proteomics is necessary as some proteins are directly responsible for stress response in plants. Various studies have proved that alterations in gene expression at the transcript level do not often lead to changes at protein level (Bogeat-Triboulot et al. 2007).

Similarly, a study by Dalcorso et al. (2013) reported that several factors such as post-transcriptional, post-translational modifications, protein folding, protein stability, protein localization, protein-protein interactions are thought to be essential determinant of protein function that determines the fate of gene translation into functional protein. Hence, a deep study of proteomics is necessary to identify target proteins responsible for detoxification/sequestration of HMs from cell (Ahsan et al. 2009). Further, a study by Kosová et al. (2011) reported that it also plays an important role in understanding physiological processes in HMs stress like stress sensing, signaling pathway that results in altered expression pattern of numerous genes responsible for stress tolerance in plants. Several studies suggest the role of defense protein in HMs stress tolerance. The level of defense proteins was enhanced to scavenge ROS that were produced due to stress and molecular chaperones are involved in re-establishment of functional protein conformation that helps to maintain the redox homeostasis in stressed plants (Zhao et al. 2011; Sharmin et al. 2012; Wang et al. 2012). Furthermore, Cobbett and Goldsbrough (2002) demonstrated that PCs belong to the metal-binding proteins family plays an important role in the sequestration of HMs during stress conditions. A study by Roth et al. (2006) demonstrated 2DE, MALDITOF-MS, LC-ESIIQTOF-MS techniques in Arabidopsis. It was revealed that out of 1,100 spots, 41 spots exhibited significant changes comprising PCs, GST, etc. Changes in these proteins help plants withstand Cd stress by moderating S assimilation. Another study by Liu et al. (2013) reported the role of protein in Hg tolerance by using 2-DE, MALDI-TOFTOF-MS techniques in *S. salsa*. The results indicate that 43 proteins result in significant changes. These proteins were related to metabolic processes of plants and changes in these proteins are linked with Hg toxicity.

6.4 Genes Involved in Signaling

Stress sensing and its transmission via signal transduction are the main steps during stress conditions that bring about alterations in physiological, molecular and cellular procedures by activating molecular response due to change in expression level of certain genes (Knight and Knight 2001). According to Sarwat (2017), the plant survival under metals stress environment, signals of metal stress are detected and transmitted via signal transduction that gradually results in activation of gene expression by initiating transcription factors. Plants' survival under stressful environment depends on their ability to signal recognition, translate these signals and take appropriate responses. Similarly, Maksymiec (2007) stated that under stress conditions several signals and proteins are expressed. This expression pattern consists of a complex signal transduction system initiated by HMs perception resulting in turning of metal responsive genes through transcription which helps to eliminate toxic impact on plants. Likewise, Cristina et al. (2010) demonstrated that when plants come across higher amounts of HMs, their gene expression pattern is altered. In various stress conditions either biotic or abiotic, MAPK cascades are considered as universal component of signal transduction. It was also reported by studies that in higher plants HMs activate MAPKs (Jonak et al. 2004; Yeh et al. 2004).

In several plant species, various candidate gene and quantitative trait loci (QTLs) have been recognized for tolerance of Zn, Fe and Cd stress that can be useful for improvement of crop via marker assisted selection or QTL pyramiding (Courbot et al. 2007; Meyer et al. 2016; Zhang et al. 2017). It has been reported in some previous studies that genetic engineering is helpful in the improvement of HMs stress (Eapen and D'Souza 2005; Farinati et al. 2010; Verma et al. 2016, 2018).

Further, Rao et al. (2011) stated that MAPK cascade together with ROS and nitric oxide production is involved in signal transduction under arsenite-mediated stress in plants. It was reported that may be OsMPK3 and OSMPK4 are the main MAPKs for signal transduction together with OsMKK4 that act as an upstream kinase of cascade. Various studies demonstrate that with the help of several omics techniques various genes are recognized to be induced under metal stress such as transcriptomic analysis in *Arabidopsis*, *Brassica* and *Lycopersicum* displayed the expression of various transcriptional factors like BHLH, AP2/ERF, BZIP and DREB during metal stress (LeDuc et al. 2006; Shameer et al. 2009; Singh et al. 2016).

Furthermore, signal transduction of HMs stress is mediated by the initiation of sensing stress signal through ion or receptors channels and additionally by messenger that are non-protein in nature like Ca, cyclic nucleotides and H^+. Various phosphatases and kinases are responsible for the transmission of stress signals that also carry out gene expression of several transcriptional factors and peptides production that detoxifies metals (Rao et al. 2011; Islam et al. 2015; Kumar and Trivedi 2016). Moreover, upon their entrance in plants, HMs stimulate different signaling pathways like MAPK signaling, Ca-dependent signaling, ROS signaling and hormone signaling that enhance the TFs and stress-responsive genes expression (Dubey et al. 2014; Kumar and Trivedi 2016).

Additionally, Huang et al. (2014) suggested that CDPKs are involved in Cr-related stress as transcription profiling of rice roots that were subjected to short and long-term Cr stress displays enhanced activity of CDPKs as per increased concentration of Cr (VI). In Arabidopsis, the changed expression pattern of At3g57530 was the main factor for the activity of Ca and CaM-dependent protein kinase (Wang et al. 2005; Srivastava et al. 2012). Also, the expression pattern of few downstream stress-responsive Transcription factors, for example, MYB, NAC and bZIP were additionally discovered to be mediated by PGPR treatment in a few plants comprising rice, chickpea and *Arabidopsis* (Srivastava et al. 2012; Tiwari et al. 2016, 2017b).

Further, various TFs, such as DREB, MYB, bZIP, ABRE, MYC, NAC and WRKY are phosphorylated by MAPKs signaling cascade, impacting the response of metal stress (Lin and Aarts 2012; Tiwari et al. 2017a). In *Medicago sativa*, distinctive MAPKs are supposed to be activated by elevated levels of Cu and Cd (Jonak et al. 2004). For instance, in another study by Liu et al. (2010) initiation of MPK3 and MPK6 through gathering of ROS was exhibited in Arabidopsis. Transcription factors phosphorylation ends up these cascades interrelating with promoters of genes and hence instigating gene expression.

Earlier, a critical function of a basic helix-loop-helix (bHLH) TF (named FER) in responses of Fe inadequacy through utilizing a tomato mutant (*Lycopersicon esculentum*). bHLH is encoded by FER and it controls the gene expression with vital functions in Fe procurement, for example, the iron-regulated transporter (IRT) and the ferric oxidase reductase (FRO) orthologs in tomato (Ling et al. 2002). The induction of these genes that are subtle to HMs stress signaling, in the presence of plant-associated microbes, indicates the complex cross-talk between plant, microbes and HMs in stress response and tolerance. Therefore, an understanding of the intricate metal stress signaling pathways and the existing cross-talk among the networks of plant-microbe-metal interaction is extremely important to elucidate the stress-responsive networks in plants.

6.5 Manipulating Metal/Metalloid Transporter Genes and Uptake System

For metal tolerance and their accumulation, metal transporters are necessary for uptake, transport and compartmentalization of metals. Therefore, metal transporter genes in plants are gaining more attention in past few decades. Natural hyperaccumulators have limited biomass for efficient phytoremediation and are constrained to specific geo-climatic environments that's why genetic engineering approaches are used to increase tolerance and accumulation of HMs in plants.

6.5.1 Ions Movement from Soil to Roots Outer Coating

According to Salt et al. (1999), root secretions inhibit the entry of HMs in root hairs via chelation. Similarly, Humphrey et al. (2007) have also observed that exogenous stress signals are detected by receptors of cell wall such as wall-associated kinases (WAKs). They transduce these signals to inside the cell through cytoplasmic kinase domain. A study by Allan and Jarrell (1989) showed that ions can be exchanged on the surface of the root through these attachment sites that help efflux and apoplastic movement.

6.5.2 Ions Movement across Cell Membrane

It was demonstrated by various studies that HMs can be absorbed with the help of various kinds of transporter protein in membrane of the cell. These proteins carry out the transport and stabilization of HMs under abiotic stress conditions.

These transporter proteins consist of NRAMP, ATPase transporter protein, cation diffusion facilitator (CDF) and ZIP families (Williams et al. 2000; Guerinot 2000).

According to Martinoia et al. (2002), plant genes that encode for transporters of HMs generally belong to large gene families, these candidate genes are prospective for renovation toward better potential of phytoremediation. Manipulation generally comprises the accumulation of HMs for the purpose of phytostabilization in roots or phytoextraction in shoots. Another purpose for transformation could be the inhibition of HMs uptake and accumulation. Numerous families of genes have been reported that codes for those proteins that function in transport system of cells, among these families, one large group of genes that encodes for proteins involved in major processes of detoxification and ion regulation is known as ATP-binding cassette (ABC).

6.5.2.1 ZIP Transporters

Liu et al. (2019) investigated Zn/Fe regulated transporter proteins in plants. According to them, these transporter proteins are members of integral membrane transporters that function in essential and non-essential metal uptake and allocation. These metal transporters are responsible for metal homeostasis critical for plant survival and growth. Similarly, Palusińska et al. (2020) reported that *NtZIP1*, *NtZIP2*, *NtZIP4* and *NtIRT1-like* are involved in the regulation of root-to-shoot translocation of Cd and Zn in *N. tobaccum*. Besides, Sasaki et al. (2014) proposed that Zn concentration was increased in root as result of 0sHMA3 overexpression that may be possibly because of increased ZIP transporter genes expression like *ZIP5*, *ZIP5*, *ZIP8*, *ZIP9* and *ZIP10*. Further, Milner et al. (2013) found out the role of *AtZIP1* and *AtZIP2* in Mn and Zn translocation from root to shoot in *A. thaliana*. These genes were seen to be expressed in root stele. It was suggested that *AtZIP1* functions in vacuolar transport, whereas *AtZIP2* is limited to plasma membrane. Mn could be remobilized from vacuole into cytoplasm in stele of root employing *AtZIP1*. Conversely, *AtZIP2* may facilitate the uptake of Mn and Zn into cells of root stele hence, it may function in translocation of Mn/Zn from stele to xylem parenchyma.

Additionally, Lin et al. (2016) reported the latest approach comprising metal transporters through species of hyper-accumulators such as *Noccaea cearulescens* that is mainly used to clean-up soil. Such as a transgene, namely NcZNT1 in *A. thaliana* encodes for Zn transporter, its overexpression may leads to hyper-accumulation of Cd and Zn in transformed plants. Moreover, Song et al. (2003) stated that metal tolerance and accumulation can be enhanced by manipulating metal transporters. Overexpression of yeast Cd factor 1 (*YCF1*) in *A. thaliana* results in improved Cd (II) and Pb (II) tolerance as well as greater amounts of metals accumulation in plants. Likewise, it was reported that YCF1 and GSH are responsible for transport of Cd into vacuole. Additionally, Arabidopsis T-DNA mutants such as cyclic nucleotide-gated ion channel 1 (*CNGC1*) that encodes for homologous protein into *NtCBP4*, moreover involved in lead tolerance. So, Zeng et al. (2015) suggested a pivotal role of *NtCBP4* and *AtCNGC1*in pathway of lead transport.

A recent study by Wu et al. (2019) reported the role of Hydrogen-rich water (HRW) in alleviating Cd stress in seedling of Chinese cabbage by using RNA-sequencing analyses. The yeast transformation system selected two ZIP encoding genes such as BcIRT1 and BcZIP2, and it was indicated that genes enhance sensitivity of various yeast strains and are capable of Zn, Cd, Mn and Fe transport in Pak choi. The results suggest that HRW decreases the uptake of Cd via restricting the expression of *BcIRT1* and *BcZIP2* transporters in response to Cd stress.

6.5.3 Movement of HMs from Root to Shoot

HMs are transported from root to shoot with the help of different transporters as describe below.

6.5.3.1 MATE Transporters

It was suggested by various studies that MATE transporters are a new class of protein family that functions in HMs movement. Researchers have shown that Ferric reductase defective 3 (*FDR3*) was found abundantly in root of plants that are hyperaccumulators of HMs such as *T. caerulescens* and *Arabidopsis halleri* (Krämer et al. 2007; Vert et al. 2009; Lin et al. 2009).

6.5.3.2 HMA Family of Transporters

Takahashi et al. (2012a, 2012b) have reported ATPase transporters, namely HMA family that were grouped on the basis of their selection of metal-substrate for various HMs such as Cu/Cd/Zn/Pb/Ag. *HMA4* carries out Zn loading in xylem, and along with *AtHMA2*, it may result in Cd translocation. Similarly, Andrés-Colás et al. (2006) stated that *AtHMA5* is involved in the removal and root-to-shoot transportation of Cu. Miyadate et al. (2011) indicated the role of toxic metal ATPase3 (namely *OsHMA3*) in their study with the help of transgenic approaches and positional cloning. Hence, *OsHMA3* was considered a gene that regulates the rate of Cd translocation from root to shoot.

Some studies reported that various metals such as boron (B), silicon (Si) and antimonite were transported by means of aquaporin channel proteins (Mosa et al. 2016a; Kumar et al. 2014). Likewise, Dubey et al. (2014) investigated various unique genes that are differentially expressed. These genes are member of various families of transporters like major facilitator superfamily antiporters. It was shown by genome-wide expression analysis that these genes are expressed in response to HMs Cd, Pb, Cr (VI) and As (V) in rice roots. Further, Wang et al. (2016) recognized Cu tolerance genes through de novo transcriptome sequencing approach in *P. ostia* plant.

6.5.4 Metal Ions Distribution to Other Organelles

When metal ions are accumulated in larger amounts, they are removed from cells by further transportation to other organelles with the help of various transporter proteins. Some transporters are discussed below.

6.5.4.1 Metal Tolerance Proteins (MTP's)

A recent study by Das et al. (2016) indicated several protein types termed as metal tolerance or transport protein (MTP), these proteins are encoded by genes that belong to family of

CDF. Extra amount of ions are excluded from the cell cytoplasm through MTP's.

6.6 Genes Involved in Metal Translocation and Sequestration at Tissue and Organelle Level

Various studies demonstrate that in order to combat HMs stress, hyperaccumulator plants have to sequester and then detoxify these toxic metals after the process of translocation. These plants sequester or detoxify HMs into vacuoles (Vögeli-Lange and Wagner 1990; Kanoun-Boulé et al. 2009; Singh et al. 2011). Numerous families of transporters are responsible for this process of sequestration and detoxification, such as *CDF, NRAMP, ABC* and *HMA*.

6.6.1 ABC Transporters

Mostly two subfamilies of *ABC* transporters are active (*MPRP* and *PDR*), are responsible for HMs transportation into the vacuole. Complexes of PC-Cd (phytochelatin-cadmium) made in cytoplasm are transported through first *ABC* transporter of vacuole, namely *HMT1* that is present on tonoplast of *S. pombe* (Ortiz et al. 1992; Kuriakose and Prasad 2008). Similarly, Fu et al. (2019) also stated in their study that OsABCG36 is responsible for Cd transportation out of the cell (i.e. efflux/detoxification/sequestration). Additionally, it was also described that ABC genes encode protein transporters and enhance toxic metals sequestration into lumen of vacuole as they are located on tonoplast of transgenic plant cells (Yazaki et al. 2006; Song et al. 2014). Further, Lu et al. (1998) revealed two transporters, namely AtMRP1 and AtMRP2 for sequestering the complexes of PC-Cd into the vacuole in *A. thaliana*.

6.6.2 Cation Efflux Transporters

According to Mäser et al. (2001), CDF is also known as cation efflux transporter due to their function in translocation and sequestration of metal ions such as Cd, Zn, Ni and Co from the cytoplasm to vacuole. Several studies also suggest *CDF* as metal tolerance proteins (*MTPs*) that enhance the movement of metallic ions from cytoplasm to other organelles or extracellular space (Haney et al. 2005; Peiter et al. 2007). Similarly, it was also supported by various studies that CDF family of transporters is responsible for metallic ions transportation from cytoplasm to vacuole (Krämer et al. 2007; Montanini et al. 2007). Likewise, Gustin et al. (2011) identified about 10 MTP genes in rice. Several studies demonstrated 12 genes in *A. thaliana* and the initial reported CDF gene was formerly known as Zn transporter 1 (*ZAT1*) gene and then changed to METAL Tolerance protein 1 (*AtMTP1*) (Van der Zaal et al. 1999; Delhaize et al., 2003). In another study, Delhaize et al. (2003) reported that these genes are responsible for Zn tolerance in *Arabidopsis*. A recent study by Das et al. (2016) reported that due to overexpressed CDF genes in plants that produce higher quantities of thiol compounds by which they can HMs sequestration by chelating them into vacuole that results in hyperaccumulation. It was also stated that the overexpressed OsMTP1 gene of rice initiated hyperaccumulation of Cd and improved tobacco tolerance during exogenous Cd stress. Further, a study by Dräger et al. (2004) on *A. halleri* demonstrated that when Zn was exogenously applied to this plant its leaves exhibited a constitutive higher expression of AhMTP1 protein. Delhaize et al. (2003) stated MTP11 and MTP8 are said to be ShMTP8 very close homologs and responsible for tolerance of Mn in Arabidopsis.

6.6.3 HMA Transporters

In addition, Morel et al. (2009) reported other transporters like HMA and NRAMP also carry out HMs sequestration from cytoplasm to vacuole. Though, in *A. thaliana* overexpressed HMAs are thought to be responsible for HMs detoxification. Ueno et al. (2011) isolated an allelic gene, namely TcHMA3. It was indicated that overexpressed TcHMA3 improves Cd tolerance and to some extent Zn tolerance in *A. thaliana*. The results suggested that TcHMA3 transporter has specificity for Cd and is located on tonoplast of cell. It also carries out toxic metals sequestration into vacuole of leaves.

6.6.4 Natural Resistance-Associated Macrophage Protein (NRAMP)

Furthermore, Belouchi et al. (1997) suggested the role of Natural resistance-associated macrophage protein (NRAMP) in HMs transportation into vacuole. In another study by Thomine et al. (2003), the role of a metal transporter, namely AtNRAMP3 belonging from NRAMP family in *A. thaliana* was discussed. It was suggested that during Fe deficiency, downregulation of AtNRAMP3 results in Zn and Mn accumulation but on the other hand its upregulation or overexpression can downregulate the accumulation of Mn in roots that was due to movement of toxic metals from vacuole to cytosol, AtNRAMP3 has greater impact on metal accumulation and IRT1 and FRO2 genes. Moreover, Anjum et al. (2014) demonstrated that the process of toxic metal detoxification in hyperaccumulator plants depends upon overexpressed patterns of genes associated with anti-oxidation activity like GSH, 0-acetylserine and cysteine. Furthermore, various studies suggested tolerance of HMs (such as Zn, Co, Ni and to some extent Cd) is due to overexpressed NgSAT in *N. goesingense* (Freeman et al. 2004; Freeman and Salt 2007).

6.7 Genes Involved in Metalloid Ligand Formation

According to Emamverdian et al. (2015), HMs accumulation in excessive amounts becomes toxic for plants as they have the ability to react with few crucial biomolecules of cells (like deoxyribonucleic acid and proteins) resulting in an increased level of ROS. Plants adapt several mechanisms in order to protect themselves from HMs stress. The fundamental components involved are chelating metals that form PCs and metal complexes known as MTs at both intra and inter-cellular levels that may result in toxic metals removal from vacuolar sequestration of ligand-metal complex.

6.7.1 Phytochelatins

Lee et al. (2003) reported one important procedure adapted by plants for HMs detoxification. It was indicated that sulfur-rich GSH is involved in short chain formation that is made up of repeated peptides in the presence of phytochelatin synthase (PCS) enzyme, these peptide chains are rich in thiols and have low molecular weight having affinity to bind with HMs in case of HMs stress. Several studies suggest that phytochelatins are produced in cytosol and are vigorously translocated into vacuole upon their conversion into metal-phytochelatins complexes of high molecular weight (Manara 2012; Song et al. 2014). Similarly, a study by Sytar et al. (2013) demonstrated that Mg ATP-dependent carrier or ATP-binding cassette (ABC) is involved in regulation of this transport. Likewise, in another study by Fidalgo et al. (2013) on *S. nigrum* L. exhibited PCs formation was increased in root part upon exposure to 200 $\mu mol{\cdot}L^{-1}$.

Further, various studies have reported that PCS enzyme action is dependent on the availability of HMs while PCs gene is constitutively expressed (Vatamaniuk et al. 2000, 2004). Additionally, several studies suggest that various PCS genes have been cloned such as TaPCS1 (wheat), AtPCS1 (*A. thaliana*), BjPCS1 (*B. juncea* L.) and OsPCS1 (rice) (Clemens et al. 1999; Vatamaniuk et al. 1999; Heiss et al. 2003). Moreover, Heiss et al. (2003) exhibited the relationship of BjPCS1 with PCS protein through CLUSTAL analysis in *A. thaliana* and *T. caerulescens*. It was revealed constitutive expression of PCs by immunoblot analysis and antiserum directed in contrast to recombinant BjPCS1. Furthermore, Clemens et al. (1999) screened plants genes that are responsible for tolerance of HMs. They recognized the role of wheat cDNA, namely *TaPCS1*, expression of this gene in *S. cerevisiae* gradually enhances Cd tolerance.

In the same way, Guo et al. (2008) worked on *A. thaliana* and reported that when AsPCS1 and GSH1 were simultaneously overexpressed, they increased the level of total phytochelatins production. The results demonstrated that the assembling of modified genes has ability to increase Cd tolerance. Moreover, Domínguez-Solís et al. (2001) stated the role of a gene in the cytosol of Arabidopsis, namely O-acetylserine (thiol) lyase (Atcys3A). This gene encodes an enzyme responsible for cysteine production as higher level of cysteine synthesis is necessary for GSH and PCs production under HMs stress.

6.7.2 Metallothioneins

According to Verkleij et al. (2003), mRNA translation leads to the formation of metallothioneins. However, Yang and Chu (2011) stated that MTs revealed higher affinity for broader range of metals like As, Zn, Cu and Cd where on the other hand PCs only detoxifies Cd. A study by Grennan (2011) on Arabidopsis revealed that isoforms of MTs from type 1, 2 and 3 are responsible for chelation of Cu but isoform from type 4 is considered as Zn binder. Similarly, another study by Garcia-Hernández et al. (1998) studied some Arabidopsis mutants and reported that MT1 may carryout Cu detoxification mainly in veins instead of mesophyll of leaves. Yang et al. (2009) described that OsMT1 (*O. sativa* metallothionein type 1) was necessary to balance Zn in root part, MT3 was primarily responsible for Cu and Zn homeostasis however MT4 carry out storage of Zn (Hegelund et al. 2012). Likewise, Pagani et al. (2012) worked on soybean and indicated that MT1, MT2 and MT3 were more liable to detoxify the toxic volume of Cd whereas MT4 has characteristics of Zn binding. Further, Zhou et al. (2014) investigated that TaMT3 (a metallothionein type 3 gene) from *T. androssowii* when engineered into tobacco leads to enhanced Cd tolerance via increasing the function of SOD that maximize ROS cleaning-up capability and may result in reduction of POD activity.

6.8 Genes Encoding Antioxidants that Detoxify Metals and Metalloids by Chemical Modification

Various studies revealed that several abiotic stresses may lead to oxidative damage as ROS and methylglyoxal (MG) are excessively accumulated. According to Hossain et al. (2012), in order to scavenge ROS and MG, higher plants adapt a refined mechanism known as the antioxidant defense system and glyoxalase system. Antioxidant defense system can reduce the impact of ROS on biomolecules (like DNA, lipids and proteins). This system is comprised of enzymatic actions that are chiefly dependent on superoxide dismutase (SOD, ascorbate peroxidase (APX), catalase (CAT) and glutathione S-transferase (GST) (Dixit et al. 2011; Hellou et al. 2012). Similarly, it was indicated by some other studies that plants when exposed to Cd, Cu and As, the overexpressed genes of SOD, CAT and APX may result in reduced level of ROS in transgenic plants and during stress condition enzymatic action of antioxidants was greatly enhanced (Lee et al. 2007; Guan et al. 2009; Gao et al. 2016).

Iannone et al. (2015) discussed the role of catalase in tolerance of Cd stress. It was revealed that upon exposure to Cd, reduced action of constitutive CAT (line CAT1AS) has the ability to activate antioxidant defense system in transgenic plants, established chiefly on enhanced action of constitutive APX and guaiacol peroxidase (GPOX). Likewise, Hossain et al. (2012) demonstrated in their studies that upon HMs entry into cells they are sequestered through various mechanisms. GSH act as vital molecule of the antioxidant defense system as well as glyoxalase system and it is responsible for controlling ROS and MG both directly and indirectly. Recently, it was revealed by studies that GSH itself as well as with its enzymes (such as GST, GPx, DHAR, etc.) are involved in protection from damage induced by ROS and MG. In addition, a study by Dixit et al. (2011) investigated that overexpressed GST gene (namely *TvGST*) is involved in tolerance of Cd stress in tobacco plants as it results in reduced Cd accumulation and lipid peroxidation in transgenic plant. Several studies described that upon exposure to Cd stress, considerably enhanced level of SOD, GPX, CAT, GST and APX was seen in transgenic plants that prove the role of GSH metabolizing enzyme in tolerance of stress (Hossain et al. 2011, 2012; Anjum et al. 2012).

In a study Ortega-Villasante et al. (2007) indicated the metabolic GSH gene expression in alfalfa seedling that was given Hg and Cd treatment, considering the GSH, AsA and homoglutathione level. By the use of RT-PCR technique, the pattern of gene expression was noticed for GR1, GR2, GS, GPX, Hgs, ECS and PCS. After Hg metal treatment of about 3hrs, GR1 and GR2 were clearly induced and showed maximum expression and then reduced afterward. Further, Mohanpuria et al. (2007) reported the transcription level of genes known as GS, GR and γ-ECS involved in biosynthesis and regeneration of GSH are enhanced upon exposure to Cd stress. Upregulation of GSH genes transcript level excluding GST in Cd stress indicates the involvement of these enzymes in tolerance of Cd stress in *C. sinensis*.

Moreover, in a study He et al. (2015) indicated that over-expressed γ-ECS gene of cytosol is responsible for Cd accumulation, transport and sequestration/detoxification through reducing ROS levels, increasing total thiols and GSH quantities in leaves and roots in transgenic plants of poplar trees. For this reason, Wen et al. (2010) stated the production of overexpressed apple spermidine synthase gene, namely *MdSPDS1* in transgenic line of *P. communis*. Upon Zn and Cd exposure, increased production and accumulation of polyamines were revealed in transgenic lines. Spermidine showed antioxidant action and is involved in binding of free metal ions. Further moreover, Faè et al. (2014) demonstrated that enhanced Cu tolerance was attained through overexpressed MtTdp2a (a DNA repair gene) that encodes tyrosol-DNA phosphodiesterase 2 in *M. truncatula* transgenic plant. Similarly, Charfeddine et al. (2017) reported that overexpressed genes that encode TFs from the drought-responsive element binding (*DREB*) family are responsible for Cu and Cd tolerance in potato.

6.9 Genes Involved in Metabolism and Physiological Response

According to Singla-Pareek et al. (2006), that plants acts as a source of transgene and provide some genes that are responsible for primary metabolism as indicator of toxic metal tolerance. In case of stress and normal metabolism, methylglyoxal is produced as a by-product that is detoxified with the help of glyoxalase cycle. Two genes, namely Indian mustard glyoxalase I (GI) and Rice glyoxalase II (GII) when transformed into tobacco plant revealed enhanced Zn, Pb and Cd tolerance and double-transgenic plants were appeared to be more tolerant than that of single gene transgenic plants. In case of Zn stress, GI-GII transgenic plants show a reduced level of lipid peroxidation and methylglyoxal accumulation along with considerably enhanced content of phytochelatin and GSH. The results revealed the positive role of glyoxalase pathway in Zn tolerance by controlling methylglyoxal level in cell as well as enhanced GSH balance.

Similarly, a study by Reed and Glick (2005) demonstrated the role of bacterial gene, namely 1-aminocyclopropane-1-carboxylic acid (ACC deaminase in HMs tolerance. It was indicated that bacteria with expressed ACC deaminase minimize the level of ethylene and harmful impact of stress by cleavage of ethylene precursor ACC. Rhizobacteria in contaminated soil with the help of expressed ACC deaminase can improve phytoremediation ability of plants. A study by Nie et al. (2002) reported the role of similar gene in transgenic *B. napus*. It was stated that plants with expressed gene accumulates more As from contaminated soil and have shown more tolerance to As (V). Likewise, another study by Zhang et al. (2008) demonstrated that concurrent and collective expression of various bacterial genes are responsible for hormone metabolism and results in most effective phytoremediation of HMs. Such as a xylem-specific promoter that drives iaaM gene of *A. tumefaciens*, that encodes a tryptophan monooxygenase catalyzing the production of indole 3 acetamide.

Conclusion and Future Prospects

Under the metal stress, numerous strategies have been developed to avoid the HMs toxicity in plants, but nowadays, science offers us new opportunities of utilizing genetic engineering to increase the plant potential for HMs uptake, mobilization and sequestration. Promising directions in genetic manipulations, such as cis- and intragenesis and gene silencing, are efficient approaches for increasing plants tolerance against HMs. In conclusion, based on the early studies, overexpression of HMs candidate genes increase plant growth and development, and remove and/or sequester HMs from soil and water. However, many such plants have limited utility for phytoremediation because of their slow growth, difficult propagation, seasonal growth and low biomass. Solutions to these problems are important and require further research. Although to date, there have been no reports exploiting cisgenesis and/or intragenesis in modulating plant responses to HMs stress, these novel practices provide an excellent opportunity to overcome concerns with transgenesis and accelerate studies on genetic manipulations toward increased efficiency of phytoremediation.

REFERENCES

Adamis, P. D., Gomes, D. S., Pinto, M. L. C., Panek, A. D. and Eleutherio, E. C. 2004. The role of glutathione transferases in cadmium stress. *Toxicol lett*, 154, 81–88.

Ahsan, N., Renaut, J. and Komatsu, S. 2009. Recent developments in the application of proteomics to the analysis of plant responses to heavy metals. *Proteomics*, 9, 2602–2621.

Ai, T. N., Naing, A. H., Arun, M., Jeon, S. M. and Kim, C. K. 2017. Expression of RsMYB1 in Petunia enhances anthocyanin production in vegetative and floral tissues. *Sci Hort*, 214, 58–65.

Ai, T. N., Naing, A. H., Yun, B. W. and Kim, C. K. 2018. Overexpression of RsMYB1 enhances heavy metal stress tolerance in transgenic petunia by elevating the transcript levels of stress tolerant and antioxidant genes. *BioRxiv*, 286849.

Alharbi, O. M., Khattab, R. A. and Ali, I. 2018. Health and environmental effects of persistent organic pollutants. *J mol liq*, 263, 442–453.

Ali, I., Aboul-Enein, H. Y. and Gupta, V. K. 2009. *Nanochromatography and Nanocapillary Electrophoresis: Pharmaceutical and Environmental Analyses*. Wiley, Hoboken.

Ali, I., Alharbi, O. M., Alothman, Z. A., Badjah, A. Y. and Alwarthan, A. 2018. Artificial neural network modelling of amido black dye sorption on iron composite nano material: kinetics and thermodynamics studies. *J mol liq*, 250, 1–8.

Ali, I., AL-Othman, Z. A. and Alwarthan, A. 2016. Molecular uptake of congo red dye from water on iron composite nano particles. *J mol liq*, 224, 171–176.

Ali, I., Alothman, Z. A. and Alwarthan, A. 2017a. Supra molecular mechanism of the removal of 17-β-estradiol endocrine disturbing pollutant from water on functionalized iron nano particles. *J mol liq*, 241, 123–129.

Ali, I., Suhail, M. and Basheer, A. A. 2017b. Advanced spiral periodic classification of the elements. *Chem int*, 3, 220–224.

Allan, A. C. and Fluhr, R. 2001. Ozone and reactive oxygen species (pp. 1–9). *eLS*, Ltd, Chichester.

Allan, D. L. and Jarrell, W. M. 1989. Proton and copper adsorption to maize and soybean root cell walls. *Plant physiol*, 89, 823–832.

Andrés-Colás, N., Sancenón, V., Rodríguez-Navarro, S., Mayo, S., Thiele, D. J., Ecker, J. R., Puig, S. and Peñarrubia, L. 2006. The Arabidopsis heavy metal P-type ATPase HMA5 interacts with metallochaperones and functions in copper detoxification of roots. *Plant j, 45*, 225–236.

Anjum, N. A., Ahmad, I., Mohmood, I., Pacheco, M., Duarte, A. C., Pereira, E., Umar, S., Ahmad, A., Khan, N.A., Iqbal, M. and Prasad, M. N. V. 2012. Modulation of glutathione and its related enzymes in plants' responses to toxic metals and metalloids—a review. *Environ exp bot*, 75, 307–324.

Anjum, N. A., Gill, S. S., Gill, R., Hasanuzzaman, M., Duarte, A. C., Pereira, E., Ahmad, I., Tuteja, R. and Tuteja, N. 2014. Metal/metalloid stress tolerance in plants: role of ascorbate, its redox couple, and associated enzymes. *Protoplasma*, 251, 1265–1283.

Arbona, V., Manzi, M., Ollas, C. D. and Gómez-Cadenas, A. 2013. Metabolomics as a tool to investigate abiotic stress tolerance in plants. *Int j mol sci*, 14, 4885–4911.

Aydinalp, C. and Marinova, S. 2009. The effects of heavy metals on seed germination and plant growth on alfalfa plant (*Medicago sativa*). *Bulg j agric sci*, 15, 347–350.

Basheer, A. A. 2018a. Chemical chiral pollution: impact on the society and science and need of the regulations in the 21st century. *Chirality*, 30, 402–406.

Basheer, A. A. 2018b. New generation nano-adsorbents for the removal of emerging contaminants in water. *J mol liq*, 261, 583–593.

Belouchi, A., Kwan, T. and Gros, P. 1997. Cloning and characterization of the OsNramp family from Oryza sativa, a new family of membrane proteins possibly implicated in the transport of metal ions. *Plant mol biol*, 33, 1085–1092.

Bogeat-Triboulot, M. B., Brosché, M., Renaut, J., Jouve, L., Le Thiec, D., Fayyaz, P., Vinocur, B., Witters, E., Laukens, K., Teichmann, T. and Altman, A. 2007. Gradual soil water depletion results in reversible changes of gene expression, protein profiles, ecophysiology, and growth performance in *Populus euphratica*, a poplar growing in arid regions. *Plant physiol*, 143, 876–892.

Bouzroud, S., Gasparini, K., Hu, G., Barbosa, M. A. M., Rosa, B. L., Fahr, M., Bendaou, N., Bouzayen, M., Zsögön, A., Smouni, A. and Zouine, M. 2020. Down regulation and loss of auxin response factor 4 function using CRISPR/Cas9 alters plant growth, stomatal function and improves tomato tolerance to salinity and osmotic stress. *Genes*, 11, 272.

Burakova, E. A., Dyachkova, T. P., Rukhov, A. V., Tugolukov, E. N., Galunin, E. V., Tkachev, A. G., Ali, I. 2018. Novel and economic method of carbon nanotubes synthesis on a nickel magnesium oxide catalyst using microwave radiation. *J mol liq*, 253, 340–346.

Castrillo, G., Sánchez-Bermejo, E., de Lorenzo, L., Crevillén, P., Fraile-Escanciano, A., Mohan, T. C., Mouriz, A., Catarecha, P., Sobrino-Plata, J., Olsson, S. and del Puerto, Y. L. 2013. WRKY6 transcription factor restricts arsenate uptake and transposon activation in Arabidopsis. *Plant cell*, 25, 2944–2957.

Charfeddine, M., Charfeddine, S., Bouaziz, D., Messaoud, R. B. and Bouzid, R. G. 2017. The effect of cadmium on transgenic potato (*Solanum tuberosum*) plants overexpressing the StDREB transcription factors. *Plant cell, tissue and organ cult*, 128, 521–541.

Chaudhary, K., Jan, S. and Khan, S. 2016. Heavy metal ATPase (HMA2, HMA3, and HMA4) genes in hyperaccumulation mechanism of heavy metals. In: *Plant Metal Interaction* (pp. 545–556). Amsterdam: Elsevier.

Cheng, Y. J., Kim, M. D., Deng, X. P., Kwak, S. S. and Chen, W. 2013. Enhanced salt stress tolerance in transgenic potato plants expressing IbMYB1, a sweet potato transcription factor. *J. Microbiol biotechnol*, 23, 1737–1746.

Clemens, S., Kim, E. J., Neumann, D. and Schroeder, J. I. 1999. Tolerance to toxic metals by a gene family of phytochelatin synthases from plants and yeast. *EMBO j*, 18, 3325–3333.

Cobbett, C. and Goldsbrough, P. 2002. Phytochelatins and metallothioneins: roles in heavy metal detoxification and homeostasis. *Annu rev plant biol*, 53, 159–182.

Courbot, M., Willems, G., Motte, P., Arvidsson, S., Roosens, N., Saumitou-Laprade, P. and Verbruggen, N. 2007. A major quantitative trait locus for cadmium tolerance in Arabidopsis halleri colocalizes with HMA4, a gene encoding a heavy metal ATPase. *Plant physiol*, 144, 1052–1065.

Cristina, M. S., Petersen, M. and Mundy, J. 2010. Mitogen-activated protein kinase signaling in plants. *Annu rev plant biol*, 61, 621–649.

DalCorso, G., Fasani, E. and Furini, A. 2013. Recent advances in the analysis of metal hyperaccumulation and hypertolerance in plants using proteomics. *Front plant sci*, 4, 280.

Das, N., Bhattacharya, S. and Maiti, M. K. 2016. Enhanced cadmium accumulation and tolerance in transgenic tobacco overexpressing rice metal tolerance protein gene OsMTP1 is promising for phytoremediation. *Plant physiol biochem*, 105, 297–309.

Delhaize, E., Kataoka, T., Hebb, D. M., White, R. G. and Ryan, P. R. 2003. Genes encoding proteins of the cation diffusion facilitator family that confer manganese tolerance. *Plant cell, 15*, 1131–1142.

Dixit, P., Mukherjee, P. K., Ramachandran, V. and Eapen, S. 2011. Glutathione transferase from Trichoderma virens enhances cadmium tolerance without enhancing its accumulation in transgenic *Nicotiana tobaccum*. *Plos one*, 6, e16360.

Domínguez-Solís, J. R., Gutiérrez-Alcalá, G., Romero, L. C. and Gotor, C. 2001. The cytosolic O-acetylserine (thiol) lyase gene is regulated by heavy metals and can function in cadmium tolerance. *J biol chem*, 276, 9297–9302.

Dräger, D. B., Desbrosses-Fonrouge, A. G., Krach, C., Chardonnens, A. N., Meyer, R. C., Saumitou-Laprade, P. and Krämer, U. 2004. Two genes encoding Arabidopsis halleri MTP1 metal transport proteins co-segregate with zinc tolerance and account for high MTP1 transcript levels. *Plant j*, 39, 425–439.

Dubey, S., Misra, P., Dwivedi, S., Chatterjee, S., Bag, S. K., Mantri, S., Asif, M.H., Rai, A., Kumar, S., Shri, M. and Tripathi, P. 2010. Transcriptomic and metabolomic shifts in rice roots in response to Cr (VI) stress. *BMC genom*, 11, 1–19.

Dubey, S. K. and Rai, L. C. 1987. Effect of chromium and tin on survival, growth, carbon fixation, heterocyst differentiation, nitrogenase, nitrate reductase and glutamine synthetase activities of Anabaena doliolum. *J plant physiol*, 130, 165–172.

Dubey, S., Shri, M., Misra, P., Lakhwani, D., Bag, S. K., Asif, M. H., Trivedi, P. K., Tripathi, R. D. and Chakrabarty, D. 2014. Heavy metals induce oxidative stress and genome-wide modulation in transcriptome of rice root. *Funct integr genomics*, 14, 401–417.

Eapen, S. and D'souza, S. F. 2005. Prospects of genetic engineering of plants for phytoremediation of toxic metals. *Biotechnol adv*, 23, 97–114.

Emamverdian, A., Ding, Y., Mokhberdoran, F. and Xie, Y. 2015. Heavy metal stress and some mechanisms of plant defense response. *The scientific world j, 2015*.

Faè, M., Balestrazzi, A., Confalonieri, M., Donà, M., Macovei, A., Valassi, A., Giraffa, G. and Carbonera, D. 2014. Copper-mediated genotoxic stress is attenuated by the overexpression of the DNA repair gene MtTdp2α (tyrosyl-DNA phosphodiesterase 2) in *Medicago truncatula* plants. *Plant cell rep*, 33, 1071–1080.

Farinati, S., DalCorso, G., Varotto, S. and Furini, A. 2010. The Brassica juncea BjCdR15, an ortholog of Arabidopsis TGA3, is a regulator of cadmium uptake, transport and accumulation in shoots and confers cadmium tolerance in transgenic plants. *New phytol*, 185, 964–978.

Fidalgo, F., Azenha, M., Silva, A. F., de Sousa, A., Santiago, A., Ferraz, P. and Teixeira, J. 2013. Copper-induced stress in *Solanum nigrum* L. and antioxidant defense system responses. *Food energy sec*, 2, 70–80.

Foyer, C. H. and Noctor, G. 2005. Redox homeostasis and antioxidant signaling: a metabolic interface between stress perception and physiological responses. *Plant cell*, 17, 1866–1875.

Freeman, J. L., Persans, M. W., Nieman, K., Albrecht, C., Peer, W., Pickering, I. J. and Salt, D. E. 2004. Increased glutathione biosynthesis plays a role in nickel tolerance in Thlaspi nickel hyperaccumulators. *Plant cell*, 16, 2176–2191.

Freeman, J. L. and Salt, D. E. 2007. The metal tolerance profile of Thlaspi goesingense is mimicked in Arabidopsis thaliana heterologously expressing serine acetyl-transferase. *BMC plant biol*, 7, 63.

Fu, S., Lu, Y., Zhang, X., Yang, G., Chao, D., Wang, Z., Shi, M., Chen, J., Chao, D. Y., Li, R. and Ma, J. F. 2019. The ABC transporter ABCG36 is required for cadmium tolerance in rice. *J exp bot*, 70, 5909–5918.

Gao, X., Ai, W. L., Gong, H., Cui, L. J., Chen, B. X., Luo, H. Y., Zhang, Z. C. and Qiu, B. S. 2016. Transgenic NfFeSOD *Sedum alfredii* plants exhibited profound growth impairments and better relative tolerance to long-term abiotic stresses. *Plant biotechnol rep*, 10, 117–128.

Gao, J., Sun, L., Yang, X. and Liu, J. X. 2013. Transcriptomic analysis of cadmium stress response in the heavy metal hyperaccumulator Sedum alfredii Hance. *PloS one*, 8, e64643.

García-Hernández, M., Murphy, A. and Taiz, L. 1998. Metallothioneins 1 and 2 have distinct but overlapping expression patterns in Arabidopsis. *Plant physiol*, 118, 387–397.

Gichner, T., Patková, Z., Száková, J. and Demnerová, K. 2004. Cadmium induces DNA damage in tobacco roots, but no DNA damage, somatic mutations or homologous recombination in tobacco leaves. *Mutat res genet toxicol environ mutagen*, 559, 49–57.

Grennan, A. K. 2011. Metallothioneins, a diverse protein family. *Plant physiol*, 155, 1750–1751.

Guan, Z., Chai, T., Zhang, Y., Xu, J. and Wei, W. 2009. Enhancement of Cd tolerance in transgenic tobacco plants overexpressing a Cd-induced catalase cDNA. *Chemosphere*, 76, 623–630.

Guerinot, M. L. 2000. The ZIP family of metal transporters. *Biochim biophys acta biomembr*, 1465, 190–198.

Guo, J., Dai, X., Xu, W. and Ma, M. 2008. Overexpressing GSH1 and AsPCS1 simultaneously increases the tolerance and accumulation of cadmium and arsenic in Arabidopsis thaliana. *Chemosphere*, 72, 1020–1026.

Gupta, V. K. and Ali, I. 2002. *Encyclopedia of Surface and Colloid Science*. Marcel Dekker, New York, pp 136–166.

Gustin, J. L., Zanis, M. J. and Salt, D. E. 2011. Structure and evolution of the plant cation diffusion facilitator family of ion transporters. *BMC evol biol*, 11, 76.

Hakeem, K. R., Sabir, M., Ozturk, M. and Mermut, A. 2015. *Soil Remediation and Plants: Prospects and Challenges* (p. 724). Elsevier, London.

Haney, C. J., Grass, G., Franke, S. and Rensing, C. 2005. New developments in the understanding of the cation diffusion facilitator family. *J ind microbiol biotechnol*, 32, 215–226.

He, J., Li, H., Ma, C., Zhang, Y., Polle, A., Rennenberg, H., Cheng, X. and Luo, Z. B. 2015. Overexpression of bacterial γ-glutamylcysteine synthetase mediates changes in cadmium influx, allocation and detoxification in poplar. *New phytol*, 205, 240–254.

Hegelund, J. N., Schiller, M., Kichey, T., Hansen, T. H., Pedas, P., Husted, S. and Schjoerring, J. K. 2012. Barley metallothioneins: MT3 and MT4 are localized in the grain aleurone layer and show differential zinc binding. *Plant physiol*, 159, 1125–1137.

Heiss, S., Wachter, A., Bogs, J., Cobbett, C. and Rausch, T. 2003. Phytochelatin synthase (PCS) protein is induced in *Brassica juncea* leaves after prolonged Cd exposure. *J exp bot*, 54, 1833–1839.

Helbig, K., Bleuel, C., Krauss, G. J. and Nies, D. H. 2008. Glutathione and transition-metal homeostasis in Escherichia coli. *J bacterial*, 190, 5431–5438.

Hellou, J., Ross, N. W. and Moon, T. W. 2012. Glutathione, glutathione S-transferase, and glutathione conjugates, complementary markers of oxidative stress in aquatic biota. *Environ sci pollut res*, 19, 2007–2023.

Hossain, M. A., Hossain, M. D., Rohman, M. M., da Silva, J. A. T. and Fujita, M. 2012. Onion major compounds (flavonoids, organosulfurs) and highly expressed glutathione-related enzymes: possible physiological interaction, gene cloning and abiotic stress response. In: *Onion Consumption and Health* (pp. 49–90). Nova Science Publishers Inc., New York, NY.

Hossain, Z., Nouri, M. Z. and Komatsu, S. 2012. Plant cell organelle proteomics in response to abiotic stress. *J proteome res*, 11, 37–48.

Hossain, M. A., Piyatida, P., Silva, J. A. T. and Fujita, M. 2012. Molecular mechanism of heavy metal toxicity and tolerance in plants: central role of glutathione in detoxification of reactive oxygen species and methylglyoxal and in heavy metal chelation. *J bot*, 2012, 1–37.

Hossain, M. A., Teixeira da Silva, J. A. and Fujita, M. 2011. Glyoxalase system and reactive oxygen species detoxification system in plant abiotic stress response and tolerance: an intimate relationship. *Abiotic stress/book*, 1, 235–266.

Huang, T. L., Huang, L. Y., Fu, S. F., Trinh, N. N. and Huang, H. J. 2014. Genomic profiling of rice roots with short-and long-term chromium stress. *Plant mol biol*, 86, 157–170.

Humphrey, T. V., Bonetta, D. T. and Goring, D. R. 2007. Sentinels at the wall: cell wall receptors and sensors. *New Phytol*, 176, 7–21.

Iannone, M. F., Groppa, M. D. and Benavides, M. P. 2015. Cadmium induces different biochemical responses in wild type and catalase-deficient tobacco plants. *Environ exp bot*, 109, 201–211.

Islam, E., Khan, M. T. and Irem, S. 2015. Biochemical mechanisms of signaling: perspectives in plants under arsenic stress. *Ecotoxicol environ saf*, 114, 126–133.

Jaspers, P. and Kangasjärvi, J. 2010. Reactive oxygen species in abiotic stress signaling. *Physiol plant*, 138,405–413.

Jonak, C., Nakagami, H. and Hirt, H. 2004. Heavy metal stress. Activation of distinct mitogen-activated protein kinase pathways by copper and cadmium. *Plant physiol*, 136, 3276–3283.

Jonak, C., Okresz, L., Bogre, L. and Hirt, H. 2002. Complexity, cross talk and integration of plant MAP kinase signalling. *Curr opin plant biol*, 5, 415–424.

Kanoun-Boulé, M., Vicente, J. A., Nabais, C., Prasad, M. N. V. and Freitas, H. 2009. Ecophysiological tolerance of duckweeds exposed to copper. *Aquat toxicol*, 91, 1–9.

Kerkeb, L. and Krämer, U. 2003. The role of free histidine in xylem loading of nickel in *Alyssum lesbiacum* and *Brassica juncea*. *Plant physiol*, 131, 716–724.

Knight, H. and Knight, M. R. 2001. Abiotic stress signalling pathways: specificity and cross-talk. *Trends plant sci*, 6, 262–267.

Kosová, K., Vítámvás, P., Prášil, I. T. and Renaut, J. 2011. Plant proteome changes under abiotic stress—contribution of proteomics studies to understanding plant stress response. *J proteomics*, 74, 1301–1322.

Krämer, U., Talke, I. N. and Hanikenne, M. 2007. Transition metal transport. *FEBS lett*, 581, 2263–2272.

Kumar, A., Kage, U., Mosa, K. and Dhokane, D. 2014. Metabolomics: a novel tool to bridge phenome to genome under changing climate to ensure food security. *Med aromat plants*, 3.

Kumar, S. and Trivedi, P. K. 2016. Heavy metal stress signaling in plants. In *Plant Metal Interaction* (pp. 585–603). Elsevier, Amsterdan.

Kuriakose, S. V. and Prasad, M. N. V. 2008. Cadmium stress affects seed germination and seedling growth in Sorghum bicolor (L.) Moench by changing the activities of hydrolyzing enzymes. *Plant growth regul*, 54, 143–156.

LeDuc, D. L., AbdelSamie, M., Móntes-Bayon, M., Wu, C. P., Reisinger, S. J. and Terry, N. 2006. Overexpressing both ATP sulfurylase and selenocysteine methyltransferase enhances selenium phytoremediation traits in Indian mustard. *Environ pollut*, 144, 70–76.

Lee, K., Bae, D. W., Kim, S. H., Han, H. J., Liu, X., Park, H. C., Lim, C. O., Lee, S. Y. and Chung, W. S. 2010. Comparative proteomic analysis of the short-term responses of rice roots and leaves to cadmium. *J plant physiol*, 167, 161–168.

Lee, S. H., Ahsan, N., Lee, K. W., Kim, D. H., Lee, D. G., Kwak, S. S., Kim, T. H. and Lee, B. H. 2007. Simultaneous overexpression of both CuZn superoxide dismutase and ascorbate peroxidase in transgenic tall fescue plants confers increased tolerance to a wide range of abiotic stresses. *J plant physiol*, 164, 1626–1638.

Lee, S., Moon, J. S., Ko, T. S., Petros, D., Goldsbrough, P. B. and Korban, S. S. 2003. Overexpression of Arabidopsis phytochelatin synthase paradoxically leads to hypersensitivity to cadmium stress. *Plant physiol*, 131, 656–663.

Li, M., Hao, P. and Cao, F. 2017. Glutathione-induced alleviation of cadmium toxicity in Zea mays. *Plant phys biochem*, 119, 240–249.

Liang, W. H., Li, L., Zhang, F., Liu, Y. X., Li, M. M., Shi, H. H., Li, H., Shang, F., Lou, C., Lin, Q. T. and Li, J. J. 2013. Effects of abiotic stress, light, phytochromes and phytohormones on the expression of OsAQP, a rice aquaporin gene. *Plant growth regul*, 69, 21–27.

Lin, Y. F. and Aarts, M. G. 2012. The molecular mechanism of zinc and cadmium stress response in plants. *Cell mol life sci*, 69, 3187–3206.

Lin, Y. F., Hassan, Z., Talukdar, S., Schat, H. and Aarts, M. G. 2016. Expression of the ZNT1 zinc transporter from the metal hyperaccumulator *Noccaea caerulescens* confers enhanced zinc and cadmium tolerance and accumulation to Arabidopsis thaliana. *PloS one*, 11, e0149750.

Lin, Y. F., Liang, H. M., Yang, S. Y., Boch, A., Clemens, S., Chen, C. C., Wu, J. F., Huang, J. L. and Yeh, K. C. 2009. Arabidopsis IRT3 is a zinc-regulated and plasma membrane localized zinc/iron transporter. *New phytol*, 182, 392–404.

Ling, H. Q., Bauer, P., Bereczky, Z., Keller, B. and Ganal, M. 2002. The tomato fer gene encoding a bHLH protein controls iron-uptake responses in roots. *Proc natl acad sci*, 99, 13938–13943.

Liu, J., Piñeros, M. A. and Kochian, L. V. 2014. The role of aluminum sensing and signaling in plant aluminum resistance. *J integr plant biol*, 56, 221–230.

Liu, X. M., Kim, K. E., Kim, K. C., Nguyen, X. C., Han, H. J., Jung, M. S., Kim, H. S., Kim, S. H., Park, H. C., Yun, D. J. and Chung, W. S. 2010. Cadmium activates Arabidopsis MPK3 and MPK6 via accumulation of reactive oxygen species. *Phytochemistry*, 71, 614–618.

Liu, X. S., Feng, S. J., Zhang, B. Q., Wang, M. Q., Cao, H. W., Rono, J. K., Chen, X. and Yang, Z. M. 2019. OsZIP1 functions as a metal efflux transporter limiting excess zinc, copper and cadmium accumulation in rice. *BMC plant biol*, 19, 283.

Liu, X., Wu, H., Ji, C., Wei, L., Zhao, J. and Yu, J. 2013. An integrated proteomic and metabolomic study on the chronic effects of mercury in *Suaeda salsa* under an environmentally relevant salinity. *PLoS one*, 8e64041.

Lu, Y. P., Li, Z. S., Drozdowicz, Y. M., Hörtensteiner, S., Martinoia, E. and Rea, P. A. 1998. AtMRP2, an Arabidopsis ATP binding cassette transporter able to transport glutathione S-conjugates and chlorophyll catabolites: functional comparisons with AtMRP1. *Plant cell*, 10, 267–282.

Maksymiec, W. 2007. Signaling responses in plants to heavy metal stress. *Acta physiol. plant*, 29, 177.

Manara, A. 2012. Plant responses to heavy metal toxicity. In *Plants and Heavy Metals*. Springer, Dordrecht.

Manavalan, L. P., Guttikonda, S. K., Phan Tran, L. S. and Nguyen, H. T. 2009. Physiological and molecular approaches to improve drought resistance in soybean. *Plant cell physiol*, 50, 1260–1276.

Martinoia, E., Klein, M., Geisler, M., Bovet, L., Forestier, C., Kolukisaoglu, U., MuÈller-RoÈber, B. and Schulz, B. 2002. Multifunctionality of plant ABC transporters–more than just detoxifiers. *Planta*, 214, 345–355.

Mäser, P., Thomine, S., Schroeder, J. I., Ward, J. M., Hirschi, K., Sze, H., Talke, I.N., Amtmann, A., Maathuis, F.J., Sanders, D. and Harper, J. F. 2001. Phylogenetic relationships within cation transporter families of Arabidopsis. *Plant physiol*, 126, 1646–1667.

Metwally, A., Safronova, V. I., Belimov, A. A. and Dietz, K. J. 2005. Genotypic variation of the response to cadmium toxicity in *Pisum sativum* L. *J exp bot*, 56, 167–178.

Meyer, C. L., Pauwels, M., Briset, L., Godé, C., Salis, P., Bourceaux, A., Souleman, D., Frérot, H. and Verbruggen, N. 2016. Potential preadaptation to anthropogenic pollution: evidence from a common quantitative trait locus for zinc and cadmium tolerance in metallicolous and nonmetallicolous accessions of *Arabidopsis halleri*. *New phytol*, 212, 934–943.

Milner, M. J., Seamon, J., Craft, E. and Kochian, L. V. 2013. Transport properties of members of the ZIP family in plants and their role in Zn and Mn homeostasis. *J exp bot*, 64, 369–381.

Miyadate, H., Adachi, S., Hiraizumi, A., Tezuka, K., Nakazawa, N., Kawamoto, T., Katou, K., Kodama, I., Sakurai, K., Takahashi, H. and Satoh-Nagasawa, N. 2011. OsHMA3, a P1B-type of ATPase affects root-to-shoot cadmium translocation in rice by mediating efflux into vacuoles. *New phytol*, 189, 190–199.

Mohanpuria, P., Rana, N. K. and Yadav, S. K. 2007. Cadmium induced oxidative stress influence on glutathione metabolic genes of *Camellia sinensis* (L.) O. Kuntze. *Environ toxicol int j*, 22, 368–374.

Montanini, B., Blaudez, D., Jeandroz, S., Sanders, D. and Chalot, M. 2007. Phylogenetic and functional analysis of the Cation Diffusion Facilitator (CDF) family: improved signature and prediction of substrate specificity. *BMC genom*, 8, 107.

Morel, M., Crouzet, J., Gravot, A., Auroy, P., Leonhardt, N., Vavasseur, A. and Richaud, P. 2009. AtHMA3, a P1B-ATPase allowing Cd/Zn/co/Pb vacuolar storage in Arabidopsis. *Plant physiol*, 149, 894–904.

Mosa, K. A., Kumar, K., Chhikara, S., Musante, C., White, J. C. and Dhankher, O. P. 2016. Enhanced boron tolerance in plants mediated by bidirectional transport through plasma membrane intrinsic proteins. *Sci rep*, 6, 21640.

Mukhopadhyay, A., Vij, S. and Tyagi, A. K. 2004. Overexpression of a zinc-finger protein gene from rice confers tolerance to cold, dehydration, and salt stress in transgenic tobacco. *Proc natl acad sci*, 101, 6309–6314.

Naing, A. H., Park, K. I., Ai, T. N., Chung, M. Y., Han, J. S., Kang, Y. W., Lim, K.B. and Kim, C. K. 2017. Overexpression of *Snapdragon delila* (Del) gene in tobacco enhances anthocyanin accumulation and abiotic stress tolerance. *BMC plant biol*, 17, 65.

Nakashima, K., Ito, Y. and Yamaguchi-Shinozaki, K. 2009. Transcriptional regulatory networks in response to abiotic stresses in Arabidopsis and grasses. *Plant physiol*, 149, 88–95.

Nematshahi, N., Lahouti, M. and Ganjeali, A. 2012. Accumulation of chromium and its effect on growth of (*Allium cepa* cv. Hybrid). *Eur j expl biol*, 2, 969–974.

Nie, L., Shah, S., Rashid, A., Burd, G. I., Dixon, D. G. and Glick, B. R. 2002. Phytoremediation of arsenate contaminated soil by transgenic canola and the plant growth-promoting bacterium Enterobacter cloacae CAL2. *Plant physiol biochem*, 40, 355–361.

Ortega-Villasante, C., Hernández, L. E., Rellán-Álvarez, R., Del Campo, F. F. and Carpena-Ruiz, R. O. 2007. Rapid alteration of cellular redox homeostasis upon exposure to cadmium and mercury in alfalfa seedlings. *New phytol*, 176, 96–107.

Ortiz, D. F., Kreppel, L., Speiser, D. M., Scheel, G., McDonald, G. and Ow, D. W. 1992. Heavy metal tolerance in the fission yeast requires an ATP-binding cassette-type vacuolar membrane transporter. *EMBO j*, 11, 3491–3499.

Osmolovskaya, N., Dung, V. V. and Kuchaeva, L. 2018. The role of organic acids in heavy metal tolerance in plants. *Biol communs*, 63, 9–16.

Ozturk, M., Ashraf, M., Aksoy, A. and Ahmad, M. S. A. 2015a. *Plants, Pollutants & Remediation*. Springer, New York.

Ozturk, M., Ashraf, M., Aksoy, A., and Ahmad, M. S. A. 2015b. *Phytoremediation for Green Energy*. Springer, New York.

Ozturk, M., Yucel, E., Gucel, S., Sakcali, S. and Aksoy, A. 2008 Plants as biomonitors of trace elements pollution in soil. In: Prasad MNV (ed.), *Trace Elements: Environmental Contamination, Nutritional Benefits and Health Implications* (pp. 723–744). Wiley, NY.

Pagani, M. A., Tomas, M., Carrillo, J., Bofill, R., Capdevila, M., Atrian, S. and Andreo, C. S. 2012. The response of the different soybean metallothionein isoforms to cadmium intoxication. *J inorg biochem*, 117, 306–315.

Palusińska, M., Barabasz, A., Kozak, K., Papierniak, A., Maślińska, K. and Antosiewicz, D. M. 2020. Zn/Cd status-dependent accumulation of Zn and Cd in root parts in tobacco is accompanied by specific expression of ZIP genes. *BMC plant biol*, 20, 37.

Peiter, E., Montanini, B., Gobert, A., Pedas, P., Husted, S., Maathuis, F. J., Blaudez, D., Chalot, M. and Sanders, D. 2007. A secretory pathway-localized cation diffusion facilitator confers plant manganese tolerance. *Proc natl acad sci*, 104, 8532–8537.

Pérez-Alfocea, F., Ghanem, M. E., Gómez-Cadenas, A. and Dodd, I. C. 2011. Omics of root-to-shoot signaling under salt stress and water deficit. *OMICS: J int biol*, 15: 893–901.

Pietrini, F., Iannelli, M. A., Pasqualini, S. and Massacci, A. 2003. Interaction of cadmium with glutathione and photosynthesis in developing leaves and chloroplasts of *Phragmites australis* (Cav.) Trin. ex Steudel. *Plant physiol*, 133, 829–837.

Pomponi, M., Censi, V., Di Girolamo, V., De Paolis, A., Di Toppi, L. S., Aromolo, R., Costantino, P. and Cardarelli, M. 2006. Overexpression of *Arabidopsis phytochelatin* synthase in tobacco plants enhances Cd 2+ tolerance and accumulation but not translocation to the shoot. *Planta*, 223, 180–190.

Rai, V. K. 2002. Role of amino acids in plant responses to stresses. *Biol plant*, 45, 481–487.

Ramos, J., Clemente, M. R., Naya, L., Loscos, J., Pérez-Rontomé, C., Sato, S., Tabata, S. and Becana, M. 2007. Phytochelatin synthases of the model legume *Lotus japonicus*. A small multigene family with differential response to cadmium and alternatively spliced variants. *Plant physiol*, 143, 1110–1118.

Rao, K. P., Vani, G., Kumar, K., Wankhede, D. P., Misra, M., Gupta, M. and Sinha, A. K. 2011. Arsenic stress activates MAP kinase in rice roots and leaves. *Arch biochem biophys*, 506, 73–82.

Raskin, I. and Ensley, B. D. 2000. *Phytoremediation of Toxic Metals*. John Wiley and Sons, New York, NY.

Reed, M. L. and Glick, B. R. 2005. Growth of canola (*Brassica napus*) in the presence of plant growth-promoting bacteria and either copper or polycyclic aromatic hydrocarbons. *Can j microbiol*, 51, 1061–1069.

Roth, U., von Roepenack-Lahaye, E. and Clemens, S. 2006. Proteome changes in Arabidopsis thaliana roots upon exposure to Cd2+. *J exp bot*, 57, 4003–4013.

Ruiz, O. N., Alvarez, D., Torres, C., Roman, L. and Daniell, H. 2011. Metallothionein expression in chloroplasts enhances mercury accumulation and phytoremediation capability. *Plant biotechnol j*, 9, 609–617.

Salt, D. E., Kato, N., Krämer, U., Smith, R. D. and Raskin, I. 1999. The role of root exudates in nickel hyperaccumulation and tolerance in accumulator and non-accumulator species of Thlaspi. In *Phytoremediation of Contaminated Soil and Water*. CRC Press LLC, Boca Raton, FL.

Sarwat, M. 2017. *Stress Signaling in Plants: Genomics and Proteomics Perspective* (Vol. 2). Springer, New York, NY.

Sasaki, A., Yamaji, N. and Ma, J. F. 2014. Overexpression of OsHMA3 enhances Cd tolerance and expression of Zn transporter genes in rice. *J exp bot*, 65, 6013–6021.

Shameer, K., Ambika, S., Varghese, S. M., Karaba, N., Udayakumar, M. and Sowdhamini, R. 2009. STIFDB—Arabidopsis stress responsive transcription factor dataBase. *Int j plant genom*, 2009.

Shanker, A. K., Cervantes, C., Loza-Tavera, H. and Avudainayagam, S. 2005. Chromium toxicity in plants. *Environ int*, 31, 739–753.

Shao, H. B., Chu, L. Y., Lu, Z. H. and Kang, C. M. 2008. Primary antioxidant free radical scavenging and redox signaling pathways in higher plant cells. *Int j biol sci*, 4, 8.

Sharma, S. and Ali, I. 2011. Adsorption of Rhodamine B dye from aqueous solution onto acid activated mango (*Magnifera indica*) leaf powder: equilibrium, kinetic and thermodynamic studies. *J Toxicol environ health sci*, 3, 286–297.

Sharmin, S. A., Alam, I., Kim, K. H., Kim, Y. G., Kim, P. J., Bahk, J. D. and Lee, B. H. 2012. Chromium-induced physiological and proteomic alterations in roots of *Miscanthus sinensis*. *Plant sci*, 187, 113–126.

Sharma, S. S. and Dietz, K. J. 2006. The significance of amino acids and amino acid-derived molecules in plant responses and adaptation to heavy metal stress. *J exp bot*, 57, 711–726.

Shiu, S. H., Shih, M. C. and Li, W. H. 2005. Transcription factor families have much higher expansion rates in plants than in animals. *Plant physiol*, 139, 18–26.

Singh, K. B., Foley, R. C. and Oñate-Sánchez, L. 2002. Transcription factors in plant defense and stress responses. *Curr opin plant biol*, 5, 430–436.

Singh, R. K., Anandhan, S., Singh, S., Patade, V. Y., Ahmed, Z. and Pande, V. 2011. Metallothionein-like gene from *Cicer microphyllum* is regulated by multiple abiotic stresses. *Protoplasma*, 248, 839–847.

Singh, S., Parihar, P., Singh, R., Singh, V. P. and Prasad, S. M. 2016. Heavy metal tolerance in plants: role of transcriptomics, proteomics, metabolomics, and ionomics. *Front plant sci*, 6, 1143.

Singla-Pareek, S. L., Yadav, S. K., Pareek, A., Reddy, M. K. and Sopory, S. K. 2006. Transgenic tobacco overexpressing glyoxalase pathway enzymes grow and set viable seeds in zinc-spiked soils. *Plant physiol*, 140, 613–623.

Siripornadulsil, S., Traina, S., Verma, D. P. S. and Sayre, R. T. 2002. Molecular mechanisms of proline-mediated tolerance to toxic heavy metals in transgenic microalgae. *Plant cell*, 14, 2837–2847.

Song, W. Y., Mendoza-Cózatl, D. G., Lee, Y., Schroeder, J. I., Ahn, S. N., Lee, H. S., Wicker, T. and Martinoia, E. 2014. Phytochelatin–metal (loid) transport into vacuoles shows different substrate preferences in barley and Arabidopsis. *Plant cell environ*, 37, 1192–1201.

Song, W. Y., Park, J., Eisenach, C., Maeshima, M., Lee, Y. and Martinoia, E. 2014. ABC transporters and heavy metals. In: *Plant ABC Transporters*. Springer, Cham.

Song, W. Y., Sohn, E. J., Martinoia, E., Lee, Y. J., Yang, Y. Y., Jasinski, M., Forestier, C., Hwang, I. and Lee, Y. 2003. Engineering tolerance and accumulation of lead and cadmium in transgenic plants. *Nat biotechnol*, 21, 914–919.

Srivastava, S., Chaudhry, V., Mishra, A., Chauhan, P. S., Rehman, A., Yadav, A., Tuteja, N. and Nautiyal, C. S. 2012. Gene expression profiling through microarray analysis in *Arabidopsis thaliana* colonized by *Pseudomonas putida* MTCC5279, a plant growth promoting rhizobacterium. *Plant signal behav*, 7, 235–245.

Stephan, U. W. and Scholz, G. 1993. Nicotianamine: mediator of transport of iron and heavy metals in the phloem?. *Physiol plant*, 88, 522–529.

Sytar, O., Kumar, A., Latowski, D., Kuczynska, P., Strzałka, K. and Prasad, M. N. V. 2013. Heavy metal-induced oxidative damage, defense reactions, and detoxification mechanisms in plants. *Acta physiol plant*, 35, 985–999.

Takahashi, H., Kawakatsu, T., Wakasa, Y., Hayashi, S. and Takaiwa, F. 2012b. A rice transmembrane bZIP transcription factor, OsbZIP39, regulates the endoplasmic reticulum stress response. *Plant cell physiol*, 53, 144–153.

Takahashi, R., Ishimaru, Y., Shimo, H., Ogo, Y., Senoura, T., Nishizawa, N. K. and Nakanishi, H. 2012a. The OsHMA2 transporter is involved in root-to-shoot translocation of Zn and Cd in rice. *Plant cell environ*, 35, 1948–1957.

Talukdar, D. and Talukdar, T. 2014. RETRACTION: Coordinated response of sulfate transport, cysteine biosynthesis, and glutathione-mediated antioxidant defense in lentil (Lens culinaris Medik.) genotypes exposed to arsenic. *Protoplasma*, 251, 839–855.

Tang, L., Mao, B., Li, Y., Lv, Q., Zhang, L., Chen, C., He, H., Wang, W., Zeng, X., Shao, Y. and Pan, Y. 2017. Knockout of OsNramp5 using the CRISPR/Cas9 system produces low Cd-accumulating indica rice without compromising yield. *Sci rep*, 7, 1–12.

Tena, G., Asai, T., Chiu, W. L. and Sheen, J. 2001. Plant mitogen activated protein kinase signaling cascades. *Curr opin plant biol*, 4, 392–400.

Thomine, S., Lelièvre, F., Debarbieux, E., Schroeder, J. I. and Barbier-Brygoo, H. 2003. AtNRAMP3, a multispecific vacuolar metal transporter involved in plant responses to iron deficiency. *Plant j*, 34, 685–695.

Tiwari, S., Lata, C., Chauhan, P. S. and Nautiyal, C. S. 2016. *Pseudomonas putida* attunes morphophysiological, biochemical and molecular responses in *Cicer arietinum* L. during drought stress and recovery. *Plant physiol biochem*, 99, 108–117.

Tiwari, S., Lata, C., Singh Chauhan, P., Prasad, V. and Prasad, M. 2017a. A functional genomic perspective on drought signalling and its crosstalk with phytohormone-mediated signalling pathways in plants. *Curr genom*, 18, 469–482.

Tiwari, S., Prasad, V., Chauhan, P. S. and Lata, C. 2017b. *Bacillus amyloliquefaciens* confers tolerance to various abiotic stresses and modulates plant response to phytohormones through osmoprotection and gene expression regulation in rice. *Front plant sci*, 8, 1510.

Tran, L. S. P., Nishiyama, R., Yamaguchi-Shinozaki, K. Shinozaki, K. 2010. Potential utilization of NAC transcription factors to enhance abiotic stress tolerance in plants by biotechnological approach. *GM crops*, 1, 32–39.

Ueno, D., Milner, M. J., Yamaji, N., Yokosho, K., Koyama, E., Clemencia Zambrano, M., Kaskie, M., Ebbs, S., Kochian, L. V. and Ma, J. F. 2011. Elevated expression of TcHMA3 plays a key role in the extreme Cd tolerance in a Cd-hyperaccumulating ecotype of *Thlaspi caerulescens*. *Plant j*, 66, 852–862.

Umezawa, T., Fujita, M., Fujita, Y., Yamaguchi-Shinozaki, K. and Shinozaki, K. 2006. Engineering drought tolerance in plants: discovering and tailoring genes to unlock the future. *Curr opin biotechnol*, 17, 113–122.

Valliyodan, B. and Nguyen, H. T. 2006. Understanding regulatory networks and engineering for enhanced drought tolerance in plants. *Curr opin plant biol*, 9, 189–195.

Van der Zaal, B. J., Neuteboom, L. W., Pinas, J. E., Chardonnens, A. N., Schat, H., Verkleij, J. A. and Hooykaas, P. J. 1999. Overexpression of a novel Arabidopsis gene related to putative zinc-transporter genes from animals can lead to enhanced zinc resistance and accumulation. *Plant physiol*, 119, 1047–1056.

Vatamaniuk, O. K., Mari, S., Lang, A., Chalasani, S., Demkiv, L. O. and Rea, P. A. 2004. Phytochelatin Synthase, a dipeptidyltransferase that undergoes multisite acylation with γ-Glutamylcysteine during Catalysis stoichiometric and site-directed mutagenic analysis of Arabidopsis thaliana pcs1-catalyzed phytochelatin synthesis. *J biol chem*, 279, 22449–22460.

Vatamaniuk, O. K., Mari, S., Lu, Y. P. and Rea, P. A. 1999. AtPCS1, a phytochelatin synthase from Arabidopsis: isolation and in vitro reconstitution. *Proc natl acad sci*, 96, 7110–7115.

Vatamaniuk, O. K., Mari, S., Lu, Y. P. and Rea, P. A. 2000. Mechanism of heavy metal ion activation of phytochelatin (PC) synthase blocked thiols are sufficient for PC synthase-catalyzed transpeptidation of glutathione and related thiol peptides. *J biol chem*, 275, 31451–31459.

Verkleij, J. A. C., Sneller, F. E. C. and Schat, H. 2003. Metallothioneins and phytochelatins: ecophysiological aspects. In *Sulphur in Plants*. Springer, Dordrecht.

Verma, P. K., Verma, S., Pande, V., Mallick, S., Deo Tripathi, R., Dhankher, O. P. and Chakrabarty, D. 2016. Overexpression of rice glutaredoxin OsGrx_C7 and OsGrx_C2. 1 reduces intracellular arsenic accumulation and increases tolerance in Arabidopsis thaliana. *Front plant sci*, 7, 740.

Verma, S., Verma, P. K., Meher, A. K., Bansiwal, A. K., Tripathi, R. D. and Chakrabarty, D. 2018. A novel fungal arsenic methyltransferase, WaarsM reduces grain arsenic accumulation in transgenic rice (*Oryza sativa* L.). *J hazard mater*, 344, 626–634.

Vert, G., Barberon, M., Zelazny, E., Séguéla, M., Briat, J. F., and Curie, C. 2009. Arabidopsis IRT2 cooperates with the high-affinity iron uptake system to maintain iron homeostasis in root epidermal cells. *Planta*, 229, 1171–1179.

Vögeli-Lange, R. and Wagner, G. J. 1990. Subcellular localization of cadmium and cadmium-binding peptides in tobacco leaves: implication of a transport function for cadmium-binding peptides. *Plant physiol*, 92, 1086–1093.

Wang, Y., Dong, C., Xue, Z., Jin, Q. and Xu, Y. 2016. De novo transcriptome sequencing and discovery of genes related to copper tolerance in Paeonia ostii. *Gene*, 576, 126–135.

Wang, Y., Hu, H., Zhu, L. Y. and Li, X. X. 2012. Response to nickel in the proteome of the metal accumulator plant *Brassica juncea*. *J plant interact*, 7, 230–237.

Wang, Y., Ohara, Y., Nakayashiki, H., Tosa, Y. and Mayama, S. 2005. Microarray analysis of the gene expression profile induced by the endophytic plant growth-promoting rhizobacteria, Pseudomonas fluorescens FPT9601-T5 in Arabidopsis. *Mol plant-microbe interact*, 18, 385–396.

Wen, X. P., Ban, Y., Inoue, H., Matsuda, N. and Moriguchi, T. 2010. Spermidine levels are implicated in heavy metal tolerance in a spermidine synthase overexpressing transgenic European pear by exerting antioxidant activities. *Transgenic res*, 19, 91–103.

Williams, L. E., Pittman, J. K. and Hall, J. L. 2000. Emerging mechanisms for heavy metal transport in plants. *Biochim biophys acta biomembr*, 1465, 104–126.

Wray, G. A., Hahn, M. W., Abouheif, E., Balhoff, J. P., Pizer, M., Rockman, M. V. and Romano, L. A. 2003. The evolution of transcriptional regulation in eukaryotes. *Mol biol evol*, 20, 1377–1419.

Wu, H., Chen, C., Du, J., Liu, H., Cui, Y., Zhang, Y., He, Y., Wang, Y., Chu, C., Feng, Z. and Li, J. 2012. Co-overexpression FIT with AtbHLH38 or AtbHLH39 in Arabidopsis-enhanced cadmium tolerance via increased cadmium sequestration in roots and improved iron homeostasis of shoots. *Plant physiol*, 158, 790–800.

Wu, X., Zhu, Z. B., Chen, J. H., Huang, Y. F., Liu, Z. L., Zou, J. W., Chen, Y. H., Su, N. N. and Cui, J. 2019. Transcriptome analysis revealed pivotal transporters involved in the reduction of cadmium accumulation in pak choi (*Brassica chinensis* L.) by exogenous hydrogen-rich water. *Chemosphere*, 216, 684–697.

Xie, Z., Nolan, T. M., Jiang, H. and Yin, Y. 2019. AP2/ERF transcription factor regulatory networks in hormone and abiotic stress responses in Arabidopsis. *Front plant sci*, 10, 228.

Xu, Z., Ge, Y., Zhang, W., Zhao, Y. and Yang, G. 2018. The walnut JrVHAG1 gene is involved in cadmium stress response through ABA-signal pathway and MYB transcription regulation. *BMC plant biol*, 18, 1–13.

Yang, Z. and Chu, C. 2011. Towards understanding plant response to heavy metal stress. *Abiotic stress plants–mech adap*, 10, 24204.

Yang, Z., Wu, Y., Li, Y., Ling, H. Q. and Chu, C. 2009. OsMT1a, a type 1 metallothionein, plays the pivotal role in zinc homeostasis and drought tolerance in rice. *Plant mol biol*, 70, 219–229.

Yazaki, K., Yamanaka, N., Masuno, T., Konagai, S., Kaneko, S., Ueda, K. and Sato, F. 2006. Heterologous expression of a mammalian ABC transporter in plant and its application to phytoremediation. *Plant mol biol*, 61, 491–503.

Yeh, C. M., Hsiao, L. J. and Huang, H. J. 2004. Cadmium activates a mitogen-activated protein kinase gene and MBP kinases in rice. *Plant cell physiol*, 45, 1306–1312.

Yin, L., Wang, S., Eltayeb, A. E., Uddin, M. I., Yamamoto, Y., Tsuji, W., Takeuchi, Y. and Tanaka, K. 2010. Overexpression of dehydroascorbate reductase, but not monodehydroascorbate reductase, confers tolerance to aluminum stress in transgenic tobacco. *Planta*, 231, 609–621.

Yuan, Y. X., Zhang, J., Wang, D. W. and Ling, H. Q. 2005. AtbHLH29 of Arabidopsis thaliana is a functional ortholog of tomato FER involved in controlling iron acquisition in strategy I plants. *Cell res*, 15, 613–621.

Zeng, H., Xu, L., Singh, A., Wang, H., Du, L. and Poovaiah, B. W. 2015. Involvement of calmodulin and calmodulin-like proteins in plant responses to abiotic stresses. *Front plant sci*, 6, 600.

Zhang, J., Chen, K., Pang, Y., Naveed, S. A., Zhao, X., Wang, X., Wang, Y., Dingkuhn, M., Pasuquin, J., Li, Z. and Xu, J. 2017. QTL mapping and candidate gene analysis of ferrous iron and zinc toxicity tolerance at seedling stage in rice by genome-wide association study. *BMC genom*, 18, 828.

Zhang, Y., Zhao, L., Wang, Y., Yang, B. and Chen, S. 2008. Enhancement of heavy metal accumulation by tissue specific co-expression of iaaM and ACC deaminase genes in plants. *Chemosphere*, 72, 564–571.

Zhao, L., Sun, Y. L., Cui, S. X., Chen, M., Yang, H. M., Liu, H. M., Chai, T.Y. and Huang, F. 2011. Cd-induced changes in leaf proteome of the hyperaccumulator plant *Phytolacca americana*. *Chemosphere*, 85, 56–66.

Zhou, B., Yao, W., Wang, S., Wang, X. and Jiang, T. 2014. The metallothionein gene, TaMT3, from Tamarix androssowii confers Cd2+ tolerance in tobacco. *Int j mol sci*, 15, 10398–10409.

Zou, J., Wang, M., Jiang, W. and Liu, D. 2006. Chromium accumulation and its effects on other mineral elements in *Amaranthus viridis* L. *Acta biol crac ser bot*, 48, 7–12.

7

Unveiling the Mechanism of Transcriptional and Epigenetic Responses in Plants under Heavy Metal Stress: An Update

Samrat Banerjee[±] **and Mehali Mitra**[±]
Department of Botany, UGC Centre for Advanced Studies, The University of Burdwan, West Bengal, India

Puja Agarwal
Constituent College in Purnea University, Purnia, Bihar, India

Sujit Roy
Department of Botany, UGC Centre for Advanced Studies, The University of Burdwan, West Bengal, India

CONTENTS

7.1 Introduction

During the years the terminology of "heavy metal" has led to create confusion. A metal is an element which is malleable and ductile, conducts electricity and has a metallic luster (Ali and Khan, 2018). Heavy metals are defined as the group of metallic elements that have a relatively high density and weights higher than about 5g/cc (Ali and Khan, 2018). Among these heavy metals, few transition metals, such as cobalt (Co), iron (Fe),, manganese (Mn), molybdenum (Mo), nickel (Ni), zinc (Zn), copper (Cu) have important role in growth and developmental processes of plants and animals as essential micronutrients and at certain threshold level they regulate various enzymatic activities that are related to plant physiological and developmental response (Salla et al. 2011; Shahid et al. 2015). These heavy metals take part in several important metabolic processes including redox reactions, electron transfer, gene expression, biosynthesis of proteins, nucleic acids, chlorophyll and secondary metabolites, lipid and carbohydrate metabolism (Parmar et al. 2013; Rengel 1999). On the other hand, there are non-essential heavy metals like cadmium (Cd), lead (Pb), mercury (Hg), chromium (Cr) and arsenic (As), which have toxic effects in plant cell (Agarwal et al. 2020) even at low concentration. Though heavy metal elements naturally present in soil, but modern agricultural practices which include use of high amount of pesticides and fertilizers resulting in sharp increase in heavy metal contamination in soil (Hossain et al. 2012; Kupper and Andresen 2016). Previous studies reportedly said that these heavy metals have density five times to that of water. Redox-active metals (Fe, Cr, Cu, Mn) directly

± Equally contributing authors.

DOI: 10.1201/9781003155089-7

activate reactive oxygen species (ROS) or free radicals via Fenton reaction resulting in DNA strand breakage, damage of proteins, plasma membrane and photosynthetic pigments which ultimately leads to disruption of cellular homeostasis. However, the non-redox active metals (Cd, Zn, Ni, Hg, aluminum (Al)) indirectly produce oxidative stress via degradation of glutathione and binding to sulfhydryl (-SH) groups of protein structure thereby inhibiting antioxidant enzymes or inducing NADPH-oxidase (Valko et al. 2005; Jozefczak et al. 2012).

Rapid industrialization and urbanization during last decade have resulted in increasing concentration of heavy metals in soil, water and air. In developing countries increased industrialization, agricultural activities, transportation, mining have largely contributed a high amount of heavy metals into the ecosystem. Also, natural processes like volcanic eruptions are responsible for heavy-metal accumulation in the ecosystem. The consumption of crops from heavy metal contaminated agricultural field poses a potential threat to both animals and humans. There are several metal elements which have essential role in plant growth and development. Heavy metal contaminated soils affect rhizospheric communities of different organism which further negatively affect ecosystem stability and environment. After entering into the plant system, the heavy metal does not undergo biodegradation rather they become bio magnified in food chain. Thus, heavy metals get accumulate in living organism and thereby causes several clinical disorders. When they are present in above their threshold level, they become toxic by the production of free radicals and ROS. High concentrations of heavy metals in soil, negatively affect the crop growth and biodiversity (Ghosh and Singh, 1997).

In plants, the accumulation of heavy metals inhibits physiological processes such as photosynthesis, cell division, nitrogen metabolism, plant-water relationship, nutrition and thus growth inhibition leading to decrease in bio-mass and death of the plant (Zenk 1996; Zornoza et al. 2002; Michalak 2006). Heavy metals also disrupt the physiological and biochemical processes through inhibition of photosynthesis and respiration and also result in decrease in essential plant pigments. In the plant cell, the heavy metals break the thiol groups disrupting the structural conformation of proteins and in addition, the heavy metals generate ROS which causes oxidative damage to the photosynthetic machinery and cellular macromolecules of plants (Dutta et al. 2018). The toxic effect of heavy metals is exerted by blocking or displacement of functional metal ions from the functional domains of proteins or enzymes (Goyer, 1997). The interaction of heavy metals with nuclear proteins and chromatin results in profound cytological and physiological abnormalities (Siddiqui 2012). In addition, several plants like rice, cotton and spinach exhibit disruption of redox homeostasis with concurrent increase in ROS production following arsenic exposure (Sharma and Dubey 2007). The elevated level of ROS in cellular micro environment disrupts structural integrity of several biomolecules such as DNA, protein, lipid and thereby alter macromolecular organization in plants. This oxidative stress, created by ROS production, leads to membrane disruption, macromolecule deterioration, lipid peroxidation, ion-leakage, DNA-double strand breakage and finally death of plants (Hossain et al. 2010; Rascio and Navari-Izzo 2011; Barconi et al. 2011; Hossain et al. 2012). The instantaneous responses include immediate changes in the transcriptional rates of thousands of responsive genes whereas the long-term processes refer to the genetic modification of epigenetic regulations (Zhang et al. 1998; Dutta et al. 2018). Together, this results in cytological, physiological and morphological aberrations in plants affecting growth and yield.

Being sessile in nature, plants develop several physiological, biochemical and molecular strategies which ultimately generate resistance in plants against heavy metals via sequestration, transportation, chelation, exclusion mechanism. The transcriptional and translational response in plants under metal stress through different TFs and non-coding RNAs exhibit a further line of stress adaptation. Various studies indicated that changes in methylation patterns in DNA resulting in alteration of gene expression and transcriptional response (Bender 2004; Peters and Schübeler 2005). In higher plants, different DNA methylase enzymes involved in stress response such as DNA methyltransferase, chromomethylase and domain rearranged methyltransferase (Wada et al. 2004). Some other stress-responsive DNA methylation enzymes also involved in metal stress regulation such as Deficient in DNA methylation 1 (DDM1) and methylcytosine binding protein (Vanyushin and Ashapkin 2011). In *Brassica napus* following Cr exposure, hypermethylation was observed at various loci (Labra et al. 2004). Plants employ various short-term and long-term strategies to develop tolerance against heavy metal toxicity, which include complicated interlinked mechanisms (Dutta et al. 2018). Plants achieve resistance to metal stress either by "avoidance" or by "tolerance" (Hossain et al. 2012). Avoidance is when plants restrict their metal uptake and tolerance is when plants survive in spite of high concentration of internal metal and metalloids. This avoidance includes reducing the metal concentration that is entering the cell by extracellular precipitation reduced uptake, biosorption to cell walls or increased efflux. In case of plants' tolerance for survival, heavy metals are chelated intracellularly by the amino acids, Glutathione (GSH), organic acids or HM binding ligands such as phytochelatins and metallothioneins, vacuole compartmentations and upregulation of antioxidant defence also take place to avoid the damaging effects caused by ROS formation (Hossain et al. 2009; Singla-Pareek et al. 2006; Leyval et al. 1997; Cobbett 2000; Hall 2002; Yang et al. 2005; Clemens 2006; Yadav 2010; Seth et al. 2012). In this chapter, we have particularly focused on the transcriptional response of plants under various heavy metal stress. Moreover, following exposure to heavy metals epigenetic modulation of stress-responsive genes also play critical role in stress management and adaptation in plants. These epigenetic responses act as memory system in plants which transmitted to the next generation, further confer another line of adaptive strategy in plants (Figure 7.1).

7.2 Transcriptional Regulation of Plants under Heavy Metal Stress

7.2.1 Transcription Factor Mediated Response

Accumulation of heavy metals in agricultural field is a matter of global concern. Consumption of fossil fuels, mining, industrial wastes and most of all anthropogenic activities (excessive use of chemical fertilizers, pesticides, insecticides, herbicides,

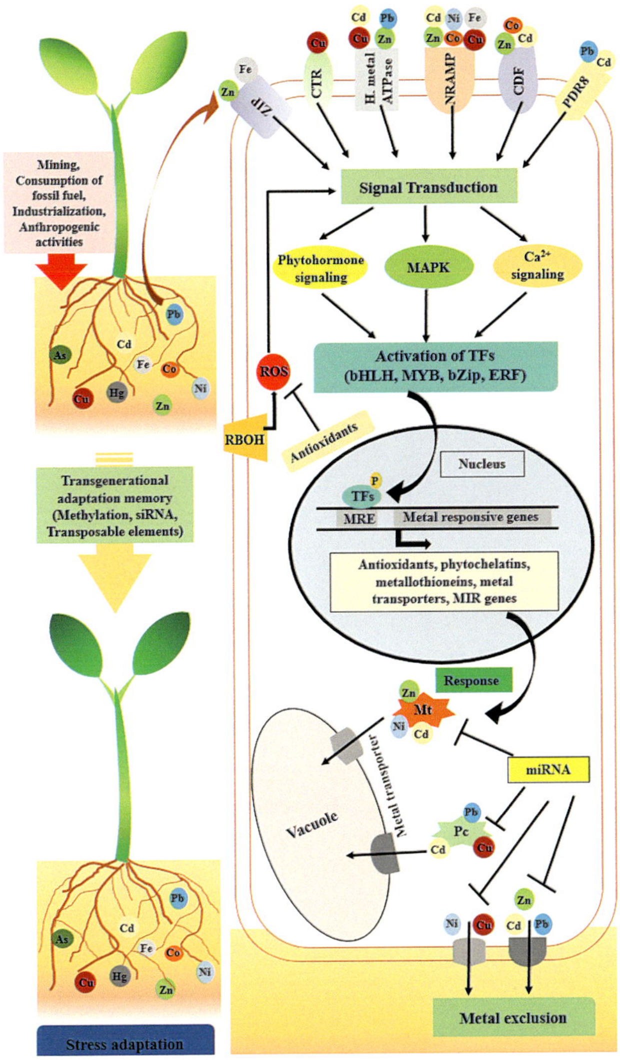

FIGURE 7.1 Schematic diagram of a transcriptional regulation and transgenerational stress adaptation of a plant cell. Plant cell uptakes heavy metals from environment via the transporter genes which eventually result in activation of transcription factors through different signaling network. In the cell the transcription factor activates metal responsive genes, activating the stress response signaling cascade. The metal responsive genes chelates the heavy metals and proceed for vacuolar sequestration, compartmentation and eventually heavy metal exclusion from the cell.

fungicides) involved with modern agricultural practices leads to increase of heavy metal contamination in rhizosphere. Metal elements with density greater than 5 gcm^{-3} are considered as heavy metals. Heavy metals impose their negative effects throughout the different stages of plant life. Non-redox active metals (Cd, Zn, Ni, Hg, Al) indirectly produce oxidative stress via inhibition of sulfhydryl group containing proteins, antioxidant enzymes or via induction of NADPH-oxidases whereas the redox-active metals (Cr, Cu, Mn, Fe) directly involved in generation of ROS resulting in genotoxic stress (Jozefczak et al. 2012). Being sedentary in nature plants can't avoid such heavy metal stress, so they develop an array of adaptive response, most of these signaling cascades regulated via induction of different classes of transcription factors (TFs). TFs regulate the expression of various stress-responsive genes and thereby ensuring defense response in plants (Yanhui et al. 2006; Dutta et al. 2018). The immediate and short-term response of heavy metals in plants are governed by different TF families. TFs, responsible for stress tolerance in plants are often targeted by plant breeders for the development of stress-resistant transgenic crops. Moreover, TFs are the most potential regulators of abiotic stress responses as they interact with the promoters of several stress-responsive genes (Figure 7.2).

Plants have evolved different molecular strategies to combat against such different redox and non-redox active metals. Different transcription factor family like WRKY, basic leucine zipper (bZip), basic helix loop helix (bHLH), MYB and ERFs are particularly involved in heavy metal homeostasis by regulating different metal stress-responsive genes (Chen et al. 2016; Dutta et al. 2018). Significant induction of genes encoding TFs such as MYB and WRKY was observed after Pb exposure. From microarray data analysis, it was suggested that two enzymes of cysteine and glutathione biosynthesis, acetylserine lyase and serineacetyl transferase were upregulated after Pb treatment (Jalmi et al. 2018). The redox-sensitive motifs of MYB TFs involved in secondary metabolite biosynthesis thereby confers another level of protection against various stresses (Myrset et al. 1993; Chen et al. 2006). Significant increase in transcript level of MYB28 and MYB72 promote Cd tolerance (Van de Mortel et al. 2008). Plants under Pb treatment, highly express different WRKY and MYB TF family genes, such as WRKY17, WRKY53, MYB39, MYB63, MYB45, MYB94, MYB93. Rapid and transient accumulation of WRKY factors under Pb stress, activate plant defense mechanism (Eulgem et al. 2000). From previous studies, it was observed that AtMYB4 positively regulates the PCS1 and MT1C genes via interacting with their promoter following Cd treatment (Agarwal et al. 2020). In rice, TF OsMYB4 when expressed in Canola plants, improves tolerance under Zn and Cu stress (Raldugina et al. 2018). Moreover, In *Arabidopsis thaliana* MYB4, MYB43, MYB48, MYB124 expression was found to be upregulated under Cd stress (Oono et al. 2014). The non-redox active metal, Cd has a more toxic impact on plant growth and development. An array of TFs like WRKY53, bZip, bHLH100, MYB4, MYB28, ERFs regulate the Cd-mediated phytotoxicity (Dutta et al. 2018). The roots of *Arabidopsis thaliana* encode different genes like AP2/ERF, HSF and MYB transcription factor genes when expose to Cd but there was no such change in expression pattern following Cu treatment (Weber et al. 2006).

In rice, an MYB TF, OsARM1 directly interacts with As transporters OsLsi1, OsLsi2, OsLsi6 regulating arsenic transport (Wang et al. 2017). In *Arabidopsis*, R2R3 MYB transcription factor, MYB49 binds with two bHLH TF, bHLH38 and bHLH101 at their promoter site and they subsequently activate the metal transporter gene IRT1 in roots. Expression of IRT1 in the roots leads to accumulation of Cd. However, basic leucine zipper TF ABI5 directly interacts with MYB49 and repress it, reducing Cd accumulation (Zhang et al. 2019). Sensitive to proton rhizotoxicity 1 (STOP1) and Al^{3+} resistance transcription factor 1 (ART1), the zinc finger family TFs are involved in enhanced Al tolerance in *Arabidopsis* and rice. In *Arabidopsis* STOP1 enhance the expression of AtALMT1, AtMATE1 and AtALS3 by binding their promoter and these transporters are

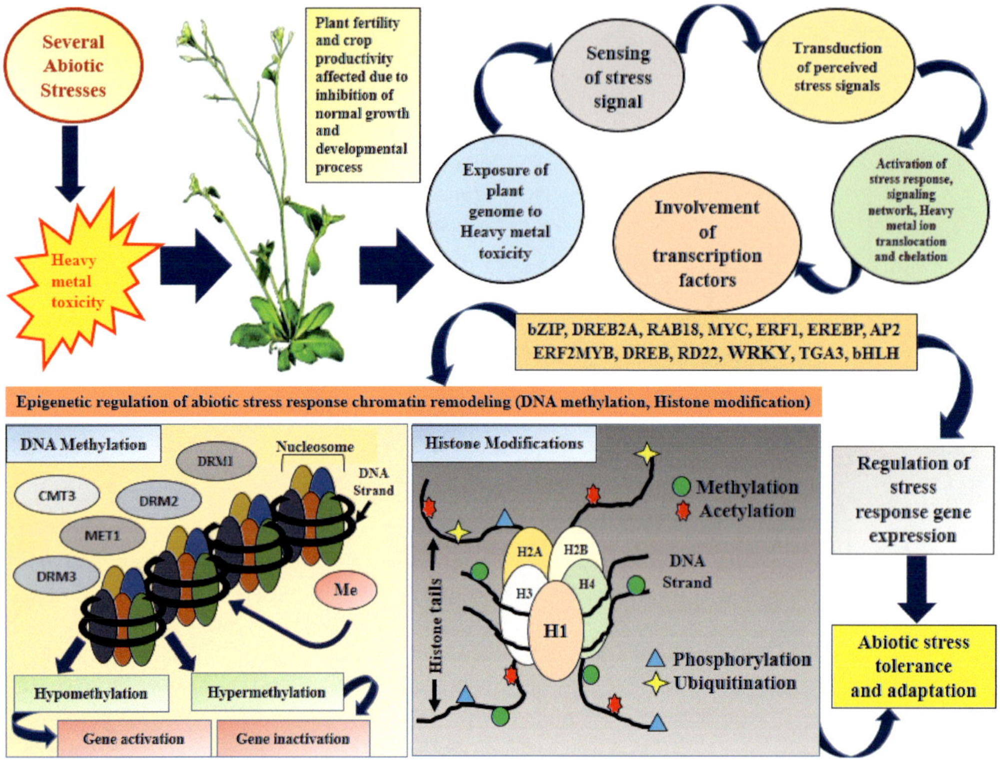

FIGURE 7.2 Schematic diagram showing the transcriptional and epigenetic stress response of plants under heavy metal toxicity. The stress signaling network activates after sensing the stress signal. The transcriptional pathway involves different transcription factor families which are involved in various signaling pathway. The epigenetic regulation involves DNA methylation and Histone Modification, resulting in alteration of chromatin structure and gene expression. Hypomethylation and hypermethylation of DNA control gene activation. Additionally, Histone Acetylation, Phosphorylation and ubiquitination play important role in epigenetic modification and gene expression.

responsible for efflux of organic anions which bind and detoxify Al^{3+} (Sawaki et al. 2009). Similarly, in rice, ART1 activates OsFRDL4, plasma membrane-associated citrate transporter in response to Al^{3+} stress (Tsutsui et al. 2011). In rice ART1 activates vacuolar transporter, ALS1, enhancing vacuolar sequestration of metal ions (Delhaize et al. 2012). The role of bZip TFs under heavy metal stress response has differential expression patterns. In *Brachypodium distachyon* most of the bZip TFs (Zip64, Zip16, Zip45) are downregulated whereas few bZip TFs (Zip39, Zip76, Zip28) are upregulated after Cu, Zn, Cd, Pb and Mn treatment (Liu and Chu, 2015; Wang et al. 2016). Overexpression of BubZip3 in Boehmeria nivea enhances root growth under metal stress (Huang et al. 2016). In plants a bZip TF, OBF5 (OCS element binding factor 5) regulates the interaction of glutathione-s-transferase with its promoter after treatment with Cd (Suzuki et al. 2001; Singh et al. 2020). Moreover, two bZip TF bZip19 and bZip23 binds with the promoter of ZNT1 transporter and results in hyperaccumulation of Zn and Cd in *Noccaea caerulescencs* (Lin et al. 2016).

In *Arabidopsis*, several bHLH proteins such as AtbHLH38, AtbHLH39, AtbHLH100, AtbHLH101 are involved in the regulation of the expression of iron-regulated transporter (IRT2), HMA3 and MTF3 under Cd stress (Wu et al. 2012). The transcriptional response of plants under Cd stress found similar with other abiotic stress response in plants. Plants under continuous Cd exposure enhance Rd29A expression via induction of DREB2A (Suzuki et al. 2001). ERF1 and ERF2 expression are also upregulated under Cd stress which further induce different dehydration responsive genes (Singh et al. 2002). Moreover, expression of different ERF genes, ERF1, ERF2 and ERF5 significantly increased following Cu treatment (Weber et al. 2006). In wheat, CBF1 and DREB1 expression get increased within 15 minutes of Cd treatment and this response would retain even after seven days. A zinc finger TF, ZAT6 in *Arabidopsis* was found to be involved in positive upregulation of GSH1, GSH2, PCS1 and PCS2 genes under Cd stress (Chen et al. 2016). Moreover, expression of ZAT12 is increased under higher Cu concentration (Opdenakker et al. 2012). Small cysteine-rich proteins, metallothioneins (MTI, MTII) are strongly induced upon exposure to lead. MTF1 specifically binds with MRE sequences of metallothionein gene promoters. The MTF1 TF also induced upon oxidative stress and hypoxia (Zhang et al. 2003). In *Brassica juncea*, an orthologue of TGA3 (TGA1A related gene 3), BjCdR15 induce the expression of various metal transporter genes which enhance the Cd accumulation and tolerance in *Arabidopsis thaliana* (Farinati et al. 2010).

In yeast cells, arsenic and cadmium lead to formation of aggregated proteins inside the cells which collectively called as unfolded protein response (UPR) (Jacobson et al. 2012). These UPRs are also responsible for generation of endoplasmic reticulum stress in plants via transcriptional modulation. In context of this different bZip TFs, such as bZip17, bZip28 and bZip60 get induced. These TFs ultimately promote the synthesis of different chaperones which regulate the unfolded protein under heavy metal stress. From expression analysis, it was observed that heat shock proteins (HSP60, HSP70, HSP90 and HSP100) get induced upon heavy metal exposure and they maintain the protein confirmation (Rhee et al. 2009). HSP70 promotes heavy metal sequestration and subsequent detoxification via regulation of metallothioneins (Ritter et al. 2010, Haap et al. 2016). In recent years, several studies on alternative splicing indicate their involvement in stress adaptation in plants (Gracz 2016; Laloum et al. 2018). Alternative splicing considered as a regulatory mechanism for the generation of mRNAs from a single copy of pre-mRNA via specific removal or retention of exons and introns. In rice two alternatively spliced, OsPCS2, OsPCS2a and OsPCS2b differentially expressed in roots and shoots under Cd stress. OsPCS2b is a truncated version of OsPCS2 lacking functional N and C terminal domains. Whereas in yeast, OsPCS2a with complete polypeptide sequence confers resistance against both Cd and As (Ramos et al. 2008). Under Cd stress, the truncated OsPCS2b highly expressed in roots than shoots (Das et al. 2017). But the requirement and the functional role of this terminated version of OsPCS2 revoked the need of further studies. Analogously, in *Sesbania rostrata* four alternatively spliced SrPCS gene is observed (Li et al. 2009). In Arabidopsis under high Zn concentration, induction of ZIF2 gene was observed which encode a tonoplast-associated transporter. Moreover, ZIF2.2, splice variant with intron retention at 5′ UTR resulting into enhanced Zn tolerance through activation of different genes associated with Zn detoxification mechanism (Remy et al. 2014).

Different signaling cascade work upstream of the TFs and regulates their activity when expose to stress. Metal transporters (CDF, NRAMP, CTR, H^+ ATPase, ABC) and signaling cascades (MAPK, Ca^{2+}, ROS, phytohormones) modulate the activity of TFs and TF mediated transcriptional regulation of stress response (Park et al. 2012; Jalmi et al. 2018). MYB TFs are the downstream target of MAPK signaling pathway under Cd or other heavy metal stress (Li et al. 2016). Plants overexpress WRKY22, WRKY25, WRKY29 TFs in a MAPK signaling-dependent manner following Cu and Cd stress (Opdenakker et al. 2012). Ethylene responsive factor, ERF104 is found to be upregulated by MPK3 and MPK6 (Bethke et al. 2009). Cd often mimics the Ca^{2+} gated ion channels. Application of Ca^{2+} on *Brassica juncea* reduces the Cd-mediated phytotoxicity and also improves the growth and development (Ahmad et al. 2015). Ca^{2+}/calmodulin also takes part in signaling under Ni or Pb stress (Ahmad et al. 2015). Transcription factor EIN3 is upregulated by MAPK after exposure to Chromium (Cr) (Trinh et al. 2014). Under Cd stress, MYB59 expression is upregulated. It was observed that MYB59 negatively regulates Ca^{2+} signaling and regulates plant growth and development (Fasani et al. 2019). Moreover, ABA also negatively regulates Cd accumulation in plants (Uraguchi et al. 2009; Fan et al. 2014).

7.3 miRNA Mediated Transcriptional Regulation of Heavy Metal Stress

Like the TFs, miRNAs are also regulating the post-transcriptional or translational gene expression. Several conserved miRNAs differentially expressed under various stress conditions. The recent advancement of molecular biology tools reveals that miRNAs involved in the maintenance of different stress-responsive genes by their inhibition of degradation (Zhou et al. 2012; Noman and Aqeel, 2017). Different molecular and experimental approaches have been applied for the identification of novel metal stress-responsive miRNAs and their potential targets (Huang et al. 2012). Thus, miRNA-based targeted gene knockout considered as a major player regulating plant growth and development under heavy metal stress.

The miRNAs are 20–24 nt long, small non-coding RNAs which regulate the plant stress response spatially or temporally (He et al. 2016). The expression of miR192 is downregulated following Cd exposure which specifically targets ABC metal transporter. Moreover, overexpression of miR192 leads to reduced seed germination and development in rice (Tang et al. 2014; He et al. 2016). Fascinatingly, various miRNAs such as miR156, miR158, miR398 and miR408 regulates the expression of metal responsive under heavy metal stress. But their expression pattern varies (either upregulated or downregulated) according to different heavy metal exposure (Noman and Aqeel 2017). Plants under Cd stress regulate the expression of metal responsive genes via miR398 and miR408. Downregulation of miR408 increases the expression of superoxide dismutase or other ROS scavengers during heavy metal stress (Ma et al. 2015). Plants under Al stress show downregulation of several miRNAs including miR156, miR171, miR395, miR396, miR398 (Lima et al. 2011). Moreover, under Pb and Cr treatment different miRNAs are upregulated. Apart from different TFs, the miRNAs are also being the downstream target of MAPK signaling pathways. Transcription factor SQUAMOSA and R2R3 MYB family proteins are also targeted by miR156/157 and miR159 respectively under Cd, Cu and Hg stress (Alonso-Peral et al. 2010). In *Ricinus communis*, miR398 is downregulated under Ni stress so that the byproduct ROS is detoxified by scavengers.

7.4 Epigenetic Regulation Not Associated with Nucleotide Modification

Epigenetics refers to the heritable changes in gene expression and function which do not involve any variation in underlying DNA sequence. The epigenetic modifications are a natural occurrence but it can be influenced by several abiotic stress factors including heavy metal toxicity which can cause severe damage to living organisms. Gene silencing associated with DNA methylation, histone modification and non-coding RNA (nc-RNA) are responsible for the epigenetic changes (Egger et al. 2004). Epigenetic modifications are transgenerationally and mitotically inheritable and provide an extra transcriptional control that regulates the way of gene expression (Fragou et al. 2011). Genes are expressed when chromatin is in extended state and the same genes are inactive when chromatin is in condensed form. The reversible epigenetic changes like methylation of cytosine in DNA, modification of lysine residues of histones and histone acetylation control the dynamic states of chromatin structure (Fragou et al. 2011; Pandey et al. 2016). Epigenetic mechanisms regulate the level of mRNA accumulation at the transcriptional level for the modulation of chromatin structure (Pandey et al. 2016). Previously, studies have revealed a correlation between chromatin remodeling and regulation of gene expression under stress conditions. Reportedly, in plants like *Arabidopsis thaliana*, *Oryza sativa* and *Zea mays* level of DNA methylation induced by abiotic stress factors modulate the expression of stress-responsive genes (Wang et al. 2009; He et al. 2010; van Dijk et al. 2010). Studies on abiotic stress-mediated epigenetic modification have focused on the modulation of chromatin at particular loci and also investigated about the proteins involved in it during periods of stress (Qiao and Fan, 2011).

7.5 DNA Methylation and Its Role in Epigenetic Regulation

DNA methylation involves the covalent binding of a methyl ($-CH_3$) group to the 5th positioned carbon atom of cytosine or the 6th positioned nitrogen atom of the adenine nucleotides. The methyl donor in the formation of 5-methylcytosine is S-adenosyl-methionine (SAM) and DNA methyltransferases catalyze the reaction (Tang and Ho, 2007). In plant genomic DNA, methylation occurs specifically in cytosines of CG dinucleotides, namely CpG dinucleotides. CpG islands are clustered formation with a length ranging from 200 bp to 5 kb (Fragou et al. 2011). This cytosine methylation is associated with transcriptionally inactive sequences and both as symmetric and asymmetric context like CpG or CpHpG (H stands for A, C or T) and CpHpH sites, respectively (Cokus et al. 2008, Pandey et al. 2016). The symmetric sites of DNA methylation, i.e. CpG and CpHpG can be maintained during DNA replication using hemi-methylated sites as substrate and thus re-establishment of the methylation occurs in the newly synthesized DNA strand. On the other hand, the asymmetric patterns of DNA methylation that is CpHpH sites cannot be maintained after DNA replication as one of two DNA copies no longer contain a methylated cytosine at its respective position. Previous studies on *Arabidopsis* genome sequencing have revealed that methylations are either located in the promoter region and in the gene sequence where both sides have an impact on gene expression (Zhang et al. 2006, Cokus et al. 2008, Yaish 2013). However, about 20% of cytosines in the *Arabidopsis* genome are methylated and 30% nucleotide in plants contain methylated cytosine reportedly (Yaish et al. 2013). According to previous reports, it has been widely accepted that the DNA methylation in promoter region is conversely related to level of gene expression whereas methylation in coding region sometimes affects level of gene expression in a positive way (Zhang et al. 2006; Zilberman et al. 2007).

Two types of cytosine DNA methyltransferases are present in plants, one type for maintenance of methylation and another type for de-novo methylation. DNA methyltransferases 1 (MET1) maintains methylation at CpG sites and chromomethylase 3 (CMT3) at CpHpG sites. These two enzymes MET1 and CMT3 are involved in maintenance of methylation. CMT3 also includes CpHpH DNA methylation with certain histone modification at specific sites (Bartee et al. 2001; Lindroth et al. 2001). Domain rearranged methyltransferases, i.e. DRM1 and DRM2 are required for methylation at asymmetric CpHpH sites (Cao et al. 2003; Chan et al. 2005). DRM1 and DRM2 promote de-novo methylation for which the site information is provided by RNA-directed DNA methylation. DRM3 plays important role in activating DRM2 (Henderson et al. 2010). In *Arabidopsis*, RNA-directed DNA methylation is controlled by DRM2 involving plant-specific RNA Polymerase V (Pol V), which physically interacts with DRM3, an evolutionarily conserved but catalytically inactive methyltransferases encoded by *Arabidopsis* genome (Zhong et al. 2015). DNA methylation has been associated with increasing methylation rate, gene silencing and hypomethylation has been linked with transcription (Choudhury et al. 2010). To explain the way DNA methylation suppresses transcription of gene, two major mechanisms have been proposed, among which, the first mechanism explains the inhibition of transcription factor binding with their respective recognition site containing CpG dinucleotides by extending the added methyl group to the major groove of DNA molecule. Another mechanism proposes that TFs could not bind to their regulatory sites because methylcytosine-binding proteins bind to methylated CpG islands and create steric hindrance (Fragou et al. 2011).

In wild-type plants, the level of methylation of few gene changes dynamically throughout plant's developmental process in response to plant's exposure to the adverse environmental conditions. DNA methylation prevents unfavorable genetic rearrangement at specific locus, thus helping plants to cope up stress conditions (Boyko et al. 2010; Chinnusamy and Zhu, 2009). Plants that are exposed to various environmental stress factors experience a chain of physiological changes to avoid damage and also to complete their life cycle to provide seeds for next generation. To a certain extent, DNA methylation controls the gene expression which is associated with a series of physiological changes (Yaish et al. 2011) such as DNA methylation regulates the expression of genes involved in flowering. The stressed plants exhibit improved stress tolerance through the processes like changes in DNA methylation patterns and transgenerational adaptation. At the transcriptional level, the sequence-specific modifications of gene expression are controlled by the small RNA molecules (Molinier et al. 2005; Boyko and Kovalchuk 2008; Carthew and Sontheimer 2009; Dutta et al. 2018). Among the abiotic stresses, heavy metals play crucial role in different stages of a plant's life but excessive dose of heavy metal leads to DNA methylation. Copper and Zinc show little changes in cytosine methylation at early stages of plant growth but at later stages extensive hypomethylation takes place. Processes like cell wall formation, amino acid metabolism and RNA processing involve these hypo-methylated genes (Cicatelli et al. 2014). DNA methylation in rice leaf tissue under heavy metal exposure, alters predominantly CHG type hypo-methylation (Pandey et al. 2016). Excess cobalt also induces altered DNA methylation. Cadmium, one of the most common heavy metals, causes reduced photosynthesis and metabolism in plants. Reportedly, DNA hyper-methylation upon cadmium treatment has been observed in sea grass (Greco et al. 2012). Thus, various reports have suggested epigenetic modification based on DNA- methylation under heavy metal stress.

7.6 DNA Methylation Associated with Cd Toxicity

The International Agency for Cancer Research and National Toxicology Program has been designated Cadmium as human carcinogen (Arita and Costa 2009). According to the previous studies, Cd exposure is associated with pancreatic and pulmonary cancers (Waalkes 2000). Cd has structural and physical similarities with Zn, so the mechanism of cadmium-induced transformation is likely to arise from zinc-dependent cellular process (Waalkes 2003). However, from previous evidences, it is clear that cadmium exposure affects DNA methylation. Cd is responsible for MT-1 gene silencing which is correlated to increased cellular DNMT activity (Takiguchi et al. 2003). Hypothetically it is been assumed that reduction of DNA methylation under cadmium stress occurs due to the inhibitory effect on expression of DNMT3 A/B genes which are responsible for DNA de-novo methylation (Doi et al. 2011). However, a different study suggested that DNA hypo-methylation occurs due to increased cell cycle rates under Cd stress (Huang et al. 2008). DNA hypomethylation occurs from low DNMT activity under dose-dependent cadmium stress whereas hyper-methylation takes place with upregulation of DNMT1 after chronic exposure to Cd stress and this hyper-methylation retains even after removal of Cd (Takiguchi et al. 2003; Benbrahim-Tallaa et al. 2007; Jiang et al. 2008).

7.7 DNA Methylation Associated with Pb Toxicity

Pb is one of the highly toxic heavy metal for animal and plant kingdoms; affecting their fitness, survival during the developmental phases. For plants lead decreases the yield of crops by disrupting the biochemical, molecular and physiological processes of plants (Babst-Kostecka et al. 2016). Impaired root growth, alteration in membrane permeability, hormonal status, changes in nutrition and delays in germination, water regime, decrease in photosynthesis mechanism and increased ROS production occurs due to Pb exposure to plants (Sengar et al. 2008; Pourrut et al. 2011). So, the Pb detoxification of plants to cope up with the adverse effects of lead toxicity is associated with binding of Pb to glutathione and amino acids, phytochelatin synthesis and sequestration into the vacuoles (Kushwaha et al. 2015). Different DNA methyltransferases are associated with DNA cytosine methylation, which occurs during flower and leaf morphogenesis, flowering time, fertility through gene silencing (Zhang et al. 2010; Shafiq and Khan 2015). The DNA methyltransferases express differentially in presence of Pb, which promotes alteration in DNA methylation pattern. From the previous work, it has been reported that DNA hypo-methylation occurs at the promoter region of some metal detoxification transporters (Shafiq et al. 2019). These transporters are associated with metal-resistant phenotype and increased gene expression (Shafiq et al. 2019). These DNA hypo-methylation occurring in the promoter region of the transporters, eventually increase the transcription rate of transporters. The increased activity of the transporters may efficiently sequestrate the accumulated heavy metals into the vacuole meanwhile some of the transporters may send the toxic heavy metals back to the apoplast. Thus, the resultant metal concentration becomes less in the cell by decreasing the impact of ROS generated from the metal exposure.

7.8 DNA Methylation Associated with Hg Toxicity

Hg has ranked among the top three potent toxic metals according to the United States Environmental Protection Agency (Schober et al. 2003) but the extent of phytotoxicity and understanding about the mechanism of Hg toxicity is still insufficient (Azevedo and Rodriguez 2012). Hg can be present in several oxidation states and can spread through different ecosystems (Boening, 2000; Clarkson and Magos 2006). The elemental Mercury (Hg^0) is widely spread through the environment and its toxic effects are less whereas the organic Hg that is methylated from inorganic Hg by living organisms are extensively harmful (Boening 2000). High level of Hg^{+2}, a product of vapor Hg, as well as organic Hg, have strong phytotoxic effect as it triggers ROS production in plant cells which leads to cell disruption (Clarkson and Magos 2006). Hg exposure to plants reduces photosynthesis, water uptake, rate of transpiration and synthesis of chlorophyll.

7.9 DNA Methylation Associated with Cr Toxicity

Cr occurs in +6, +3 and 0 valance states. Industries that produce stainless steel, ferrochrome and chromate use Cr extensively. The toxic effects of Cr have been well studied but its effect on DNA methylation and histone modification is still an emerging field for the epigenetic researchers. The cytotoxicity due to chromium exposure is thought to be genotoxic as reportedly Cr leads to oxidative stress, DNA-protein crosslinks, DNA double-strand breaks and formation of DNA Cr adducts. According to Rozhon et al. (2008), 3mM Cr (VI) treatment increases global DNA methylation level in *Medicago sativa* seedlings. Cr exposure induces gene silencing in promoter region of gpt transgene in plant cells (Labra et al. 2004). All the previous reports have suggested that Cr induces gene transcription or silencing by inducing alterations in DNA methylation and histone modification.

7.10 DNA Methylation Associated with As Toxicity

As contamination occurs either in Arsenite (As^{III}) or Arsenate (As^{V}) form among which Arsenate is dominant in soil. As is already present on earth's surface and can easily contaminate the ground water. Apart from that, industrial factories release arsenic contaminated waters in environment. Anthropologic and geogenic activities of humans, such as mining, smelting, pesticide application, fossil fuel burning release a huge amount of arsenic in the environment as well as soil. According to US Environmental Protection Agency (USEPA), Arsenic ranked top in the list of hazardous substances (Tang et al. 2016). At higher concentrations, As is toxic for most of the plants. As contamination in the soil leads to arsenic accumulation in food crops like rice (Meharg and Rahman, 2003; Zhu et al. 2008). Rice consumes a large proportion of As from soil and its very hazardous for human as rice is used as one of the main food supplements in China and Asian countries. Arsenic-mediated DNA methylation involves several mechanisms leading to epigenetic changes. To be more specific the effect of As alters the regulation of DNA methyltransferases. As causes whole-genome demethylation and localized gene-specific demethylation by blocking the activity of the enzymes. However, according to previous reports DNA hyper-methylation occurs due to As stress but that is difficult to explain mechanistically.

7.11 Histone Modifications upon Heavy Metal Exposure

Chromatin which is an array of nucleosomes is formed by condensed linear DNA. Each nucleosome is consisting of an octamer, wrapped by a DNA strand and the octamer is made up of two copies of four histones, H2A, H2B, H3 and H4 (Tang and Ho, 2007). The most susceptible part of histone to get modified is the tail that consists of the flexible N-terminus. Epigenetic modification includes histone acetylation, methylation, ubiquitination, phosphorylation, glycosylation, carboxylation and ADP-ribosylation. Histone acetylation reduces their overall positive charge and thus reducing their affinity to bind with negatively charged DN (Choudhury et al. 2010). Histone de-acetylation and methylation are associated with heterochromatic state and interrupting gene transcription by gene silencing (Tang and Ho, 2007). Histone protein acetylation has important roles in the transcription regulation process as acetylases activate transcription process by enhancing the initiation and elongation stages. Thus the core histones of nucleosome undergo reversible acetylation which is catalyzed by two types of enzymes, i.e. histone acetyltransferase (HAT) and histone deacetylase (HDAC) and this acetylation takes place at selected lysine residues of N-terminal histone tails which is associated with actively transcribed genes (Spencer and Davie 1999; De Ruijter et al. 2003; Tian et al. 2005). Among heavy metals, Nickel has been confirmed for having an inhibitory effect on acetylation of all the histone cores. Reportedly, Nickel suppresses the acetylation of H2A, H2B and H4 via the action of HAT which is associated with silencing of gene expression (Yan et al. 2003; Kang et al. 2004; Moggs and Orphanides 2004). In case of histone H4, Nickel inhibits acetylation by binding to histidine 18 which is located in the conserved N-terminal tail of the histone (Zoroddu et al. 2000, 2002). Nickel exposure also causes cleavage of an octapeptide through truncation of histone H2A, eventually altering the chromatin structure (Karaczyn et al. 2003; Moggs and Orphanides 2004).

Mono, di- and tri-methylation occurs at lysine and arginine residues of histone H3 and H4 (Choudhury et al. 2010). Site-specific methylation either contributes in activation or in repression of gene expression and thus methylation has important roles in many biological processes (Feng et al. 2002). From the previous studies, it has been seen that Nickel chloride induces di-methylation of the 9th lysine residue of histone H3, confirming the influence of heavy metal in the alteration of histone pattern by histone methylation (Chen et al. 2006; Ke et al. 2006).

Histone phosphorylation occurs when negatively charged γ-phosphate, donated by ATP or GTP, added to serine or threonine residues of histone. Histone phosphorylation, mediated by kinases, disrupts the electrostatic interactions between the DNA and the histones by which open chromatin state is produced (Choudhury et al. 2010). Phosphorylation is a cell-cycle-dependent process occurring largely in the M-phase (Schroeder-Reiter et al. 2003; Kallin and Zhang, 2004). Phosphorylation of histone H3 at 10th and 28th serine residues responds to various environmental stress factors (Burkhart et al. 2007). Several heavy metals change the status of phosphorylation of histone proteins. Arsenite exposure causes rapid phosphorylation of 10th serine residue of histone H3 (He et al. 2003, Moggs and Orphanides 2004).

Histone ubiquitination takes place in the lysine residues of histone H1, H2A, H2B and H3 among which histone H2A and H2B are reversibly ubiquitinated. Ubiquitination takes place at specific lysine residues and is linked to transcriptionally active genes. According to previous studies ubiquitination level of H2A and H2B change upon exposure to Ni (Karaczyn et al. 2006) whereas soluble nickel compounds increase the ubiquitination level of the 2 core histones (Ke et al. 2006, Pawlak and Deckert 2007).

The epigenetic modulation in plants involves a methyl-CpG binding protein (MBD) which has a specific domain binding to DNA-methyl group. This domain of MBD protein works as a platform for attachment of other chromatin remodeling proteins, involved in the regulation of gene expression (Berg et al. 2003; Springer and Kaeppler 2005; Yaish 2013). In *Arabidopsis,* AtMBD protein bind to DDM1 (Decrease in DNA-methylation 1) and PRMT1 (Arginine methyltransferase 1) (Scebba et al. 2007; Zemach et al. 2008, Yaish 2013). These DDM1 and PRMT1 proteins control gene expression in plants by modifying chromatin structure. Studies involving an *Arabidopsis* atmbd9 loss-of-function mutant line revealed that the mutant line leads to decreased histone acetylation along with increased DNA methylation at the FLC locus (Peng et al. 2006, Yaish et al. 2009, Yaish 2013). Similarly, study of ddm1 loss-of-function mutant line of *Arabidopsis* displayed increased H3K4 methylation and increased H3K9 methylation along with decreased DNA methylation (Gendrel et al. 2002, Yaish 2013).

Such histone modifications are controlled by different biotic and abiotic stresses. Alteration in the chromatin structure alters the ability of plants to cope up with the adverse environmental conditions as chromatin structure is very important for maintaining stability of genome. According to previous studies, among the heavy metals, Ni exposure leads to de-methylation of histone H3K4 leading to modification in chromatin structure and eventually silencing of the transgene (Chen et al. 2006; Sun et al. 2013; Dutta et al. 2018). Nickel exposure has also been shown to be responsible for de-methylation of H3K9 leading to ubiquitination of HA and H2B histone. In maize seedlings, expansion in cell cycle duration occurs under heavy metal stress, as hyper- and hypo-methylation of specific lysine residues of H3 and H4 histone tails takes place at the promoter region of cell cycle regulatory genes leading to regulation of expression of those genes which eventually leads to growth inhibition of plants (Zhao et al. 2014; Dutta et al. 2018). Cr induces formation of di- and tri-methyl H3K4 and decreases the activity of histone de-methylase. Cr has also been shown to induce de-acetylation of histones and inhibit phosphorylation repressing gene expression (Fragou et al. 2011). Studies on epigenetic modification mediated by copper are limited but it has only been linked to decreased histone acetylation which could lead to gene silencing (Fragou et al. 2011). However, Cd has only been shown to be responsible for DNA methylation so far. But evidences from previous studies suggest that exposure to Cd stress leads to rapid transcription activation due to heritable changes in the structure of chromatin (Martinez-Zamudio and Ha 2011).

7.12 Transgenerational Adaptive Response of Plants under Heavy Metal Stress

Like the immunogens in animal system, plants also carry such stress memory in their genome. In previous studies, it is also described as defense priming in which a plant rapidly responds to a certain stress condition than the previous one. This defense priming is also transmitted across generations, result into better performance of the offspring and plant survival. Though the information related to parental, specifically maternal response to different abiotic stress and their subsequent transmission to next progenies is considerably lacking. Ecologically significant variation in plants generates transgenerational phenotypic plasticity. When parent plant population gains a specific trait for environmental stress adaptation, their offspring exhibit enhanced growth and stress tolerance under the same stress. It was observed that Arabidopsis seedlings of F0 and F1 generation when expose to heat stress (30°C) the adaptive transgenerational response noticed even after two generations (Whittle et al. 2009). Dyer et al. (2010) studied the transgenerational adaptive response exhibit by offsprings of *Aegilops triuncialis* under drought, heavy metal and nutrient stress. Several epigenetic markers are associated with phenotypic plasticity and their modulation of expression is observed under various environmental stress. DNA methylation is considered as a potent adaptive strategy of plants for such stress (Henderson and Jacobsen 2007; Hauser et al. 2011). In rice, the transgenerational changes in methylation pattern were reported under various heavy metal treatment such as Cu, Hg, Cr and Cd along with the reduction of root and shoot length (Ou et al. 2012). With the increase in Cu, Cr and Cd concentration, hypomethylation at specific loci occurred. RT-PCR expression study revealed that two transposable elements, Tos17A and Tos17B show upregulated expression pattern under heavy metal stress which transgenerationally maintained in two successive generations (Ou et al. 2012; Cong et al. 2019). The stress memory in plants under Hg^{2+} stress associated with methylation of two copies of Tos17 is transduced to the offsprings. Moreover, several transporter genes in rice (OsHMAs) and their methylation pattern in parental generation under different heavy metal exposure, successfully transmitted and maintained in the next generation which shows better adaptivity and phenotypic plasticity when expose to similar condition (Cong et al. 2019). Under low humidity, methylation loci coincide with bHLH transcription factor (Tricker et al. 2013a). The core of abiotic stress adaptation and their transgenerational transmission is mainly regulated by RNA-dependent DNA methylation (RdDM) via alteration of gene expression and chromatin dynamics (Bond and Baulcombe 2015). MET1, DRM2, CMT3, NRPD1, NRPE1 are the key component of DNA methylation machinery, respond rapidly under metal stress (Ou et al. 2012). Plethora studies on DNA methylation pattern show that CG, CHG and CHH are the major target of methylation machinery which may be modified in subsequent generations (Vanyushin and Ashapkin 2009; Ou et al. 2012). The imprinted maternal effect of adaptive stress response in plants also regulated by non-coding RNAs (siRNA) (Kinoshita and Seki 2004). During male and female gametogenesis siRNAs act as mobile signal molecules develop transgenerational memory during mitotic and meiotic processes (Hauser et al. 2011). Arabidopsis seedlings treated with Ni^{2+}, Cd^{2+}, Cu^{2+} show reversible increase in homologous recombination frequency up to five generations (Rahavi et al. 2011). The stochastic transgenerational response associated with increased frequency of HR under heavy metal stress needs to be further studied.

Conclusion

The non-biodegradable nature of heavy metals in soil make them a potential threat for plants and humans after entering into the food chain. Their bio-magnification property results into several health hazards to animals and humans. The accumulation of heavy metals in plants occurs through various transporters of root system from contaminated soil and wastewater. It was observed that highest accumulation of heavy metals is observed in leafy vegetables and lowest amount is observed in legumes (Alexander et al. 2006). Intracellular vacuoles accumulate a significant amount of metal ions when grow in heavy metal contaminated soils. These heavy metals disrupt the structural and functional stability of different biomolecules which leads to permanent alteration of physiological and morphological characteristics of plants. Moreover, heavy metals affect major metabolic processes like photosynthesis, respiration, nutrient uptake of several crop plants.

Several studies described the possible phytoremediation mediated approaches and employment of various biosorption strategies which would be beneficial for improvement of plant resistance against heavy metals (Gupta et al. 2015; Ali et al. 2018). But recent advancement in molecular biology tools and omics approaches revealed that plants equipped with different transcriptional and epigenetic machineries which help them to combat against different life-threatening abiotic stresses. Many genes related to heavy metal uptake, sequestration, translocation and chelation are transcriptionally regulated (Dutta et al. 2018). Different families of TFs and non-coding RNAs play crucial role in this transcriptional regulation of stress-responsive genes. Moreover, epigenetic stress-responsive regulators and their subsequent transgenerational transmission helps plants to adapt stress tolerance. These stress-responsive gene regulators should allow us to generate crop plants which are more suitable for cultivation in heavy metal contaminated soils.

Acknowledgments

The authors gratefully acknowledge Council of Scientific and Industrial Research, Govt. of India, (Ref. No. 38(1417)/16/EMR-II, dated: 17/05/2016 to SR), UGC, Govt. of India (Start-Up research grant No. F.30-141/2015 (BSR) and SERB, DST, Govt of India (Ref. No. ECR/2016/000539 to SR) for providing financial supports for performing research related to the topic discussed in this chapter. We apologize to those authors whose work could not be cited due to space limitation.

REFERENCES

Agarwal P, Mitra M, Banerjee S, Roy S (2020) MYB4 transcription factor, a member of R2R3-subfamily of MYB domain protein, regulates cadmium tolerance via enhanced protection against oxidative damage and increases expression of PCS1 and MT1C in Arabidopsis. Plant Sci 297:0168–9452.

Ahmad A, Hadi F, Ali N (2015) Effective phytoextraction of cadmium (Cd) with increasing concentration of total phenolics and free proline in Cannabis sativa (L) plant under various treatments of fertilizers, plant growth regulators and sodium salt. Int J Phytoremed 17:56–65.

Alexander PD, Alloway BJ, Dourado AM (2006) Genotypic variations in the accumulation of Cd, Cu, Pb and Zn exhibited by six commonly grown vegetables. *Environmental pollution*, 144(3), 736–745.

Ali H, Khan E (2018) What are heavy metals? Long-standing controversy over the scientific use of the term 'heavy metals'—proposal of a comprehensive definition. Toxicol Environ Chem 100:6–19.

Ali H, Khan E, Sajad MA (2013) Phytoremediation of heavy metals–concepts and applications. *Chemosphere*, 91(7), 869–881.

Alonso-Peral, MM, Li J, Li Y, Allen RS, Schnippenkoetter W, Ohms S et al. (2010) The microRNA159-regulated GAMYB-like genes inhibit growth and promote programmed cell death in Arabidopsis. Plant Physiol 154:757–771.

Arita A, Costa M (2009) Epigenetics in metal carcinogenesis: nickel, arsenic, chromium and cadmium. Metallomics 1:222–228.

Azevedo R, Rodriguez E (2012) Phytotoxicity of mercury in plants: a review. J Botany 1–6. doi:10.1155/2012/848614.

Babst-Kostecka AA, Waldmann P, Frérot H, Vollenweider P (2016) Plant adaptation to metal polluted environments—Physiological, morphological, and evolutionary insights from *Biscutella laevigata*. Environ Exp Bot 127, 1–13.

Barconi D, Bernardini G, Santucci A (2011) Linking protein oxidation to environmental pollutants: redox proteome approaches. J Proteomics 74(11):2324–2337.

Bartee L, Malagnac F, Bender J (2001) Arabidopsis cmt3 chromomethylase mutations block non-CG methylation and silencing of an endogenous gene. Genes Dev 15(14):1753–1758.

Benbrahim-Tallaa L, Waterland RA, Dill AL, Webber MM, Waalkes MP (2007) Tumor suppressor gene inactivation during cadmium-induced malignant transformation of human prostate cells correlates with overexpression of de novo DNA methyltransferase. Environ Health Perspect 115:1454–9.

Bender J (2004) DNA methylation and epigenetics. Annu Rev Plant Biol 55:41–68.

Berg A, Meza TJ, Mahic M, Thorstensen T, Kristiansen, K, Aalen RB (2003) Ten members of the Arabidopsis gene family encoding methyl-CpG-binding domain proteins are transcriptionally active and at least one, AtMBD11, is crucial for normal development. Nucleic Acids Res 31(18):5291–5304.

Bethke G, Unthan T, Uhrig JF, Pöschl Y, Gust AA, Scheel D et al. (2009) Flg22 regulates the release of an ethylene response factor substrate from MAP kinase 6 in Arabidopsis thaliana via ethylene signaling. Proc Natl Acad Sci USA 106:8067–8072.

Boening DW (2000) Ecological effects, transport, and fate of mercury: a general review. Chemosphere 40(12): 1335–1351.

Bond DM, Baulcombe DC (2015) Epigenetic transitions leading to heritable, RNA-mediated de novo silencing in Arabidopsis thaliana. Proc Natl Acad Sci USA 112:917–922.

Boyko A, Blevins T, Yao Y, Golubov A, Bilichak A, Ilnytskyy Y, Hollander J, Meins F Jr, Kovalchuk I (2010) Transgenerational adaptation of Arabidopsis to stress requires DNA methylation and the function of Dicer-like proteins. PLoS One 5(3):e9514.

Boyko A, Kovalchuk I (2008) Epigenetic control of plant stress response. Environ Mol Mutagen 49: 61–72.

Burkhart BA, Kennett SB, Archer TK (2007) Osmotic stress-dependent repression is mediated by histone H3 phosphorylation and chromatin structure. J Biol Chem 282: 4400–4407.

Cao X, Aufsatz W, Zilberman D, Mette MF, Huang MS, Matzke M, Jacobsen SE (2003) Role of the DRM and CMT3 methyltransferases in RNA-directed DNA methylation. Curr Biol 13:2212–2217.

Carthew RW, Sontheimer EJ (2009) Origins and mechanisms of miRNAs and siRNAs. Cell 136:642–655.

Chan SWL, Henderson IR, Jacobsen SE (2005) Gardening the genome: DNA methylation in Arabidopsis thaliana. Genetics 6:351–360.

Chen H, Ke Q, Kluz T, Yan Y, Costa M (2006) Nickel ions increase histone H3 lysine 9 dimethylation and induce transgene silencing. Mol Cell Biol 26:3728–3737.

Chen J, Yang L, Yan X, Liu Y, Wang R, Fan, T et al. (2016) Zinc-Finger transcription factor ZAT6 positively regulates cadmium tolerance through the glutathione-dependent pathway in Arabidopsis. Plant Physiol 171:707–719.

Chinnusamy V, Zhu JK (2009) Epigenetic regulation of stress responses in plants. Curr Opin Plant Biol 12(2):133–139.

Choudhury S, Cui Y, Klaassen CD (2010) Molecular targets of epigenetic regulation and effectors of environmental influences. Toxicol Appl Pharmacol 245:378–393.

Cicatelli A, Todeschini V, Lingua G, Biondi S, Torrigiani P, Castiglione, S2014) Epigenetic control of heavy metal stress response in mycorrhizal versus non-mycorrhizal poplar plants. Environ. Sci. Pollut. Res. Int. 21(3):1723–1737.

Clarkson TW, Magos L (2006) The toxicology of mercury and its chemical compounds. Critical Reviews in Toxicology 36(8):609–662.

Clemens S (2006) Toxic metal accumulation, responses to exposure and mechanisms of tolerance in plants, Biochimie 88(11):1707–1719.

Cobbett CS (2000) Phytochelatins and their roles in heavy metal detoxification. Plant Physiol 123(3):825–832.

Cokus SJ, Feng S, Zhang X, Chen Z, Merriman B, Haudenschild CD, Pradhan S, Nelson SF, Pellegrini M, Jacobsen SE (2008) Shotgun bisulphite sequencing of the Arabidopsis genome reveals DNA methylation patterning. Nature. 452(7184):215–219.

Cong W, Miao Y, Xu L et al. (2019) Transgenerational memory of gene expression changes induced by heavy metal stress in rice (Oryza sativa L.). BMC Plant Biol 19:282.

Das N, Bhattacharya S, Bhattacharyya S et al. (2017) Identification of alternatively spliced transcripts of rice phytochelatin synthase 2 gene OsPCS2 involved in mitigation of cadmium and arsenic stresses. Plant Mol Biol 94:167–183.

De Ruijter AJ, Van Gennip AH, Caron HN, Kemp S, Van Kuilenburg AB (2003) Histone deacetylase (HDACs): Characterization of the classical HDAC family. Biochem. J 370: 737–749.

Delhaize E, Ma JF, Ryan PR (2012) Transcriptional regulation of aluminium tolerance genes. Trends Plant Sci 17(2012):341–348.

Doi T, Puri P, McCann A, Bannigan J, Thompson J (2011) Epigenetic effect of cadmium on global de novo DNA hypomethylation in the cadmium-induced ventral body wall defect (VBWD) in the chick model. Toxicol Sci 120:475–480.

Dutta S, Mitra M, Agarwal P, Mahapatra K, De S, Sett U, Roy S (2018) Oxidative and genotoxic damages in plants in response to heavy metal stress and maintenance of genome stability. Plant Signaling & Behavior 1–49.

Dyer AR, Brown CS, Espeland EK, Mckay JK, Meimberg H, Rice KJ (2010) The role of adaptive trans-generational plasticity in biological invasions of plants. Evol Appl 3:179–192.

Egger G, Liang G, Aparicio A, Jones PA (2004) Epigenetics in human disease and prospects for epigenetic therapy. Nature 429(6990):457–63. Doi: 10.1038/nature02625. PMID: 15164071.

Eulgem T, Rushton PJ, Robatzek S, Somssich IE (2000) The WRKY superfamily of plant transcription factors. Trends Plant Sci 5(5):199–206.

Fan SK, Fang XZ, Guan MY, Ye YQ, Lin XY, Du ST, Jin CW (2014) Exogenous abscisic acid application decreases cadmium accumulation in Arabidopsis plants, which is associated with the inhibition of IRT1- mediated cadmium uptake. Front Plant Sci 5(2014):721.

Farinati S, DalCorso G, Varotto S, Furini A (2010) The brassica juncea BjCdR15, an ortholog of arabidopsis TGA3, is a regulator of cadmium uptake, transport and accumulation in shoots and confers cadmium tolerance in transgenic plants. New Phytol 185(2010):964–978.

Fasani E, DalCorso G, Costa A, Zenoni S, Furini A (2019) The Arabidopsis thaliana transcription factor MYB59 regulates calcium signalling during plant growth and stress response. Plant Mol Biol 99(6):517–534.

Feng Q, Wang H, Ng HH, Erdjument-Bromage H, Tempst P, Struhl K, Zhang Y (2002) Methylation of H3-lysine 79 is mediated by a new family of HMTases without a SET domain. Curr. Biol. 12:1052–1058.

Fragou D, Fragou A, Kouidou S, Njau S, Kovatsi L (2011) Epigenetic mechanisms in metal toxicity. Toxicology Mechanisms and Methods, 21(4):343–352.

Gendrel AV, Lippman Z, Yordan C, Colot V, Martienssen RA (2002) Dependence of heterochromatic histone H3 methylation patterns on the Arabidopsis gene DDM1. Science 297(5588):1871–1873. doi:10.1126/science.1074950.

Ghosh, M, Singh, Satendra. (2005). A Review on Phytoremediation of Heavy Metals and Utilization of Its By-products. Applied Ecology and Environmental Research. 3. 10.15666/aeer/0301_001018.

Goyer RA. (1997). Toxic and essential metal interactions. Annu Rev Nutr 17:37–50.

Gracz J (2016) Alternative splicing in plant stress response. BioTechnologia J Biotechnol Comput Biol Bionanotechnol 97(1):9–17.

Greco M, Chiappetta A, Bruno L, Bitonti MB (2012) In Posidoniaoceanica cadmium induces changes in DNA methylation and chromatin patterning. J. Exp. Bot 63(2):695–709.

Gupta, Vinod, Nayak, Arunima, Agarwal, Shilpi. (2015). Bioadsorbents for remediation of heavy metals: Current status and their future prospects. Environmental Engineering Research. 20. 1-18. 10.4491/eer.2015.018.

Haap T, Schwarz S, Köhler HR (2016) Metallothionein and Hsp70 trade-off against one another in Daphnia magna cross-tolerance to cadmium and heat stress. Aquat. Toxicol 170:112–119.

Hall JL (2002) Cellular mechanisms for heavy metal detoxification and tolerance. J Exp Botany 53(366):1–11.

Hauser MT, Aufsatz W, Jonak C, Luschnig C (2011) Transgenerational epigenetic inheritance in plants. Biochim Biophys Acta 1809:459–468.

He G, Zhu X, Elling AA, Chen L, Wang X, Guo L, Liang M, He H, Zhang H, Chen F, Qi Y, Chen R, Deng XW (2010) Global epigenetic and transcriptional trends among two rice subspecies and their reciprocal hybrids. Plant Cell 22:17–33.

He X, Zheng W, Cao F, Wu F (2016) Identification and comparative analysis of the microRNA transcriptome in roots of two contrasting tobacco genotypes in response to cadmium stress. Sci Rep 6:32805.

He Z, Ma WY, Liu G, Zhang Y, Bode AM, Dong Z (2003) Arsenite-induced phosphorylation of histone H3 at serine 10 is mediated by Akt1, extracellular signal-regulated kinase 2, and p90ribosomal S6kinase 2 but not mitogen- and stress activated protein kinase 1. J Biol Chem 278:10588–10593.

Henderson IR, Deleris A, Wong W, Zhong X, Chin HG, Horwitz GA, Kelly KA, Pradhan S, Jacobsen SE (2010) The de-novo cytosine methyltransferase DRM2 requires intact UBA domains and a catalytically mutated paralog DRM3 during RNA directed DNA methylation in Arabidopsis thaliana. PLoS Genet 6(10).

Henderson IR, Jacobsen SE (2007) Epigenetic inheritance in plants. *Nature* 447, 418–424.

Hossain MA, Hasanuzzaman M, Fujita M (2009) Stress-induced changes of methylglyoxal level and glyoxalase I activity in pumpkin seedlings and cDNA cloning of glyoxalase I gene. Austr J Crop Sci 3(2):53–64.

Hossain MA, Hasanuzzaman M, Fujita M (2010) Up-regulation of antioxidant and glyoxalase systems by exogenous glycinebetaine and proline in mung bean confer tolerance to cadmium stress. Physiol Mol Biol Plants 16(3):259–272.

Hossain MA, Piyatida P, da Silva JAT, Fujita M (2012) Molecular mechanism of heavy metal toxicity and tolerance in plants: central role of glutathione in detoxification of reactive oxygen species and methylglyoxal and in heavy metal chelation. J Botany 1–37.

Huang C, Zhou J, Jie Y, Xing H, Zhong Y, She W et al. (2016) A ramie (Boehmeria nivea) bZIP transcription factor BnbZIP3 positively regulates drought, salinity and heavy metal tolerance. Mol Breed 36:120.

Huang D, Zhang Y, Qi Y, Chen C, Ji W (2008) Global DNA hypomethylation, rather than reactive oxygen species (ROS), a potential facilitator of cadmium-stimulated K562 cell proliferation. Toxicol Lett 179:43–47.

Huang J, Zhang Y, Peng JS, Zhong C, Yi HY, Ow DW et al. (2012) Fission yeast HMT1 lowers seed cadmium through phytochelatin-dependent vacuolar sequestration in Arabidopsis. Plant Physiol 158:1779–1788.

Jacobson T, Navarrete C, Sharma SK, Sideri TC, Ibstedt S, Priya S et al. (2012) Arsenite interferes with protein folding and triggers formation of protein aggregates in yeast. J Cell Sci 125:5073–5083.

Jalmi SK, Bhagat PK, Verma D, Noryang S, Tayyeba S, Singh K, Sharma D, Sinha AK (2018) Traversing the links between heavy metal stress and plant signaling. Front Plant Sci 9:12.

Jiang G, Xu L, Song S, Zhu C, Wu Q, Zhang L et al. (2008) Effects of long-term low-dose cadmium exposure on genomic DNA methylation in human embryo lung fibroblast cells. Toxicology 244:49–55.

Jozefczak M, Remans T, Vangronsveld J, Cuypers A. Glutathione is a key player in metal-induced oxidative stress defences. International Journal of MolecularSciences 13(2012): 3145–3175.

Kallin E, Zhang Y (2004) Chromatin remodelling. In Lennarz WJ, Lane MD (eds) Encyclopaedia of biological chemistry. Academic Press, Elsevier Inc, 456–462.

Kang J, Lin C, Chen J, Liu Q (2004) Copper induces histone hypoacetylation through directly inhibiting histone acetyltransferase activity. Chem. Biol. Interact. 148: 115–123.

Karaczyn AA, Bal W, North SL, Bare RM, Hoang VM, Fisher RJ, Kasprzak KS (2003) The octapeptide end of the C-terminal tail of histone H2a is cleaved off cells exposed to carcinogenic nickel (II). Chem. Res. Toxicol. 16: 1555–1559.

Karaczyn AA, Golebiowski F, Kasprzak KS (2006) Ni(II) affects ubiquitination of core histones H2B and H2A. Exp. Cell Res. 312: 3252–3259.

Ke Q, Davidson T, Chen H, Kluz T, Costa M (2006) Alterations of histone modifications and transgene silencing by nickel chloride. Carcinogenesis 27: 1481–1488.

Kupper H, Andresen E (2016) Mechanisms of metal toxicity in plants. Metallomics 8:269–285.

Kushwaha A, Rani R, Kumar S, Gautam A (2015) Heavy metal detoxification and tolerance mechanisms in plants: Implications for phytoremediation. Environ. Rev 24: 39–51.

Labra M, Grassi F, Imazio S, Di Fabio T, Citterio S, Sgorbati S, Agradi E (2004) Genetic and DNA-methylation changes induced by potassium dichromate in *Brassica napus* L. Chemosphere 54:1049–1058.

Laloum T, Martín G, Duque P (2018) Alternative splicing control of abiotic stress responses. Trends Plant Sci 23(2):140–150.

Leyval C, Turnau K, Haselwandter K (1997) Effect of heavy metal pollution on mycorrhizal colonization and function: physiological, ecological and applied aspects. Mycorrhiza 7(3): 139–153.

Li AM, Yu BY, Chen FH, Gan HY, Yuan JG, Qiu R, Huang JC, Yang ZY, Xu ZF (2009) characterization of the Sesbania rostrata phytochelatin synthase gene: alternative splicing and function of four isoforms. Int J Mol Sci 10:3269–3282.

Li S, Gao J, Yin K, Wang R, Wang C, Petersen M et al. (2016) MYB75 Phosphorylation by MPK4 is required for light-induced anthocyanin accumulation in Arabidopsis. Plant Cell 16:130.

Lima J, Arenhart RA, Margis-Pinheiro M, Margis R (2011) Aluminum triggers broad changes in microRNA expression in rice roots. Genet Mol Res 10:2817–2832.

Lin XY, Ye YQ, Fan SK, Jin CW, Zheng SJ (2016) Increased sucrose accumulation regulates iron-deficiency responses by promoting auxin signaling in Arabidopsis Plants. Plant Physiology 170:907–920.

Lindroth AM, Cao X, Jackson JP, Zilberman D, McCallum CM, Henikoff S, Jacobsen SE (2001) Requirement of CHROMOMETHYLASE3 for maintenance of CpXpG methylation. Science 292:2077–2080.

Liu X, Chu Z (2015) Genome-wide evolutionary characterization and analysis of bZIP transcription factors and their expression profiles in response to multiple abiotic stresses in Brachypodium distachyon. BMC Genomics 16:227.

Ma C, Burd S, Lers A (2015) miR408 is involved in abiotic stress responses in Arabidopsis. Plant J 84:169–187.

Martinez-Zamudio R, Ha HC (2011) Environmental epigenetics in metal exposure. Epigenetics, 6(7):820–827. doi:10.4161/epi.6.7.16250.

Meharg AA, Rahman M (2003) Arsenic contamination of Bangladesh paddy field soils: Implications for rice contribution to arsenic consumption. Environ. Sci. Technol 37: 229–234.

Michalak A (2006) Phenolic compounds and their antioxidant activity in plants growing under heavy metal stress. Polish J Environ Stud 15(4):523–530.

Moggs JG, Orphanides G (2004) The role of chromatin in molecular mechanisms of toxicity. Toxicol. Sci. 80: 218–224.

Molinier J, Oakeley EJ, Niederhauser O, Kovalchuk I, Hohn B (2005) Dynamic response of plant genome to ultraviolet radiation and other genotoxic stresses. Mutat. Res. 2005; 571: 235–47.

Myrset AH, Bostad A, Jamin N, Lirsac PN, Toma F, Gabrielsen OS (1993) DNA and redox state induced conformational changes in the DNA-binding domain of the Myb oncoprotein. EMBO J 12:4625–4633.

Navari-Izzo F (1998) Thylakoid-bound and stromal antioxidative enzymes in wheat treated with excess copper. Physiologia Plantarum, 104(4):630–638.

Noman A, Aqeel M (2017) miRNA- based heavy metal homeostasis and plant growth. Environ Sci Pollut Res Int 24:10068–10082.

Oono Y, Yazawa T, Kawahara Y, Kanamori H, Kobayashi F, Sasaki HH, Mori S, Wu J, Handa H, Itoh T, Matsumoto T et al. (2014) Genomewide transcriptome analysis reveals that cadmium stress signaling controls the expression of genes in drought stress signal pathways in rice. PLoS ONE 9(5):e96946.

Opdenakker K, Remans T, Vangronsveld J, Cuypers A (2012) Mitogen-activated protein (MAP) kinases in plant metal stress: regulation and responses in comparison to other biotic and abiotic stresses. Int J Mol Sci 13:7828–7853.

Ou X, Zhang Y, Xu C et al. (2012) Transgenerational inheritance of modified DNA methylation patterns and enhanced tolerance induced by heavy metal stress in rice (Oryza sativa L.). PLoS One 7(9):e41143.

Pandey G, Sharma N, Sahu PP, Prasad M (2016) Chromatin-Based Epigenetic Regulation of Plant Abiotic Stress Response. Curr Genomics 17(6):490–498. Doi: 10.2174/1389202917666160520103914. PMID: 28217005; PMCID: PMC5282600.

Park J, Song WY, Ko D, Eom Y, Hansen TH, Schiller M et al. (2012) The phytochelatin transporters AtABCC1 and AtABCC2 mediate tolerance to cadmium and mercury. Plant J 69(2012):278–288.

Parmar P, Dave B, Sudhir A, Panchal K, Subramanian RB (2013) Physiological, biochemical and molecular response of plants against heavy metals stress. Int J Curr Res 5(1):080–089.

Pawlak S, Deckert J (2007) Histone modifications under environmental stress. Biological Lett 44:65–73.

Peng M, Cui Y, Bi YM, Rothstein SJ (2006) AtMBD9: a protein with a methyl-CpG-binding domain regulates flowering time and shoot branching in Arabidopsis. Plant J Cell Mol Biol 46(2):282–296.

Peters AH, Schübeler D (2005) Methylation of histones: playing memory with DNA. Curr Opin Cell Biol 17:230–238.

Pourrut B, Shahid M, Dumat C, Winterton P, Pinelli E (2011) Lead Uptake, Toxicity, and Detoxification in Plants. In Whitacre DM (ed) Reviews of environmental contamination and toxicology. Springer: New York, NY, USA, 213:113–136.

Qiao W, Fan L (2011) Epigenetics, a mode for plants to respond to abiotic stresses. Front. Biol. (Beijing) 6:477–481.

Rahavi MR, Migicovsky Z, Titov V, Kovalchuk I (2011) Transgenerational adaptation to heavy metal salts in Arabidopsis. Front Plant Sci 2:91.

Raldugina GN, Maree M, Mattana M et al. (2018) Expression of rice *OsMyb4* transcription factor improves tolerance to copper or zinc in canola plants. Biol Plant 62:511–520.

Ramos J, Naya L, Gay M, Abian J, Becana M (2008) Functional characterization of an unusual phytochelatin synthase, LjPCS3, of Lotus japonicus. Plant Physiol 148:536–545.

Rascio N, Navari-Izzo F (2011) Heavy metal hyperaccumulating plants: how and why do they do it? And what makes them so interesting? Plant Sci 180(2):169–181.

Remy E, Cabrito TR, Batista RA, Hussein MA, Teixeira MC, Athanasiadis A et al. (2014) Intron retention in the 5' UTR of the novel ZIF2 transporter enhances translation to promote zinc tolerance in Arabidopsis. PLoS Genet 10:e1004375.

Rengel Z (1999) Physiological mechanisms underlying differential nutrient efficiency of crop genotypes. In:Rengel Z (ed) Mineral nutrition of crops—fundamental mechanisms and implications. Food Products Press 1:227–265.

Rhee JS, Raisuddin S, Lee KW, Seo JS, Ki JS, Kim IC et al. (2009) Heat shock protein (Hsp) gene responses of the intertidal copepod Tigriopus japonicus to environmental toxicants. Comp Biochem Physiol C Toxicol Pharmacol 149:104–112.

Ritter A, Ubertini M, Romac S, Gaillard F, Delage L, Mann A et al. (2010) Copper stress proteomics highlights local adaptation of two strains of the model brown alga Ectocarpus siliculosus. Proteomics 10:2074–2088.

Rozhon W, Baubec T, Mayerhofer J, Mittelsten Scheid O, Jonak C (2008) Rapid quantification of global DNA methylation by isocratic cation exchange high-performance liquid chromatography. Anal Biochem 375:354–360.

Salla V, Hardaway CJ, Sneddon J (2011) Preliminary investigation of Spartina alterniflora for phytoextraction of selected heavy metals in soils from Southwest Louisiana. Microchem J 97:207–212.

Sawaki Y, Iuchi S, Kobayashi Y, Kobayashi Y, Ikka T, Sakurai N et al. (2009) STOP1 regulates multiple genes that protect Arabidopsis from proton and aluminum toxicities. Plant Physiol 150:281–294.

Scebba F, De Bastiani M, Bernacchia G, Andreucci A, Galli A, Pitto L (2007) PRMT11: a new Arabidopsis MBD7 protein partner with arginine methyltransferase activity. Plant J 52(2):210–222.

Schober SE, Sinks TH, Jones RL, Bolger PM, McDowell M, Osterloh JES, Garrett RA, Canady CF, Dillon Y Sun (2003) Blood mercury levels in US children and women of childbearing age, 1999–2000. Jama 289(13):1667–1674.

Schroeder-Reiter E, Houben A, Wanner G (2003) Immunogold labelling of chromosomes for scanning electron microscopic: A closer look at phosphorylated histone H3 in mitotic metaphase chromosomes of Hordeum vulgare. Chromosome Res. 11: 585–596.

Sengar RS, Gautam M, Sengar RS, Garg SK, Sengar K, Chaudhary R (2008) Lead stress effects on physiobiochemical activities of higher plants. Rev. Env. Contam. Toxicol. 196:73–93.

Seth CS, Remans T, Keunen E et al. (2012) Phytoextraction of toxic metals: a central role for glutathione. Plant, Cell and Environment 35(2):334–346.

Shafiq S, Khan AR (2015) Plant Epigenetics and Crop Improvement. In Barh D, Khan MS, Davies E (eds) PlantOmics: The omics of plant science. Springer: New Delhi, India, 157–179.

Shafiq S, Zeb Q, Ali A, Sajjad Y, Nazir R, Widemann E, Liu L (2019) Lead, cadmium and zinc phytotoxicity alter DNA methylation levels to confer heavy metal tolerance in wheat. Int J Mol Sci 20(19):4676. doi: 10.3390/ijms20194676.

Shahid M, Khalid S, Abbas G, Shahid N, Nadeem M, Sabir M, Aslam M, Dumat C (2015) Heavy metal stress and crop productivity. In Hakeem KR (ed) Crop production and global environmental issues. Springer International Publishing 1–25.

Sharma P, Dubey RS (2007) Involvement of oxidative stress and role of antioxidative defense system in growing rice seedlings exposed to toxic concentrations of aluminium. Plant Cell Rep 26(11):2027–2038.

Siddiqui S (2012) Lead induced genotoxicity in Vigna mungo var. HD94. J Sau Soc Agric Sci 11:107–112.

Singh K, Foley RC, Oñate-Sánchez L (2002) Transcription factors in plant defense and stress responses. Curr Opin Plant Biol 5:430–436.

Singh P, Siddiqui H, Sami F, Arif Y, Bajguz A, Hayat S (2020) Cadmium: A Threatening Agent for Plants. Plant Resp Soil Pollution 59–88.

Singla-Pareek SL, Yadav SK, Pareek A, Reddy MK, Sopory SK (2006) Transgenic tobacco overexpressing glyoxalase pathway enzymes grow and set viable seeds in zinc-spiked soils. Plant Physiol 140(2):613–623.

Spencer VA, Davie JR (1999) Role of covalent modifications of histones in regulating gene expression. Gene 240: 1–12.

Springer NM, Kaeppler SM (2005) Evolutionary divergence of monocot and dicot methyl-CpG-binding domain proteins. Plant Physiol 138(1):92–104.

Sun H, Shamy M, Costa M (2013) Nickel and epigenetic gene silencing. Genes. 4: 583–95. doi: 10.3390/genes4040583.

Suzuki N, Koizumi N, Sano H (2001) Screening of cadmium-responsive genes in Arabidopsis thaliana. Plant Cell Environ 24(2001):1177–1188.

Takiguchi M, Achanzar WE, Qu W, Li G, Waalkes MP (2003) Effects of cadmium on DNA-(Cytosine-5) methyltransferase activity and DNA methylation status during cadmium-induced cellular transformation. Exp Cell Res 286:355–365.

Tang M, Mao D, Xu L, Li D, Song S, Chen C (2014) Integrated analysis of miRNA and mRNA expression profiles in response to Cd exposure in rice seedlings. BMC Genomics 15:835.

Tang WY, Ho SM (2007) Epigenetic reprogramming and imprinting in origins of disease. Rev Endocr Metab Disord 8:173–182.

Tang Z, Lv Y, Chen F, Zhang W, Rosen BP, Zhao FJ (2016) Arsenic Methylation in *Arabidopsis thaliana* expressing an algal arsenite methyltransferase gene increases arsenic phytotoxicity. Agri Food Chem 64(13):2674–2681.

Tian L, Fong MP, Wang JJ, Wei NE, Jiang H, Doerge RW, Chen ZJ (2005) Reversible histone acetylation and deacetylation mediate genome-wide, promoter-dependent and locus specific changes in gene expression during plant development. Genetics 169:337–345.

Tricker PJ, Lopez CM, Gibbings G, Hadley P, Wilkinson MJ (2013) Transgenerational, dynamic methylation of stomata genes in response to low relative humidity. Int J Mol Sci 14:6674–6689.

Trinh NN, Huang TL, Chi WC, Fu SF, Chen CC, Huang HJ (2014) Chromium stress response effect on signal transduction and expression of signaling genes in rice. Physiol Plant 150:205–224.

Tsutsui T, Yamaji N, Ma J (2011) Identification of a cis-acting element of ART1, a C2H2-type zinc-finger transcription factor for aluminum tolerance in rice. Plant Physiol 156:925–931.

Uraguchi S, Mori S, Kuramata M, Kawasaki A, Arao T, Ishikawa S (2009) Root-to-shoot Cd translocation via the xylem is the major process determining shoot and grain cadmium accumulation in rice. J Exp Bot 60:2677–2688.

Valko M, Morris H, Cronin MTD (2005) Metals, toxicity and oxidative stress. Curr Med Chem 12(10):1161–1208.

van de Mortel JE, Schat H, Moerland PD, Ver Loren van Themaat, E, Van Der Ent S, Blankestijn H et al. (2008) Expression differences for genes involved in lignin, glutathione and sulphate metabolism in response to cadmium in Arabidopsis thaliana and the related Zn/Cd hyperaccumulator Thlaspi caerulescens. Plant Cell Environ 31:301–324.

van Dijk K, Ding Y, Malkaram S, et al. Dynamic changes in genome-wide histone H3 lysine 4 methylation patterns in response to dehydration stress in Arabidopsis thaliana. BMC Plant Biology. 2010 Nov;10:238. DOI: 10.1186/1471-2229-10-238. PMID: 21050490; PMCID: PMC3095321.

Vanyushin BF, Ashapkin VV (2009) DNA methylation in plants. Nova Science Publishers Inc, NY.

Vanyushin BF, Ashapkin VV (2011) DNA methylation in higher plants: past, present and future. Biochim Biophys Acta 1809:360–368.

Waalkes MP (2000) Cadmium carcinogenesis in review. J Inorg Biochem 79:241–244.

Waalkes MP (2003) Cadmium carcinogenesis. Mutat Res 533:107–120.

Wada Y, Miyamoto K, Kusano T et al. (2004) Association between up-regulation of stress responsive genes and hypomethylation of genomic DNA in tobacco plants. Mol Genet Genomics 271:658–666.

Wang FZ, Chen MX, Yu LJ, Xie LJ, Yuan LB, Qi H et al. (2017) OsARM1, an R2R3 MYB transcription factor, is involved in regulation of the response to arsenic stress in rice. Front Plant Sci 30:1868.

Wang H, Wang H, Shao H, Tang X (2016) Recent advances in utilizing transcription factors to improve plant abiotic stress tolerance by transgenic technology. Front Plant Sci 7:67.

Wang X, Elling AA, Li X, Li N, Peng Z, He G, Sun H, Qi Y, Liu XS, Deng XW (2009) Genome-Wide and organ-specific landscapes of epigenetic modifications and their relationships to mRNA and small RNA transcriptomes in maize. Plant Cell 21:1053–1069.

Weber M, Trampczynska A, Clemens S (2006) Comparative transcriptome analysis of toxic metal responses in Arabidopsis thaliana and the Cd2+- hypertolerant facultative metallophyte Arabidopsis halleri. Plant Cell Environ 29:950–963.

Whittle CA, Otto SP, Johnston MO, Krochko JE (2009) Adaptive epigenetic memory of ancestral temperature regime in Arabidopsis thaliana. Botany 87:650–657.

Wu H, Chen C, Du J, Liu H, Cui Y, Zhang Y, He Y, Wang Y, Chu C, Feng Z et al. (2012) Co-overexpression FIT with AtbHLH38 or AtbHLH39 in Arabidopsis-enhance Cadmium tolerance via increased cadmium sequestration in roots and improved iron homeostasis of shoots. Plant Physiol 158:790–800.

Yadav SK (2010) Heavy metals toxicity in plants: an overview on the role of glutathione and phytochelatins in heavy metal stress tolerance of plants. South African Journal of Botany 76 (2):167–179.

Yaish M.W. (2013) DNA Methylation-Associated Epigenetic Changes in Stress Tolerance of Plants. In: Rout G., Das A. (eds) Molecular Stress Physiology of Plants. Springer, India. https://doi.org/10.1007/978-81-322-0807-5_17.

Yaish MW, Colasanti J, Rothstein SJ (2011) The role of epigenetic processes in controlling flowering time in plants exposed to stress. J Exp Bot 62(11): 3727–3735.

Yaish MW, Peng M, Rothstein SJ (2009) AtMBD9 modulates Arabidopsis development through the dual epigenetic pathways of DNA methylation and histone acetylation. Plant J 59(1):123–135.

Yan Y, Kluz T, Zhang P, Chen HB, Costa M (2003) Analysis of specific lysine histone H3 and H4 acetylation and methylation status in clones of cells with a gene silenced by nickel exposure. Toxicol. Appl. Pharmacol. 190: 272–277.

Yang XE, Jin XF, Feng Y, Islam E (2005) Molecular mechanisms and genetic basis of heavy metal tolerance/hyperaccumulation in plants. J Integr Plant Biol 47(9): 1025–1035.

Yanhui C, Xiaoyuan Y, Kun H, Meihua L, Jigang L, Zhaofeng G et al. (2006) The MYB transcription factor superfamily of Arabidopsis: expression analysis and phylogenetic comparison with the rice MYB family. Plant Mol Biol 60:107–124.

Zemach A, Gaspan O, Grafi G (2008) The three methylCpG-binding domains of AtMBD7 control its subnuclear localization and mobility. J Biol Chem 283(13):8406–8411.

Zenk M (1996) heavy metal detoxification in higher plants–a review. Gene 179(21), 21–30.

Zhang B, Georgiev O, Hagmann M et al. (2003) Activity of metal-responsive transcription factor 1 by toxic heavy metals and H2O2 in vitro is modulated by metallothionein. Mol Cell Biol 23(23):8471–8485.

Zhang M, Kimatu JN, Xu K, Liu B (2010) DNA cytosine methylation in plant development. J. Genet. Genom 37:1–12.

Zhang P, Wang R, Ju Q, Li W, Tran LSP, Xu J (2019) The R2R3-MYB Transcription Factor MYB49 Regulates Cadmium Accumulation Plant Physiology 180(1):529–542.

Zhang W, Bone JR, Edomondson DG, Turner BM, Roth SY (1998) Essential and redundant functions of histone acetylation revealed mutation of target lysine and loss of the Gcn5 p acetyltransferase. EMBO J 17:3155–3167. doi:10.1093/emboj/17.11.3155.

Zhang X, Yazaki J, Sundaresan A, Cokus S, Chan SW, Chen H, Henderson IR, Shinn P, Pellegrini M, Jacobsen SE, Ecker JR (2006) Genome-wide high-resolution mapping and functional analysis of DNA methylation in Arabidopsis. Cell 126(6):1189–1201. doi:10.1016/j.cell.2006.08.003.

Zhao L, Wang P, Hou H, Zhang H, Wang Y, Yan S, Huang Y, Li H, Tan J, Hu A et al. (2014) Transcriptional regulation of cell cycle genes in response to abiotic stresses correlates with dynamic changes in histone modifications in maize. PLoS ONE 9(8):e106070.doi:10.1371/journal.pone.0106070.

Zhong X, Hale CJ, Nguyen M, Ausin I, Groth M, Hetzel J, Vashisht AA, Henderson IR, Jacobsen Wohlschlegel JA (2015) Domains rearranged methyltransferasE3 controls DNA methylation and regulates RNA polymerase V transcript abundance in Arabidopsis. Proc Natl Acad Sci 112(3): 911–916.

Zhou ZS, Zeng HQ, Liu ZP, Yang ZM (2012) Genome-wide identification of Medicago truncatula microRNAs and their targets reveals their differential regulation by heavy metal. Plant Cell Environ 35:86–99.

Zhu YG, Sun GX, Lei M, Teng M, Liu YX, Chen NC, Wang LH, Carey AM, Deacon C, Raab A, Meharg AA, Williams PN (2008) High percentage inorganic arsenic content of mining impacted and nonimpacted Chinese rice. Environ. Sci. Technol 42:5008–5013.

Zilberman D, Gehring M, Tran RK, Ballinger T, Henikoff S (2007) Genome-wide analysis of Arabidopsis thaliana DNA methylation uncovers an interdependence between methylation and transcription. Nat Genet 39 (1):61–69.

Zornoza P, vÁzQuEz S, EstEBan E, FErnÁnDEz-pascual M, carpEna R (2002) Cadmium-stress in nodulated white lupin: strategies to avoid toxicity. Plant Physiol Biochem 40:1003.

Zoroddu MA, Kowalik-Jankowska T, Kozlowski H, Molinari H, Salnikow K, Broday L, Costa M (2000) Interaction of Ni(II) and Cu(II) with a metal binding sequence of histone H4: AKRHRK, a model of the H4 tail. Biochem. Biophys Acta 1475:163–168.

Zoroddu MA, Schinocca L, Kowalik-Jankowska T, Kozlowski H, Salnikow K, Costa M (2002) Molecular mechanisms in nickel carcinogenesis: modelling Ni(II) binding site in histone H4. Environ. Health Persp 110:719–723.

8

Emerging Roles of Osmoprotectants in Heavy Metal Stress Tolerance in Plants

Akbar Hossain
Bangladesh Wheat and Maize Research Institute, Dinajpur, Bangladesh

Zahoor Ahmad
Department of Botany, University of Central Punjab, Bahawalpur Campus, Punjab Group of Colleges, Pakistan

Muhammad Adeel
Institute of Soil and Environmental Science, University of Agriculture, Faisalabad, Punjab, Pakistan

Md. Atikur Rahman
Species Research Center, Bangladesh Agricultural Research Institute (BARI), Bogra, Bangladesh

Md. Jahangir Alam
On-Farm Research Division, Bangladesh Agricultural Research Institute, Gaibandha, Bangladesh

Sharif Ahmed
International Rice Research Institute, Bangladesh Office, Dhaka, Bangladesh

Tariq Aftab
Department of Botany, Aligarh Muslim University, Aligarh, India

CONTENTS

Abbreviations

Al	Aluminum
AsA	Ascorbic acid
BADH	Betaine aldehyde dehydrogenase
Cd	Cadmium
COD	Choline oxidase
Cr	Chromium
CS	Compatible solutes
Cu	Copper
DNA	Deoxy ribonucleic acid
EL	Electrolyte leakage

DOI: 10.1201/9781003155089-8

Fe	Iron
GB	Glycine betaine
GR	Glutathione reductase
GSH	Glutathione
H2O2	Hydrogen peroxide
Hg	Mercury
HMs	Heavy metals
MDA	Malondialdehyde
MG	Methylglyoxal
Mn	Manganese
MTs	Metallothioneins
Ni	Nickel
OSP	Osmoprotectants
PAs	Polyamines
PCs	Phytochelatins
ROS	Reactive oxygen species
SA	Salicylic acid
UV	Ultraviolet radiation
Zn	Zinc

8.1 Introduction

Among the abiotic stresses, heavy metals (HMs) toxicity is of wide importance, as urbanization and industrialization caused a major concern for environmentalists through the inclusion of HMs in the environment (Kohli et al. 2017; Masindi and Muedi 2018). At the toxic level, these elements generate reactive oxygen species (ROS) (Ghori et al. 2019) by interacting with cellular molecules (Ruciniska-Sobkowiak, 2010; Shahid et al. 2014), that results in reduction in the growth and productivity of plants (Ali et al. 2020). The activity of HMs is divided in two categories based on their physicochemical properties that are redox-active and non-redox active groups. Iron (Fe), manganese (Mn), copper (Cu) and chromium (Cr) are the examples of redox-active group and aluminum (Al), zinc (Zn), mercury (Hg), nickel (Ni) and cadmium (Cd) are the examples of the non-redox active group (Valko et al. 2005; Jozefczak et al. 2012). The redox metals can directly generate oxidative injury via undergoing HaberWeiss and Fenton reactions, which leads to the aforementioned production of ROS or oxygen free radicals species in plants, resulting in cell homeostasis disruption, DNA strand breakage, defragmentation of proteins, or cell membrane and damage to photosynthetic pigments, which may trigger cell death (Schutzendübel and Polle 2002; Flora 2009). In contrast, non-redox active metals indirectly inflict oxidative stress via multiple mechanisms including glutathione depletion, binding to sulfhydryl groups of proteins (Valko et al. 2005), inhibiting antioxidative enzymes, or inducing ROS-producing enzymes like NADPH oxidases (Bielen et al. 2013).

For mitigating the worst effects of HMs on plants, the use of osmoprotectants (OSP) is a very interesting approach (Hanson and Scott 1980). Compatible solutes (CS) are also involved in the osmoregulation in plants. CS are usually produced in plants that do not disturb cellular metabolism even at high concentration, these are low molecular weight organic compounds. Much attention has been given to their role as the adaptive measure against abiotic stresses. Apart from osmotic adjustment, in enzyme protection, membrane protection and mitigation of free radicals, some sulfur and nitrogen-containing compounds, amino acids (proline, glycine, alanine, etc.) and sugars (sucrose, fructose, trehalose, etc.) play imperative roles (Handa et al. 2018).

For example, glycine betaine (GB), that is an osmoprotectant is very useful in minimizing the oxidative stress by minimizing the uptake of HMs, under HMs stress it also improves nutrient uptake, antioxidant enzyme activities, photosynthesis and growth of plants. Besides, GB also regulates to activate the glutathione (GSH), ascorbic acid (AsA) and glutathione reductase (GR) in plants under HM stress (Ali et al. 2020). In higher plants, GB is synthesized in chloroplast from serine through ethanolamine, choline and betaine aldehyde (Hanson and Scott 1980). The various plant generates GB under stressful conditions. GB is a water-soluble, non-toxic and environmentally safe compound. Under different stress conditions, GB is engaged in the fortification of plants, like under HMs (Ali et al. 2015; Farooq et al. 2016; Jabeen et al. 2016; Xalxo et al. 2017; Yao et al. 2018), drought (Gupta et al. 2013; Yang et al. 2016; Liu et al. 2017) and salinity (Tian et al. 2017; Khan et al. 2018) stress. GB application, by protecting the photosynthesis process, hunting ROS and stabilizing the proteins, lowers HMs stress on plants (Nusrat et al. 2014). In *Lemna gibba*, exogenously applied GB reduced Cd toxicity (Duman et al. 2011).

Proline also acts as an elevator of some biotic and abiotic stress like drought, salinity, HMs, ultraviolet radiation (UV) and some of the oxidative stress (Szabados and Savoure, 2010). It also plays an important role in the adaptation of plants at stressed condition, recovery from the stressed condition as well as signaling to plants under the unfavorable condition for plants (Fidalgo et al. 2013). Several studies demonstrated that the metabolism and accumulation of proline are important for plant growth and development as well as to survive against adverse environmental conditions (Hong et al. 2000; Mattioli et al. 2008; Szekely et al. 2008).

Similar to other OSP, salicylic acid (SA) also plays an important role as phytohormone for regulating the metabolic pathways. SA when applied at an optimal dose, enhances tolerance of plants to HMs stress through regulating several metal-chelating compounds, secondary metabolites, osmolytes and components of antioxidative defense cascade (Kohli et al. 2017).

Considering the important aspects of OSP, the present chapter provides a comprehensive overview of the OSP-based survival adaptability of plants against HMs-induced oxidative stress in plants in for sustainability of crop production in the modern era of climate change.

8.2 Consequences of Heavy Metals Stress in Plants

Unprecedented competition among nations to become self-reliant in economy indulges adoption of high mechanical power run agriculture, mining and growing industrialization like a frog's umbrella (Shahid et al., 2015) causing the emission of HMs into the environment resulting pollution of air, water and soil. Plants perpetuating in a biome having a high concentration of HMs in soil, air and water means leaving them in stress called HMs toxicity (Tiwari and Lata 2018). Tolerance of plants my generate in HMs toxicity by the genesis of different complex biomolecules or physiological mechanisms (Panuccio et al., 2009; Hassan et al., 2017) but those mechanisms will fail under heavy exposure of plants to these metals finally paying death toll. HMs are the elements that have a specific gravity greater than 5 (As – 5.7, Zn – 6.9, Fe – 7.8, Cd – 8.6, Ni – 8.9, Cu – 9.0, Pb – 11.3, Au – 19.4, Pt – 21.4 g/cm^3) (Islam, 2014).

High densities, toxic in nature and non-degradability are the basic properties of HMs and there are at least 38 HMs where 13 are common (Cd, Cr, cobalt (Co), Cu, iron (Fe), Hg, Mn, molybdenum (Mo), Ni, lead (Pb), arsenic (As), tin (Sb), Zn) (Islam, 2014). Sources of HMs to pollute arable soil may be agricultural (animal wastes-cattle manure/poultry manure, pesticides–herbicides/insecticides/fungicides, fertilizers – SP/TSP (containing Cd), irrigation – As) and non-agricultural (municipality wastes, sewage sludge, industrial wastes, combustion of fuels) (Islam 2014; Table 8.1).

Some HMs (Zn, Cu, Mo, Mn, Co and Ni) are essential for plants to complete their life cycle by taking part in different processes, plant growth, metabolism, physiology and senescence (Ghori et al. 2019) but in high concentration are toxic to plants (Salla et al. 2011; Shahid et al. 2015). Also, HMs polluted soils bring plants liable to absorb higher quantities of As, Pb, Cd, Hg, Cr, Al and Be which will reduce growth and development and eventually yield of crops (Xiong et al. 2014; Pierart et al. 2015). Destruction of redox balance of plant cells by producing superoxide anion radical (O^{2-}), H_2O_2 and hydroxyl radical (OH^-) are common on HMs stressed environment but to overcome this stress dilemma plants may develop unique defense mechanisms like preventing plants exposure to HMs of soil, reducing the uptake of metals from entering into the cells by cellular and root exudates (Ghori et al. 2019), impounding of metal into vacuoles, fixing into phytochelatins (PCs)/metallothioneins (MTs) and stimulation of various antioxidants (Shahid et al. 2015). Plants show affinity to build a symbiotic relationship with other companions to address such kind of stress for accumulating metals in the rhizosphere area preventing their availability and uptake by the plants (Ghori et al. 2019). HMs are generally sequestered in the root cell wall by the blockage in Casparian strip (Ghori et al. 2019). During the toxicity of specific HMs, oxidative stress is profound in plants and synthesis of stress-related proteins (heat-shock proteins), hormones, signaling molecules, antioxidants are initiated (Ghori et al. 2019). Plants show visible symptoms upon HMs toxicity like chlorosis, stunted growth, root browning, the decline in growth and death (Ozturk et al. 2008) but invisible effects are inference with photosynthesis and respiration, alteration of the integrity of cell membranes, cleavage of DNA strands and homeostasis (Hossain et al. 2012a,b, Rascio and Navari-Izzo 2011). Due to higher toxicity potential, repeated existence and exposure potential to flora and fauna As, Pb, Cd and Hg are considered as the most toxic HMs (ATSDR 2003).

TABLE 8.1

Sources of Heavy Metal Elements

Source	Heavy Metals
Fertilizers (as impurities)	Cd, Pb, Zn, Mo
Sewage sludge	Cd, Ni, Cu, Pb, Zn
Animal manure	Cu, Zn, As
Composts	Cd, Cu, Ni, Pb, Zn
Irrigation water	As
Fossil fuels	Pb
Wood preservations	As, Cu, Cr
Chlorine manufacture	Hg
Batteries	Pb, Zn, Ni, Hg, Sb
Paints & pigments	Pb, Cr, As, Cd, Sb
Pharmaceuticals	Ag, As, Cu, Sb, Se, Zn
Cables, contacts, semi-conductors	Cu, Zn, Pb, Sn, Co, Mo, As

As releasing from heavily occurred parent materials or irrigation with As contaminated water or bringing from any other external sources, it remains in high concentration in soil solution in the form arsenate As (V) and arsenite As (III) resulting in higher uptake by crops affecting crop productivity, yield and accumulated in plant tissues even in grains thus contaminating entire food chain (Verma et al. 2016). All two species of As (arsenate and arsenite) are toxic to plants and animals and considered as foremost environmental pollutants (V) (Tripathi et al. 2007; Verma et al. 2016). As (III) can easily disturb plant metabolic functions by binding proteins with sulfhydryl groups and impeding normal functioning of proteins (Verma et al. 2016). It can produce reactive oxygen species-ROS (superoxide anion radical (O^{2-}), H_2O_2 and hydroxyl radical (OH^-)) hindering respiration of plants by binding to vicinal thiols in pyruvate dehydrogenase and 2-oxo-glutarate dehydrogenase and induce intrachromosomal homologous recombination indirectly as a mutagen (Helleday et al. 2000). As (V) is less toxic than As (III) can affect energy metabolism by interfering in oxidative phosphorylation and ATP synthesis (Carbonell et al. 1998; Verma et al. 2016).

Exposure of plants to widely and evenly distributed lead leads to acute toxicity to plants by several harmful effects in biological systems as Pb^{2+} is non-biodegradable in nature (Bradham et al. 2006; Pehlivan et al. 2009). Pb can impede the growth and development and biological processes of plants, such as seed germination, root elongation, seedling development, transpiration, chlorophyll biosynthesis and cell division (Pourrut et al. 2011; Kumar et al. 2017). It can react with active groups of different metabolic enzymes and phosphate groups of ADP or ATP consequently impairing in cell membrane permeability and also can replace essential ions causing phytotoxicity (Pourrut et al. 2011; Kumar et al. 2017). With the help of over-production of ROS, Pb can hinder with ATP production but encourage lipid peroxidation and DNA damage (Tiwari and Lata 2018).

Cd is treated as the major phytotoxic HMs and as most soluble elements entered easily into the plant system making degraded food chain (Buchet et al.1990). It interferes with the carbohydrate metabolism, Calvin cycle, phosphorus metabolism, nitrogen (N) metabolism, sulfur (S) metabolism and CO_2 fixation (Sandalio et al. 2001; Verma and Dubey 2001; Sharma and Dubey 2006; Gill and Tuteja 2011) after altering several enzymatic activities even at low concentration. Exposure of plants to the Cd-toxicity enhances stunted growth, inhibition of photosynthesis, leaf epinasty, chlorosis and alterations in chloroplast ultrastructure. It affects pollen germination and tube growth, induction of lipid peroxidation and disruption of antioxidant machinery (Gill and Tuteja 2011). Cd toxicity produces brown root tip of plants and reduces Fe (III) reductase resulting Fe (II) deficiency, inhibits chlorophyll synthesis and retardance of enzymes synthesis for CO_2 fixation leading to reduced photosynthesis. Cd has an antagonistic effect on uptake of Ca, P, K, Mg and water by plants and reduces the synthesis of nitrate reductase which will reduce the translocation and assimilation of nitrate. Cd toxicity damages nucleolus enhancing chromosomal fragmentation and abnormality (Ghori et al. 2019). Decomposition of mitochondria, hindering plant respiration, binding to sulfur-containing groups (sulfhydryl) of various structural proteins leading to misfolding of proteins resulting interference with redox reactions especially in the electron transport chain and increasing proline amino acid synthesis are common in Cd toxicity. Cd-treated fenugreek plants show significant inhibitory effect (Alaraidh et al. 2018) and biochar can be used to reduce availability and root uptake of Cd in plants (Rizwan et al. 2018).

Mercury in low concentration is not harmful to higher plants but in higher concentration as Hg^{2+} can accumulating into the higher plant system (Israr et al. 2006; Yadav 2010) and easily interfere with the normal activities of plant cells and enhance physiological disorders (Ortega-Villasante et al. 2005; Zhou et al. 2007). On the other hand, Hg^{2+} has an affinity to bind with the proteins of water channels of plants causing the closure of stomata of leaf resulting inhibition of water flow in plants (Zhang and Tyerman 1999; Zhou et al. 2008). It can even hamper the activities of mitochondria (Zhou et al. 2008). Hg can exist in several forms in the external environment such as elemental (Hg^0), associated (HgS, $ClHg_2$, Hg_2Cl_2) and organic (CH_3–Hg) in addition to the most prominently occurred inorganic (Hg^{2+}) form (Zhou et al. 2008). These mercuric ions are testified to stimulate the generation of ROS resulting in oxidative stress in crops leading to disruption of bio-membrane lipids and cellular metabolism which may increase the degree of stress in plants by increasing the activities of antioxidant enzymes like SOD, POD or APX (Cargnelutti et al. 2006; Zhou et al. 2007). Hg has no beneficial effect on plants (Hameed et al. 2017) but can persist in soil by adhering in clay, sulfides, organic matter and chelated and precipitated as hydroxide, carbonate, sulfide and phosphate. It may be converted to methylated form by an anaerobic bacterium (Tangahu et al. 2011) and then absorbed by plants and expressing phytotoxicity. Genomic template stability of plants is highly influenced by mercury treatment (Malar et al. 2015).

Zn is an essential element for plants as it is involved in enzymatic reactions, pollen formation, disease resistance mechanisms and carbohydrate, auxin and protein metabolism. Though it is required in a trace amount but any deficiency may bring severe yield loss (Aziz et al. 2016) and in excess concentration may lead to a terrible consequence. In excess, Zn acts as a toxic element as like as HMs and alters enzymatic reactions enhancing growth retardance, senescence and chlorosis of younger leaves (Ghori et al. 2019). On the other hand, it has a synergistic effect on uptake of some other HMs (Cu and Mn) by plants resulting in extending exacerbation of the consequences of stress (Nagajyoti et al. 2010). Zn toxicity has an antagonistic effect on P uptake by plants resulting production of purplish-red color leaf (Lee et al. 1996), thickened and blunt roots and have restricted cell division and elongation. It also affects photosystems I and II hampering photosynthesis eventually reducing ATP synthesis (Van Assche and Clijsters 1990). Mg^{2+} is required for carbon fixation in plants may be inhibited by Zn toxicity because Zn as a divalent ion displaces Mg^{2+} in RuBisCO (ribulose-1, 5-bis-phosphate-carboxylase/oxygenase) (Ghori et al. 2019). Exposure of poplar leaves in Zn toxicity may enhance Zn accumulation in xylem and parenchyma cells, increasing the number of calcium oxalate crystals in leaves and amplified free calcium occurred (Todeschini et al. 2011).

Cu is an essential element for plants as it plays important roles in plant systems such as carbon assimilation, ATP synthesis and constituent part of plastocyanin and cytochrome oxidase but in high concentration, Cu may be dire to plants as it may cause oxidative stress, damage to cell membranes and interfere with various biochemical pathways (Yadav 2010). Hindrance in seed germination, reduction of seedling length and number of roots of the *Solanum melongena* were caused when exposed in Cu toxicity (Neelima and Reddy 2002).

Cr toxicity effect of plants may be evident by chlorosis, growth retardation, wilting of top and injury of roots (Ozturk et al. 2015), hindering the activity of amylases (Zeid 2001), interfering with photosynthesis, affecting physiological processes, producing ROS (Yadav 2010), alteration of synthesis of essential plant pigments for photosynthesis and anthocyanin (Boonyapookana et al. 2002), encouraging the production of glutathione and AsA (Shanker et al. 2005) and induces production of PCs and histidine (Schmfger 2001).

Nickle occurs in the structure of glyoxalases, ureases, methyl-CoM reductase, superoxide dismutases, peptide deformylases and some hydrogenases enzymes. On the other hand, ureolysis, methane biogenesis, acidogenesis, maintaining the cellular redox state, stress tolerance/defense, optimum nitrogen use efficiency (Vatansever et al. 2017) and hydrogen metabolism require Ni (Ghori et al. 2019). Under the toxic level of Ni for plants, necrosis, chlorosis, nutrient deficiency, disturbed function of cell membrane, disrupted water balance, a decline in water content of monocot and dicot plants, reduction in plants growth, reduction in photosynthetic ability, oxidative stress induction, inhibition of nitrogen metabolism, reduction in mitotic activities, interference with the uptake of other metals are common phenomena.

In addition to the above discussion, if the HMs can successfully enter into the nucleus of plant cells can cause promutagenic damage that encompasses inter-molecular and intra-molecular cross-linkage of DNA and proteins, DNA base modifications, rearrangements, DNA strand breakage and depurination by producing oxidative species. It develops promutagenic adduct 8-oxoG (7,8-dihydro-8-oxoguanine) that miss-pairs with adenine in the absence of DNA repair and in consequence developing transversion of C to T (Cunningham 1997; Kasprzak 1995).

8.3 Osmoprotectants and Their Emerging Role to Survive against Heavy Metals Stress

HMs stress is considered as one of the major environmental stresses when they appear in soil solution in a high concentration. The high concentration of HMs in soil solution forces plants to uptake through root system; leads to cause serious disorders in plants as well as in human beings by plant foodstuffs. The adverse effect of HMs are linked to the inhibition of vital physiochemical processes of plants such as photosynthesis, nutrients and uptake of water, lead to excessive production of various ROS. Generally, oxidative stress occurs when the generation of ROS exceeds that of the scavenging potential of the antioxidants defence system (Figure 8.1). Although, tolerant plants could survive against HMs stress through following a range of potential mechanisms which are linked with the decontamination of HMs (Ghori et al. 2019). The physiological adaptation strategies of plants are controlled by various OSP. OSP are the small organic compounds having low molecular weight, primarily amassed in plant cells and provide to maintain the osmotic pressure in various stressed conditions (Yancey et al. 1982).

These are supporting no net charge at physiological pH and are non-toxic even at high absorption due to its high solubility. OSP are also indicated as a scavenger of ROS, having matron-like action and assist in metabolic detoxification (Serraj and Sinclair 2002). Moreover, OSP romp an intrinsic role in balancing proteins and membranes during oxidative damage in stressed condition (Yancey 1994; Bohnert and Jensen 1996). Usually. OSP are gathered in cytosol and chloroplast of plant cells but sometimes noticed to in other few plant parts. OSP, which have a functional role are hinted in agriculture. OSP demonstrate intensified survival rates of plants in high stressed conditions due to the regulation of toxic ROS.

OSP have three major classes viz., amino acids (e.g., proline), polyols and sugars (e.g., mannitol, d-ononitol, trehalose, fructants) and quaternary ammonium compounds (e.g., glycine betaine) as reported by Yancey (1994). Amino acids and their derivatives, polyols and sugars, methylamines and methyl-sulphonium compounds are absorbed when organic osmolytes play a vital role in minimizing the harmful effects of abiotic stresses. Proline, GB and mannitol are generally traced in plants among these OSP. Abiotic stress (high or low temperature, drought, salinity, HMs and UV) has some negative effects on physical, biochemical and molecular activities of plants (Atkinson and Urwin 2012; Suzuki et al. 2014; Singh et al. 2015; Zhu 2016; Tripathi et at. 2016a, 2016b, 2017a, 2017b; Jeandroz and Lamotte 2017). Under this condition, plant cells accumulate some organic molecules (OSP) which protect plats from this adverse situation (Mahajan and Tuteja, 2005) as well as protect them from oxidative damage (Burg and Ferraris, 2008), also protecting DNA, protein and enzymes due to the activation of some polyamines (Pas), amino acids and their derivatives (Rodríguez et al. 2005; Groppa and Benavides 2008; Roychoudhury et al. 2011; Paul et al. 2017). OSP are also

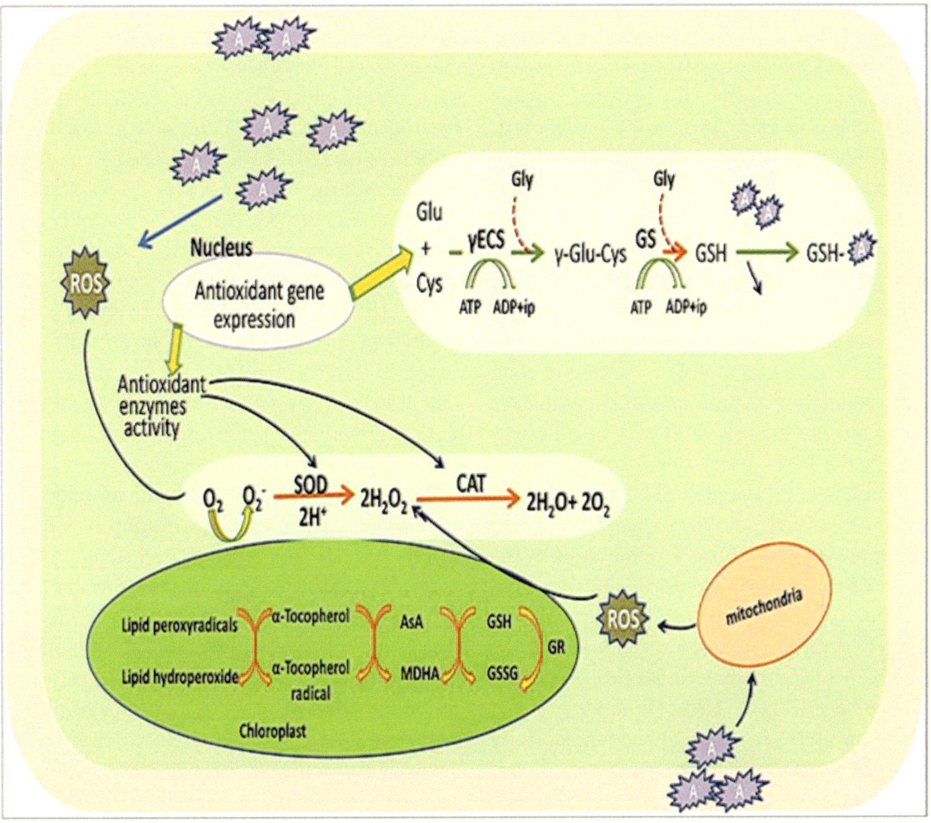

FIGURE 8.1 Heavy metal induced-oxidative stress, tolerance and detoxification mechanisms in the plant cell. Source: Adapted and modified from Singh et al. (2016) with permission.

a significant role for the retaining of topsoil bacteria populations. Besides, it also assists gene expression with respect to environmental variations. The important role of some OSP has been discussed in this section.

8.3.1 Proline

Proline, an amino acid, commonly associated with metabolic osmolyte with high water-soluble, involved in the cell wall formation, scavenger of free radical, antioxidant, stabilizer and maintain osmotic pressure in plant cells under stressed or adverse condition (Seregin et al. 2003; Pavlikova et al. 2007; Zarei et al. 2012; Yancey 2005). Proline also flourishes the embryo, elongation of the stem, as well as shifting from the vegetative to the reproductive stage of plants (Burritt 2012); source of energy, carbon and organic nitrogen at the stressed condition for plants (Tyagi and Sairam 2004). Proline also plays a significant role during abiotic stress-induced oxidative stress. For example, HMs induced proline accumulation in plants mainly due to the water balance disorder is excessive metal stress is accountable for the introduction of proline (Clemens 2006). In this perspective, proline act as an osmoprotectant for the survival of plants in HMs stressed condition. Proline also plays a significant role as a protein stabilizer and HMs chelator (Mishra and Dubey 2006); scavenger of ROS, that is triggered in HMs stress condition (Mourato et al. 2012).

Under HMs stressed condition, proline increases antioxidant enzymatic activities, protection, control the redox homeostasis, regulate the intracellular pH as well as the reorganizing of chlorophyll (Rastgoo et al. 2011; Mourato et al. 2012). The activities of proline in response to HMs depend on the concentration, plant organ as well as the specific HMs (Emamverdian et al. 2015). There is a linear relationship between proline augmentation in plant cell and HMs in *Cynara scolymus* L (Karimi et al. 2012). In other, Gohari et al. (2012) reported that the concentration of proline increased in the rapeseed roots (*Brassica napus* L.) when Pb concentration increased (100–400 μM) but not distinct to the areal parts of the plants. Proline accumulation is more (two folds higher) in the roots compared to leaves at higher concentration of HMs (Sharma and Dietz 2006; Zengin and Kirbag 2007) but no significant differences at a lower concentration of metals as reported by Nikolic et al. (2008). However, the formation of proline–metal complexes due to the induction of proline in plant roots may be an effective approach to nullify the HMs toxicity compared to allow them in the above-ground parts of the plant. The enhancement of HMs concentration increases the proline in plant cell up to a certain level, which suppresses the proline accumulation of the HMs with the increase of metals beyond a particular threshold as noticed by Ruscitti et al. (2011). Besides, foliar application of proline in plant parts is an effective approach to avoid the HMs toxicity and also increase the protective mechanism in plants against the adverse effects of HMs (Emamverdian et al. 2015). The application of proline can enhance the activity of antioxidative enzymes, photosynthetic parameters and carbonic anhydrase in chickpea on HMs (Cd) stressed condition as well as enhanced the HMs tolerance of the plant (Hayat et al. 2013) and detoxified nickel (Ni) on pea (*Pisum sativum* L.) by decreasing electrolyte leakage and lipid peroxidation, increasing PA biosynthetic enzymes activities and endogenous compatible solutes (Shahid et al. 2014).

8.3.2 Glycine Betaine

GB is composed of ammonium compound, commonly found in plants, involved in many morphological and biological activities in plants. It also helps stabilize the structure of protein and enzymes as well as to maintain the integrity of cell membrane under stressed condition (Sakamoto and Murata 2000). GB has the capability to alleviate HM stress for plants. GB also behaved as OSP, that inhibited the ROS and free radical production in plants (Fariduddin et al. 2013), changing of gene expressions (Tiwari and Lata, 2018; Hasan et al. 2017) by creating additional antioxidants enzymes SOD, POD and CAT, as well as scavenge the unexpected ROS under HMs stress condition (Moharramnejad et al. 2015). GB also enhances plant growth by escalating chlorophyll content and repair oxidative damage (Demidchik 2015), as well work as a safeguard of photosynthetic pigments (Ahmad et al. 2020; Kurepin et al. 2017; Anjum et al. 2011).

GB successfully illuminates HMs tolerance in plants and keeps the cells free from the HMs toxicity. GB can improve the antioxidant system in plants and nullify the adverse effects of HMs stressed condition (Jabeen et al. 2016; Zouari et al. 2018). Foliar application of GB can effectively alleviate HM toxicity and improve plant growth, root and shoot weight, cell division along with dry matter production in plants (Ali et al. 2015; Farooq et al. 2016; Bhatti et al. 2013). Reports available that, foliar application of GB and successfully nullify Cd toxicity in cotton (Farooq et al. 2016), tobacco (He et al. 2019; Bai et al. 2015), ryegrass (Lou et al. 2015) and wheat (Rasheed et al. 2014) under Cd stressed condition by the decreasing of hydrogen peroxide (H_2O_2), electrolyte leakage (EL) and malondialdehyde (MDA) level in plants in HMs stressed situation. Similarly, GB can also protect plants from Pb toxicity of perennial ryegrass (*Lolium perenne* L.) (Islam et al. 2009); harmal (*Peganum harmala* L.) (Mahdavian et al. 2016), olive trees (Zouari et al. 2018) and cotton (Ali et al. 2013; Hossain et al. 2010) by enhancing photosynthetic activities and chlorophyll synthesis (Bharwana et al. 2014), as well as protect juvenile tissues of plants. The exogenous application of GB can effectively avoid Cr toxicity in wheat (Ali et al. 2015), due to the activation of some antioxidant enzymes like SOD, CAT, POX, etc.

The rate of photosynthesis, the ascent of sap, mineral nutrition, water uptake, the functioning of stomata and slowdown of chlorophyll biosynthesis are affected by HMs stress which is familiar to the plant in relation to water (Cuypers et al. 2009; Pourrut et al. 2011). GB resists the thylakoid membrane when it is found massive in the chloroplast (Genard et al. 1991). The cytoplasmic dehydration is constrained contributing to GB and retains turgor pressure in leaves of plant yielded to water-deficient state in sunflower (Iqbal et al. 2008). GB can effectively improve the leaf expansion, turgidity and photosynthetic pigments in sorghum (Ibrahim and Aldesuquy, 2003). The lethal impacts of HMs by improving the resistance of plants under stressful conditions are lessened by GB. GB also provide to improve the nutrient content, especially potassium (K^+)

and sodium (Na^+) in root and shoot of plants (Anjum et al. 2011), which ensure plant growth, carbohydrate production, metabolism, development of seed and fruit, chlorophyll synthesis in plants (Tripathi et al. 2015).

8.3.3 Phytochelatins

PCs, a low-molecular-weight short-chain peptide synthesized from sulfur-rich glutathione, act as an osmoprotectant have the affinity of HMs binding at toxic levels (Lee et al. 2003; Chen et al. 2008; Wang et al. 2009, Shukla et al. 2013; Gupta et al. 2013). PCs are also treated as detoxifying of HMs in plants as well as salinity, heat and UV stressed conditions (Zagorchev et al. 2013) and worked as biomarkers for preliminary diagnosis of HMs stress in plants (Saba et al. 2013). PCs can fix both the cations and anions of some HMs like Cd, Cu, Ag, Zn, Pb, Ni and Ar (Gupta et al. 2013; Cobbett 2000), of which Cd^{2+} is the most effective (induced 4–6 times higher PCs) compared to Cu^{2+} and Zn^{2+} (Thangavel et al. 2007; Kotrba et al. 1999). PCs can be produced and accumulated in roots and shoots of the plants. Out of them, roots had accumulated two-fold higher PCs than that of shoots exposed to Cd intoxication (Yurekli and Kucukbay 2003). PCs enhanced in plant roots at a high level of Cu (200 $\mu mol \cdot L^{-1}$) in *Solanum nigrum* L. and protect the plants from Cu toxicity by the restriction of Cu movement from root to the areal parts (Fidalgo et al. 2013), and As toxicity in rice (Batista et al. 2014) by the formation of As-PC complexed that inhibited the transportation of As form soil or roots to the plant parts. Conversely, PCs accumulated in plant leaves and functioned as HMs protectors factored on time variables. The synthesis of PCs in plant leaves detoxified Cd in *Brassica juncea* (Heiss et al. 2003) and in maize (Szalai et al. 2013) because of the three times more PCs accumulation in leaves compared to roots. It seems that the production, accumulation and transportation of PCs vary on plant species and the level of tolerance of HMs in plants. Reports are available that Cd detoxification in *Sedum alfredii* and wheat when PCs accumulated in plant roots and shoots, respectively (Chen et al. 2008; Hentz et al. 2012) resulted that the enzymatic biosynthesis of PCs along with the antioxidative mechanism detoxified Cd in *Brassica chinensis* L.

8.3.4 Metallothioneins

MTs is a cytoplasmic metal-binding protein having low molecular weight, commonly found in fungi, mammals, invertebrates and some of prokaryotes and eukaryotes (Du et al. 2012; Macovei et al. 2010; Cai and Ma 2003). MTs are mainly deal with Cu, Zn, Cd and As detoxification in plants (Yang and Chu 2011). The role of the activities of MTs in plants varying on their structural and molecular characteristic (Leszczyszyn et al. 2013). In plants, MTs are contributed to detoxification of HMs by the homeostasis of intracellular metal ions and cellular sequestration as well as transport of metals (Guo et al. 2013; Kohler et al. 2004). Moreover, MTs also contributed to maintaining the redox level (Macovei et al. 2010), the growth and proliferation of plant cell (Grennan 2011), repair the plasma membrane (Mishra and Dubey 2006) and damaged deoxyribonucleic acid (DNA) (Grennan, 2011), as well act as ROS scavenger (Wong et al. 2004). The production and activity of MTs mainly depend on both the exogenous and endogenous factors like heat, drought, osmotic stress, nutrient deficiency, hormones, tissue senescence, viral infections and other plant injuries (Manara, 2012; Du et al. 2012; Yang and Chu 2011). MTs are divided into four subgroups as suggested by (Huang and Wang, 2009). These are MT1 (mostly occurs in roots), MT2 (mainly expressed in shoots), MT3 (occurs in leaves during fruit ripening) and MT4 (found in seeds) (Yang and Chu 2011; Kohler et al. 2004). Each subgroup (MT1 to MT4) is further divided and known as isoforms. All the four types of MTs and their isoforms play a significant role in nullifying HMs through chelating of HMs, may be varied depending on metal types and their ability to bind the HMs (Du et al. 2012). The MT isoforms (1a, 2a and 2b) chelated Cu but type 4 (4a and 4b) act as a Zn chelator in *Arabidopsis* (Grennan 2011), but MT1 to MT3 detoxified Cd and MT4 showed the Zn binding characters in soybean (Pagani et al. 2012). MTs can induce a decrease of ROS and improve antioxidant activity in plant cell under HMs stressed conditions.

8.3.5 Glyoxalase

Glyoxalase, a common output in excessive HMs stress due to the over deposition of methylglyoxal (MG) in plant cell can protect the DNA and enzymatic damage or interact with other major components of plants. It also included the direct and indirect control of MG and their activities in plant cells (Yadav et al. 2005a, 2005b; Pareek et al. 2006; Hossain et al. 2009, 2012a, 2012b). One of the important strategies of avoiding HMs toxicity in plants is cellular dismissal. A great number of HMs are found in diffusion space outside of the plasma membrane (apoplastic space) in the plant roots. Reports are available that, Al toxicity had less symplastic in sensitive wheat cultivar due to the exclusion proteins in the eviction of toxic HMs from the diffusion space of plant roots.

8.3.6 Salicylic Acid

SA plays a significant physiological role (e.g., morphology, development, flowering and stomatal closure) of plants (Miura and Tada 2014, Mohsenzadeh et al. 2011); improved growth, leaf number, stem diameter (Hussein et al. 2007); seed germination, cell growth and nodulation in leguminous crops (Vlot et al. 2009). It also helps regulate photosynthesis, respiration, antioxidants' defense mechanism in HMs stressed condition (Das and Roychoudhury 2014; Miura and Tada 2014; Mohsenzadeh et al, 2011), as well as save from several HMs (Pb, Hg and Cd) to plants (Ghani et al. 2015; Gondor et al. 2016; Zhou et al. 2009). In addition, the application of SA with some phytohormones (auxin, gibberellins, cytokinins, ethylene, brassinosteroids) can contribute an important role against HMs toxicity in Plants (Khokon et al. 2011). The mode of action of SA varies on the concentration and method of application (Khokon et al. 2011; Alaey et al. 2011; Divi et al. 2010). SA increased the leaf area and dry weight production in corn and soybean (Khan et al. 2013); increase the shelf life of rose under HMs stressed condition. SA mediated some enzyme activities (CAT and SOD) in barely leaves in Zn, Cu and Mn stressed condition (Song et al. 2014); carbohydrate metabolism in maize plant in Cr stressed condition (Islam et al. 2016); mitigation of Cd toxicity in mustard, improves growth, yield and nullifying oxidative damage in *Brassica campestris* in Pb toxicity condition (Hasanuzzaman et al. 2019), as well as reduce ROS and stability of cell membranes by strengthening antioxidant defense system in plants under HMs stressed condition (Faraz et al. 2019).

Photosynthesis is one of the key mechanisms in plant cells. SA treated as a vital photosynthesis regulator that influences on enzymes related to photosynthesis, stomatal conductivity and chlorophyll content in plants (Rivas-San Vicente and Plasencia 2011). The synthesis of photosynthetic pigments (chlorophyll and carotenoids) may be hampered due to the deposition of HMs in various plant parts (Sidhu et al. 2016, 2017); can be avoided by SA under HM stress condition (Shi et al. 2009). Several studies reported that SA can deduce the HM activity on plants. SA enhanced Chla, Chlb and carotenoid content in barley under Pb stress (Arshad et al. 2017), reduce Cd deposition in tomato (Guo et al. 2018). The protective role of SA might be due to the blockage of Ca channels in roots as well as detoxify ROS by the activation of antioxidant in plant cells (Arshad et al. 2017; Rivas-San Vicente and Plasencia 2011). However, it was found that PSI and PSII showed a negative effect of HMs toxicity (Babu et al. 2010; Chugh and Sawhney 1999). In other, Khan et al. (2007) noticed that PSI is more active than that of PSII in HMs stressed condition; but convinced the activity at high concentration, which debilitate the transpiration rate and stomatal conductance (Yusuf et al. 2012).

8.3.7 Polyamines

In the regulation of signaling network of plants and different developmental processes, PAs and ethylene play an important role as phytohormones. Especially in abiotic stress adaptation and tolerance, PAs and ethylene play an indirect or direct role. In response to HMs stress, the interplay between PAs and ethylene is due to a common precursor of both these hormones that is S-adenosylmethionine. The reported pathways for the biosynthesis of ethylene and PAs, in the literature, are showing differential modulations. However, synergism between the biosynthesis of ethylene and PAs could help to maintain optimum cellular metabolism, modulate major signaling compounds, increasing water (nutrients) uptake and improving the ratio of root to shoot under HM stress (Asgher et al. 2018).

8.3.8 Brassinosteroids

Under cadmium stress, the activity of several enzymes (carbonic anhydrase and nitrate reductase), leaf water potential and photosynthetic parameters are reduced significantly in two varieties (Sarvodya and K-25) of tomato, to a greater extent in Sarvodya than K-25. Anyhow, both brassinosteroids (HBL/EBL) application and metal treatment enhanced the proline contents as well as antioxidant enzyme activity. Besides, brassinosteroids applied exogenously not only lowered the harmful effects of mental stress but also improved the antioxidant defense system and the activity of photosynthetic machinery in both cultivars (Hasan et al. 2011).

8.4 Osmoprotectants Engineering against Heavy Metals Stress

HMs contamination and its toxicity is a major threat to the environment as well as crop production of many areas in the world. They badly affect plant growth, development and productivity by the production of excessive ROS which alters cellular metabolism by disrupting various macromolecules resulting in sometimes ultimately lead to plant die or accumulate HMs to reproductive parts of plants (Pongrac et al. 2009). Most likely, ions of HMs stay in the cytoplasm of plant and cause oxidative stress, therefore, regulation of ROS is crucial to avoid cytotoxicity and oxidative damage (Hossain et al. 2012b; Ghori et al. 2019). Accumulation of some of the compatible organic solutes known as OSP can protect plants from oxidative damages by directly scavenging for ROS or protecting proteins and membranes from damage by high concentrations of inorganic ions (Slama et al. 2015). Although certain plants, bacteria and marine algae have evolved several adaptations to such abiotic stresses, many major crops cannot synthesize the special OSP that are naturally accumulated by stress-tolerant organisms. In plant cells, OSP are primarily accumulated in cytosol and chloroplast but are also reported to be distributed in few other organelles. OSP serve to raise osmotic pressure in the cytoplasm and can also stabilize proteins and membranes when plant faced different abiotic stress including HMs. The application of various OSP is a well-known approach to mitigate the negative effects of HMs on plants. GB, proline and polyamines are the most studied osmoregulation used as OSP and their level varies considerably among different plants (Chen et al. 2014). Application of GB on plants under HMs stress successfully improves antioxidant enzyme activities, nutrients uptake, minimizes excessive HMs uptake and oxidative stress, resulting in increased photosynthesis and plant growth (Nusrat et al. 2014). In addition, GB activates the adjustment of GR, AsA and GSH contents in plants under HMs stress (Chen and Murata 2002).

Several strategies including genetic engineering have been successfully applied to generate plants which are able to grow in HMs contaminated soils and accumulate or tolerate this stress. The plants engineered for OSP to cope with HMs stress have been an important area of plant scientists (Chen and Murata 2002). In contrast to traditional breeding, genetic engineering appears to be an attractive alternative with respect to identify osmoprotectant-related genes and their associated biosynthesis pathways and the possibility of direct introduction them (Diamant et al. 2001). Hence genetic engineering approaches would be useful to manipulate these OSP biosynthetic pathways for accumulating such molecules that act by scavenging ROS, reducing lipid peroxidation, maintaining protein structure and functions (Yamada et al. 2005). Many crops are engineered using different OSP and the level of tolerance exhibited by these engineered crops varies greatly (Huang et al. 2000). Ashraf et al. (2010) reported genetic engineering creates transgenic plants which contain various genes for the GB biosynthesis pathway, which ultimately develops increased resistance to different abiotic stress. They also reported betaine aldehyde dehydrogenase (BADH) and choline oxidase

(COD) are two main enzymes which are chiefly used to create GB synthesis in that plant which does not normally synthesize GB, a very crucial function in plants under stressful conditions. Increase in the accumulation of GB through genetic engineering could be effectively exploited as an essential implement to mitigate the HMs stress confrontation in plants. Engineered plants predominately accumulate GB in the chloroplast, and display tolerance to different types of abiotic stresses including HMs (Park et al. 2007). Transgenic studies over the years have demonstrated that plants which generate additional GB have improved the resistance to HMs stress (Kishitani et al. 2000; Khan et al. 2009). Qin et al. (2017) reported that the enzymes containing BADH transgenic soybeans have shown a 6–17% increase in germination index, POX activity enhanced by 1–7% and decreased in contents of MDA by 1.5–13% in comparison to the control treatment. Sharma and Dietz (2006) reported HMs induced proline accumulation and its protective role in transgenic plants. Transgenic *Chlamydomonas reinhardtii* expressing P5CS isolated from moth bean accumulated 80% more proline than the wild type under Cd toxicity (Siripornadulsil et al. 2002). Transgenic *Oryza sativa* cultivar overexpressing the betaine synthesis gene BADH showed improved Cd tolerance compared to non-transgenic cultivars (Shao et al. 2008).

The accumulation of GB through the utilization of a genetic engineering approach can successfully enhance tolerance against stress Co, Cu and Fe (Chen and Murata 2011; Ali et al. 2015; Farooq et al. 2016; Jabeen et al 2016). In developing engineering transgenic plants with different abiotic stress including HMs tolerance that utilized genes encoding OSP, and other stress-related functional proteins so far, much progress has been made. Biosynthetic accumulation of GB, proline and other osmoprotectant genes in several transgenic crop plants have shown some improvement in HMs stress tolerance. However, the success of these genes has been limited in the sense that most of the transgenic plants have been developed and tested under controlled (laboratory) conditions. In addition, most of the studies transgenic plants are evaluated at the early growth stage and are exposed to stress condition for a short period and such short exposure experiments at early growth stages may not predict the response of the plant in their whole lifecycle under realistic field conditions.

Conclusion

From the discussion of the chapter, it may be concluded that HMs stress is one of the most important abiotic stresses which cause poisonous effects in the plant process. The adverse effect of HMs are linked to the inhibition of vital physiochemical processes of plants, such as photosynthesis, nutrients and uptake of water, lead to excessive production of various ROS. Although tolerant plants could survive against HMs stress via several mechanisms such as immobilization and exclusion of HMs in plant cells' plasma membrane, efflux and limit the transport of toxic HMs ions in plant cells, compartmentalization and metal chelation via tonoplast transporters, etc. These physiological adaptation strategies of plants generally controlled by various OSP such as sulfur- and nitrogen-containing compounds, amino acids (proline, glycine, GB, SA), alanine, etc.) and sugars (sucrose, fructose, trehalose, etc.). The present chapter provides a broad overview of the OSP-based survival adaptability of plants against HMs. The understanding of survival mechanisms will be helpful for the sustainability of crop production under changing climate.

REFERENCES

Ahmad, R., Ali, S., Abid, M., Rizwan, M., Ali, B., Tanveer, A., Ghani, M.A. 2020. Glycinebetaine alleviates the chromium toxicity in *Brassica oleracea* L. by suppressing oxidative stress and modulating the plant morphology and photosynthetic attributes. Environ Sci Pollut Res. 27, 1101–1111.

Alaey, M., Babalar, M., Naderi, R., Kafi, M. 2011. Effect of pre- and postharvest salicylic acid treatment on physiochemical attributes in relation to vase-life of rose cut flowers. Postharvest Biol Tech. 61, 91–94.

Alaraidh, I. A., A. A. Alsahli., and E. S. A. Razik. 2018. Alteration of antioxidant gene expression in response to heavy metal stress in Trigonella foenum-graecum L. S Afr J Bot. 115, 90–93.

Ali, Q., Anwar, F., Ashraf, M., Saari, N., Perveen, R. 2013. Ameliorating effects of exogenously applied proline on seed composition, seed oil quality and oil antioxidant activity of maize (*Zea mays* L.) under drought stress. Int J Mol Sci. 14, 818–835.

Ali, S., Abbas, Z., Seleiman, M. F., Rizwan, M., YavaŞ, İ., Alhammad, B. A., … Kalderis, D. 2020. Glycine betaine accumulation, significance and interests for heavy metal tolerance in plants. Plants. 9(7), 896.

Ali, S., Chaudhary, A., Rizwan, M., Anwar, H. T., Adrees, M., Farid, M., Irshad, M. K. Hayat, T. and Anjum, S.A. 2015. Alleviation of chromium toxicity by glycinebetaine is related to elevated antioxidant enzymes and suppressed chromium uptake and oxidative stress in wheat (*Triticum aestivum* L.). Environ Sci Pollut Res. 22, 10669–10678.

Anjum, S.A., Farooq, M., Wang, L.C., Xue, L.L., Wang, S.G., Wang, L., Chen, M. 2011. Gas exchange and chlorophyll synthesis of maize cultivars are enhanced by exogenously-applied glycinebetaine under drought conditions. Plant Soil Environ. 57, 326–331.

Arshad, T., Maqbool, N., Javed, F., Wahid, A., Arshad, M.U. 2017. Enhancing the defensive mechanism of lead affected barley (*Hordeum vulgare* L.) genotypes by exogenously applied salicylic acid. J Agric Sci. 9, 139–146.

Asgher, M., Khan, M.I.R., Anjum, N.A., Verma, S., Vyas, D., Per, T.S., Masood, A., Khan, N.A., 2018. Ethylene and polyamines in counteracting heavy metal phytotoxicity: a crosstalk perspective. J Plant Growth Regul. 37(4), 1050–1065.

Ashraf, M., Akram, N.A., Arteca, R.N., Foolad, M.R. 2010. The physiological, biochemical and molecular roles of brassinosteroids and salicylic acid in plant processes and salt tolerance. Crit Rev Plant Sci. 29, 162–190.

Atkinson, N.J., Urwin, P.E. 2012. The interaction of plant biotic and abiotic stresses: from genes to the field. J Exp Bot. 63(10), 3523–3543.

ATSDR. 2003. Agency for Toxic Substances and Disease Registry. Available at: http://www.atsdr.cdc.gov/.

Aziz, M.A., Ahmad, H.R., Corwin, D.L., Sabir, M., Ozturk, M., Hakeem, K.R. 2016. Influence of farmyard manure on retention and availability of nickel, zinc and lead in metal-contaminated calcareous loam soils. J Environ Eng Landsc Manag. 25(3), 289–296.

Babu, N.G., Sarma, P.A., Attitalla, I.H., Murthy, S. 2010. Effect of selected heavy metal ions on the photosynthetic electron transport and energy transfer in the thylakoid membrane of the cyanobacterium, Spirulina platensis. Acad J Plant Sci. 3, 46–49.

Bai, X.Y., Dong, Y.J., Wang, Q.H., Xu, L.L., Kong, J., Liu, S. 2015. Effects of lead and nitric oxide on photosynthesis, antioxidative ability, and mineral element content of perennial ryegrass. Biol Plantarum. 59, 163–170.

Batista, B.L., Nigar, M., Mestrot, A. 2014. Identification and quantification of phytochelatins in roots of rice to long-term exposure: evidence of individual role on arsenic accumulation and translocation. The J Expt Bot. 65(6), 1467–1479.

Bharwana, S.A., Ali, S., Farooq, M.A., Iqbal, N., Hameed, A., Abbas, F., Ahmad, M.S.A. 2014. Glycine betaine-induced lead toxicity tolerance related to elevated photosynthesis, antioxidant enzymes suppressed lead uptake and oxidative stress in cotton. Turk J Bot. 38, 281–292.

Bhatti, K.H., Anwar, S., Nawaz, K., Hussain, K., Siddiqi, E.H., Sharif, R.U., Talat, A., Khalid, A. 2013. Effect of exogenous application of glycinebetaine on wheat (*Triticum aestivum* L.) Under Heavy Metal Stress. Middle-East J Sci Res. 14, 130–137.

Bielen, A., Remans, T., Vangronsveld, J., Cuypers, A. 2013. "The influence of metal stress on the availability and redox state of ascorbate, and possible interference with its cellular functions. Int J Mol Sci 14(3), 6382–6413.

Bohnert, H.J., Jensen, R.G. 1996. Strategies for engineering water stress tolerance in plants. Trends Biotechnol. 14, 89–97.

Boonyapookana, B., Upatham, E.S., Kruatrachue, M., Pokethitiyook, P., Singhakaew, S. 2002. Phytoaccumulation and phytotoxicity of cadmium and chromium in duckweed Wolffia globosa. Int J Phytoremed 4, 87–100.

Bradham, K.D., Dayton, E.A., Basta, N.T., Schroder, J., Payton, M.,Lanno, R.P. 2006. Effect of soil properties on lead bioavailability and toxicity to earthworms. Environ Toxicol. Chem. 25, 769–775. doi: 10.1897/04-552R.1.

Buchet, J.P., Lauwerys, R., Roels, H., Bernard, A., Bruaux, P., Claeys, F. et al. 1990. Renal effects of cadmium body burden of the general population. Lancet. 336, 699–702. doi: 10.1016/0140-6736(90)92201-R.

Burg, M.B. Ferraris, J.D. 2008. Intracellular organic osmolytes: function and regulation. J Biol Chem 283, 7309–7313.

Burritt, D.J. 2012. Proline and the Cryopreservation of Plant Tissues: Functions and Practical Applications, in Current Frontiers in Cryopreservation, I. Katkov, Ed., pp. 415–426, InTech.

Cai, Y., Ma, L.Q. 2003. Metal tolerance accumulation and detoxication in plants with emphasis on arsenic in terrestrial plants, in Proceedings of the ACS Symposium Series 835 on Biogeochemistry of Environmentally Important Trace Elements, Y. Cai and O. C. Braids, Eds., pp. 95–114, American Chemical Society.

Carbonell, A.A., Aarabi, M.A., Delaune, R.D., Grambrell, R.P., Patrick, W.H. 1998. Arsenic in wetland vegetation: availability, phytotoxicity, uptake and effects on plants growth and nutrition. Sci Total Environ. 217, 189–199. doi: 10.1016/S0048-9697(98)00195-8.

Cargnelutti, D., Tabaldi, L.A., Spanevello, R.M., de Oliveira Jucoski, G., Battisti, V., Redin, M. et al. 2006. Mercury toxicity induces oxidative stress in growing cucumber seedlings. Chemosphere 65, 999–1006. doi: 10.1016/j.chemosphere.2006.03.037.

Chen, L., Guo, Y., Yang, L., Wang, Q. 2008. Synergistic defensive mechanism of phytochelatins and antioxidative enzymes in Brassica chinensis L. against Cd stress. Chin Sci Bull. 53(10), 1503–1511.

Chen, T.H. and Murata, N. 2011. Glycinebetaine protects plants against abiotic stress: mechanisms and biotechnological applications. Plant Cell Environ. 34(1), 1–20.

Chen, T.H.H., Murata, N. 2002. Enhancement of tolerance of abiotic stress by metabolic engineering of betaines and other compatible solutes. Curr Opin Plant Biol. 5, 250–257.

Chen, X.M., Lung, C.S., Du, Y.Z., Chye, L.M. 2014. Engineering plants to tolerate abiotic stresses. Biocatal Agric Biotechnol. 3, 81–87.

Chugh, L.K., Sawhney, S.K. 1999. Photosynthetic activities of Pisum sativum seedlings grown in presence of cadmium. Plant Physiol Biochem. 37, 297–303.

Clemens, S. 2006. Toxic metal accumulation, responses to exposure and mechanisms of tolerance in plants. Biochimie. 88(11), 1707–1719.

Cobbett, C.S. 2000. Phytochelatins and their roles in heavy metal detoxification. Plant Physiol. 123(3), 825–832.

Cunningham R. P. 1997. DNA repair: caretakers of the genome? Curr Biol. 7, 576–579

Cuypers, A., Smeets, K., Vangronsveld, J. 2009. Heavy metal stress in plants, in Plant Stress Biology: From Genomics to Systems Biology, H. Hirt, Ed., Wiley-VCH Verlag GmbH & Co. KGaA: Weinheim, Germany.

Das, K., Roychoudhury, A. 2014. Reactive oxygen species (ROS) and response of antioxidants as ROS-scavengers during environmental stress in plants. Front. Environ Sci. 2, 53–78.

Demidchik, V. 2015. Mechanisms of oxidative stress in plants: from classical chemistry to cell biology. Environ Exp Bot. 109, 212–228.

Diamant, S., Eliahu, N., Rosenthal, D., Goloubinof, P. 2001. Chemical chaperones regulate molecular chaperones *in vitro* and in cells under combined salt and heat stresses. J Biol Chem. 276, 39586–39591.

Divi, U.K., Rahman, T., Krishna, P. 2010. Brassinosteroid-mediated stress tolerance in Arabidopsis shows interactions with abscisic acid, ethylene and salicylic acid pathways. BMC Plant Biol. 10, 151.

Du, J., Yang, J.L., Li, C.H. 2012. Advances in metallotionein studies in forest trees. Plant OMICS. 5(1), 46–51.

Duman, F., Aksoy, A., Aydin, Z., Temizgul, R. 2011. Effects of exogenous glycinebetaine and trehalose on cadmium accumulation and biological responses of an aquatic plant (Lemna gibba L.). Water Air Soil Pollut. 217, 545–556.

Emamverdian, A., Ding, Y., Mokhberdoran, F., Xie, Y. 2015. Heavy metal stress and some mechanism of plant defense response—A review. The Sci World J. 2015, 1–18. http://dx.doi.org/10.1155/2015/756120.

Faraz, A., Faizan, M., Sami, F., Siddiqui, H., Hayat, S. 2019. Supplementation of salicylic acid and citric acid for alleviation of cadmium toxicity to *Brassica juncea*. J. Plant Growth Regul. 2019, 1–15.

Fariduddin, Q., Varshney, P., Yousuf, M., Ali, A., Ahmad, A. 2013. Dissecting the role of glycine betaine in plants under abiotic stress. Plant Stress. 7, 8–18.

Farooq, M. A., Ali, S., Hameed, A., Bharwana, S. A., Rizwan, M., Ishaque, W., Farid, M., Mahmood, K., Iqbal, Z. 2016. Cadmium stress in cotton seedlings: physiological, photosynthesis andoxidative damages alleviated by glycinebetaine. South Afr J Bot, 104, 61–68.

Fidalgo, F., Azenha, M., Silva, A.F. 2013. Copper-induced stress in *Solanum nigrum* L. and antioxidant defense system response. Food and Energy Sec. 2(1), 70–80.

Flora, S.J.S. 2009. Structural, chemical and biological aspects of antioxidants for strategies against metal and metalloid exposure. Oxid Med Cell Long, 2(4), 191–206.

Genard, H., LeSaos, J., Hillard, J., Tremolieres, A., Boucaud, J. 1991. Effect of salinity on lipid composition, glycine betaine content and photosynthetic activity in chloroplasts of Suaeda maritime. Plant Physiol. Biochem. 29, 421–427.

Ghani, A., Khan, I., Ahmed, I., Mustafa, I., Abdur, R., Muhammad, N. 2015. Amelioration of lead toxicity in *Pisum sativum* (L.) by foliar application of salicylic acid. J Environ Anal Toxicol. 5(10), 4172–91.

Ghori, N.H., Ghori, T., Hayat, Imadi, S.R., Gul, A., Altay, V., Ozturk, M. 2019. Heavy metal stress and responses in plants. Int J Environ Sci Technol 16, 1807–1828.

Gill, S. S., and N. Tuteja. 2011. Cadmium stress tolerance in crop plants: probing the role of sulfur. Plant Signal. Behav. 6, 215–222. doi: 10.4161/psb.6.2.14880.

Gohari, M., Habib-Zadeh, A.R., Khayat, M. 2012. Assessing the intensity of tolerance to lead and its effect on amount of protein and proline in root and aerial parts of two varieties of rape seed (*Brassica napus* L.). J Basic Appl Sci Res. 2(1), 935–938.

Gondor, O.K., Pál, M., Darkó, É., Janda, T., Szalai, G. 2016. Salicylic acid and sodium salicylate alleviate cadmium toxicity to different extents in maize (*Zea mays* L.). PLoS ONE. 11, e0160157.

Grennan, A.K. 2011. Metallothioneins, a diverse protein family. Plant Physiol. 155(4), 1750–1751.

Groppa, M.D., Benavides, M.P. 2008. Polyamines and abiotic stress: recent advances. Amino Acids 34, 35–45.

Guo, J., Zhou, R., Ren, X., Jia, H., Hua, L., Xu, H., Lv, X., Zhao, J., Wei, T. 2018. Effects of salicylic acid, Epibrassinolide and calcium on stress alleviation and Cd accumulation in tomato plants. Ecotoxic Environ Safety. 157, 491–496.

Guo, J.L., Xu, L.P., Su, Y.C. 2013. ScMT2-1-3, a metallothionein gene of sugarcane, plays an important role in the regulation of heavy metal tolerance/accumulation. BioMed Res Inter. 2013, Article ID 904769.

Gupta, D., Vandenhove, H., Inouhe, M. 2013. Role of phytochelatins in heavy metal stress and detoxification mechanisms in plants, in Heavy Metal Stress Plants, pp. 73–94, Springer, Berlin.

Gupta, N., Thind, S.K., Bains, N.S. 2013. Glycine betaine application modifies biochemical attributes of osmotic adjustment in drought-stressed wheat. Plant Growth Regul. 72, 221–228.

Hameed, A., T. N. Qadri., M. Zaffar., T. O. Siddiqi., M. Ozturk., V. Altay., and P. Ahmad. 2017. Biochemical and nutritional responses of Abelmoschus esculentus L. exposed to mercury contamination. Fresenius Environ Bull. 26(10):5814–5823

Handa, N., Kohli, S.K., Kaur, R., Sharma, A., Kumar, V., Thukral, A.K., Arora, S. and Bhardwaj, R., 2018. Role of compatible solutes in enhancing antioxidative defense in plants exposed to metal toxicity, in Plants under Metal and Metalloid Stress, Springer, Singapore, pp. 207–228.

Hanson, A.D., Scott, N.A. 1980. Betaine synthesis from radioactive precursors in attached, water-stressed barley leaves. Plant Physiol. 66, 342–348.

Hasan, M.K., Cheng, Y., Kanwar, M.K., Chu, X.Y., Ahammed, G.J., Qi, Z.Y. 2017. Responses of plant proteins to heavy metal stress—A review. Front Plant Sci. 8, 1492.

Hasan, S. A., Hayat, S., Ahmad, A. 2011. Brassinosteroids protect photosynthetic machinery against the cadmium induced oxidative stress in two tomato cultivars. Chemosphere. 84(10), 1446–1451.

Hasanuzzaman, M., Matin, M.A., Fardus, J., Hasanuzzaman, M., Hossain, M.S., Parvin, K. 2019. Foliar application of salicylic acid improves growth and yield attributes by upregulating the antioxidant defense system in Brassica campestris plants grown in lead-amended soils. Acta Agrobot. 2019, 72, doi:10.5586/aa.1765.

Hassan, T.U., Bano, A., Naz, I. 2017. Alleviation of HMs toxicity by the application of plant growth promoting rhizobacteria and effects on wheat grown in saline sodic field. Int J Phytoremed. 19, 522–529. doi: 10.1080/15226514.2016.1267696.

Hayat, S., Hayat, Q., Alyemeni, M.N., Ahmad, A. 2013. Proline enhances antioxidative enzyme activity, photosynthesis and yield of *Cicer arietinum* L. exposed to cadmium stress. Acta Botanica Croatica. 72(2), 323–335.

He, X., Richmond, M.E., Williams, D.V., Zheng, W., Wu, F. 2019. Exogenous glycine betaine reduces cadmium uptake and mitigates cadmium toxicity in two tobacco genotypes differing in cadmium tolerance. Int J Mol Sci. 20, 1612.

Heiss, S., Wachter, A., Bogs, J., Cobbett, C., Rausch, T. 2003. Phytochelatin synthase (PCS) protein is induced in *Brassica juncea* leaves after prolonged Cd exposure. J Expt Bot. 54(389), 1833–1839.

Helleday, T., Nilsson, R., Jenssen, D. 2000. Arsenic (III) and heavy metal ions induce intrachromosomal homologous recombination in the hprt gene of V79 Chinese hamster cells. Environ Mol Mutagen. 35, 114–122. doi: 10.1002/(SICI)1098-2280(2000)35:2<114::AID-EM6>3.0.CO;2-Q.

Hentz, S., McComb, J., Miller, G., Begonia, M., Begonia, G. 2012. Cadmium uptake, growth and phytochelatin contents of *Triticum aestivum* in response to various concentrations of cadmium. World Environ. 2(3), 44–50.

Hong, Z., Lakkineni, K., Zhang, Z., Verma, D.P.S. 2000. Removal of feedback inhibition of delta1- pyrroline-5-carboxylate synthetase results in increased proline accumulation and protection of plants from osmotic stress. Plant Physiol. 122, 1129–1136.

Hossain, M.A., Hasanuzzaman, M., Fujita, M. 2010. Up-regulation of antioxidant and glyoxalase systems by exogenous glycine betaine and proline in mung bean confer tolerance to cadmium stress. Physiol Mol Biol Plants. 16, 259–272.

Hossain, M.A., Hossain, M.D., Rohman, M.M, da Silva, J.A.T., Futija, M. 2012a. Onion major compounds (flavonoids, organosulfurs) and highly expressed glutathione-related enzymes: possible physiological interaction, gene cloning and abiotic stress response, in Onion Consumption and Health, C. B. Aguirre and L. M. Jaramillo, Eds., Nova, New York.

Hossain, M.A., Hossain, M.Z., Futija, M. 2009. Stress-induced changes of methylglyoxal level and glyoxalase I activity in pumpkin seedlings and cDNA cloning of glyoxalase I gene. Aust J Crop Sci. 3(2), 53–64.

Hossain, M.A., Piyatida, P., da Silva, J.A.T., Fujita, M. 2012b. Molecular mechanism of heavy metal toxicity and tolerance in plants: central role of glutathione in detoxification of reactive oxygen species and methylglyoxal and in heavy metal chelation. J Bot. 2012:1–37, doi:10.1155/2012/872875

Huang, G.Y., Wang, Y.S. 2009. Expression analysis of type 2 metallothionein gene in mangrove species (*Bruguiera gymnorrhiza*) under heavy metal stress. Chemosphere. 77(7), 1026–1029.

Huang, J., Hirji, R., Adam, L., Rozwadowski, K.L., Hammerlindl, J.K., Keller, W.A., Selvaraj G. 2000. Genetic engineering of glycinebetaine production toward enhancing stress tolerance in plants: metabolic limitations. Plant Physiol, 122, 747–756.

Hussein, M., Balbaa, L., Gaballah, M. 2007. Salicylic acid and salinity effects on growth of maize plants. Res J Agricul Biol Sci. 3, 321–328.

Ibrahim, A.H., Aldesuquy, H.S. 2003. Glycine betaine and shikimic acid—induced modification in growth criteria, water relation and productivity of droughted sorghum bicolor plants. Phyton Horn. 43, 351–361.

Iqbal, N., Ashraf, M., Ashraf, M.Y. 2008. Glycinebetaine, an osmolyte of interest to improve water stress tolerance in sunflower (*Helianthus annuus* L.): water relations and yield. South Afr J Bot. 74, 274–281.

Islam, F., Yasmeen, T., Arif, M.S., Riaz, M., Shahzad, S.M., Imran, Q., Ali, I. 2016. Combined ability of chromium (Cr) tolerant plant growth promoting bacteria (PGPB) and salicylic acid (SA) in attenuation of chromium stress in maize plants. Plant Physiol Biochem. 108, 456–467.

Islam, M.R. 2014. Heavy metals and their impacts on agriculture and environment. A class lecture arranged by BARC, Bangladesh (31 March 2014).

Islam, M.M., Anamul-Hoque, M., Okuma, E., Banu, N.A., Shimoishi, Y., Nakamura, Y., Murata, Y. 2009. Exogenous proline and glycinebetaine increase antioxidant enzyme activities and confer tolerance to cadmium stress in cultured tobacco cells. J Plant Physiol. 166, 1587–1597.

Israr, M., Sahi, S., Datta, R., Sarkar, D. 2006. Bioaccumulation and physiological effects of mercury in Sesbania drummondii. Chemosphere 65, 591–598. doi: 10.1016/j.chemosphere.2006.02.016.

Jabeen, N., Abbas, Z., Iqbal, M., Rizwan, M., Jabbar, A., Farid, M., Ali, S., Ibrahim, M., Abbas, F. 2016. Glycinebetaine mediates chromium tolerance in mung bean through lowering of Cr uptake and improved antioxidant system. Arch Agron Soil Sci. 62, 648–662.

Jeandroz, S., Lamotte, O. 2017. Plant responses to biotic and abiotic stresses: lessons from cell signaling. Front Plant Sci. 8, 1772.

Jozefczak, M., Remans, T., Vangronsveld, J., Cuypers, A. 2012. Glutathione is a key player in metal-induced oxidative stress defenses. Int J Mol Sci. 13(3), 3145–3175.

Karimi, L.N., Khanahmadi, M., Moradi, B. 2012. Accumulation and phytotoxicity of lead in *Cynara scolymus*. Indian J Sci Technol. 5(11), 3634–3641.

Kasprzak, K.S. 1995. Possible role of oxidative damage in metal induced carcinogenesis. Cancer Invest. 13, 411–430.

Khan, A., Khan, S., Khan, M.A., Aamir, M., Ullah, H.J., Nawab, J., Rehman, I.U., Shah, J. 2018. Heavy metals effects on plant growth and dietary intake of trace metals in vegetables cultivated in contaminated soil. Int J Environ Sci Technol. 16, 2295–2304.

Khan, M.I.R., Iqbal, N., Masood, A., Per, T.S., Khan, N.A. 2013. Salicylic acid alleviates adverse effects of heat stress on photosynthesis through changes in proline production and ethylene formation. Plant Sign Behav. 2013, 8, e26374.

Khan, M.S., Yu, X., Kikuchi, A., Asahina, M., Watanabe, K.N. 2009. Genetic engineering of glycine betaine biosynthesis to enhance abiotic stress tolerance in plants. Plant Biotechnol, 26, 125–134.

Khan, N., Samiullah, Singh, S., Nazar, R. 2007. Activities of antioxidative enzymes, sulphur assimilation, photosynthetic activity and growth of wheat (*Triticum aestivum*) cultivars differing in yield potential under cadmium stress. J Agron Crop Sci. 193, 435–444.

Khokon, M.A.R., Okuma, E., Hossain, M.A., Munemasa, S., Uraji, M., Nakamura, Y., Mori, I.C., Murata, Y. 2011. Involvement of extracellular oxidative burst in salicylic acid-induced stomatal closure in Arabidopsis. Plant Cell Environ. 34, 434–443.

Kishitani, S., Takanami, T., Suzuki, M., Oikawa, M., Yokoi, S., Ishitani, M. 2000. Compatibility of glycinebetaine in rice plants: Evaluation using transgenic rice plants with a gene for peroxisomal betaine aldehyde dehydrogenase from barley. Plant Cell Environ. 23, 107–114.

Kohler, A., Blaudez, D., Chalot, M., Martin, F. 2004. Cloning and expression of multiple metallothioneins from hybrid poplar. New Phytologist. 164(1), 83–93.

Kohli, S.K., Handa, N., Kaur, R., Kumar, V., Khanna, K., Bakshi, P., Singh, R., Arora, S., Kaur, R. and Bhardwaj, R., 2017. Role of salicylic acid in heavy metal stress tolerance: insight into underlying mechanism, in Salicylic Acid: A Multifaceted Hormone, Springer, Singapore, pp. 123–144.

Kotrba, P., Macek, T., Ruml, T. 1999. Heavy metal-binding peptides and proteins in plants. A review. Coll Czechoslovak Chem Commun. 64(7), 1057–1086.

Kumar, B., Smita, K., Flores, L.C. 2017. Plant mediated detoxification of mercury and lead. Arabian J Chem. 10, S2335–S2342. doi: 10.1093/pcp/pcu117.

Kurepin, L.V., Ivanov, A.G., Zaman, M., Pharis, R.P., Hurry, V., Hüner, N.P. 2017. Interaction of glycine betaine and plant hormones: Protection of the photosynthetic apparatus during abiotic stress, in Photosynthesis: Structures, Mechanisms, and Applications, Springer: Cham, Switzerland, pp. 185–202.

Lee, C.W., Choi, J.M., Pak, C.H. 1996. Micronutrient toxicity in seed germination (Pelargonium x Hortorum Bailey). J Am Soc Hortic Sci. 121, 77–82.

Lee, S., Moon, J.S., Ko, T.S., Petros, D., Goldsbrough, P.B., Korban, S.S. 2003. Overexpression of Arabidopsis phytochelatin synthase paradoxically leads to hypersensitivity to cadmium stress. Plant Physiol. 131(2), 656–663.

Leszczyszyn, O.I., Imam, H.T., Blindauer, C.A. 2013. Diversity and distribution of plant metallothioneins: a review of structure, properties and functions. Metallomics. 5(9), 1146–1169.

Liu, N., Lin, S., Huang, B. 2017. Differential effects of glycine betaine and spermidine on osmotic adjustment and antioxidant defense contributing to improved drought tolerance in creeping bentgrass. J Am Soc Hortic Sci. 142, 20–26.

Lou, Y., Yang, Y., Hu, L., Liu, H., Xu, Q. 2015. Exogenous glycinebetaine alleviates the detrimental effect of Cd stress on perennial ryegrass. Ecotoxicology. 24, 1330–1340.

Macovei, A., Ventura, L., Donà, M., Faè, M., Balestrazzi, A., Carbonera, D. 2010. Effects of heavy metal treatments on metallothionein expression profiles in white poplar (*Populus alba* L) cell suspension cultures. Analele Universitatii din Oradea-Fascicula Biologie. 18(2), 274–279.

Mahajan, S., Tuteja, N. 2005. Cold, salinity and drought stresses: an overview. Arch Biochem Biophys. 444, 139–158.

Mahdavian, K., Ghaderian, S.M., Schat, H. 2016. Pb accumulation, Pb tolerance, antioxidants, thiols, and organic acids in metallicolous and non-metallicolous Peganum harmala L. under Pb exposure. Environ Exp Bot. 126, 21–31.

Malar, S., Sahi, S.V., Favas, P.J.C., Venkatachalam, P. 2015. Mercury heavy-metal-induced physiochemical changes and genotoxic alterations in water hyacinths [Eichhornia crassipes (Mart.)]. Environ Sci Pollut Res. 22(6), 4597–4608.

Manara, A. 2012. Plant responses to heavy metal toxicity, in Plants and Heavy Metals, A. Furini, Ed., Springer Briefs in Molecular Science, pp. 27–53, Springer, Dordrecht, Netherlands.

Masindi, V., Muedi, K. 2018. Environmental contamination by heavy metals, in Heavy Metals, Hosam M. El-Din and F. Saleh Refaat Aglan, Eds., IntechOpen, Rijeka, Croatia.

Mattioli, R., Marchese, D., D'Angeli, S., Altamura, M.M., Costantino, P., Trovato, M. 2008. Modulation of intracellular proline levels affects flowering time and in florescence architecture in Arabidopsis. Plant Mol. Biol. 66, 277–288.

Mishra, S., Dubey, R.S. 2006. Heavy metal uptake and detoxification mechanisms in plants. International J Agril. Res. 1(2), 122–141.

Miura, K., Tada, Y. 2014. Regulation of water, salinity, and cold stress responses by salicylic acid. Front. Plant Sci. 5, 4. doi.10.3389/fpls.2014.00004

Moharramnejad, S., Sofalian, O., Valizadeh, M., Asgari, A., Shiri, M. 2015. Proline, glycine betaine, total phenolics and pigment contents in response to osmotic stress in maize seedlings. J Biosci Biotechnol. 4, 313–319.

Mohsenzadeh, S., Shahrtash, M., Mohabatkar, H. 2011. Interactive effects of salicylic acid and silicon on some physiological responses of cadmium-stressed maize seedlings. Iranian J Sci Tech. (Sciences) 35, 57–60.

Mourato, M., Reis, R., Martins, L.L. 2012. Characterization of plant antioxidative system in response to abiotic stresses: a focus on heavy metal toxicity, in Advances in Selected Plant Physiology Aspects, G. Montanaro and B. Dichio, Eds., InTech, Vienna, Austria, pp. 23–44. http://hdl.handle.net/10400.5/4410.

Nagajyoti, P.C., Lee, K.D., Sreekanth, T.V.M. 2010. Heavy metals, occurrence and toxicity for plants: a review. Environ Chem Lett. 8(3), 199–216.

Neelima, P, Reddy, K.J. 2002. Interaction of copper and cadmium with seedlings growth and biochemical responses in Solanum melongena. Environ Pollut Technol. 1, 285–290.

Nikolic, N., Kojic, D., Pilipovic, A. et al. 2008. Responses of hybrid poplar to cadmium stress: photosynthetic characteristics, cadmium and proline accumulation, and antioxidant enzyme activity. Acta Biologica Cracoviensia Series Botanica. 50(2), 95–103.

Nusrat, N., Shahbaz, M., Perveen, S. 2014. Modulation in growth, photosynthetic efficiency, activity of antioxidants and mineral ions by foliar application of glycine betaine on pea (*Pisum sativum* L.) under salt stress. Acta Physiol Plant. 36, 2985–2998.

Ortega-Villasante, C., Rellán-Álvarez, R., Del Campo, F.F., Carpena-Ruiz, R.O., Hernández, L.E.2005. Cellular damage induced by cadmium and mercury in Medicago sativa. J Exp Bot. 56, 2239–2251. doi: 10.1093/jxb/eri223.

Ozturk, M., Ashraf, M., Aksoy, A., Ahmad, M.S.A. and Hakeem, K.R. (Eds.). 2015. Plants, Pollutants and Remediation. Springer, Dordrecht. pp. 1–404. doi: 10.1007/978-94-017-7194-8.

Ozturk, M., Yucel, E., Gucel, S., Sakcali, S., Aksoy, A. 2008. Plants as biomonitors of trace elements pollution in soil, in Trace Elements: Environmental Contamination, Nutritional Benefits and Health Implications, M. N. V. Prasad, Ed., Wiley, New York, pp. 723–744.

Pagani, M.A., Tomas, M., Carrillo, J. et al. 2012. The response of the different soybean metallothionein isoforms to cadmium intoxication. J Inorganic Biochem. 117, 306–315.

Panuccio, M.R., Sorgona, A., Rizzo, M., Cacco, G. 2009. Cadmium adsorption on vermiculite, zeolite and pumice: batch experiment studies. J Environ Manage. 90, 364–374. doi: 10.1016/j.jenvman.2007.10.005.

Pareek, S.L.S., Yadav, S.K., Pareek, A., Reddy, M.K., Soproy, S.K. 2006. Transgenic tobacco overexpressing glyoxalase pathway enzymes grow and set viable seeds in zinc spiked soils. Plant Physiol. 140(2), 613–623.

Park, E.J., Jeknic, Z., Pino, M. T., Murata, N., Chen, T. H. 2007. Glycinebetaine accumulation is more effective in chloroplasts than in the cytosol for protecting transgenic tomato plants against abiotic stress. Plant Cell Environ. 30, 994–1005.

Paul, S., Roychoudhury, A., Banerjee, A. et al. 2017. Seed pretreatment with spermidine alleviates oxidative damages to different extent in the salt (NaCl)-stressed seedlings of three Indica rice cultivars with contrasting level of salt tolerance. Plant Gene. 11, 112–123.

Pavlikova, D., Pavlik, M., Staszkova, L., Tlustos, P., Szakova, J., Balik, J. 2007. The effect of potentially toxic elements and sewage sludge on the activity of regulatory enzyme glutamate kinase. Plant, Soil Environ. 53(5), 201–206.

Pehlivan, E., Özkan, A.M., Dinc, S., Parlayici, S. 2009. Adsorption of Cu2+ and Pb2+ ion on dolomite powder. J Hazard Mater. 167, 1044–1049. doi: 10.1016/j.jhazmat.2009.01.096.

Pierart, A., Shahid, M., Séjalon-Delmas, N., Dumat, C. 2015. Antimony bioavailability: knowledge and research perspectives for sustainable agricultures. J Hazard Mater. 289, 219–234. doi: 10.1016/j.jhazmat.2015.02.011.

Pongrac, P., Zhao, F.J., Razinger, J., Zrimec, A., Regvar, M. 2009. Physiological responses to Cd and Zn in two Cd/Zn hyperaccumulating Thlaspi species. Environ Exp Bot. 66, 479–486.

Pourrut, B., Jean, S., Silvestre, J., Pinelli, E. 2011. Lead-induced DNA damage in Vicia faba root cells: potential involvement of oxidative stress. Mutat Res. 726, 123–128. doi: 10.1016/j.mrgentox.2011.09.001.

Pourrut, B., Pohu, A.L., Pruvot, C., Garçon, G., Verdin, A., Waterlot, C., Bidar, G., Shirali, P., Douay, F. 2011. Assessment of flash-aided phytostabilisation of highly contaminated soils after an 8-year field trial Part 2. Influence on plants. Sci. Total Environ. 409, 4504–4510.

Qin, D., Zhao, C.L., Liu, X.Y., Wang, P.W. 2017. Transgenic soybeans expressing betaine aldehyde dehydrogenase from Atriplex canescens show increased drought tolerance. Plant Breed. 136, 699–709.

Rascio, N., Navari-Izzo, F. 2011. Heavy metal hyperaccumulating plants: how and why do they do it? And what makes them so interesting? Plant Sci. 180(2), 169–181.

Rasheed, R., Ashraf, M.A., Hussain, I., Haider, M.Z., Kanwal, U., Iqbal, M. 2014. Exogenous proline and glycinebetaine mitigate cadmium stress in two genetically different spring wheat (*Triticum aestivum* L.) cultivars. Braz J. Bot. 37, 399–406.

Rastgoo, L., Alemzadeh, A., Afsharifar, A. 2011. Isolation of two novel isoforms encoding zinc- and copper-transporting P1B-ATPase from Gouan (*Aeluropus littoralis*). Plant Omics J. 4(7), 377–383.

Rivas-San Vicente, M., Plasencia, J. 2011. Salicylic acid beyond defence: its role in plant growth and development. J Exp Bot. 62, 3321–3338.

Rizwan, M., Ali, S., Abbas, T., Adrees, M., Zia-ur-Rehman, M., Ibrahim, M., Abbas, F., Qayyum, M.F., Nawaz, R. 2018. Residual effects of biochar on growth, photosynthesis and cadmium uptake in rice (Oryza sativa L.) under Cd stress with different water conditions. J Environ Manag. 206, 676–683.

Rodríguez, M., Canales, E., Borrás-Hidalgo, O. 2005. Molecular aspects of abiotic stress in plants. Biotecnol Apl. 22, 1–10.

Roychoudhury, A., Basu, S., Sengupta, D.N. 2011. Amelioration of salinity stress by exogenously applied spermidine or spermine in three varieties of Indica rice differing in their level of salt tolerance. J Plant Physiol. 168, 317–328.

Ruciniska-Sobkowiak, R. 2010. Oxidative stress in plants exposed to heavy metals. Postepy Biochem. 56, 191–200.

Ruscitti, M., Arango, M., Ronco, M., Beltrano, J. 2011. Inoculation with mycorrhizal fungi modifies praline metabolism and increases chromium tolerance in pepper plants (*Capsicum annuum* L.). Brazilian J Plant Physiol. 23(1), 15–25.

Saba, H., Jyoti, P., Neha, S. 2013. Mycorrhizae and phytochelators as remedy in heavy metal contaminated land remediation. Int Res J Environ Sci. 2(1), 74–78.

Sakamoto, A., Murata, N. 2000. Genetic engineering of glycinebetaine synthesis in plants: current status and implications for enhancement of stress tolerance. J Exp Bot. 51, 81–88.

Salla, V., Hardaway, C.J., Sneddon, J. 2011. Preliminary investigation of Spartina alterniflora for phytoextraction of selected heavy metals in soils from Southwest Louisiana. Microchem J. 97, 207–212. doi: 10.1016/j.microc.2010.09.005.

Sandalio, L.M., Dalurzo, H.C., Gomez, M., Romero-Puertas, M.C., del Rio, L.A. 2001. Cadmium-induced changes in the growth and oxidative metabolism of pea plant. J Exp Bot. 52, 2115–2126. doi: 10.1093/jexbot/52.364.2115.

Schmfger, M.E.V. 2001. Phytochelatins: complexation of metals and metalloids, studies on the phytochelatin synthase, Ph.D. Thesis, Munich University of Technology (TUM), Munich.

Schutzendübel, A., Polle, A. 2002. Plant responses to abiotic stresses: heavy metal-induced oxidative stress and protection by mycorrhization. The J Exp Bot. 53(372), 1351–1365.

Seregin, I.V., Kozhevnikova, A.D., Kazyumina, E.M., Ivanov, V.B. 2003. Nickel toxicity and distribution in maize roots. Russian J Plant Physiol. 50(5), 711–717.

Serraj, R., Sinclair, T.R. 2002. Osmolyte accumulation: can it really help increase crop yield under drought conditions? Plant Cell Environ. 25, 333–341.

Shahid, M., Khalid, S., Abbas, G., Shahid, N., Nadeem, M., Sabir, M. et al. 2015. Heavy metal stress and crop productivity, in Crop Production and Global Environmental Issues, K. R. Hakeem, Ed., Springer International Publishing, Cham, pp. 1–25.

Shahid, M., Pourrut, B., Dumat, C., Nadeem, M., Aslam, M., Pinelli, E. 2014. Heavy-metal-induced reactive oxygen species: Phytotoxicity and physicochemical changes in plants. Rev Environ Contamin Toxicol. 232, 1–44.

Shahid, M.A., Balal, R.M., Pervez, M. A. et al. 2014. Exogenous proline and proline enriched *Lolium perenne* leaf extract protects against phytotoxic effects of nickel and salinity in *Pisum sativum* by altering polyamine metabolism in leaves. Turkish J Bot. 38(5), 914–926.

Shanker, A.K., Cervantes, C., Loza-Tavera, H. and Avudainayagam, S., 2005. Chromium toxicity in plants. Environ Int. 31(5), 739–753.

Shao, G., Chen, M., Wang, W. and Zhang, G. 2008. The effect of salinity pretreatment on Cd accumulation and Cd-induced stress in BADH-transgenic and nontransgenic rice seedlings. J Plant Growth Regul. 27, 205–210.

Sharma, P., Dubey, R.S. 2006. Cadmium uptake and its toxicity in higher plants, in Cadmium Toxicity and Tolerance in Plants, N. A. Khan and Samiullah, Eds., Narosa Publishing House, New Delhi, pp. 64–86.

Sharma, S.S., Dietz, K.J. 2006. The significance of amino acids and amino acid-derived molecules in plant responses and adaptation to heavy metal stress. The J Expert Bot. 57(4), 711–726.

Shi, G., Cai, Q., Liu, Q., Wu, L. 2009. Salicylic acid-mediated alleviation of cadmium toxicity in hemp plants in relation to cadmium uptake, photosynthesis, and antioxidant enzymes. Acta Physiol Plant. 31, 969–977.

Shukla, D., Tiwari, M., Tripathi, R.D., Nath, P., Trivedi, P.K. 2013. Synthetic phytochelatins complement a phytochelatin deficient *Arabidopsis* mutant and enhance the accumulation of heavy metal(loid)s. Biochem Biophys Res Commun. 434(3), 664–669.

Sidhu, G.P.S., Singh, H.P., Batish, D.R., Kohli, R.K. 2016. Effect of lead on oxidative status, antioxidative response and metal accumulation in *Coronopus didymus*. Plant Physiol Biochem. 105, 290–296.

Sidhu, G.P.S., Singh, H.P., Batish, D.R., Kohli, R.K. 2017. Tolerance and hyperaccumulation of cadmium by a wild, unpalatable herb *Coronopus didymus* (L.) Sm. (Brassicaceae). Ecotox Environ Safety. 135, 209–215.

Singh, S., Parihar, P., Singh, R., Singh, V.P., Prasad, S.M. 2016. Heavy metal tolerance in plants: role of transcriptomics, proteomics, metabolomics, and ionomics. Front Plant Sci. 6, 1143. https://doi.org/10.3389/fpls.2015.01143.

Singh, S., Srivastava, P.K., Kumar, D. et al. 2015. Morpho-anatomical and biochemical adapting strategies of maize (Zea mays L.) seedlings against lead and chromium stresses. Biocatal Agric Biotechnol. 4(3), 286–295.

Siripornadulsil, S., Traina, S., Verma, D. P. S. and Sayre, R. T. 2002. Molecular mechanisms of proline-mediated tolerance to toxic heavy metals in transgenic microalgae. Plant Cell. 14, 2837–2847.

Slama, I., Abdelly, C., Bouchereau, A., Flowers, T., Savoure, A. 2015. Diversity, distribution and roles of osmoprotective compounds accumulated in halophytes under abiotic stress. Ann Bot. 115, 433–447. doi: 10.1093/aob/mcu239.

Song, W.Y., Yang, H.C., Shao, H.B., Zheng, A.Z., Brestic, M. 2014. The alleviative effects of salicylic acid on the activities of catalase and superoxide dismutase in malting barley (*Hordeum vulgare* L.) seedling leaves stressed by heavy metals. CLEAN–Soil, Air, Water. 42, 88–97.

Suzuki, N., Rivero, R.M., Shulaev, V. et al. 2014. Abiotic and biotic stress combinations. New Phytol. 203, 32–43.

Szabados, L., Savoure, A. 2010. Proline: a multifunctional amino acid. Trends Plant Sci. 15(2), 89–97.

Szalai, G., Krantev, A., Yordanova, R., Popova, L.P., Janda, T. 2013. Influence of salicylic acid on phytochelatin synthesis in *Zea mays* during Cd stress. Turkish J Bot. 37(4), 708–714.

Szekely, G., Abraham, E., Cseplo, A., Rigo, G., Zsigmond, L., Csiszar, J., Ayaydin, F., Strizhov, N., Jasik, J., Schmelzer, E., Koncz, C., Szabados, L. 2008. Duplicated P5CS genes of Arabidopsis play distinct roles in stress regulation and developmental control of proline biosynthesis. Plant J. 53, 11–28.

Tangahu, B.V., Sheikh Abdullah, S.R., Basri, H., Idris, M., Anuar, N., Mukhlisin, M. 2011. A review on heavy metals (As, Pb, and Hg) uptake by plants through phytoremediation. Int J Chem Eng. 2011, 1–31.

Thangavel, P., Long, S., Minocha, R. 2007. Changes in phytochelatins and their biosynthetic intermediates in red spruce (*Picea rubens* Sarg.) cell suspension cultures under cadmium and zinc stress. Plant Cell, Tissue Organ Cult. 88(2), 201–216.

Tian, F., Wang, W., Liang, C., Wang, X., Wang, G., Wang, W. 2017. Over accumulation of glycine betaine makes the function of the thylakoid membrane better in wheat under salt stress. Crop J. 5, 73–82.

Tiwari, S., Lata, C. 2018a. Heavy metal stress, signaling, and tolerance due to plant-associated microbes: an overview. Front Plant Sci. 9(April), 1–12. https://doi.org/10.3389/fpls.2018.00452.

Tiwari, S., Lata, C. 2018b. Heavy Metal Stress, Signaling, and Tolerance Due to Plant-Associated Microbes: An Overview. Front. Plant Sci. 9, 452.

Todeschini, V., Lingua, G., D'Agostino, G., Carniato, F., Roccotiello, E., Berta, G. 2011. Effects of high zinc concentration on poplar leaves: a morphological and biochemical study. Environ Exp Bot. 71(1), 50–56.

Tripathi, A., Liu, S., Singh, P.K. et al. 2017a. Differential phytotoxic responses of silver nitrate (AgNO3) and silver nanoparticle (AgNps) in Cucumis sativus L. Plant Gene. 11, 255–264.

Tripathi, A., Tripathi, D.K., Chauhan, D.K., Kumar, N. 2016a. Chromium (VI)-induced phytotoxicity in river catchment agriculture: evidence from physiological, biochemical and anatomical alterations in Cucumis sativus (L.) used as model species. Chem Ecol. 32(1), 12–33.

Tripathi, D.K., Shweta, S.S., Yadav, V. et al. 2017b. Silicon: a potential element to combat adverse impact of UV-B in plants, in UV-B Radiation: From Environmental Stressor to Regulator of Plant Growth, V. P. Singh, S. Singh, S. M. Prasad and P. Parihar, Eds., vol. 1, Wiley Blackwell, Oxford, pp. 175–195.

Tripathi, D.K., Singh, S., Singh, S. et al. 2016b. Silicon as a beneficial element to combat the adverse effect of drought in agricultural crops, in Water Stress and Crop Plants: A Sustainable Approach, P. Ahmad, Ed., Wiley Blackwell, Oxford, pp. 682–694.

Tripathi, D.K., Singh, S., Singh, S., Mishra, S., Chauhan, D.K., Dubey, N.K. 2015. Micronutrients and their diverse role in agricultural crops: Advances and future prospective. Acta Physiol Plant. 37, 1–14.

Tripathi, R.D., Srivastava, S., Mishra, S., Singh, N., Tuli, R., Gupta, D.K. et al. 2007. Arsenic hazards: strategies for tolerance and remediation by plants. Trends Biotechnol. 25, 158–165. doi: 10.1016/j.tibtech.2007.02.003.

Tyagi, A., Sairam, R.K. 2004. Physiology and molecular biology of salinity stress tolerance in plants. Curr Sci. 86(3), 407–420.

Valko, M., Morris, H., Cronin, M.T.D. 2005. Metals, toxicity and oxidative stress. Curr Med Chem. 12(10), 1161–1208.

Van Assche, F., Clijsters, H. 1990. Effects of metals on enzyme activity in plants. Plant Cell Environ. 13, 195–206. doi: 10.1111/j.1365-3040.1990.tb01304.x

Vatansever, R., Ozyigit, I.I., Filiz, E. 2017. Essential and beneficial trace elements in plants, and their transport in roots: a review. Appl Biochem Biotechnol. 181(1), 464–482.

Verma, S., Dubey, R.S. 2001. Effect of Cadmium on soluble sugars and enzymes of their metabolism in rice. Biol Plant. 44, 117–123. doi: 10.1023/A:1017938809311.

Verma, S., Verma, P.K., Meher, A.K., Dwivedi, S., Bansiwal, A.K., Pande, V. et al. 2016. A novel arsenic methyltransferase gene of Westerdykella aurantiaca isolated from arsenic contaminated soil: phylogenetic, physiological, and biochemical studies and its role in arsenic bioremediation. Metallomics 8, 344–353. doi: 10.1039/c5mt00277j.

Vlot, A.C., Dempsey, D.M.A., Klessig, D.F. 2009. Salicylic acid, a multifaceted hormone to combat disease. Ann Rev Phytopathol. 47, 177–206.

Wang, H.C., Wu, J.S., Chia, J.C., Yang, C.C., Wu, Y.J., Juang, R.H. 2009. Phytochelatin synthase is regulated by protein phosphorylation at a threonine residue near its catalytic site. J Agricul Food Chem. 57(16), 7348–7355.

Wong, H.L., Sakamoto, T., Kawasaki, T., Umemura, K., Shimamoto, K. 2004. Down-regulation of metallothionein, a reactive oxygen scavenger, by the small GTPase OsRac1 in rice. Plant Physiol. 135(3), 1447–1456.

Xalxo, R., Yadu, B., Chakraborty, P., Chandrakar, V., Keshavkant, S. 2017. Modulation of nickel toxicity by glycinebetaine and aspirin in Pennisetum typhoideum. Acta Biol. Szeged. 61, 163–171.

Xiong, T., Leveque, T., Shahid, M., Foucault, Y., Mombo, S., Dumat, C. 2014. Lead and cadmium phytoavailability and human bioaccessibility for vegetables exposed to soil or atmospheric pollution by process ultrafine particles. J Environ Qual. 43, 1593–1600. doi: 10.2134/jeq2013.11.0469.

Yadav, S.K. 2010. Heavy metals toxicity in plants: an overview on the role of glutathione and phytochelatins in heavy metal stress tolerance of plants. South Afr J Bot. 76, 167–179. doi: 10.1016/j.sajb.2009.10.007.

Yadav, S.K., Pareek, S.L.S., Ray, M., Reddy, M.K., Soproy, S.K. 2005a. Transgenic tobacco plants overexpressing glyoxalase enzymes resist an increase in methyl-glyoxal and maintain higher reduced glutathione levels under salinity stress. FEBS Lett. 579(27), 6265–6271.

Yadav, S.K., Pareek, S.L.S., Ray, M., Reddy, M.K., Soproy, S.K. 2005b. Methylglyoxal levels in plant under salinity stress are dependent on glyoxalase I and glutathione. Biochem Biophys Res Commun. 337(1), 61–67.

Yamada, M., Morishita, H., Urano, K., Shiozaki, N., Yamaguchi-Shinozaki, K., Shinozaki, K. and Yoshiba, Y. 2005. Effects of free proline accumulation in petunias under drought stress. J Exp Bot. 56, 1975–1981.

Yancey, P.H. 1994. Compatible and counteracting solutes, in Cellular and Molecular Physiology of Cell Volume Regulation, K.Strange, Ed., CRC, Boca Raton, pp. 81–109.

Yancey, P.H. 2005. Organic osmolytes as compatible, metabolic and counteracting cytoprotectants in high osmolarity and other stresses. J Exp Biol. 208, 2819–2830.

Yancey, P.H., Clark, M.E., Hand, S.C., Bowlus, R.D., Somero, G.N. 1982. Living with water stress: evolution of osmolyte systems. Sci. 217, 1214–1222.

Yang, N., Wang, C.L., He, W.P., Qu, Y.Z., Li, Y.S. 2016. Photosynthetic characteristics and effects of exogenous glycine of Chorispora bungeana under drought stress. Photosynthetica. 54, 459–467.

Yang, Z., Chu, C. 2011. Towards understanding plant response to heavy metal stress, in Abiotic Stress in Plants—Mechanisms and Adaptations, InTech, Shanghai, China, pp. 59–78.

Yao, W., Xu, T., Farooq, S.U., Jin, P., Zheng, Y. 2018. Glycine betaine treatment alleviates chilling injury in zucchini fruit (Cucurbita pepo L.) by modulating antioxidant enzymes and membrane fatty acid metabolism. Postharvest Biol Technol. 144, 20–28.

Yurekli, F., Kucukbay, Z. 2003. Synthesis of phytochelatins in Helianthus annuus is enhanced by cadmium nitrate. Acta Botanica Croatica. 62(1), 21–25.

Yusuf, M., Fariduddin, Q., Varshney, P., Ahmad, A. 2012. Salicylic acid minimizes nickel and/or salinity-induced toxicity in Indian mustard (*Brassica juncea*) through an improved antioxidant system. Environ Sci Poll Res. 19, 8–18.

Zagorchev, L., Seal, C.E., Kranner, I., Odjakova, M. 2013. A central role for thiols in plant tolerance to abiotic stress. Int J Mol Sci. 14(4), 7405–7432.

Zarei, S., Ehsanpour, A.A., Abbaspour, J. 2012. The role of over expression of P5CS gene on proline, catalase, ascorbate peroxidase activity and lipid peroxidation of transgenic tobacco (*Nicotiana tabacum* L.) plant under in vitro drought stress. J Cell Mol Res. 4(1), 43–49.

Zeid, I.M. 2001. Responses of Phaseolus vulgaris to chromium and cobalt treatments. Biol Plant. 44, 111–115.

Zengin, F.K., Kirbag, S. 2007. Effects of copper on chlorophyll, proline, protein and abscisic acid level of sunflower (*Helianthus annuus* L.) seedlings. J Environ Biol. 28(3), 561–566.

Zhang, W.H., Tyerman, S.D. 1999. Inhibition of water channels by HgCl2 in intact wheat root cells. Plant Physiol. 120, 849–858. doi: 10.1104/pp.120.3.849.

Zhou, Z.S., Guo, K., Elbaz, A.A., Yang, Z.M. 2009. Salicylic acid alleviates mercury toxicity by preventing oxidative stress in roots of Medicago sativa. Environ Exp Bot. 65, 27–34.

Zhou, Z.S., Huang, S.Q., Guo, K., Mehta, S.K., Zhang, P.C., Yang, Z.M. 2007. Metabolic adaptations to mercury-induced oxidative stress in roots of Medicago sativa L. J Inorg Biochem. 101, 1–9. doi: 10.1016/j.jinorgbio.2006.05.011.

Zhou, Z.S., Wang, S.J., Yang, Z.M. 2008. Biological detection and analysis of mercury toxicity to alfalfa (Medicago sativa) plants. Chemosphere. 70, 1500–1509. doi: 10.1016/j.chemosphere.2007.08.028.

Zhu, J.K. 2016. Abiotic stress signaling and responses in plants. Cell. 167, 313–324.

Zouari, M., Elloumi, N., Labrousse, P., Ben Rouina, B., Ben Abdallah, F., Ben Ahmed, C. 2018. Olive trees response to lead stress: Exogenous proline provided better tolerance than glycine betaine. South Afr J Bot. 118, 158–165.

9

Uptake and Transformation of Heavy Metals/Metalloids in Plants

Muhammad Ashar Ayub
Institute of Soil and Environmental Sciences, University of Agriculture Faisalabad, Punjab, Pakistan
Horticultural Sciences Department, Plant Root Biology Lab, Indian River Research and Education Center, Institute of Food and Agriculture Sciences, University of Florida, Fort Pierce, Florida, USA

Muhammad Zia ur Rehman
Institute of Soil and Environmental Sciences, University of Agriculture, Faisalabad, Punjab, Pakistan

Wajid Umar
Institute of Environmental Science, Hungarian University of Agriculture and Life Sciences, Gödöllő, Hungary

Asad Jamil, Muhamad Zohaib Aslam, and Hamaad Raza Ahmad
Institute of Soil and Environmental Sciences, University of Agriculture, Faisalabad, Punjab, Pakistan

Zahoor Ahmad
Department of Botany, University of Central Punjab, Bahawalpur Campus, Punjab Group of Colleges, Pakistan

Ayesha Siddique
Institute of Soil and Environmental Sciences, University of Agriculture, Faisalabad, Punjab, Pakistan

Julio E. Quinones III
Horticultural Sciences Department, Plant Root Biology Lab, Indian River Research and Education Center, Institute of Food and Agriculture Sciences, University of Florida, Fort Pierce, Florida, USA

CONTENTS

DOI: 10.1201/9781003155089-9

9.1 Introduction

Heavy metal (HM) is a broad term referring to elements having specific gravity greater than 4–5 gcm^{-3} (Koller and Saleh 2018). Regarding agroecosystem, organisms must deal with two kinds of HMs, heavy metal pollutants (HMPs) and heavy metal nutrients (HMNs) with later ones becoming toxic at higher concentrations (Figure 9.1). Anthropogenic soil pollution, of various of contaminants and sources, has become an emerging concern of modern-day communities. Soil being an ultimate pollutant sink, is also the main source of re-entry of these pollutants into the human food chain (Ali et al. 2019; Ayub et al. 2020; Farooqi et al. 2021). Among various environmental pollutants, inorganic pollutants are primarily persistent in the soil and thus have become of main concern. The HMPs in this regard are very important. Among HMNs, iron (Fe), zinc (Zn), nickel (Ni), copper (Cu), manganese (Mn), and molybdenum (Mo) are included which are metallic micronutrients of plants (Mertz 1981; Balaguer et al. 1993) which can become toxic at higher concentrations. In the category of HMPs, lead (Pb), arsenic (As)/metalloid, mercury (Hg), cadmium (Cd), and chromium (Cr) are important (Duffus 2002). Sources of HMNs are very diverse and mainly include agrochemical inputs added by farmers, but their unchecked addition, in conjunction with the use of untreated wastewater and animal farm wastes can lead to accumulations resulting toxic for plants in cases of elements such as Fe, Zn, Cu (Ahmad et al. 2020) while same can be a big source of HMPs (Alloway 2013; Yan et al. 2018). These HMPs and excessive HMNs are capable of disturbing normal geochemistry, thus can compromise health of agro-ecosystems.

In soil, the chemistry of HMs is very diverse, and must be considered to predict interactions with edaphic and biotic factors. Normally, plant availability of HMNs is in large part controlled by soil pH, redox potential, and soil clay content (Najafi-Ghiri et al. 2013; Riaz et al. 2020). However, availability becomes non-significant if HMNs persist in above permissible limits. For HMPs, chemistry is diverse, yet most of the HMPs are mobile and readily available for plant uptake (Wuana and Okieimen 2011; Ahmad et al. 2016). Unfortunately, plants cannot sequester infinite amounts of HMPs and excess HMs in solution are carried off, contaminating the systems they end up in. Translocation of HMs from soil to plant is a diverse and complex process involving various active and passive (inter and intra-cellular) pathways in plant uptake and eventual biomass assimilation (Thakur et al. 2016; Chandra et al. 2018). These signal cascades must be understood to properly track, and control the influx of these HMPs, and HMNs in the human food chain. The consequence of improper HMPs biomass accumulation in the human food chain is responsible for various diseases and disorders (Monisha et al. 2014).

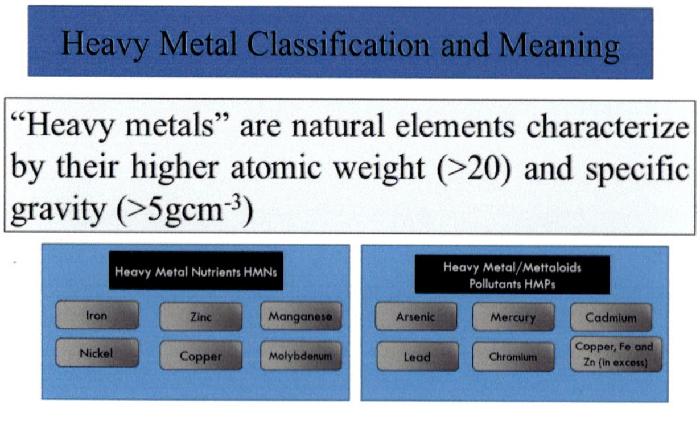

FIGURE 9.1 General introduction of HMs.

9.2 Soil Chemistry, Uptake, and Translocation of HMs: HMNs vs HMPs

There are many phases and portions of HMs in the soil, of which a small fraction is available to plants (bioavailable). Among the HMs, some are more mobile (e.g. Cd, and Zn). Others such as Pb are less mobile. The main source of these metals to plants is the soil solution. Solubility, and solid-liquid phase transition of these metal elements is controlled by various soil physicochemical properties as well as plant factors (rhizosphere chemistry, root exudates production, etc.). For HMNs, plants tend to adopt various methods to enhance bioavailability via siderophore and carboxylate production, in addition to active acidification of the rhizosphere. Furthermore, the presence of diverse soil microbial communities plays a critical role in the bioavailability of HMs in soil (Thakur et al. 2016). Figure 9.2 presents visual representation of fate of HMs in soil–plant system.

9.2.1 Micronutrients (HMNs): Fate in Soil and Role in Plants

Micronutrients are an integral part of plants homeostasis and metabolism, without which, plants cannot complete said metabolic functions, and ultimately their life cycle. The only source of HMNs to plants is the soil, but the fate of HMNs in the soil depends on the chemistry and the chemical nature of the HMN in question. Different kinds of metal compounds (carbonates, phosphates, sulfates, chlorides, etc.), precipitates, and chelates significantly control soil HMNs chemistry, their uptake, translocation, and assimilation in plants (Nakandalage and Seneweera 2018; Yujun et al. 2002; Riaz et al. 2020).

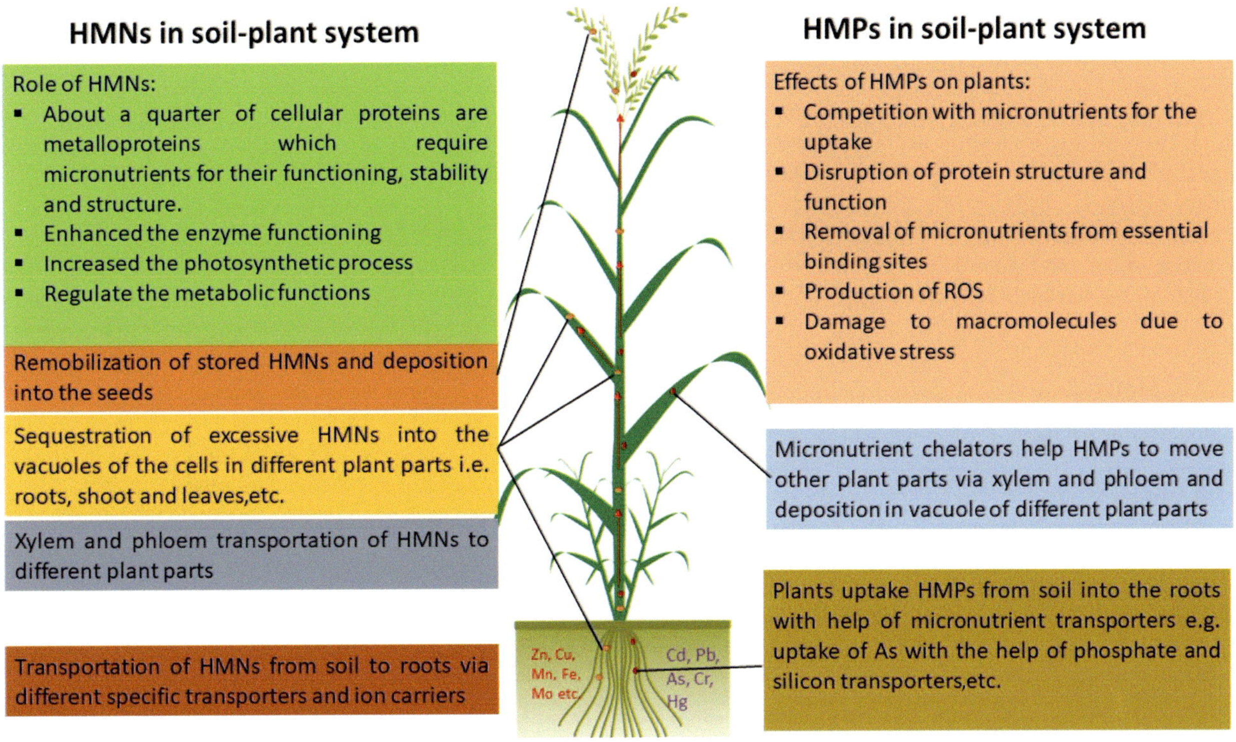

FIGURE 9.2 Fate of HMNs and HMPs in soil–plant system.

9.2.1.1 Zinc

Plants absorb zinc as Zn^{+2} and $ZnOH^{+}$ from soil, and its availability is controlled by various factors such as soil pH, clay and minerals adsorption sites, and soil organic matter content. Major sources of Zn in the soil are Zn fertilizers, oxides, and minerals such as sphalerites. Said sources are widely used for diverse purposes (Bradley et al. 2012; Sandstead 2015). Zn is essential to plant and animal metabolism, as it is involved in the synthesis of various macromolecules. The metabolic functions Zn are essential for cell growth, photosynthesis, gene expression via controlling enzymatic activities (cofactor for > 100 enzymes) and promoting cellular function. Deficiency of Zn causes stunted growth, chlorosis, and disturbed physiology. In contrast excess bioavailable Zn in soils leads to its higher accumulation in plants, causing various enzymatic dysfunctions (Alloway 2009; Cakmak 2000, 2009).

9.2.1.2 Copper

Copper is an important plant nutrient needed for plant metabolism. Although there is a remarkable amount of Cu present in the earth's soil/crust (Riedel 2008) but it is limited in its plant bioavailable form due to ionic transformations. Soil redox potential, pH, clay, and organic matter contents control the solid-liquid transition of bioavailable Cu in soil (Sharma et al. 2009). Total bioavailable soil Cu content can range between 700 and 4000 ppm in soil (Printz et al. 2016). Coppers uptake from the soil involves active use of plant's physiological machinery(active transport) (Sun et al. 2014). It is an important plant nutrient involved in catalytic activities, protein synthesis, redox reactions, and electron transfer in plants. It is involved in ATP synthesis, metalloprotein production, and cell organelle homeostasis. Copper deficiency in plants leads to cell death, at sub-lethal concentrations it leads to oxidative damage to plant cells (Gupta et al. 2008; Cook et al. 1998; Vinit-Dunand et al. 2002; Priyanka et al. 2019).

9.2.1.3 Iron

Iron comprises about 5 percent of the earth's crust, and is another important plant micronutrient and soil constituent with average soil contents of 40 ppm. It is the second most abundant metal in the earth's crust after Al, but its bioavailable concentration in alkaline soil is limited. Fe in the soil can be found in mineral ores such as hematite, magnetite, and goethite (Broadley et al. 2012; Masunaga and Fong 2018; Riaz et al. 2020; Naveed et al. 2020). It is involved in the various metabolic process of plants such as photosynthesis, respiration, redox reactions and chlorophyll, metalloproteins, and phytohormone production. Iron deficiency causes interveinal chlorosis in plants (Bertini and Rosato 2007; Abadía et al. 2011; McGrath et al. 2014; Rout and Sahoo 2015).

9.2.1.4 Nickel

Nickel the fifth most abundant element in the earth's crust (after O, Fe, Mg, and Si). Nickel exists in soil solution as Ni ions (–1, +1, +3, and +4) with Ni^{+1} being the most abundant form. The geochemistry of Ni in the soil is closely associated with Cu and Fe and usually, these three elements co-exist in more than 100 minerals and said minerals are used worldwide for various purposes. Naturally, it reaches topsoil via fires depositing ash, dust deposition, and volcanic eruptions while anthropogenic sources such as fossil fuel and incineration of sludge/waste also contribute significantly (Mulrooney and Hausinger 2003; Kabata-Pendias 2004; Cempel and Nikel 2006; Harasim and Filipek 2015). Nickel is involved in N-assimilation (as it is the integral part of urease enzyme), redox-reaction homeostasis, methane biogenesis, hydrogen metabolism, acetogenesis, synthesis of phytoalexins, plant growth promotion, and disease prevention (Sirko and Polonica 2000; Fageria 2001; Fageria et al. 2002; Mulrooney and Hausinger 2003).

9.2.1.5 Molybdenum

Molybdenum is an integral micronutrient for plants and animals. Naturally, Mo is found in mineral forms such as molybdenite, powellite, wulfenite, and ferrimolybdite. Its average soil content can range between 2 and 2.3 ppm. Among all HMNs, Mo is the only nutrient whose availability increases with an increase in alkalinity due to the formation of soluble hydroxides. Its bioavailable form is MnO_4^{+2} whose chemistry can be altered by sulfur and Cu intake by plants. Its wide use in fertilizer, pigments, corrosion-resistance provision to steel, catalytic activation, lubrication, and metallic alloy formation makes its non-point anthropogenic sources widespread (Fageria 2001; Fageria et al. 2002; Tallkvist and Oskarsson 2015). Mo is required by the plant for various roles such as disease resistance, seedling growth promotion, and seed quality. Mo is also an integral part of the N-fixing enzyme (nitrate-reductase) and is involved in redox reactions. Usually, the inactive form Mo remains in organisms as a complex with organic proteins called molybdoenzymes; as found in plants, animals, and prokaryotes. Its deficiency in plants leads to leaves withering and death of leaf tips. Leguminous plants may also suffer from N deficiency as Mo-catalyzed N fixation processes stop (Williams and Fraústo da Silva 2002; Kaiser et al. 2005; Dimkpa and Bindraban 2016).

9.2.1.6 Manganese

Manganese is found in soil as Mn^{+2} (soluble) and Mn (III, IV insoluble). Its normal concentration in soil (Mn^{+2}) is in-between 0.01 and 1 ppm and said concentration is primarily under the control of soil pH and redox potential. Manganese deficiency hampers plant growth, while its toxicity disrupts plant physiology, enzyme activity, and mineral homeostasis via initiation of oxidative stress (Rengel 2000; Graham 2004; Dučić and Polle 2005; Watmough et al. 2007; Masunaga and Fong 2018). In soil, its bioavailable fraction is present in soil solution as well as on soil particle exchange sites. It is readily up taken by plants and is used in protein synthesis, photosystem activation, chlorophyll synthesis, catalytic activation, primary/secondary metabolite production, and disease resistance.

Proper bioavailable Mn and subsequent plant uptake is important for the various roles it plays in plant cellular physiology. At pH > 6.5, Mn availability starts decreasing; its deficiency causes interveinal chlorosis, cell death, and retarded plant growth (Millaleo et al. 2010; Broadley et al. 2012; McGrath et al. 2014).

9.2.2 Heavy Metal/Metalloid Pollutants

9.2.2.1 Cadmium

Cadmium is a well-reported toxin of agricultural ecosystems with radial availability in soil and can be up taken into edible portion of plants (grains and tissue) in above permissible limits (Rehman et al. 2020a, 2021). It is a potent toxin for human as well and can be cause of cardiovascular, pulmonary, and skeletal diseases (Reeves and Chaney 2008; Hu et al. 2016) and crops grown on contaminated soils are major source of Cd inclusion in human food (Azhar et al. 2019; Rehman et al. 2020b). Soil contamination with Cd is a potential threat in various countries worldwide and can be an upcoming lethal pollutant (Yang et al. 2017; Ishikawa et al. 2012; Kim et al. 2016; USFDA 2017). Among various sources of Cd input in agriculture soils, wastewater application and agrochemical excessive uses are leading causes, other sources can be industrial activities and atmospheric depositions (Rehman et al. 2019; Azhar et al. 2019; Ayub et al. 2020).

9.2.2.2 Lead

In the category of HMPs, Pb is another potential toxic element which is non-degradable in soil and has significant bioavailability to crop plants (Nagajyoti et al. 2010). Total Pb contents in the earth's crust can be up to 20 ppm and are characterized as second most potent toxic pollutant after arsenic (As) because it has widespread toxicity to all organisms. Due to earth geochemistry, lead's presence in pedosphere is widespread and other major sources include erosion, volcanic eruption, and anthropogenic activities like application of agrochemicals (Yokel and Delistraty, 2003; Gottesfeld et al. 2018). Once reaching soil solution phas, Pb can disrupt plant physiology as its permissible limit for plants is 2 ppm while for soil is 50–300 ppm (WHO 1996; Inglezakis et al. 2014) and above this limit, plant morpho-physiological and biochemical attributes start disrupting and severe damping is observed due to severe production of ROS (Kushwaha et al. 2018). The widespread effect of Pb toxicity on plant's germination, growth, mineral acquisition, oxidative stress, enzyme deactivation, water stress, and carbon assimilation disruption is well reported (Zulfiqar et al. 2019).

9.2.2.3 Chromium

Another HMP, chromium's soil chemistry is also closely dependent upon soil parent material, weathering pace, rock, and sediments type as well as anthropogenic activities like application of untreated industrial effluents to irrigate crops. Among anthropogenic sources, Cr deposition in soil, dumping of contaminated solid waste and liquids, slag/sludge or Cr plating baths are important. In soil, Cr is found in two states as Cr (III) and Cr (VI) and in presence of moisture or in water it can also transform to other species (III to VI). Cr (VI) is most persistent form in earth crust and can stay in place for years if soil doesn't have reducing power. In comparison between Cr (VI) and Cr (III), Cr (VI) is more mobile and toxic to plants (Kimbrough et al. 1999; Rosas et al. 1989; Mallhi et al. 2019; Farid et al. 2020).

9.2.2.4 *Arsenic-Metalloid*

Due to high persistence in soil and water, and potential toxicity to human, (As) has been classified a threat globally with a lot of research going on for its possible tracking and management. The As has been reported to be found in ground water and drinking water like reported in Bangladesh and posses' severe threat to human. In soil, (As) chemistry is totally controlled by soil physico-chemical properties, organic carbon microbial community, and mineral composition (Abbas et al. 2018; Niazi et al. 2018; Hussain et al. 2020). In soil solution phase, As is found in two states (V and III) which are found primarily in surface and subsurface soil. The chemistry of As in soil is primarily controlled by soil pH and redox potential and plant and microbial activity may change it into methylated species (Fendorf et al. 2010).

9.2.2.5 *Mercury*

Mercury is a liquid metal which is extracted from cinnabar, cordierite, Livingstone (mercury antimony sulfosalt), meta-cinnabar, and native Hg and can be found in soil as Hg^0, Monovalent, or divalent forms. Hg is involved in formation of various organic and inorganic compounds and found about 0.05 ppm in the earth's crust. Though it's very rare but found in all types of rocks as well as in soil near mines. Upon toxic accumulation in plant, Hg can disrupt normal functioning of crop (Beckers and Rinklebe 2017).

9.3 Micronutrients (HMNs) Deficiency and Toxicity: Plant Cellular/Molecular Responses

As disused above all HMNs are very crucial for plant growth and must be up taken by plants in adequate amount along with avoidance of their toxicity. For this reason, plants tend to behave via modulating physiological, cellular, and molecular aspects to avoid HMNs deficiency as well as toxicity as disused ahead. All HMNs are actively involved in plant cellular processes and metabolism and their adequate concentration is must be acquired by the plant. The soil fate of HMNs is the main factor controlling their availability to plants and varying plant species respond to soil HMs contents distinctively (Rout and Das, 2009). Under HMNs deficiency, plants tend to release root exudates to increase solubility while under toxicity intercellular sequestration via chelators, such as amino acids, metallothionein's (MT), ferritins happen. Moreover, the plant also tends to modulate metabolic pathways to escape HMNs deficiency and toxicity to the maximum extent (Hall 2002; Cho et al. 2003). HMNs toxicity response is rapidly shown via disturbed photosystem, decreased carotenoids and chlorophyll production, decrease CO_2 fixation via RUBISCO, and disrupting ATP production and plant tend to escape this via every source and way possible (Monnet et al. 2001; Hall 2002; Nagajyoti et al. 2010; Todeschini et al. 2011; Maleva et al. 2012; Ghori et al. 2019).

9.3.1 Zinc: Homeostasis under Deficiency and Toxicity

Zn being one of the most important HMNs is an important limiting factor of modern-day agriculture and its content in soil and plants must be managed properly. Under Zn deficiency and toxicity plant tends to escape (Henriques et al. 2012) via modulating Zn bioavailability (Cakmak et al. 2011), controlling its uptake into plant tissue (Broadley et al. 2007), and assimilation management (Marschner, 1995). Under Zn, deficiency plant tends to acquire more while under its surplus it tries to stop its uptake and in this process plant root play a vital role via modulating Zn transporter genes like AtNASA2,4 (Robson 1994; van de Mortel et al. 2006; Klatte et al. 2009). Plant root exudates/siderophores like 2-deoxymugenic acid) (Rengel 1997), malate and mugineic acids (Gao et al. 2009) which are helpful in efficient Zn absorption under deficiency conditions (Impa and Johnson-Beebout 2012). Zinc translocation in a plant is controlled by Zn which is transported and is under the control of genes in plants and can be modulated according to Zn availability (deficiency vs toxicity). A major class of transporter in this regard are natural resistance-associated macrophage proteins (NRAMPs) involving AtNARMP3,4 activate under Zn deficiency conditions (Thomine et al. 2000, 2003; Grotz and Guerinot 2006) while PIB-ATPase likes HMA1-4 help in exclusion of excessive Zn under toxic conditions (Eren and Argüello 2004; Kim et al. 2009). The AtHMA1 is found in the chloroplast, AtHMA2 gene codes ATPase present in the plasma membrane, AtHMA3 is found in hyperaccumulators and AtHMA4 helps in inter tissue translocation of Zn (Eren and Argüello 2004; Hassan and Aarts 2011; Waters and Sankaran 2011). Similarly, other classes of Zn transporters are also present in plants for effective Zn homeostasis like cation diffusion facilitator (CDF) (Gustin et al. 2011) and iron-regulated-transporter like protein (IRT-ZIP) (Colangelo and Guerinot 2004; Palmer and Guerinot 2009; Song et al. 2010).

9.3.2 Copper: Homeostasis under Deficiency and Toxicity

All plants have specialized Cu transporters which help them translocate a significant amount of Cu under Cu deficiency conditions like CTRs and COPT1-6. The presence and expression of these genes vary in the plant body as COPT4 in roots, COPT1,2,6 in leaves, and COPT3,5 present predominantly in the stem which help inter-plant translocation of Cu. The potential decrease in Cu concentration significantly alters the expression of these genes and so does affect long-distance transport of Cu (Singh et al. 2011; Shin et al. 2012; Carrió Seguí 2017). Intercellular and tissue moment of Cu is done by yellow strip-like proteins such as HMA5 which is ATP dependent and involved in the efflux of Cu from plant's apoplast (Liang et al. 2012; Garcia et al. 2014). For proper functioning and synthesis of plastocyanin, Cu is needed in chloroplast cells to where it is transported via Cu/Zn SOD (Ravet and Pilon 2013). Cu being involved in major physiological processes of a plant can cause a disturbance in photosystem, chlorosis, and even death of plant tissue if not provided in adequate amount. Cu deficiency in a plant also triggers many genes involved in the production of proteins and chemical complexes needed for normal light harvesting and carbon fixation and cellular metabolism and bioproduction of lignin needed for cell protection (Yruela 2005; Raldugina et al. 2016). It has been reported that decreased availability of Cu from the system can lead to an electron transport chain and may increase oxidative stress in tinder plants (Thomas et al. 2016; Ameh and Sayes 2019).

On contrary to this, Cu toxicity negatively alters plant physiology, growth, and metabolism as reported in various research works (Gong et al. 2019; Marques et al. 2019; Jaime-Pérez et al. 2019) but proper toxicity mechanisms of Cu are still under investigations and need better understanding. The major influence of Cu toxicity appears as plant root death (Cambrollé et al. 2013) and imbalanced nutrition uptake (Sheldon and Menzies 2005). Excessive Cu contents can disturb cell wall structure and rupture the epidermis and cortex (Kopittke et al. 2011), decrease plant growth and root morphology (Marques et al. 2019). As molecular toxicity response, plant's pigment production decreases and protein metabolism gets disturbed completely resulting in severe oxidative stress and reactive oxygen species production (Silva et al. 2018; Jaime-Pérez et al. 2019; Nazir et al. 2019).

9.3.3 Iron: Homeostasis under Deficiency and Toxicity

As reported earlier, Fe makes a major part of the earth's crust but its bioavailable concentration in soil is very low and plants must tend to adopt every possible mode in their approach to absorb needed Fe from soil (Morrissey and Guerinot 2009; Hindt and Guerinot 2012). Under Fe deficiency, plant root tends to release H^+ into rhizosphere thus lowering soil pH which can increase soil Fe bioavailability (Guerinot and Yi 1994; Santi et al. 2005, 2008). Another way which helps plant achieve this objective is the release of iron-reducing enzymes (Fe-chelate-reductase) which can help in active transformation of Fe (III) to Fe (II) which is more mobile and available for plant's absorption (Waters et al. 2002; Vasconcelos et al. 2006; Ishimaru et al. 2005). The bivalent ionic form of Fe (Fe^{+2}) is taken up by plants into roots via special transporters (IRT1) located on the epidermal layer of plant roots and are specialized in Fe absorption under deficit conditions (Vert et al. 2002; Dinneny et al. 2008). Another form of Fe deficiency management is the release of Fe chelators (Fe-siderophores) which help retain Fe bioavailable for a longer period (Conte and Walker 2011) production of which is controlled by nicotianamine present in the plant cell (Hindt and Guerinot 2012). Besides Fe solubilization, Fe-siderophore complex can also directly be uptaken by plant roots via transporters like YS1, OsYSL15 (Curie et al. 2001, 2009; Lee et al. 2009; Inoue et al. 2009). Plant during the process of adaption and evolution has developed specific genes like 21 NAS, TaNAAT, Ta DMAS1, HuNAAT, HuDMAS1, TaDMAS1/2, and TaNAAT1 involved in active translocation of Fe under deficient conditions (Beasley et al. 2017). For Fe-Ps complex translocation, phyto siderophores-efflux-transporter- TOM1 is also present in plants (Nozoye et al. 2011) which can be helpful in both Fe deficiency and toxicity conditions. Under Fe toxicity, the plant tends to release Fe (II) into the rhizosphere via active efflux or assimilate it into the plant's cell vacuole thus making it is potentially inactive and also can mitigate Fe mediated oxidative damage in the cell via the production of antioxidants (Briat et al. 2010; Deng et al. 2010; Gallie 2013; Wang et al. 2019).

9.3.4 Nickel: Homeostasis under Deficiency and Toxicity

Ni was categorized as an essential plant nutrient in the early '70s (Dixon et al. 1975) but very little work on its actual fate in the soil-plant system is done and so does very little is known about it (Bai et al. 2006). It was initially reported to be an essential element for leguminous plants by Eskew et al. (1984) and later got known for essentiality for non-leguminous plants by Brown et al. (1987). Now it is well known that Ni is needed for the enzyme (glyoxalases, urease, superoxide dismutase, and hydrogenase) activation and makes about 10 ppm of plant dry body mass (Küpper and Kroneck, 2007). Maximum availability of Ni in the soil is in the range of acidic to neutral pH thus plant must adopt different techniques to escape its deficiency in soil (Nkrumah et al. 2019). Ni is an integral part of the urease enzyme and EU2-3 protein genes involved in the activation of urease (Seregin and Kozhevnikova 2006). Despite being that much essential for plants, the higher concentration of Ni can be toxic to plants as can disrupt mineral homeostasis and induce oxidative stress in plant's cell (Sreekanth et al. 2013) subsequently affecting plant growth and development (Gajewska et al. 2006; Ahmad et al. 2007). Antagonistic interaction of Ni and Fe can be used in the potential reversal of Ni toxicity. At the molecular level, Ni toxicity disrupts the Ca binding capability of photosystem-oxygen-evolution complex (Solymosi et al. 2004) and initiate ROS flux into the cell body (Boominathan and Doran 2002) due to decreased antioxidants production this decrease plant growth (Gomes-Junior et al. 2006).

9.3.5 Molybdenum: Homeostasis under Deficiency and Toxicity

Mo is taken up from the soil in the form of molybdate and its valency in the soil varies from 0 to 6 and 6+ being the most common valence form of transition metal ion. It is an integral part of many enzymes and acts as a co-factor needed for their activation. The bioavailability of Mo in soil depends upon soil pH and redox potential (Williams and Frausto da Silva 2002; Bittner and Mendel 2010; Mendel and Kruse 2012; Bittner 2014). Upon experiencing Mo deficiency, the plant's physiology can be severely damped as Mo-enzymes have a unique role in plant's metabolism as nitrate-reductase, sulfite oxidase, xanthine dehydrogenase, aldehyde oxidase, mitochondrial amidoxime reductase. These enzymes can be classified further based upon the type of Mo compounds present in them. The Mo-redox potential, enzyme activation, growth promotion and help in N-fixation are key role which gets damped under Mo deficiency (Mendel and Haensch 2002; Rana et al. 2020). The Mo deficiency causes mottling, leaf cupping, discoloration, and leaf turgor pressure loss (Hewitt and Bolle-Jones 1952a, 1952b; Graham and Stangoulis 2005). Most legumes have Mo-enzymes involved in N-fixation and Mo deficiency leads to disruptive N-fixation thus compromising legume growth and N-efficiency (Chatterjee and Nautiyal 2001) and exogenous application of Mo can reverse these effects (Rana et al. 2020). On the other hand, just like all other HMNs, the toxicity of Mn can also occur but Mo limits in the soil are yet to decide and a lot of work needs to be done to Mn sufficiency/ toxicity fringe determination (McGrath et al. 2010).

9.3.6 Manganese: Homeostasis under Deficiency and Toxicity

Mn is a very important micronutrient and plays a vital role in plants. Its chemistry in the soil is predominantly controlled by soil properties and redox potential while the plant's root activity can also affect its bioavailability in the rhizosphere (Porter et al. 2004; Kluwer et al. 2010). The main role of Mn in a plant's body is enzyme activation, metal-enzyme creation helping in the electron transport chain and carbon assimilation, and very minute concentration is required for all these processes (Millaleo et al. 2010; Andresen et al. 2018). It is also vital in the provision of defense and tolerance in plants against diseases and pest stress (Eaton 2015; Yano and Yachandra 2014). The deficiency of Mn is one of the main limiting reasons for the growth and yield of wheat and barley as its involvement in a phenomenon called winterkill (Hebbern et al. 2005; Jiang 2006). The main cellular process initiated in Mn deficiency is the degradation of chlorophyll in plants leading to interveinal chlorosis as well as involved in light spot disease and is predominately observed in saline soils (Papadakis et al. 2007; George et al. 2014; Hernandez-Apaolaza 2014; Schmid et al. 2014). The oxidative stress experienced by metallic-enzyme complexes is another molecular response toward Mn deficiency which ultimately decreases plant growth and increases susceptibility to abiotic and biotic stresses (Heine et al. 2011; Moradtalab et al. 2018; Saidi et al. 2012; Ihnatowicz et al. 2014). Toxicity of Mn in plants is species-dependent but overall response of plant is disruptive nutrients acquisition, imbalanced metabolism, and plant growth (St. Clair and Lynch 2005; Husted et al. 2009; Broadley et al. 2012; Zhao et al. 2017).

Higher Mn concentrations are toxic to plant, but this again depends upon plant type (Husted et al. 2009; Broadley et al. 2012) as Mn inter and intracellular accumulation in plants highly varies. The Mn toxicity in plants can disturb nutrient acquisition (St. Clair and Lynch 2005), cause imbalance in chlorophyll biosynthesis (Subrahmanyam and Rathore 2001), cause a decrease in cell division and elongation (Zhao et al. 2017) as well as cause oxidative damage to cell (Fecht-Christoffers et al. 2006). The toxic levels of Mn are not fixed and vary remarkably from crop to crop (200 ppm for maize and 5300 ppm for hyperaccumulator like sunflower). Some plant has specialized transporters which can help in effective uptake, re-distribution and potential detoxification of Mn hyper concentrations like OsNram3, AtNram1, AtZIP1/2, HvNramp5 and HvIRT1 (Edwards and Asher 1981; Shao et al. 2017).

9.4 Heavy Metal Pollutants Toxicity: Cellular and Molecular Level Homeostasis and Plant Response

Major HMPs posse severe toxicity to plants and plants need to evolve some mechanisms to escape or tolerate these stresses. One of the ways in which plant responds to HMPs stress is upregulating and triggering its genes encoding stress suppressing proteins. Once making their way into plant cell, HMPs tend to initiate oxidative stress resulting interference in homeostasis of nascent protein. This makes plant go for regulation of gene expression coding MTs and phytochelatins which bind the HMPs in cytosol which is then compartmentalized into vacuole. These proteins also help plant cell revive from oxidative damage caused by HMPs (Hassan et al. 2017).

As discussed above, HMPs can be easily taken up by plant and translocated into different tissues where they end up entering into plant cell causing excessive surge of ROS promoting DNA and organelle damage, disrupting membrane integrity, and curbing protein activity (Li et al. 2016a; Tamás et al. 2014). To avoid these catastrophic effects of HMPs, plants have evolved various kinds of strategies resulting in metal detoxification. First such strategy is prevention or reducing metal uptake via development of apoplastic barriers, binding them to cell wall or to cellular exudates thus restricting their entry into plant cell (Manara 2012; Hasan et al. 2015). If the concentration of HMPs is too much high and still some of metal contents end up into plant cell, then plants have network of storage and detoxification sites via metal-chelators assisted translocation into vacuole where non-active HMPs are dumped (Zhao and Chengcai, 2011). Below mentioned are ways in detail.

9.4.1 Phytochelatins and ATPase Mediated Homeostasis

These complex molecules are cysteine-rich-oligomers which are involved in cell protection against HMPs stress and commonly called phytochelatins or PC and are only effective in early stress stages (Ashraf et al. 2010; Pochodylo and Aristilde, 2017). The PCs are first protective stage in plant cell for potential nullification of HMPs mediated stress via active detoxification, but action of PCs is very limited as they only decrease free circulation of HMPs inside plant cell (Hasan et al. 2016). Gene involved in coding of PC in Arabidopsis is AtPCS1 and shows hypersensitivity toward Cd and Zn but the production of PC is higher compared to parent wild type plant (Lee et al. 2003). As PCs are only involved in chelation of metals that's why most of the times if excessive PCs are produced then plant tend to accumulate higher amount of HMPs without improving HMPs tolerance in plants (Pomponi et al. 2006). Investigations have shown that besides HMPs chelation, PCs are involved in HMNs homeostasis, antioxidant stimulation, and S-metabolism (Furini 2012).

Chelation of HMPs inside plant cell via PCs is the first step as these complexes have to be translocated in vacuoles via transporters which are primarily ATP-driven pumps (V-ATPase and V-PPase) (Sharma et al. 2016). Major classes of ATPase involved in HMPs translocation (ATPase-HMAs) are ABC transporters, Zn-Fe-permease (ZIP), and NRAMPs (Xu et al. 2015; Sharma et al. 2016). Another class of Cd transporter in plant cell are IRT-1 belonging to ZIP family beside those NRAMP-5 has been reported to translocate Mn and Cd in rice roots, while HMA2,3 were found in plasma membrane and tonoplast, respectively (Clemens and Ma, 2016). Another class of transporters ABCC1/2 has been found to be effective detoxification of As and Cd in Arabidopsis as well as help in homeostasis of HMNs working in synergism of PCs (Xu et al. 2015; Sharma et al. 2016; Song et al. 2014; Zhang et al. 2017).

HMPs/HMNs toxicity management in plants	Management of HMNs deficiency in plants
▪ Deposition of excessive metals to cell vacuoles of different plant parts ▪ Activation of antioxidant enzymes, such as SOD, POD, CAT, APX,etc. ▪ Chelation of metals with amino acids (histidine and proline) and amines and with organic acids (malate, oxalate,etc.) to reduce the toxicity. ▪ Increase in the activity of glutathione reductase to reduce the glutathione disulphide under HMs stress. ▪ Production of α-tocopherol to scavenge lipid peroxidase and ROS. ▪ Increase the phenolic compound production to chelate the metal ions. ▪ Activation of heavy metal ATPase and cation diffusion facilitators to exclude the HMs from cytoplasm. ▪ Introduction of NRAMPs transporters. ▪ Enhancing gene expression to confer the oxidative stress. ▪ Increase in mineral nutrient uptake to overcome metal stress.	▪ Induction of micronutrient acquisition related genes ▪ Increase in the phenolic compounds and GSH ▪ Maintenance of high SOD activity ▪ Higher AsA concentration to maintain the redox balance ▪ Activation of antioxidant defense system ▪ Activation of microRNAs ▪ Inducing transcription factors by plants

FIGURE 9.3 Summarized modes of plant behavior under HMs deficiency and toxicity.

9.4.2 Metallothionein's Mediated Homeostasis

Like PCs this class of cysteine-rich metal-binding proteins also helps in immobilization, sequestration, and detoxification of HMPs (Capdevila and Atrian, 2011). The MTs are involved in net homeostasis of all HMNs and HMPs as well as help plant overcome oxidative damage (Hossain et al. 2012)

9.4.3 Repairing Cellular Proteins

Upon onset of HMPs mediated toxicity, denaturation, and disintegration of protein is a major issue especially metalloproteins involved in plant cellular physiology (Tamás et al. 2014). To overcome this issue, plants tend to produce a special class of proteins called HSPs involved in maintaining cellular homeostasis by helping folding of non-active proteins thus preventing protein degradation (Hüttner et al. 2012; Park and Seo 2015).

9.5 Modes of Overcoming HMPs Toxicity in Field Crops

To overcome HMPs mediated stress and decreased crop production a wide range of practical and viable options have been provided in literature like application of various inorganic (gypsum, illite, nanoparticles, silicon) and organic (biochar, press mud, FYM, etc.) amendments (Sohail et al. 2019; Azhar et al. 2019; Rehman et al. 2019, 2020a,b; 2021). Authors encourage all readers to work on various aspects of these amendments in potential recovery and sustainable safe produce production from HMs contaminated soils.

9.6 Summary

The HMs are special elements with wide range enclosed as essential nutrients as well as pollutants. Plants tend to escape HMNs deficiency as well as toxicity of HMNs and HMPs via various ways out of which modulation of cellular and molecular mechanisms are most important as HMs upon accumulation above permissible limits initiate cellular oxidative damage causing irreversible damage to the plant. A lot of work in proper investigation of physiological and molecular levels in this regard is still needed. Figure 9.3 presents a summary of plant behavior under HMs deficiency and toxicity.

REFERENCES

Abadía, J., Vázquez, S., Rellán-Álvarez, R. (2011) Towards a knowledge-based correction of iron chlorosis. Plant Physiol Biochem 49:471–482. https://doi.org/10.1016/j.plaphy.2011.01.026.

Abbas, G., Murtaza, B., Bibi, I., Shahid, M., Niazi, N.K., Khan, M.I., Amjad, M., Hussain, M. (2018) Arsenic uptake, toxicity, detoxification, and speciation in plants: physiological, biochemical, and molecular aspects. Int J Environ Res Public Health 15:59.

Ahmad, H.R., Aziz, T., Zia-ur-Rehman, M., Sabir, M., Khalid, H. (2016) "Sources and composition of waste water: threats to plants and soil health." In Soil science: Agricultural and environmental prospectives. Springer, Cham, pp. 349–370.

Ahmad, H.R., Sabir, M., ur Rehman, M.Z., Aziz, T., Maqsood, M.A., Ayub, M.A., Shahzad, A. (2020) "Wastewater irrigation-sourced plant nutrition: concerns and prospects." In Plant micronutrients. Springer, Cham, pp. 417–434.

Ali, H., Khan, E., Ilahi, I. (2019) Environmental chemistry and ecotoxicology of hazardous heavy metals: Environmental persistence, toxicity, and bioaccumulation. J Chem 2019.

Alloway, B.J. (2009) Soil factors associated with zinc deficiency in crops and humans. Environ Geochem Health 31:537–548. https://doi.org/10.1007/s10653-009-9255-4.

Alloway, Brian J. (2013) "Sources of heavy metals and metalloids in soils." In Heavy metals in soils. Springer, Dordrecht, pp. 11–50.

Ameh, T., Sayes, C.M. (2019) The potential exposure and hazards of copper nanoparticles: a review. Environ Toxicol Pharmacol 71:103220.

Andresen, E., Peiter, E., Küpper, H. (2018) Trace metal metabolism in plants. J Exp Bot 69:909–954. doi: 10.1093/jxb/erx465.

Ashraf, M., Qztürk, M.A., Ahmad, M.S.A. (2010) Plant adaptation and phytoremediation. Springer, New York, NY.

Ayub, M.A., Usman, M., Faiz, T., Umair, M., ul Haq, M.A., Rizwan, M., Ali, S., ur Rehman, M.Z. (2020) "Restoration of degraded soil for sustainable agriculture." In Soil health restoration and management. Springer, Singapore, pp. 31–81.

Azhar, M., ur Rehman, M.Z., Ali, S., Qayyum, M.F., Naeem, A., Ayub, M.A., ul Haq, M.A., Iqbal, A., Rizwan, M. (2019) Comparative effectiveness of different biochars and conventional organic materials on growth, photosynthesis and cadmium accumulation in cereals. Chemosphere 227:72–81.

Bai, C., Reilly, C., Wood, B.W. (2006) Ni deficiency disrupts metabolism of ureides, amino acids, and organic acids of young pecan foliage. Plant Physiol 140:433–443.

Balaguer, J., Almendro, M.B., Gomez, I., Navarro Pedreño, J., Mataix, J. (1993) Tomato growth and yield affected by nickel presented in the nutrient solution. Int Symp Water Quality Quantity Greenhouse 458:269–272.

Beasley, J.T., Bonneau, J.P., Johnson, A.A.T. (2017) Characterisation of the nicotianamine aminotransferase and deoxymugineic acid synthase genes essential to strategy II iron uptake in bread wheat (Triticum aestivum L.). PLoS ONE 12:e0177061. doi: 10.1371/journal.pone.0 177061.

Beckers, F., Rinklebe, J. (2017) Cycling of mercury in the environment: Sources, fate, and human health implications: A review. Crit Rev Environ Sci Technol 47(9):693–794.

Bertini, I., Rosato, A. (2007) From genes to metalloproteins: a bioinformatic approach. Eur J Inorg Chem 2007:2546–2555. https://doi.org/10.1002/ejic.200700186.

Bittner, F., Mendel, R.R. (2010) "Cell biology of molybdenum." In: Hell, R., Mendel, R.R. (Eds.), Cell biology of metals and nutrients. Plant cell monography, Vol. 17. Springer-Verlag, Berlin, Heidelberg, pp. 119–143.

Boominathan, R., Doran, P.M. (2002) Ni-induced oxidative stress in roots of the Ni hyperaccumulator, Alyssum bertolonii. New Phytol 156(2):205–215.

Briat, J.F., Duc, C., Ravet, K., Gaymard, F. (2010) Ferritins and iron storage in plants. Biochimica Et Biophysica Acta-Gen Sub 1800:806–814. doi:10.1016/j.bbagen.2009.12.003.

Broadley, M., Brown, P., Cakmak, I. (2012) "Function of nutrients." In Marschner's mineral nutrition of higher plants. Academic Press, London, pp. 191–248.

Broadley, M.R. (2012) Marschner's mineral nutrition of higher plants. Elsevier/Academic Press.

Broadley, M.R., White, P.J., Hammond, J.P., Zelko, I., Lux, A. (2007) Zinc in plants. New Phytol 173:677–702.

Brown, P.H., Welch, R.M., Cary, E.E. (1987) Ni: A micronutrient essential for higher plants. Plant Physiol 85:801–803.

Cakmak, I. (2000) Tansley review no. 111: possible roles of zinc in protecting plant cells from damage by reactive oxygen species. New Phytol 146:185–205.

Cakmak, I. (2009) Enrichment of fertilizers with zinc: an excellent investment for humanity and crop production in India. J Trace Elem Med Boroniol 23:281–289.

Cakmak, I., Pfeiffer, W., Mcclafferty, B. (2011) Biofortification of durum wheat with zinc and iron. Cereal Chem 87:10–20.

Cambrollé, J., García, J. L., Ocete, R., Figueroa, M. E., Cantos, M. (2013) Growth and photosynthetic responses to copper in wild grapevine. Chemosphere 93(2):294–301.

Capdevila, M., Atrian, S. (2011) Metallothionein protein evolution: a miniassay. J Biol Inorg Chem 16, 977–989. doi: 10.1007/s00775-011-0798-3.

Cempel, M., Nikel, G. (2006) Nickel: a review of its sources and environmental toxicology. Polish J Environ Stud 15:375–382.

Chandra., R., Kumar, V., Singh, K. (2018) "Hyperaccumulator versus nonhyperaccumulator plants environment waste management." In: Chandra, R., Dubey, N.K., Kumar, V. (Eds.), Phytoremediation of environmental pollutants. CRC Press, Boca Raton, pp. 14–35.

Chatterjee, C., Nautiyal, N. (2001) Molybdenum stress affects viability and vigour of wheat seeds. Journal of Plant Nutrition 24:1377–1386.

Cho, M., Chardonnens, A.N., Dietz, K.J. (2003) Differential heavy metal tolerance of Arabidopsis halleri and Arabidopsis thaliana: a leaf slice test. New Phytol 158:287–293.

Clemens, S., Ma, J. F. (2016) Toxic heavy metal and metalloid accumulation in crop plants and foods. Annu Rev Plant Biol 67:489–512. doi: 10.1146/annurev-arplant-043015-112301.

Colangelo, E., Guerinot, M. (2004) Put the metal to the petal: metal uptake and transport throughout plants. Curr Opin Plant Biol 9:322–330.

Conte, S.S., Walker, E.L. (2011) Transporters contributing to iron trafficking in plants. Mol Plant 4:464–476.

Cook, C.M., Kostidou, A., Vardaka, E., Lanaras, T. (1998) Effects of copper on the growth, photosynthesis and nutrient concentrations of Phaseolus plants. Photosynthetica 34:179–193.

Curie, C., Cassin, G., Counch, D., Divol, F., Higuchi, K., Le Jean, M., Misson, J., Schikora, A., Czernic, P., Mari, S. (2009) Metal movement within the plant: contribution of nicotianamine and yellow stripe 1-like transporters. Ann Bot 103:1–11.

Curie, C., Panaviene, Z., Loulergue, C., Dellaporta, S.L., Briat, J.-F., Walker, E.L. (2001) Maize yellow stripe1 encodes a membrane protein directly involved in Fe(III) uptake. Nature 409:346–349.

Deng, D., Wu, S.C., Wu, F.Y., Deng, H., Wong, M.H. (2010) Effects of root anatomy and Fe plaque on arsenic uptake by rice seedlings grown in solution culture. Environ Pollut 158:2589–2595. doi:10.1016/j.envpol.2010.05.015.

Dimkpa, C.O., Bindraban, P.S. (2016) Fortification of micronutrients for efficient agronomic production: a review. Agron Sustain Dev 36:1–26. https://doi.org/10.1007/s13593-015-0346-6.

Dinneny, J.R., Long, T.A., Wang, J.Y., Jung, J.W., Mace, D., Pointer, S., Barron, C., Brady, S.M., Schiefelbein, J., Benfey, P.N. (2008) Cell identity mediates the response of Arabidopsis roots to abiotic stress, Science 320:942–945.

Dixon, N.E., Gazzola, C., Blakeley, R.L., Zerner, R. (1975) Jack bean urease. A metalloenzyme. A simple biological role for Ni. J Am Chem Soc 97:4131–4133.

Dučić, T., Polle, A. (2005) Transport and detoxification of manganese and copper in plants. Brazilian J Plant Physiol 17:103–112. https://doi.org/10.1590/S1677-04202005000100009.

Duffus, J.H. (2002). "Heavy metals" a meaningless term? (IUPAC Technical Report). Pure Appl Chem 74(5): 793–807.

Eaton, E.T. (2015). "Manganese." In: Barker, A.V., Pilbeam, D. J. (Eds.), Handbook of plant nutrition. CRC Press, pp. 427–485.

Edwards, D.G., Asher, C.J. (1981) "Tolerance of crop and pasture species to manganese toxicity." In: Scaife, A. (Ed.), Proceedings of the ninth international plant nutrition colloquium. Warwick University, pp. 145–150.

EFSA Panel on Contaminants in the Food Chain (CONTAM). (2011) Statement on tolerable weekly intake for cadmium. EFSA J 9(2):1975.

Eren, E., Argüello, J. (2004) Arabidopsis HMA2, a divalent heavy metal-transporting PIB-Type ATPase, is involved in cytoplasmic Zn2+ homeostasis. Plant Physiol 136:3712–3723.

Eskew, D.L., Welch, R.M., Norvell, W.A. (1984) Ni in higher plants: further evidence for an essential role. Plant Physiol 76:691–693.

Fageria, N.K., Baligar, C., Clark, R.B. (2002) Micronutrients in crop production. Adv Agron 77:185–268.

Fageria, V.D. (2001) Nutrient interactions in crop plants. J Plant Nutr 24:1269–1290. https://doi.org/10.1081/PLN-100106981.

Farid, M., Ali, S., Rizwan, M., Yasmeen, T., Arif, M.S., Riaz, M., Saqib, M., Ayub, M.A. (2020) Combined effects of citric acid and 5-aminolevulinic acid in mitigating chromium toxicity in sunflower (Helianthus annuus L.) grown in Cr spiked soil. Pakistan J Agri Sci 57(2):477–488.

Farooqi, Z.U.R., Ahmad, Z., Ayub, M.A., Umar, W., Nadeem, M., Fatima, H., Shoukat, A., Ali, T., Ashraf, M.I. (2021) Threats to Arable land of the world: current and future perspectives of land use. Examining International Land Use Policies, Changes, and Conflicts, pp. 186–209.

Fecht-Christoffers, M.M., Führs, H., Braun, H.P., Horst, W.J. (2006) The role of hydrogen peroxide-producing and hydrogen peroxide-consuming peroxidases in the leaf apoplast of cowpea in manganese tolerance. Plant Physiol 140:1451–1463. doi: 10.1104/pp.105.070474.

Fendorf, S., Nico, P.S., Kocar, B.D., Masue, Y., Tufano, K.J. (2010) "Arsenic chemistry in soils and sediments." In Developments in soil science. Elsevier, Vol. 34, pp. 357–378.

Furini, A. (2012) Plants and heavy metals. Netherlands, Springer.

Gajewska, E., Skłodowska, M., Słaba, M., Mazur, J. (2006) Effect of nickel on antioxidative enzyme activities, proline and chlorophyll contents in wheat shoots. Biologia Plantarum 50(4):653–659.

Gallie, D.R. (2013) The role of l-ascorbic acid recycling in responding to environmental stress and in promoting plant growth. J. Exp. Bot 64:433–443. doi:10.1093/jxb/ers330.

Gao, X., Zhang, F., Hoffland, E. (2009) Malate exudation by six aerobic rice genotypes varying in zinc uptake efficiency. J Environ Qual 38(6):2315–2321.

Garcia, L., Welchen, E., Gonzalez, D.H. (2014). Mitochondria and copper homeostasis in plants. Mitochondrion 19:269–274.

George, T.S., French, A.S., Brown, L.K., Karley, A.J., White, P.J., Ramsay, L., Daniell, T.J. (2014) Genotypic variation in the ability of landraces and commercial cereal varieties to avoid manganese deficiency in soils with limited manganese availability: is there a role for root–exuded phytases? Physiol Plant 151:243–256. https://doi.org/10.1111/ppl.12151.

Ghori, N.H., Ghori, T., Hayat, M.Q., Imadi, S.R., Gul, A., Altay, V., Ozturk, M. (2019) Heavy metal stress and responses in plants. Int J Environ Sci Technol 16(3):1807–1828.

Gomes-Junior, R.A., Moldes, C.A., Delite, F.S., Gratão, P.L., Mazzafera, P., Lea, P.J., Azevedo, R.A. (2006) Nickel elicits a fast antioxidant response in Coffea arabica cells. Plant Physiol Biochem 44(5-6):420–429.

Gong, Q., Wang, L., Dai, T., Zhou, J., Kang, Q., Chen, H., Li, K., Li, Z. (2019) Effects of copper on the growth, antioxidant enzymes and photosynthesis of spinach seedlings. Ecotoxicol Environ Safety 171:771–780.

Gottesfeld, P., Were, F.H., Adogame, L., Gharbi, S., San, D., Nota, M.M., Kuepouo, G. (2018) Soil contamination from lead battery manufacturing and recycling in seven African countries. Environ Res 161:609–614.

Graham, R.C. (2004) Soil mineralogy with environmental applications. Vadose Zo J 3:724. https://doi.org/10.2136/vzj2004.0724.

Graham, R.D., Stangoulis, J.R.C. (2005) Molybdenum and disease. In: Datnoff, L., Elmer, W., Huber, D. (Eds.), Mineral Nutrition and Plant Diseases. APS Press, St Paul, MN.

Grotz, N., Guerinot, M.L. (2006) Molecular aspects of Cu, Fe and Zn homeostasis in plants. Biochim Biophys Acta 1763:595–608.

Guerinot, M.L., Yi, Y. (1994) Iron: nutritious, noxious, and not readily available. Plant Physiol 104:815–820.

Gupta, U.C., Wu, K., Liang, S. (2008) Micronutrients in soils, crops, and livestock. Earth Sci Front 15:110–125. https://doi.org/10.1016/s1872-5791(09)60003-8.

Gustin, J.L., Zanis, M.J., Salt, D.E. (2011) Structure and evolution of the plant cation diffusion facilitator family of ion transporters. BMC Evol Biol 11:76.

Hall, J.L. (2002) Cellular mechanisms for heavy metal detoxification and tolerance. J Exp Bot 53:1–11.

Harasim, P., Filipek, T. (2015) Nickel in the environment. J Elem 20:525–534. https://doi.org/10.5601/jelem.2014.19.3.651.

Hasan, M., Ahammed, G. J., Yin, L., Shi, K., Xia, X., Zhou, Y. Yu, J., Zhou, J. (2015) Melatonin mitigates cadmium phytotoxicity through modulation of phytochelatins biosynthesis, vacuolar sequestration, and antioxidant potential in *Solanum lycopersicum* L. Front Plant Sci 6:601. doi: 10.3389/fpls.2015.00601.

Hasan, M., Cheng, Y., Kanwar, M.K., Chu, X.Y., Ahammed, G.J., Qi, Z.Y. (2017) Responses of plant proteins to heavy metal stress—a review. Front Plant Sci 8:1492.

Hasan, M.K., Liu, C., Wang, F., Ahammed, G.J., Zhou, J., Xu, M.X., Yu, J.Q., Xia, X.J. (2016) Glutathione-mediated regulation of nitric oxide, S-nitrosothiol and redox homeostasis confers cadmium tolerance by inducing transcription factors and stress response genes in tomato. Chemosphere161:536–545. doi: 10.1016/j.chemosphere.2016.07.053.

Hassan, Z., Aarts, M. (2011) Opportunities and feasibilities for biotechnological improvement of Zn, Cd or Ni tolerance and accumulation in plants. Environ Exp Bot 72:53–63.

Hebbern, C.A., Pedas, P., Schjoerring, J.K., Knudsen, L., Husted, S. (2005) Genotypic differences in manganese efficiency: field experiments with winter barley (Hordeum vulgare). Plant Soil 272:233–244.

Heine, G., Max, J.F.J., Führs, H., Moran-Puente, D.W., Heintz, D., Horst, W.J. (2011) Effect of manganese on the resistance of tomato to Pseudocercospora fuligena. J Plant Nutr Soil Sci 174:827–836. doi: 10.1002/jpln.201000440.

Henriques, A.R., Chalfun-Junior, A., Aarts, M. (2012) Strategies to increase zinc deficiency tolerance and homeostasis in plants. Brazilian J Plant Physiol 24(1):3–8.

Hernandez-Apaolaza, L. (2014) Can silicon partially alleviate micronutrient deficiency in plants? A review. Planta 240:447–458. https://doi.org/10.1007/s00425-014-2119.

Hewitt, E.J., Bolle-Jones, E.W. (1952a) Molybdenumas a plant nutrient. II. The effects of molybdenum deficiency on some horticultural and agricultural crop plants in sand culture. J Hortic Sci 27:257–265.

Hewitt, E.J., Bolle-Jones, E.W. (1952b) Molybdenum as a plant nutrient. I. The influence of molybdenum on the growth of some Brassica crops in sand culture. J Hortic Sci 27:245–256.

Hindt, M.N., Guerinot, M.L. (2012) Getting a sense for signals: regulation of the plant iron deficiency response. Biochimica et Biophysica Acta (BBA)-Mol Cell Res 1823(9):1521–1530.

Hossain, M.A., Piyatida, P., da Silva, J.A.T., Fujita, M. (2012) Molecular mechanism of heavy metal toxicity and tolerance in plants: central role of glutathione in detoxification of reactive oxygen species and methylglyoxal and in heavy metal chelation. J. Bot 2012:872875. doi: 10.1155/2012/872875.

Hussain, M.M., Wang, J., Bibi, I., Shahid, M., Niazi, N.K., Iqbal, J., Mian, I.A., Shaheen, S.M., Bashir, S., Shah, N.S., Hina, K. (2020) Arsenic speciation and biotransformation pathways in the aquatic ecosystem: The significance of algae. J Hazard Mater 403:124027.

Husted, S., Thomsen, M.U., Mattsson, M., Schjoerring, J. (2009) Influence of nitrogen and sulphur form on manganese acquisition by barley (Hordeum vulgare). Plant Soil 268:309–317.

Hu, Y., Cheng, H., Tao, S. 2016. The challenges and solutions for cadmium-contaminated rice in China: a critical review. Environ Int 92, 515–532.

Hüttner, S., Veit, C., Schoberer, J., Grass, J., Strasser, R. (2012) Unraveling the function of *Arabidopsis thaliana* OS9 in the endoplasmic reticulum-associated degradation of glycoproteins. Plant Mol Biol 79:21–33. doi: 10.1007/s11103-012-9891-4.

Ihnatowicz, A., Siwinska, J., Meharg, A. A., Carey, M., Koornneef, M., Reymond, M. (2014) Conserved histidine of metal transporter AtNRAMP1 is crucial for optimal plant growth under manganese deficiency at chilling temperatures. New Phytol 202:1173–1183. doi: 10.1111/nph.12737.

Impa, S.M., Johnson-Beebout, S.E. (2012). Mitigating zinc deficiency and achieving high grain Zn in rice through integration of soil chemistry and plant physiology research. Plant and Soil 361(1-2): 3–41.

Inglezakis, V.J., Zorpas, A.A., Karagiannidis, A., Samaras, P., Voukkali, I., Sklari, S. (2014) European Union legislation on sewage sludge management. Fresenius Environ Bull 23(2A):635–639.

Inoue, H., Kobayashi, T., Nozoye, T., Takahashi, M., Kakei, Y., Suzuki, K., Nakazono, M., Nakanishi, H., Mori, S., Nishizawa, N.K. (2009) Rice OsYSL15 is an iron-regulated iron(III)-deoxymugineic acid transporter expressed in the roots and is essential for iron uptake in early growth of the seedlings. J Biol Chem 284:3470–3479.

Ishikawa, S., Ishimaru, Y., Igura, M., Kuramata, M., Abe, T., Senoura, T., Hase, Y., Arao, T., Nishizawa, N.K., Nakanishi, H. (2012). Ion-beam irradiation, gene identification, and marker-assisted breeding in the development of low-cadmium rice. Proc Natl Acad Sci 109(47):19166–19171.

Ishimaru, Y., Suzuki, M., Kobayashi, T., Takahashi, M., Nakanishi, H., Mori, S., Nishizawa, N.K. (2005) OsZIP4, a novel zinc-regulated zinc transporter in rice. J Exp Bot 56(422):3207–3214.

Jaime-Pérez, N., Kaftan, D., Bína, D., Bokhari, S. N. H., Shreedhar, S., Küpper, H. (2019) Mechanisms of sublethal copper toxicity damage to the photosynthetic apparatus of Rhodospirillum rubrum. Biochimica et Biophysica Acta (BBA)-Bioenerg 1860(8):640–650.

Jiang, W.Z. (2006) Mn use efficiency in different wheat cultivars. Environ Exp Bot 57:41–50.

Kabata-Pendias, A. (2004) Soil–plant transfer of trace elements—an environmental issue. Geoderma 122:143–149. https://doi.org/10.1016/j.geoderma.2004.01.004.

Kaiser, B.N., Gridley, K.L., Brady, J.N. (2005) The role of molybdenum in agricultural plant production. Ann Bot 96:745–754. https://doi.org/10.1093/aob/mci226.

Kim, Y.Y., Choi, H., Segami, S., Cho, H.T., Martinoia, E., Maeshima, M. Lee Y. (2009) AtHMA1 contributes to detoxification cation of excess Zn(II) in Arabidopsis. Plant J 58:737–753.

Kim, S.C., Kim, H.S., Seo, B.H., Owens, G., Kim, K.R. (2016) Phytoavailability control based management for paddy soil contaminated with Cd and Pb: Implications for safer rice production. Geoderma 270:83–88.

Kimbrough, D.E., Cohen, Y., Winer, A.M., Creelman, L., Mabuni, C. (1999) A critical assessment of chromium in the environment. Crit Rev Environ Sci Technol 29(1):1–46.

Klatte, M., Schuler, M., Wirtz, M., Fink-Straube, C., Hell, R., Bauer, P. (2009) The analysis of Arabidopsis nicotianamine synthase mutants reveals functions for nicotianamine in seed iron loading and iron deficiency responses. Plant Physiol 150:257–271.

Koller, M., Saleh, H.M. (2018). Introductory chapter: introducing heavy metals. Heavy Metals. 1: 3–11.

Kopittke, P.M., Blamey, F.P.C., McKenna, B.A., Wang, P., Menzies, N.W. (2011) Toxicity of metals to roots of cowpea in relation to their binding strength. Environ Toxicol Chem 30(8):1827–1833.

Küpper, H., Kroneck, P.M.H. (2007) "Nickel in the environment and its role in the metabolism of plants and cyanobacteria." In: Sigel, A., Sigel, H., Sigel, R.K.O. (Eds.), Metal ions in life sciences. John Wiley and Sons Ltd., Chichester, UK, Vol. 2, pp. 31–62.

Kushwaha, A., Hans, N., Kumar, S., Rani, R. (2018) A critical review on speciation, mobilization and toxicity of lead in soil-microbe-plant system and bioremediation strategies. Ecotoxicol Environ Safety 147:1035–1045.

Lee, S., Chiecko, J.C., Kim, S.A., Walker, E.L., Lee, Y., Guerinot, M.L., An, G. (2009) Disruption of OsYSL15 leads to iron inefficiency in rice plants, Plant Physiol 150:786–800.

Lee, S., Moon, J. S., Ko, T.-S., Petros, D., Goldsbrough, P. B., Korban, S.S. (2003) Overexpression of *Arabidopsis* phytochelatin synthase paradoxically leads to hypersensitivity to cadmium stress. Plant Physiol 131:656–663. doi: 10.1104/pp.014118.

Li, M., Ahammed, G. J., Li, C., Bao, X., Yu, J., Huang, C. Yin, H., Zhou, J. (2016a) Brassinosteroid ameliorates zinc oxide nanoparticles-induced oxidative stress by improving antioxidant potential and redox homeostasis in tomato seedling. Front Plant Sci 7:615. doi: 10.3389/fpls.2016.00615.

Liang, Z.D., Tsai, W.B., Lee, M.Y., Savaraj, N., Kuo, M.T. (2012) Specificity protein 1 (spl) oscillation is involved in copper homeostasis maintenance by regulating human high-affinity copper transporter 1 expression. Mol Pharmacol 81(3):455–464.

Maleva, M.G., Nekrasova, G.F., Borisova, G.G., Chukina, N.V., Ushakova, O.S. (2012). Effect of heavy metal on photosynthetic apparatus and antioxidant status of elodea. Russian J Plant Physiol 59:190–197. doi: 10.1134/S1021443712020069.

Mallhi, A.I., Chatha, S.A.S., Shafaqat, A., Tahsin, G., Muhammad, R., Muhammad, Z.U.R., Shahid, M.R., Ayub, M.A., Mallhi, Z.I. (2019) Glutathione alleviates chromium stress in Helianthus annuus irrigated with tannery wastewater. Int J Agri Biol 22(6):1376–1384.

Manara, A. (2012). Plant responses to heavy metal toxicity. In: Furini, A. (Ed.), Plants and heavy metals. Springer, Dordrecht, Heidelberg, New York, London, pp. 27–53.

Marques, D.M., da Silva, A.B., Mantovani, J.R., Magalhães, P.C., de Souza, T.C. (2019). Root morphology and leaf gas exchange in *Peltophorum dubium* (Spreng.) Taub. (Caesalpinioideae) exposed to copper-induced toxicity. South Af J Bot 121:186–192.

Marschner H. (1995) Mineral nutrition of higher plants, 2nd edition. Academic Press, London.

Masunaga, T., Fong, J.D.M. (2018). Strategies for increasing micronutrient availability in soil for plant uptake. In: Plant Micronutrient Use Efficiency. Academic Press, pp. 195–208.

McGrath, J.M., Spargo, J., Penn, C.J. (2014) Soil fertility and plant nutrition. Encycl Agric Food Syst 5:166–184. https://doi.org/10.1016/B978-0-444-52512-3.00249-7.

Mendel, R.R., Haensch, R. (2002) Molybdoenzymes and molybdenum cofactor in plants. J Exp Bot 53:1689–1698.

Mendel, R.R., Kruse, T. (2012). Cell biology of molybdenum in plants and humans. Biochimica et Biophysica Acta (BBA)-Mol Cell Res 1823(9):1568–1579.

Mertz, W. (1981). The essential trace elements. Science 213(4514):1332–1338.

Millaleo R., Reyes-Díaz M., Ivanov A.G. (2010) Manganese as essential and toxic element for plants: transport, accumulation and resistance mechanisms. J Soil Sci Plant Nutr 10:476–494.

Millaleo, R., Reyes-Díaz, M., Ivanov, A.G., Mora, M.L., Alberdi, M. (2010) Manganese as essential and toxic element for plants: transport, accumulation and resistance mechanisms. J Soil Sci Plant Nutri 10(4):470–481. https://doi.org/10.4067/s0718-95162010000200008.

Monisha, J., Tenzin, T., Naresh, A., Blessy, B.M., Krishnamurthy, N.B. (2014) Toxicity, mechanism and health effects of some heavy metals. Interdisciplinary Toxicology 7(2):60–72.

Monnet, F., Vaillant, N., Vernay, P., Coudret, A., Sallanon, H., Hitmi, A. (2001). Relationship between PSII activity, CO2 fixation, and Zn, Mn and Mg contents of Lolium perenne under zinc stress. J. Plant Physiol 158:1137–1144. doi: 10.1078/S0176-1617(04)70140-6.

Moradtalab, N., Weinmann, M., Walker, F., Höglinger, B., Ludewig, U., Neumann, G. (2018) Silicon improves chilling tolerance during early growth of maize by effects on micronutrient homeostasis and hormonal balances. Front Plant Sci 9:1–17. https://doi.org/10.3389/fpls.2018.00420.

Morrissey, J., Guerinot, M.L. (2009) Iron uptake and transport in plants: the good, the bad, and the ionome. Chem Rev 109:4553–4567. https://doi.org/10.1021/cr900112r.

Mulrooney, S.B., Hausinger, R.P. (2003) Nickel uptake and utilization by microorganisms. Iron MS Microbiol Rev 27:239–261. https://doi.org/10.1016/S0168-6445(03)00042-1.

Naeem, A., Zafar, M., Khalid, H., Zia-ur-Rehman, M., Ahmad, Z., Ayub, M.A., Qayyum, M.F. (2019) "Cadmium-induced imbalance in nutrient and water uptake by plants." In Cadmium toxicity and tolerance in plants. Academic Press, pp. 299–326.

Nagajyoti, P.C., Lee, K.D., Sreekanth, T.V.M. (2010) Heavy metals, occurrence and toxicity for plants: a review. Environ Chem Lett 8(3):199–216.

Najafi-Ghiri, M., Ghasemi-Fasaei, R., Farrokhnejad, E. (2013) Factors affecting micronutrient availability in calcareous soils of Southern Iran. Arid Land Res Manag 27(3):203–215.

Nakandalage, N., & Seneweera, S. (2018). Micronutrients use efficiency of crop-plants under changing climate. In: Hossain, M.A., Kamiya, T., Burritt, D.J., Tran, L.S.P. and Fujiwara, T. Plant micronutrient use efficiency. Academic Press, London, pp. 209–224.

Naveed, M., Khalid, H., Ayub, M.A., ur Rehman, M.Z., Rizwan, M., Rasul, A., ul Haq, M.A. (2020) Biofortification of cereals with zinc and iron: recent advances and future perspectives. Res Use Eff Agri: 615–646.

Nazir, F., Hussain, A., Fariduddin, Q. (2019). Hydrogen peroxide modulate photosynthesis and antioxidant systems in tomato (Solanum lycopersicum L.) plants under copper stress. Chemosphere 230:544–558.

Niazi, N.K., Bibi, I., Shahid, M., Ok, Y.S., Burton, E.D., Wang, H., Shaheen, S.M., Rinklebe, J., Lüttge, A. ((2018)) Arsenic removal by perilla leaf biochar in aqueous solutions and groundwater: an integrated spectroscopic and microscopic examination. Environ Poll 232:31–41.

Nkrumah, P.N., Echevarria, G., Erskine, P.D., Chaney, R.L., Sumail, S., van der Ent, A. (2019) Effect of nickel concentration and soil pH on metal accumulation and growth in tropical agromining "metal crops." Plant Soil 443(1-2):27–39.

Nozoye, T., Nagasaka, S., Kobayashi, T., Sato, Y., Uozumi, N., Nakanishi, H., Nishizawa, N.K. (2011) The phytosiderophore efflux transporter TOM2 is involved in metal transport in rice. J Biol Chem 290:27688–27699. doi: 10.1074/jbc.M114.635193.

Palmer, C., Guerinot, M. (2009) Facing the challenges of Cu, Fe and Zn homeostasis in plants. Nature Chem Biol 5:333–340.

Papadakis, I.E., Giannakoula, A., Therios, I.N., Bosabalidis, A.M., Moustakas, M., Nastou, A. (2007) Mn-induced changes in leaf structure and chloroplast ultrastructure of *Citrus volkameriana* (L.) plants. J Plant Physiol 164:100–103.

Park, J., Song, W.Y., Ko, D., Eom, Y., Hansen, T.H., Schiller, M. Lee, T.G., Martinoia, E., Lee, Y. (2012) The phytochelatin transporters AtABCC1 and AtABCC2 mediate tolerance to cadmium and mercury. Plant J. 69:278–288. doi: 10.1111/j.1365-313X.2011.04789.x.

Pochodylo, A.L., Aristilde, L. (2017) Molecular dynamics of stability and structures in phytochelatin complexes with Zn, Cu, Fe, Mg, and Ca: Implications for metal detoxification. Environ Chem Lett 1–6. doi: 10.1007/s10311-017-0609-3.

Pomponi, M., Censi, V., Di Girolamo, V., De Paolis, A., Di Toppi, L.S., Aromolo, R. Costantino, P., Cardarelli, M. (2006) Overexpression of *Arabidopsis* phytochelatin synthase in tobacco plants enhances Cd^{2+} tolerance and accumulation but not translocation to the shoot. Planta 223:180–190. doi: 10.1007/s00425-005-0073-3.

Porter, G.S., Bajita-Locke, J.B., Hue, N.V., Strand, D. (2004) Manganese solubility and phytotoxicity affected by soil moisture, oxygen levels, and green manure additions. Commun Soil Sci Plan 35(1–2):99–116.

Printz, B., Lutts, S., Hausman, J.F., Sergeant, K. (2016) Copper trafficking in plants and its implication on cell wall dynamics. Front Plant Sci 7:1–16. https://doi.org/10.3389/fpls.2016.00601.

Priyanka, N., Geetha, N., Ghorbanpour, M., Venkatachalam, P. (2019) Role of engineered zinc and copper oxide nanoparticles in promoting plant growth and yield: present status and future prospects. Academic Press, London, pp. 183–201.

Rana, M., Bhantana, P., Sun, X. C., Imran, M., Shaaban, M., Moussa, M., Hamzah Saleem, M., Elyamine, A., Binyamin, R., Alam, M. and Afzal, J. (2020) Molybdenum as an essential element for crops: an overview. Int J Sci Res Growth 24:18535.

Ravet, K., Pilon, M. (2013). Copper and iron homeostasis in plants: the challenges of oxidative stress. Antiox Redox Signal 19(9):919–932.

Reeves, P.G., Chaney, R.L. (2008). Bioavailability as an issue in risk assessment and management of food cadmium: A review. Sci Total Environ 398(1–3):13–19.

Rehman, M.Z., Batool, Z., Ayub, M.A., Hussaini, K.M., Murtaza, G., Usman, M., Naeem, A., Khalid, H., Rizwan, M., Ali, S. (2020a) Effect of acidified biochar on bioaccumulation of cadmium (Cd) and rice growth in contaminated soil. Environ Technol Innov 19:101015.

Rehman, M.Z., Khalid, H., Rizwan, M., Ali, S., Sohail, M.I., Usman, M., Umair, M. (2019) Inorganic amendments for the remediation of cadmium-contaminated soils. In Cadmium tolerance in plants. Academic Press, Elsevier, Cambridge, MA, pp. 113–141.

Rehman, M.Z., Waqar, M., Bashir, S., Rizwan, M., Ali, S., El Baroudy, A.A.E.F., Khalid, H., Ayub, M.A., Usman, M., Jahan, S. (2021) Effect of biochar and compost on cadmium bioavailability and its uptake by wheat–rice cropping system irrigated with untreated sewage water: a field study. Arabian J Geosci 14(2):1–12.

Rehman, M.Z., Zafar, M., Waris, A.A., Rizwan, M., Ali, S., Sabir, M., Usman, M., Ayub, M.A., Ahmad, Z. (2020b) Residual effects of frequently available organic amendments on cadmium bioavailability and accumulation in wheat. Chemosphere 244:125548.

Rengel, Z. (1997) "Root exudation and microflora populations in rhizosphere of crop genotypes differing in tolerance to micronutrient deficiency." In Plant nutrition for sustainable food production and environment. Springer, Dordrecht, pp. 243–248.

Rengel, Z. (2000) Manganese uptake and transport in plants. Met Ions Biol Syst 37:57–87. https://doi.org/10.1201/9781482289893-13.

Riaz, M.U., Ayub, M.A., Khalid, H., ul Haq, M.A., Rasul, A., ur Rehman, M.Z., Ali, S. (2020) "Fate of micronutrients in alkaline soils." In Resources use efficiency in agriculture. Springer, Singapore, pp. 577–613.

Riedel, G.F. (2008) "Copper production and usage chemistry of copper." In Ecotoxicology. Elsevier, Amsterdam, pp. 778–783.

Robson, A.D. (1994) Zinc in soils and plants. Springer, New York.

Rosas, I., Belmomt, R., Baez, A., Villalobos-Pietrini, R. (1989) Some aspects of the environmental exposure to chromium residues in Mexico. Water Air Soil Poll 48(3-4):463–475.

Rout, G.R., Das, P. (2009). "Effect of metal toxicity on plant growth and metabolism: I. Zinc." In Sustainable agriculture. Springer, Dordrecht, pp. 873–884.

Rout, G.R., Sahoo, S. (2015) Role of iron in plant growth and metabolism. Rev Agric Sci 3:1–24. https://doi.org/10.7831/ras.3.1.

Saidi, M.N., Jbir, R., Ghorbel, I., Namsi, A., Drira, N., Gargouri-Bouzid, R. (2012) Brittle leaf disease induces an oxidative stress and decreases the expression of manganese-related genes in date palm (Phoenix dactylifera L.). Plant Physiol Biochem 50:1–7. https://doi.org/10.1016/j.plaphy.2011.09.016.

Santi, S., Cesco, S., Varanini, Z., Pinton, R. (2005) Two plasma membrane H+-ATPase genes are differentially expressed in iron-deficient cucumber plants. Plant Physiol Biochem 43:287–292.

Santi, S., Schmidt, W. (2009) Dissecting iron deficiency-induced proton extrusion in Arabidopsis roots. New Phytol 183(4):1072–1084.

Schmid, N.B., Giehl, R.F., Döll, S., Mock, H.P., Strehmel, N., Scheel, D., Kong, X., Hider, R.C. and von Wirén, N. (2014) Feruloyl-CoA 6′-Hydroxylase1-dependent coumarins mediate iron acquisition from alkaline substrates in Arabidopsis. Plant Physiology, 164(1), 160–172.

Seguí, À.C., Blasco, L.P. (2017). *Arabidopsis thaliana COPT transporters function in hormone metabolism and in metal stress* (Doctoral dissertation, Universitat de València).

Seh, E. (2002) "Metal permeability, transport and efflux in plants." In: Prasad, M.N.V., Strazalka, K. (Eds.), Physiology and Biochemistry of Metal Toxicity and Tolerance in Plants. London. Kluwer Academic Publishers.

Seregin, I.V., Kozhevnikova, A.D. (2006) Physiological role of nickel and its toxic effects on higher plants. Russ J Plant Physiol 53:285–308.

Sharma, S.S., Dietz, K.J., Mimura, T. (2016) Vacuolar compartmentalization as indispensable component of heavy metal detoxification in plants. Plant Cell Environ 39:1112–1126. doi: 10.1111/pce.12706.

Sharma, S.K., Sehkon, N.S., Deswal, S., John, S. (2009). Transport and fate of copper in soils. Int J Civil Environ Eng 1(1):19–39.

Sheldon, A.R., Menzies, N.W. (2005) The effect of copper toxicity on the growth and root morphology of Rhodes grass (Chloris gayana Knuth.) in resin buffered solution culture. Plant Soil 278(1-2):341–349.

Shin, L.J., Lo, J.C., Yeh, K.C. (2012). Copper chaperone antioxidant protein1 is essential for copper homeostasis. Plant Physiol 159(3):1099–1110.

Silva, J.C., Echeveste, P., Lombardi, A. T. (2018) Higher biomolecules yield in phytoplankton under copper exposure. Ecotoxicol Environ Safety 161:57–63.

Singh, S., Korripally, P., Vancheeswaran, R., Eapen, S. (2011). Transgenic Nicotiana tabacum plants expressing a fungal copper transporter gene show enhanced acquisition of copper. Plant Cell Rep 30(10):1929–1938.

Sirko, A., Brodzik, R. (2000). Plant ureases: roles and regulation. Acta Biochim Polon 47(4), 1189–1195.

Sohail, M.I., Waris, A.A., Ayub, M.A., Usman, M., ur Rehman, M.Z., Sabir, M., Faiz, T. (2019) "Environmental application of nanomaterials: A promise to sustainable future." In Comprehensive analytical chemistry. Elsevier Amsterdam The Netharland, Vol. 87, pp. 1–54.

Solymosi, K., Lenti, K., Myśliwa-Kurdziel, B., Fidy, J., Strzałka, K., Böddi, B. (2004) Hg2+ reacts with different components of the NADPH: protochlorophyllide oxidoreductase macrodomains. Plant Biol 6(3):358–368.

Song, W.Y., Choi, K.S., Kim do, Y., Geisler, M., Park, J., Vincenzetti, V., Schellenberg, M. Kim, S.H., Lim, Y.P., Noh, E.W., Lee, Y. (2010) Arabidopsis PCR2 is a zinc exporter involved in both zinc extrusion and long-distance zinc transport. Plant Cell 22:2237–2252.

Song, W.Y., Mendoza-Cozatl, D.G., Lee, Y., Schroeder, J.I., Ahn, S.N., Lee, H.S. Wicker, T., Martinoia, E. (2014) Phytochelatin–metal (loid) transport into vacuoles shows different substrate preferences in barley and *Arabidopsis*. Plant Cell Env 37:1192–1201. doi: 10.1111/pce.12227.

Sreekanth, T.V.M., Nagajyothi, P.C., Lee, K.D., Prasad, T.N.V.K.V. (2013) Occurrence, physiological responses and toxicity of nickel in plants. Int J Environ Sci Technol 10:1129–1140.

St. Clair, S.B., Lynch, J.P. (2005). Element accumulation patterns of deciduous and evergreen tree seedlings on acid soils: implications for sensitivity to manganese toxicity. Tree Physiol 25:85–92. doi: 10.1093/treephys/25.1.85.

Sun, X.H., Yu, G., Li, J.T. (2014) A heavy metal-associated protein (AcHMA1) from the halophyte, Atriplex canescens (Pursh) Nutt., confers tolerance to iron and other abiotic stresses when expressed in Saccharomyces Cerevisiae. Int J Mol Sci 15:14891–14906. https://doi.org/10.3390/ijms150814891.

Tallkvist, J., Oskarsson, A. (2015) Molybdenum. In: Handbook on the Toxicology of Metals. Academic Press, London, pp. 1077–1089.

Tamás, M.J., Sharma, S.K., Ibstedt, S., Jacobson, T., Christen, P. (2014) Heavy metals and metalloids as a cause for protein misfolding and aggregation. Biomolecules 4:252–267. doi: 10.3390/biom4010252.

Thakur, S., Singh, L., Ab Wahid, Z., Siddiqui, M.F., Atnaw, S.M., Din, M.F.M. (2016) Plant-driven removal of heavy metals from soil: uptake, translocation, tolerance mechanism, challenges, and future perspectives. Environ Monitor Assess 188(4):206.

Thomas, G., Andresen, E., Mattusch, J., Hubáček, T., Küpper, H. (2016) Deficiency and toxicity of nanomolar copper in low irradiance—a physiological and metalloproteomic study in the aquatic plant *Ceratophyllum demersum*. Aquatic Toxicol 177:226–236.

Thomine, S., Lelièvre, F., Debarbieux, E., Schroeder, J.I., Barbier-Brygoo, H. (2003). AtNRAMP3, a multispecific vacuolar metal transporter involved in plant responses to iron deficiency. Plant J 34(5):685–695.

Thomine, S., Wang, R., Ward, J.M., Crawford, N.M., Schroeder, J.I. (2000). Cadmium and iron transport by members of a plant metal transporter family in Arabidopsis with homology to Nramp genes. Proceed Natl Acad Sci 97(9):4991–4996.

Todeschini, V., Lingua, G., D'Agostino, G., Carniato, F., Roccotiello, E., Berta, G. (2011) Effects of high zinc concentration on poplar leaves: a morphological and biochemical study. Environ Exp Bot 71(1):50–56.

US Food and Drug Administration (USFDA). (2017) Total diet study elements results summary statistics-market baskets 2006 through 2013. Available online at https://www.fda.gov/food/total-diet-study/analytical-results-total-diet-study.

van de Mortel, J.E., Almar Villanueva, L., Schat, H., Kwekkeboom, J., Coughlan, S., Moerland, P.D. van Themaat, E.V.L., Koornneef, M., Arts, M.G. (2006) Large expression differences in genes for iron and zinc homeostasis, stress response, and lignin biosynthesis distinguish roots of Arabidopsis thaliana and the related metal hyperaccumulator Thlaspi caerulescens. Plant Physiol 42:1127–1147.

Vasconcelos, M., Eckert, H., Arahana, V., Graef, G., Grusak, M.A., Clemente, T. (2006) Molecular and phenotypic characterization of transgenic soybean expressing the *Arabidopsis ferric* chelate reductase gene, FRO2. Planta 224:1116–1128.

Vert, G., Grotz, N., Dedaldechamp, F., Gaymard, F., Guerinot, M.L., Briat, J.-F., Curie, C. (2002) IRT1, an Arabidopsis transporter essential for iron uptake from the soil and plant growth. Plant Cell 14:1223–1233.

Vinit-Dunand, F., Epron, D., Alaoui-Sossé, B., Badot, P.M. (2002) Effects of copper on growth and on photosynthesis of mature and expanding leaves in cucumber plants. Plant Sci 163:53–58. https://doi.org/10.1016/S0168-9452(02)00060-2.

Wang, M., Kawakami, Y., Bhullar, N.K. (2019) Molecular analysis of iron deficiency response in hexaploid wheat. Front Sustain Food Syst 3:67.

Waters, B., Sankaran, R. (2011) Moving micronutrients from the soil to the seeds: genes and physiological processes from a biofortification perspective. Plant Sci 180:562–574.

Watmough, S.A., Eimers, M.C., Dillon, P.J. (2007) Manganese cycling in Central Ontario forests: response to soil acidification. Appl Geochem 22:1241–1247. https://doi.org/10.1016/j.apgeochem.2007.03.039.

Williams, R.J.P., Fraústo da Silva, J.J.R. (2002) The involvement of molybdenum in life. Boroniochem Biophys Res Commun 292:293–299. https://doi.org/10.1006/bbrc.2002.6518.

World Health Organization (WHO). (1996) Permissible limits of heavy metals in soil and plants. Geneva, Switzerland.

Wuana, R.A., Okieimen, F.E. (2011) Heavy metals in contaminated soils: a review of sources, chemistry, risks and best available strategies for remediation. Int Schol Res Notice 2011. https://doi.org/10.5402/2011/402647

Xu, L., Wang, Y., Liu, W., Wang, J., Zhu, X., Zhang, K. Yu, R., Wang, R., Xie, Y., Zhang, W. and Gong, Y. (2015). *De novo* sequencing of root transcriptome reveals complex cadmium-responsive regulatory networks in radish (*Raphanus sativus* L.). Plant Sci 236:313–323. doi: 10.1016/j.plantsci.2015.04.015.

Yan, X., Liu, M., Zhong, J., Guo, J., Wu, W. (2018) How human activities affect heavy metal contamination of soil and sediment in a long-term reclaimed area of the Liaohe River Delta, North China. Sustainability 10(2):338.

Yang, Y., Wang, M., Chen, W., Li, Y., Peng, C. (2017) Cadmium accumulation risk in vegetables and rice in southern China: insights from solid-solution partitioning and plant uptake factor. J Agri Food Chem 65(27):5463–5469.

Yano, J., Yachandra, V. (2014) Mn4Ca cluster in photosynthesis: where and how water is oxidized to dioxygen. Chem Rev 114:4175–4205.

Yokel, J., Delistraty, D.A. (2003) Arsenic, lead, and other trace elements in soils contaminated with pesticide residues at the Hanford site (USA). Environ Toxicol Int J 18(2):104–114.

Yruela, I. (2005) Copper in plants. Brazilian J Plant Physiol 17(1):145–156.

Zhang, J., Hwang, J. U., Song, W. Y., Martinoia, E., Lee, Y. (2017) Identification of amino acid residues important for the arsenic resistance function of Arabidopsis ABCC1. FEBS Lett 591:656–666. doi: 10.1002/1873-3468.12576.

Zhao, J., Wang, W., Zhou, H., Wang, R., Zhang, P., Wang, H., Pan, X., Xu, J. (2017) Manganese toxicity inhibited root growth by disrupting auxin biosynthesis and transport in Arabidopsis. Front Plant Sci 8:272.

Zhao, L., Sun, Y. L., Cui, S. X., Chen, M., Yang, H. M., Liu, H. M., Chai, T.Y., Hunag, F. (2011) Cd-induced changes in leaf proteome of the hyperaccumulator plant *Phytolacca americana*. Chemosphere 85:56–66. doi: 10.1016/j.chemosphere.2011.06.029.

Zulfiqar, U., Farooq, M., Hussain, S., Maqsood, M., Hussain, M., Ishfaq, M., Ahmad, M. and Anjum, M.Z., 2019. Lead toxicity in plants: Impacts and remediation. J Environ Manage 250:109557.

10

Physiological, Biochemical, and Molecular Mechanisms of Gasotransmitter-Mediated Heavy Metal Stress Tolerance in Plants

Vivek Kumar, Akash Hidangmayum, and Ankita Singh
Institute of Agriculture Sciences, Banaras Hindu University, Varanasi, UP, India

Rekha Sodani
College of Agriculture, Agriculture University, Nagaur, Jodhpur, Rajasthan, India

Basant Kumar Dadrwal, Navneet Kumar, Surendra Kumar Chaudhary, and Brijesh Kumar Chaudhary
Institute of Agriculture Sciences, Banaras Hindu University, Varanasi, UP, India

S.P. Kushwaha
Chandra Shekhar Azad University of Agriculture & Technology, Kanpur, UP, India

Jyoti Chauhan
Institute of Agriculture Sciences, Banaras Hindu University, Varanasi, UP, India

Hirdayesh Anuragi
ICAR-Central Agroforestry Research Institute, Jhansi, UP, India

Udit Nandan Mishra
Faculty of Agriculture, Sri Sri University, Cuttack, Odisha, India

Devidutta Lenka
Orissa University of Agriculture and Technology, Bhubaneswar, Odisha, India

Rajesh Kumar Singhal
ICAR-Indian Grassland and Fodder Research Institute, Jhansi, UP, India

CONTENTS

10.1 Introduction

Anthropogenic interruption in natural biosphere severely affects living organisms including plants. Expedite production of emerging contaminants, such as heavy metals (HMs) from industries, mining, intensive agriculture and municipal waste have not only havoc the natural assets but also affects the equilibrium in living organism. HMs include both essential and non-essential elements for growth, physiological function and development of plants, such as Fe, Zn, Cu, Mn, Mo, Ni, Cd, Cr, Pb, Co, As, Ag and Hg. Supply of these elements in accurate concentration plays a crucial role in plants but excess can lead to hazardous, and retardation of plant growth (Ghori et al. 2019). Various transporter family such as Zn transporter family members (ZIP, ZRT), copper cation transporter (COPT), yellow stripe-like (YSL) transporter, nodulin-26-like intrinsic protein (NIP) aquaporin's and high-affinity Ca^{2+} channels, helps in transport of these elements by

DOI: 10.1201/9781003155089-10

plants roots surface (Terrón-Camero et al. 2019). After entering in root cells, metals are immobilized in cell vacuole or translocate through apoplast towards upper side of the xylem. Bioaccumulation and biomagnification of HMs interact with several cellular biomolecules leading to extreme generation of reactive oxygen species (ROS), such as hydrogen peroxide (H_2O_2), superoxide radicals (O_2^-), hydroxyl free radicals (OH^-), singlet oxygen (O_2^*) (Emamverdian et al. 2015). These ROS may cause serious complications in plants which include chlorosis, lipid peroxidation, protein degradation, denaturation of cellular membrane and redox imbalance. Plants evolved various tricks for tolerating HMs toxicity and maintain cellular homeostasis. Mechanism such as reduction in uptake through production of exudates compound, chelation *via* initiation of organic compounds, phytochelatins (PCs), polysaccharides and metallothioneins (MTs); subcellular compartmentalization for sequestration of metal ions is primarily used (Maksymiec 2007). Plant can bind the metals to the cell wall with electrostatic interaction and minimize their explosion and retain them within cellulosic and lignin structure (Memon and Schröder 2009; Loix et al. 2017). Other defence during HMs toxicity includes generation of stress defensive protein, production and osmolyte accumulation, enhanced SOD (superoxide dismutase) activities, APX (ascorbate peroxidase), CAT (catalase) and peroxide like antioxidant enzymes (Most and Papenbrock 2015), hormones and antioxidant biosynthesis and functioning and signalling compound activation such as phytohormones (jasmonic acid (JA), abscisic acid (ABA), ethylene, salicylic acid (SA)), mineral (Ca^{2+}) and gaseous molecule (NO, H_2S, CO, H_2, CH_4). Gasotransmitters either synthesized endogenously in plant cells or receive externally from other organisms and exogenous application that communicate chemical signals in plants and induce biochemical, physiological and molecular changes in cells. These gasotransmitters also act as cytoprotective molecules and have stress acclimatization properties (Terrón-Camero et al. 2019), which induce transcriptional changes in cells and is very effective in the regulation of hormone level and ROS production. These gasotransmitter molecules also detoxify ROS in cell; enhanced synthesis and activity of enzymes, thus improve antioxidant properties, increase osmolyte level, and stable redox homeostasis of cell and maintaining cellular processes. Role of these gasotransmitters as signalling molecules in HMs stress tolerance is studied in many crop plants (Hancock and Whiteman 2016). Several reports indicated the crosstalk network of these gaseous molecules between themselves and to plant hormones. Hydrogen peroxide (H_2O_2), nitric oxide (NO), Hydrogen sulphide (H_2S), carbon dioxide (CO_2) and plant hormone-mediated signalling pathways initiated as a complicated web and trigger H_2S accompanied lateral root formation in tomato through H_2O_2 accumulation (Mei et al. 2017). NO triggered H_2S synthesis improved antioxidant enzymes interaction and accumulation, ameliorated osmotic stress and increased accumulation of water-soluble solids (Khan et al. 2017). H_2S and NO have a vital role in senescence, seed physiology, program cell death (PCD), closing of stomata, fruit ripening and root development (Mishina et al. 2007; Corpas et al. 2018). These compounds also check the various essential processes in antagonistic or synergistic ways to enhanced plant tolerance. The present chapter covers the process/pathways of gasotransmitters biosynthesis, their biological function and crosstalk with other signalling molecule and role in post-transition modification activities in plants.

10.2 Gasotransmitters' Biosynthesis and Metabolism

In plants, Gasotransmitters synthesis is accomplished through various catalytic and non-catalytic pathways. Organic and inorganic nitrite sources, NO synthase (NOS) and nitrate reductase (NR) activity, nitrogen oxide reaction with plant metabolites, mitochondrial nitrite reduction, microbial metabolism are the furthermost probable origin of NO biosynthesis in plants (Stöhr and Ullrich 2002; Besson-Bard et al. 2008; Moreau et al. 2010; Wang et al., 2010; Kolbert et al. 2019a). External stimuli such as abiotic and biotic stresses and nitrate level in soil are responsible for modulation of NO biosynthesis (Lipka and Müller 2014; Astier et al. 2018). Under hypoxic and low oxygen stress conditions, NO productions regulate by non-symbiotic haemoglobin1 (nsHb1); and enhanced NO production through NR activity, protect plants under stress (Gupta et al. 2017). Non-symbiotic haemoglobin also involves in establishing symbiosis and increase in nodule formation with concomitant reduced yielding of NO (Shimoda et al. 2009; Hichri et al. 2015). GSNO reductase (GSNOR) enzyme also regulates NO levels with respect to HMs stress (Frungillo et al. 2014). GSNOR activity during different metal stress conditions, such as As, Cd, Cu depends on treatment strength (Rodríguez-Ruiz et al. 2019; Hu et al. 2019). Soil nitrogen level affects the NO formation in nutrient exchange zone of roots, where nitrate encouraged the formation of LR with endogenous NO production (Zhang and Forde 2000; Forde 2002).

Heme oxygenase (HO)-dependent pathway involves in CO generation in plants, oxidative conversion of heme in CO, Fe^{2+} and biliverdin catalyzed through HOs (Bilban et al. 2008). HO genes in plants can be separated in two subfamilies that is *HO1* and *HO2*, and it has been identified that *HO2* subfamily does not involve in CO producing activity in plants. *AtHO1* (a plastid localized HO recombinant protein) catalyzed CO formation in-vitro condition (Muramoto et al. 2002). Other non-enzymatic sources of CO production include ureide metabolism and lipid peroxidation, along with it also release during breaking of alpha-methylene bridge of heme (Bilban et al. 2008, Zilli et al. 2014). Oxidative damages caused by the HMs stress (Chen and Yang 2011; Chen et al. 2018), salt stress (Ling et al. 2009) and drought stress (Liu et al. 2010), generate CO and interact with other signalling molecules under stress conditions (Lin et al. 2014; Xie et al. 2013). Growth regulators auxin, NO, and JA regulates CO biosynthesis and their subsequent plant signalling (Wang and Liao 2016). CO also affects various biological functions like germination (Dekker and Hargrove 2002), stomatal closures and root development (Cao et al. 2007). Like NO, CO also relieving HMs stress in plants (Zheng et al. 2011).

Endogenous H_2S biosynthesis is regulated by cysteine synthase, D-cysteine desulphydrase (DCD), cysteine desulphurases, L-cysteine desulphydrase (LCD), sulphite reductase (SiR), cyanoalanine synthase and *O*-acetylserine (thiol) lyase (OASTL) (González-Gordo et al. 2020), in which cysteine desulphydrase generates higher amount of H_2S along with pyridoxal 5′-phosphate (PLP) as cofactor (Papenbrock et al. 2007). Photosynthetic sulphate assimilation is activated by SiR and cytosolic L-DES that responsible for chloroplast H_2S biosynthesis (Fu et al. 2013). *O*-acetylserine (thiol) lyase catalyzes the inorganic sulphur into cysteine and produces free H_2S molecules. Nine *O*-acetylserine (thiol) lyase enzymes also have been recognized in *Arabidopsis* which are found in the double membrane-bound organelles (mitochondria, chloroplasts) and cytosol. Among the OASTL-A1, OASTL-B and OASTL-C are the primary Cys synthases in *Arabidopsis* (Heeg et al. 2008; Álvarez et al. 2010). H_2S modulates growth and defence mechanisms via signalling in the plants. H_2S regulates germination, stomatal action, growth of roots and their architecture and photosynthesis (Shivaraj et al. 2020). Likewise, H_2S priming (NaHS) alleviate Al stress in barley, wheat and other important agricultural crops with increasing antioxidant enzymes. It also upregulates H^+-ATPase expression and induced citrate transporter (HvAACT1) in barley (Chen et al. 2013). Arsenic stress could enhance ROS formation in pea seedlings thus damaging lipid and protein molecules, and cellular membranes (Most and Papenbrock 2015). H_2S-Cys system having able to protect Arabidopsis from Cr stress relying on glutathione production and synthesis of PCs (Fang et al. 2017).

H_2 molecule is produced by hydrogenase activity irrespective of the genus (Stephenson and Stickland 1931; Sanadze 1961). Abiotic stresses, such as cold, salt, HM also trigger H_2 production in various crops such as rice and alfalfa (Cui et al. 2013). Phytohormones like ABA, ethylene, JA might be involved in inducing H_2 production (Zeng et al. 2013).

CH_4 is mainly produced in the rice field and from other crops under stress like salt, low light, high temperature and ultraviolet radiation (Yao et al. 2019). Polyethylene glycol (PEG) treatment enhanced production of CH_4 in maize (Han et al. 2017). HMs stresses also increase its production in alfalfa (Samma et al. 2017; Gu et al. 2018). CH_4 treatment decrease metal accumulation and alleviated Al toxicity through enhancement of antioxidant enzyme activities and redox homeostasis (Cui et al. 2017). It also alleviated Cu toxicity by reducing thiobarbituric acid reactive substances (TBARS) content and improving amylase enzyme activity along with sugar content in alfalfa seedlings (Samma et al. 2017).

10.3 Gasotransmitters Crosstalk with Plant Growth Regulators in Plant Growth and Development

Gasotransmitters able to integrate with endogenous plant growth regulators (PGRs) and other external signal molecules that govern main molecular mechanisms and reprogrammed internal function of plants. ABA catabolism occurs through 8′-hydroxylation pathway catalyzed by the cytochrome P450 enzyme ABA 8′-hydroxylase (encoded by CYP707A family) and those having a major role in germination and seed development. Reduction in ABA sensitivity can be deciphered by NO and *CYP707A2* induction with subsequent accumulation of protein (Okamoto et al. 2006). NO-deficient mutants '*nia1nia2*' and '*nia1nia2noa1-2*' expression has remarkable impact on inhibition of ABA-mediated dormancy of seeds (Gniazdowska et al. 2007). NO also influences the level of sensing and signalling of ABA, that also relies upon on the pyrabactin resistance (PYR)/pyrabactin 1 related (PYR)/regulatory element of ABA receptor (RCAR), protein phosphatase 2c (PP2C), as well as SNF1-related protein kinase2 (SnRK2) (Yoshida et al. 2006; Ma et al. 2009; Park et al. 2009; Umezawa et al. 2009; Vlad et al. 2009). Nitration modification of tyrosine moiety inhibited PYR/PYL/RCAR receptor and activate PP2C sequentially inactivates SnRKs. SnRK downregulation by NO is brought about by damage to seeds and inhibits stomatal opening. UP-regulating the RCAR family also up-regulates the PP2C transcription by NO application in the nia1nia2 mutant (Castillo et al. 2015). CH_4 mediated adventitious root (AR) development is depend on induction of CsDNAJ-1, CsCDPKs, CsCDC6 (cell division related gene) and CsAux22D, CsAux22B (Auxin signalling gene) genes, and EIN2 (cell cycle), which facilitated by NO signalling (Novikova et al. 2017; Qi et al. 2017).

H_2S affects genes encoding ion channel protein and transcription factors (WRKY) with respect to ABA (Jin et al. 2017). The interaction of ABA and H_2S could increase stomatal aperture in *aba3* and *abi1* mutant of ABA (Jin et al. 2013). JA is also responsible for stomatal movement with enhancing LCD activity along with H_2S content. Methyl jasmonate (MeJA) pre-treatment encourages in the decrease of MDA and H_2O_2 with Cd injury alleviation with aids of H_2S (Tian et al. 2017). The pre-treatment of H_2S enhanced SA content that could help in acquiring Cd tolerance in Arabidopsis (Qiao et al. 2015). An H_2S and ethylene interaction are observed during ripening of the fruits like banana and kiwi, both combinations delay ripening processes (Gao et al. 2013, Ge et al. 2017). Combined treatment of ethylene (ET), NO and H_2S reduced ROS (H_2O_2, O^{2-}) and MDA content (Hancock and Whiteman 2016). Ethylene induces the indigenous H_2S production in stomatal guard cells helping in stomatal closing. H_2S involves in repression of 1-aminocyclopropane-1-carboxylic acid synthase (ACS) and 1-aminocyclopropane-1-carboxylic acid oxidase (ACO) gene expression responsible for ET biosynthesis (Jia et al. 2018). H_2S delayed PCD in the GA-treated seeds aleurone layer and also accelerated amylase production (Zhang et al. 2015). The important role of H_2S during plant growth and HM stress is represented in Figure 10.1.

Like NO, CO also crosstalks with phytohormones like IAA, ABA, ET and GA (Melo et al. 2016; Kapczynski et al. 2017). CO involve in IAA-induced lateral root development in tomato plants via altering biosynthesis (Guo et al. 2008). IAA activated CO signalling, which triggers signal transduction, and helps in AR formation in cucumber (Xuan et al. 2008). CO also involves in H_2S induces cytoprotective role during GA induced PCD (Xie et al. 2014). At the time of root development, 'CO heme oxygenase' and 'NO' are associated with jasmonic acid, auxin and ethylene signalling pathway (Ivanchenko et al. 2008; Terrón-Camero et al. 2019).

10.4 Heavy Metal Induced Gasotransmitter Crosstalk

During abiotic stresses, phytohormones decide plant adaptation mechanisms. In various studies exhibited that NO, H_2, CH_4, H_2S and CO closely relates with PGRs under normal and

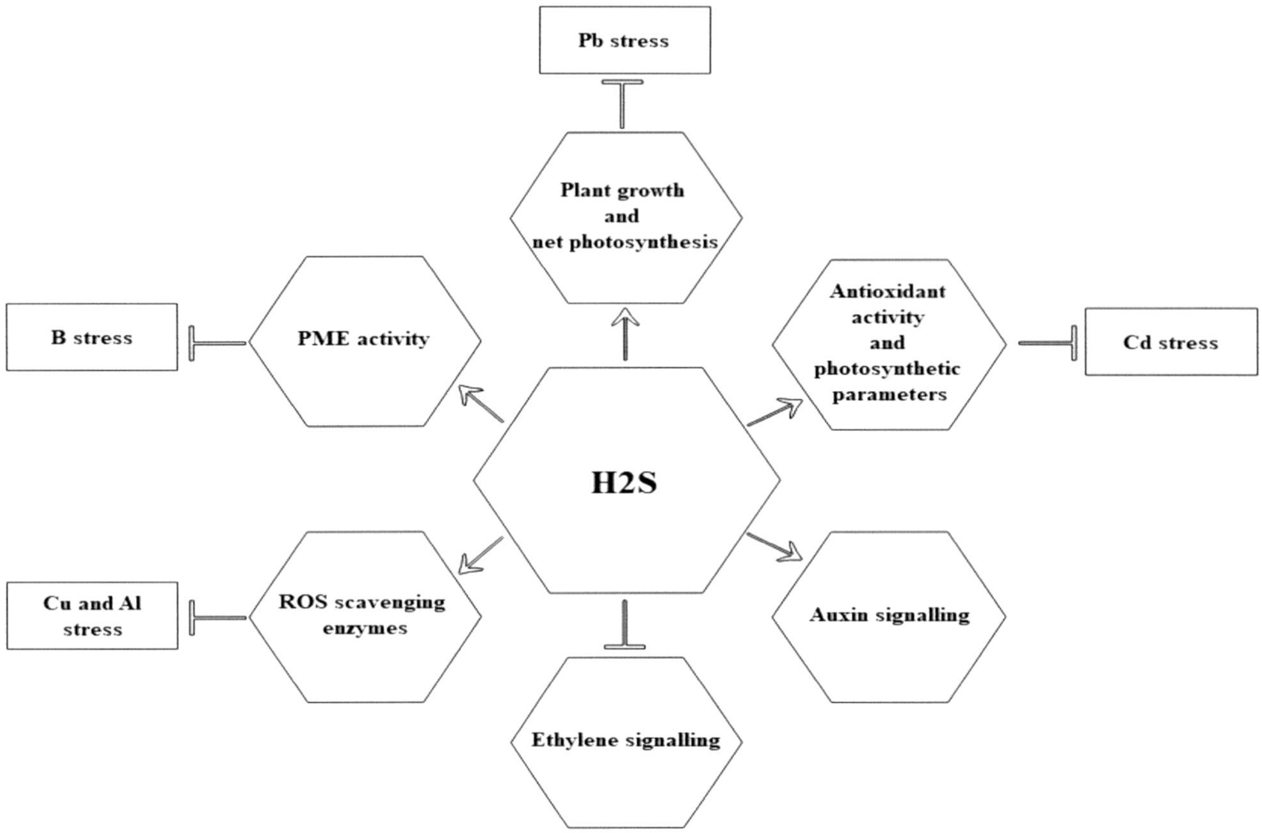

FIGURE 10.1 Represents the role of H_2S during plant growth and heavy metal stress tolerance.

Symbol: ⊥: Inhibition → Induction.

stress conditions during the growth of plant (Freschi 2013). H_2S can suppress chlorophyll loss and delay senescence (Wei et al. 2017). In an experiment, Brassica leaves fumigated with H_2S and found decreased ET production with increased antioxidant activity and lower rate of senescence (Al Ubeed et al. 2019). Exogenous NO application reduced ET burst in sold pea leaves (Leshem et al. 1998) and NO can neutralize ABA induces senescence in rice leaves (Hung and Kao 2003). Likewise, NO induces the H_2O_2 production that enhanced the production of antioxidant enzymes and gene expression and provide Al tolerance in wheat (Sun et al. 2018). Exogenous H_2S application in wheat alleviates the toxic effects of metal stress by inducing an antioxidant system (Zhang et al. 2008). Either NO or H_2S treatment significantly reduces Cr accumulation in the roots, indicate that NO or H_2S involve in Cr uptake by the root system. Other findings show NO or H_2S improves Fe uptake in strawberry plants by alleviating leaf chlorosis (Kaya et al. 2019; Kaya and Ashraf 2018). In the pea, studied the Cd effect on NO and ROS with an alternation in SA level, JA and ET (Rodríguez-Serrano et al. 2009). Co-treatment of H_2S-ET in tomato showed that the H_2S inhibits ET synthesis upregulation and genes under signal transduction that results in suppression of ET induced petiole abscission (Liu et al. 2020). NaHS treatment during LR development increases auxin content with simultaneous increase in number and length of ARs (Zhang et al. 2009). H_2S treatment decline Cd-induced rice oxidative stress (Mostofa et al. 2015) and in *Brassica napus* (Ali et al. 2014) by maintaining cellular structure, enhancing photosynthetic parameters and antioxidant activities in leaves and root.

Improve plant growth, chlorophyll content, photosynthesis along with antioxidant activity while reducing H_2O_2, electrolyte leakage, MDA content and the endogenous level of Pb in cotton by H_2S treatment (Bharwana et al. 2014). The higher amount of boron (B) in the growth medium results in root inhibition and pectin methylesterase (PME) up-regulation (Wang et al. 2010). H_2S application in wheat alleviates the repressive effect of Cu and Al by increasing the activity of ROS scavenging enzymes, i.e. SOD, CAT and APX (Zhang et al. 2008).

NO and ABA crosstalk during adverse climatic conditions, thus ameliorate plant adaptive responses, like antioxidant defence and closure of stomata (Neill et al. 2008). NO and auxin signalling pathways are intensively interact to regulate growth and development and plant root morphology (Sanz et al. 2015). The NO application in Cd-stressed *Medicago truncatula* plant reduced auxin degradation by suppressing the activities of IAA oxidase, thereby viewing another interaction mechanism between NO and auxin during Cd tolerance (Xu et al. 2010). Another positive crosstalk found between auxin and NO during aluminium toxicity that sodium nitroprusside (SNP) increases IAA content, GA accumulation and promote apical root growth in wheat and rye induce Al tolerance (He et al. 2012). Cytokinin (CK) stimulates NO biosynthesis and NO inhibits CK signalling. Some literature has shown CK

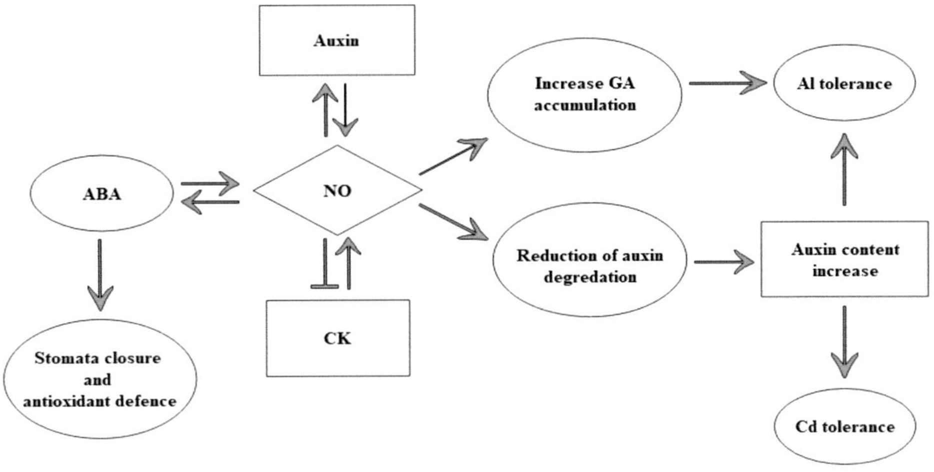

FIGURE 10.2 Represents the NO and crosstalk with PGRs in heavy metal stress tolerance.

treatment increase NO production but, other reports found low NO level or unchanged after treatment of CK in mutant/transgenic plants (Freschi 2013). Endogenous CK content directly affected by NO due to peroxynitrite reaction, a NO derivative, with zeatin, so decreasing availability of this CK (Liu et al. 2013). The interaction of PGRs during HMs is represented in Figure 10.2.

10.5 Molecular Understanding of Gasotransmitters Signalling under Heavy Metal Stress

Gasotransmitters act as second messenger molecule which is able to crosstalk with each other, and affects various physiological responses, such as germination, stomata movement, root development and provides protection against stress condition (Arc et al. 2013). The H_2S and cysteine cycle increases the MTs and PCs protein expression thereby increasing antioxidant activity and assisting in dealing with Cd^{2+} toxicity (Jia et al. 2016). H_2S promoted oxidative stress resistance by regulating ROS and NADPH production in maize (Kharbech et al. 2017). Treatment with sodium hypochlorite activates CAT and SOD and decreases Cu accumulation and the extent of Cu-induced damage (Zhang et al. 2008). H_2 rich water counteracts the negative impact of Cd, Hg and Al stress on seedling. This is probably due to enhancement of internal antioxidant activities (up-regulation of SOD, POD, APX, SOD, POD and APX1/2 genes) in root tissue (Cui et al. 2013, 2014; Zhao et al. 2017). H_2 can maintain redox homeostasis of a cell by equalizing GSH content and regulating miRNA under HM stress (Wu et al. 2015; Xu et al. 2017). NO participate in downstream H_2 signalling cascade under abiotic stress such as high salinity (Chen et al. 2017; Su et al. 2018) through the up-regulation of antioxidant enzymes with NR activity (Zhu et al. 2016; Su et al. 2018). The number of scientific publications is evidence that AR formation follows CO, CH_4 and H_2. HO1 acts as downstream mechanism but is involved with regulating H_2S-induced AR formation *via* DNAJ-1 and CDPK1/5 genes expression in cucumber (Lin et al. 2012). CO-produced stomatal closure in *Vicia faba*, possibly mediated via O_2-signalling (She and Song 2008). CH_4 and H_2 interact to enhance antioxidant activity in cells under stress (Zhu et al. 2016). Both NO and H_2S participate in the minimizing of oxidant stress and the regulation of glutathione, which may help in the counterbalancing of oxidative damage (Cui et al. 2011; Fu et al. 2013; Singh et al. 2015). NO-induced H_2S synthesizing enzyme activities, e.g. LCD and DCD activities in wheat seedling and also improves antioxidant enzyme CAT, APX, POD and SOD activities (Khan et al. 2017). NO could be regulating the upstream of H_2S under HM stress by enhancing antioxidants with and/or without simultaneous decreasing MDA (Shi et al. 2014). Likewise, NO and H_2S regulates post-translational modifications (PTMs) for instance persulphidation and S-nitrosation (Aroca et al. 2018). Exogenously applied NO induced ATPase and PPase activity in the plasma membrane and the tonoplast in tomato crop developing under Cu stress condition. NO provides tolerance to HMs to plant by encouraging various biochemical and physiological molecular changes in cells. The activity of NADPH oxidase might potential to enhanced H_2O_2 generation under HM toxicity that changes redox homeostasis and increases antioxidant activity through plant signalling and gene expression (Rentel and Knight 2004). These gasotransmitters also involve in PTMs under HMs stress and facilitate regulatory role in gene regulation and protein alteration, cellular metabolic pathways, such as photorespiration, hormonal and ROS regulation (Terrón-Camero et al. 2019). The role of gasotransmitters crosstalk during HMs stress is represented in Figure 10.3.

Protein NADP-isocitrate dehydrogenase balance the metabolic, reprogramming function and redox homeostasis during maturation (Begara-Morales et al. 2013). Chromatin remodelling (Han et al. 2015) and histone demethylation (Zhou et al. 2015) are associated with the ABA signalling pathway under stress conditions (Nonogaki 2014). The reprogramming expression induced by NO is responsible for chromatin's PTM state (Imran et al. 2018). The basic hormonal equilibrium is NO-regulated leading to reprogramming of the signalling cascades. The pathway combines the development of NR-dependent NO with the regulation of ERFVII-mediated

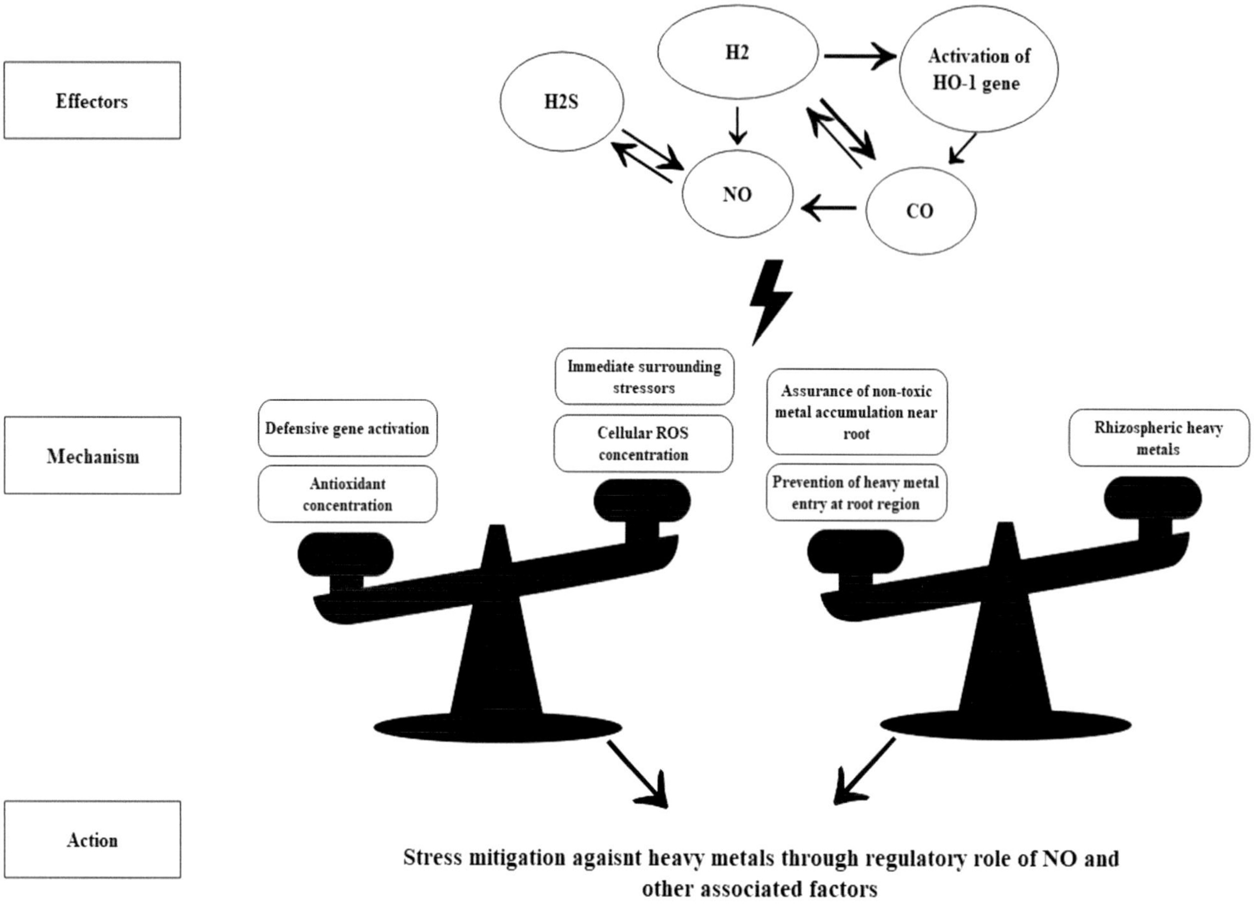

FIGURE 10.3 Represents the mechanism of NO and other gasotransmitters in heavy metal stress tolerance.

chromatin-remodelling ATPase BRAHMA, resulting in genetic reprogramming that regulates growth and stress responses to boost plant survival (Zhang et al. 2018). NtHb1 overexpression in tobacco plants has shown how downregulating NO levels can increase Cd tolerance via decreased accumulation of Cd (Lee and Hwang 2015; Bahmani et al. 2019). Co-expressed transcriptome analysis discloses expression of multiple pathways. Biosynthetic suppression of NO in the endothelium *via* the eNOS co-factor BH_4 is the direct result of endothelium oxidative stress, conceivably because the BH_4 is readily oxidized in the presence of higher levels of ROS (Antoniades et al. 2007). Protein function control by NO has been well-researched after PTMs like nitrosation and S-nitrosylation. Parani et al. (2004) reported that 422 plant defence-associated genes along with many other differentially expressed genes (genes for oxidative stress resistance and transcription contributing factors in phytohormone pathways) under concentrations of SNP sodium nitroprusside (SNP). Likewise, numerous other experiments have exhibited the transcriptional regulation of specific genes by NO (Kawakita et al. 2010). Now bioinformatics methodologies make it possible to determine which genes are responsive to NO. Aux/IAA auxin-responsive transcriptional repressors and ARF transcriptional activators play a chief role in Arabidopsis responses to auxin. Expression of 'Aux/IAA 2-11' (AT5G43700), 'Aux/IAA 14' (AT4G14550) and 'Aux/IAA 29' (AT4G32280) transcriptional repressors were reduced in response to 1 mM CysNO treatment. It has been observed that Aux/IAA repressors bind transcriptional activators to ARF. The ARF is unveiled to trigger Auxin-responsive genes upon ubiquitin-dependent degradation of repressors (Nanao et al. 2014). A reciprocal adjustment in the redox level follows different abiotic and biotic stresses (Foyer and Noctor 2005). When ROS accumulate, it alters the disturbed DNA methylation. For example, high sodium nitroprusside (NO donor) in the rice plant leads to hypomethylation of DNA resulting in growth inhibition. In relation, it is feasible to retain the effects that gene expression has on DNA methylation towards the next generation (Ou et al. 2015). So, therefore, gasotransmitter possess important roles during post-transcriptional and/or post-translational thresholds to coordinate the appropriate sensing pathways to fight with the stress conditions.

Conclusions

As plants are sessile in nature, they need to have a sophisticated coordination among receptors and transducers (signalling molecules) in order to maintain optimum fitness during stressful situations. In these consequences, the gasotransmitters are novel emerging chemical compounds, have a major role during coordination of signalling mechanism. Evidence was observed that gasotransmitters drastically affect the growth and development of plants, particularly in synergistic or antagonistic ways. Understanding the complex signalling mechanism among gasotransmitters and PGRs molecules is a turning point for pharmaceutical drug research. This chapter explores the importance of plant gasotransmitters in plant signalling and stress resistance, identifying their biosynthesis, metabolism and functions. Gasotransmitters are essential for plant growth and development as they play a key role in the regulation and expression of multiple genes and proteins throughout plant development. After research on abiotic stresses and plant development, we discussed how abiotic stresses affect plant development. Gasotransmitter transducers-PGRs are integral part of plants and modulation in them is what protects the plant against abiotic stress. In the last chapter, several past studies were discussed concerning the. PGR–gasotransmitter interrelationships and how these substances ameliorate the HMs by decreasing oxidative stress and strengthening the antioxidant system. At the end of the report, we concentrated on HM stress mitigation and their molecular mechanisms. However, there is a dearth of research findings for molecular mechanisms of gasotransmitters in HMs tolerance, but some present studies suggested that these compounds have a great importance in post-translational alterations, protein re-modelling and improving the stress tolerance genes.

REFERENCES

Al Ubeed, H. M. S., Wills, R.B.H., Bowyer, M.C. and Golding, J. B. 2019. Interaction of the hydrogen sulphide inhibitor, propargylglycine (PAG), with hydrogen sulphide on postharvest changes of the green leafy vegetable, pak choy. *Postharv Biol Tech*, 147, 54–58.

Ali, B., Gill, R. A., Yang, S., Gill, M. B., Ali, S., Rafiq, M. T. and Zhou, W. 2014. Hydrogen sulfide alleviates cadmium-induced morpho-physiological and ultrastructural changes in Brassica napus. *Ecotoxicol Environ Saf*, 110, 197–207.

Álvarez, C., Calo, L., Romero, L. C., García, I. and Gotor, C. 2010. An O-acetylserine (thiol) lyase homolog with L-cysteine desulfhydrase activity regulates cysteine homeostasis in Arabidopsis. *Plant Physiol*, 152, 656–669.

Antoniades, C., Shirodaria, C., Crabtree, M., Rinze, R., Alp, N., Cunnington, C. and Ratnatunga, C. 2007. Clinical perspective. *Circulation*, 116, 2851–2859.

Arc, E., Sechet, J., Corbineau, F., Rajjou, L. and Marion-Poll, A. 2013. ABA crosstalk with ethylene and nitric oxide in seed dormancy and germination. *Front Plant Sci*, 4, 63.

Aroca, A., Gotor, C. and Romero, L. C. 2018. Hydrogen sulfide signaling in plants: emerging roles of protein persulfidation. *Front Plant Sci*, 9, 1369.

Astier, J., Gross, I. and Durner, J. 2018. Nitric oxide production in plants: an update. *J Expl Bot*, 69, 3401–3411.

Bahmani, R., Kim, D., Na, J. and Hwang, S. 2019. Expression of the tobacco non-symbiotic class 1 hemoglobin gene Hb1 reduces cadmium levels by modulating Cd transporter expression through decreasing nitric oxide and ROS level in Arabidopsis. *Front Plant Sci*, 10, 201.

Begara-Morales, J.C., Sánchez-Calvo, B., Chaki, M., Valderrama, R., Mata-Pérez, C., López-Jaramillo, J., Padilla, M.N., Carreras, A., Corpas, F. J. and Barroso, J. B. 2013 Dual regulation of cytosolic ascorbate peroxidase (APX) by tyrosine nitration and S-nitrosylation. *J Exp Bot*, 65, 527–538.

Besson-Bard, A., Courtois, C., Gauthier, A., Dahan, J., Dobrowolska, G., Jeandroz, S. and Wendehenne, D. 2008. Nitric oxide in plants: production and cross-talk with Ca2+ signaling. *Mol Plant*, 1, 218–228.

Bharwana, S. A., Ali, S., Farooq, M. A., Ali, B., Iqbal, N., Abbas, F. and Ahmad, M. S. A., 2014. Hydrogen sulfide ameliorates lead-induced morphological, photosynthetic, oxidative damages and biochemical changes in cotton. *Environ Sci Pol Res*, 21, 717–731.

Bilban, M., Haschemi, A., Wegiel, B., Chin, B. Y., Wagner, O. and Otterbein, L. E. 2008. Heme oxygenase and carbon monoxide initiate homeostatic signaling. *J Mol Med, 86*, 267–279.

Cao, Z. Y., Xuan, W., Liu, Z. Y., Li, X. N., Zhao, N., Xu, P. and Shen, W. B. 2007. Carbon monoxide promotes lateral root formation in rapeseed. *J Integr Plant Biol*, 49, 1070–1079.

Castillo, M. C., Lozano-Juste, J., González-Guzmán, M., Rodriguez, L., Rodriguez, P. L. and León, J. 2015. Inactivation of PYR/PYL/RCAR ABA receptors by tyrosine nitration may enable rapid inhibition of ABA signaling by nitric oxide in plants. *Sci Signal*, 8, 1–10.

Chen, J., Wang, W. H., Wu, F. H., You, C. Y., Liu, T. W., Dong, X. J. and Zheng, H. L. 2013. Hydrogen sulfide alleviates aluminum toxicity in barley seedlings. *Plant and Soil*, 362, 301–318.

Chen, J. and Yang, Z. M. 2011. Enhancement of tolerance of Indian mustard (*Brassica juncea*) to mercury by carbon monoxide. *J Hazard Mat*, 186, 1823–1829.

Chen, Q., Gong, C., Ju, X., Zhu, Z., Shen, W., Shen, Z. and Cui, J. 2018. Hemin through the heme oxygenase 1/ferrous iron, carbon monoxide system involved in zinc tolerance in *Oryza Sativa L. J. Plant Growth Regul*, 37, 947–957.

Chen, Z., Chen, M. and Jiang, M. 2017. Hydrogen sulfide alleviates mercury toxicity by sequestering it in roots or regulating reactive oxygen species productions in rice seedlings. *Plant Physiol Biochem*, 111, 179–192.

Corpas, F.J., Freschi, L., Rodriguez-Ruiz, M., Mioto, P.T., Gonzalez-Gordo, S. and Palma, J.M. 2018. Nitro-oxidative metabolism during fruit ripening. *J Exp Bot*, 69, 3449–3463.

Cui, W., Cao, H., Yao, P., Pan, J., Gu, Q., Xu, S., Wang, R., Ouyang, Z., Wang, Q. and Shen, W. 2017. Methane enhances aluminum resistance in alfalfa seedlings by reducing aluminum accumulation and reestablishing redox homeostasis. *Biometals*, 30, 719–732.

Cui, W., Fang, P., Zhu, K., Mao, Y., Gao, C., Xie, Y. and Shen, W. 2014. Hydrogen-rich water confers plant tolerance to mercury toxicity in alfalfa seedlings. *Ecotoxicol Environ Saf*, 105, 103–111.

Cui, W., Fu, G., Wu, H. and Shen, W. 2011. Cadmium-induced heme oxygenase-1 gene expression is associated with the depletion of glutathione in the roots of *Medicago sativa*. *Biometals*, 24, 93–103.

Cui, W., Gao, C., Fang, P., Lin, G. and Shen, W. 2013. Alleviation of cadmium toxicity in Medicago sativa by hydrogen-rich water. *J. Hazard Mater*, 260, 715–724.

Dekker, J. and Hargrove, M. 2002. Weedy adaptation in Setaria spp. V. Effects of gaseous environment on giant foxtail (*Setaria faberii*) (Poaceae) seed germination. *Am J Bot*, 89(3),410–416.

Emamverdian, A., Ding, Y., Mokhberdoran, F. and Xie, Y. 2015. Heavy metal stress and some mechanisms of plant defense response. *Sci World J*, 2015.

Fang, H., Liu, Z., Long, Y., Liang, Y., Jin, Z., Zhang, L. and Pei, Y. 2017. The Ca2+/calmodulin2-binding transcription factor TGA 3 elevates LCD expression and H2S production to bolster Cr6+ tolerance in Arabidopsis. *The Plant Journal*, 91, 1038–1050.

Forde, B. G. 2002. Local and long-range signaling pathways regulating plant responses to nitrate. *Ann Rev Plant Biol*, 53(1), 203–224.

Foyer, C. H. and Noctor, G. 2005. Redox homeostasis and antioxidant signaling: a metabolic interface between stress perception and physiological responses. *The Plant Cell*, 17, 1866–1875.

Freschi, L. 2013. Nitric oxide and phytohormone interactions: current status and perspectives. *Front Plant Sci*, 4, 1–22.

Frungillo, L., Skelly, M. J., Loake, G. J., Spoel, S. H. and Salgado, I. 2014. S-nitrosothiols regulate nitric oxide production and storage in plants through the nitrogen assimilation pathway. *Nat Comm*, 5, 1–10.

Fu, P., Wang, W., Hou, L. and Liu, X. 2013. Hydrogen sulfide is involved in the chilling stress response in Vitis vinifera L. *Acta Societatis Botanicorum Poloniae*, 82.

Gao, S. P., Hu, K. D., Hu, L. Y., Li, Y. H., Han, Y., Wang, H. L. and Zhang, H. 2013. Hydrogen sulfide delays postharvest senescence and plays an antioxidative role in fresh-cut kiwifruit. *HortScience*, 48, 1385–1392.

Ge, Y., Hu, K. D., Wang, S. S., Hu, L. Y., Chen, X. Y., Li, Y. H. and Zhang, H. 2017. Hydrogen sulfide alleviates postharvest ripening and senescence of banana by antagonizing the effect of ethylene. *PLoS One*, 12, e0180113.

Ghori, N. H., Ghori, T., Hayat, M. Q., Imadi, S. R., Gul, A., Altay, V. and Ozturk, M. 2019. Heavy metal stress and responses in plants. *Int J Environ Sci Tech*, 16, 1807–1828.

Gniazdowska, A., Dobrzynska, U., Babanczyk, T. and Bogatek, R. 2007. Breaking the apple embryo dormancy by nitric oxide involves the stimulation of ethylene production. *Planta*, 225, p. 1051–1057.

González-Gordo, S., Palma, J.M. and Corpas, F.J. 2020. Appraisal of H2S metabolism in Arabidopsis thaliana: In silico analysis at the subcellular level. *Plant Physiol Biochem*, 155, 579–588.

Gu, Q., Chen, Z., Cui, W., Zhang, Y., Hu, H., Yu, X., Wang, Q. and Shen, W. 2018. Methane alleviates alfalfa cadmium toxicity via decreasing cadmium accumulation and reestablishing glutathione homeostasis. *Ecotoxicol Environ Saf*, 147, 861–871.

Guo, K., Xia, K. and Yang, Z. M. 2008. Regulation of tomato lateral root development by carbon monoxide and involvement in auxin and nitric oxide. *J Exp Bot*, 59, 3443–3452.

Gupta, D. K., Pena, L. B., Romero-Puertas, M. C., Hernández, A., Inouhe, M. and Sandalio, L. M. 2017. NADPH oxidases differentially regulate ROS metabolism and nutrient uptake under cadmium toxicity. *Plant, Cell & Environ*, 40, 509–526.

Han, B., Duan, X., Wang, Y., Zhu, K., Zhang, J., Wang, R., Hu, H., Qi, F., Pan, J. and Yan, Y. 2017. Methane protects against polyethylene glycol-induced osmotic stress in maize by improving sugar and ascorbic acid metabolism. *Sci Rep*, 7, 46185.

Han, S. K., Wu, M. F., Cui, S. and Wagner, D. 2015. Roles and activities of chromatin remodeling ATPases in plants. *The Plant J*, 83, 62–77.

Hancock, J. T. and Whiteman, M. 2016. Hydrogen sulfide signaling: interactions with nitric oxide and reactive oxygen species. *Ann New York Acad Sci*, 1365, 5–14.

He, H. Y., He, L.F., Gu, M. H. and Li, X. F. 2012. Nitric oxide improves aluminum tolerance by regulating hormonal equilibrium in the root apices of rye and wheat. *Plant Sci*, 183, 123–130.

Heeg, C., Kruse, C., Jost, R., Gutensohn, M., Ruppert, T., Wirtz, M. and Hell, R. 2008. Analysis of the Arabidopsis O-acetylserine (thiol) lyase gene family demonstrates compartment-specific differences in the regulation of cysteine synthesis. *The Plant Cell*, 20, 168–185.

Hichri, I., Boscari, A., Castella, C., Rovere, M., Puppo, A. and Brouquisse, R. 2015. Nitric oxide: a multifaceted regulator of the nitrogen-fixing symbiosis. *J Exp Bot*, 66, 2877–2887.

Hu, Y., Lu, L., Tian, S., Li, S., Liu, X., Gao, X. and Lin, X. 2019. Cadmium-induced nitric oxide burst enhances Cd tolerance at early stage in roots of a hyperaccumulator Sedum alfredii partially by altering glutathione metabolism. *Sci Tot Env*, 650, 2761–2770.

Hung, K. T. and Kao, C. H. 2003. Nitric oxide counteracts the senescence of rice leaves induced by abscisic acid. *J Plant Physiol*, 160, 871–879.

Imran, Q. M., Hussain, A., Lee, S. U., Mun, B. G., Falak, N., Loake, G. J. and Yun, B. W. 2018. Transcriptome profile of NO-induced Arabidopsis transcription factor genes suggests their putative regulatory role in multiple biological processes. *Sci Rep*, 8, 1–14.

Ivanchenko, M.G., Muday, G.K. and Dubrovsky, G. 2008. Ethylene–auxin interactions regulate lateral root initiation and emergence in *Arabidopsis thaliana*. *Plant J*, 55, 335–347.

Jia, H., Chen, S., Liu, D., Liesche, J., Shi, C., Wang, J. and Li, J. 2018. Ethylene-induced hydrogen sulfide negatively regulates ethylene biosynthesis by persulfidation of ACO in tomato under osmotic stress. *Front Plant Sci*, 9, 1517.

Jia, H., Wang, X., Dou, Y., Liu, D., Si, W., Fang, H. and Li, J. 2016. Hydrogen sulfide-cysteine cycle system enhances cadmium tolerance through alleviating cadmium-induced oxidative stress and ion toxicity in Arabidopsis roots. *Sci Rep*, 6, 1–14.

Jin, X., Liao, W.B., Yu, J.H., Ren, P.J., Dawuda, M.M., Wang, M., Niu, L.J., Li, X.P. and Xu, X.T. 2017. Nitric oxide is involved in ethylene-induced adventitious rooting in marigold (*Tagetes erecta* L.). *Can J Plant Sci*, 97, 620–631.

Jin, Z., Wang, Z., Ma, Q., Sun, L., Zhang, L., Liu, Z. and Pei, Y. 2017. Hydrogen sulfide mediates ion fluxes inducing stomatal closure in response to drought stress in Arabidopsis thaliana. *Plant and Soil*, 419, 141–152.

Jin, Z., Xue, S., Luo, Y., Tian, B., Fang, H., Li, H. and Pei, Y. 2013. Hydrogen sulfide interacting with abscisic acid in stomatal regulation responses to drought stress in Arabidopsis. *Plant Physiol Biochem*, 62, 41–46.

Kawakita, K., Shahjahan, M. M. and Takemoto, D. 2010. Plant defense-related activities of NO producing elicitor candidates on potato and *Nicotiana benthamiana*. In *Nitric Oxide-Biology and Chemistry*, USA: Academic Press Inc Elsevier Science.

Kaya, C., Akram, N. A. and Ashraf, M. 2019. Influence of exogenously applied nitric oxide on strawberry (Fragaria × ananassa) plants grown under iron deficiency and/or saline stress. *Physiologia Plantarum*, 165, 247–263.

Kaya, C. and Ashraf, M. 2018. The mechanism of hydrogen sulfide mitigation of iron deficiency-induced chlorosis in strawberry (Fragaria × ananassa) plants. *Protoplasma*, 256(2), 1–12.

Kepczynski, J., Cembrowska-Lech, D. and Sznigir, P. 2017. Interplay between nitric oxide, ethylene, and gibberellic acid regulating the release of *Amaranthus retroflexus* seed dormancy. *Acta Physiol Plant*, 39, 254.

Khan, M. N., Mobin, M., Abbas, Z. K. and Siddiqui, M. H. 2017. Nitric oxide-induced synthesis of hydrogen sulfide alleviates osmotic stress in wheat seedlings through sustaining antioxidant enzymes, osmolyte accumulation and cysteine homeostasis. *Nitric Oxide*, 68, 91–102.

Kharbech, O., Houmani, H., Chaoui, A. and Corpas, F. J. 2017. Alleviation of Cr(VI)-induced oxidative stress in maize (*Zea mays* L.) seedlings by NO and H2S donors through differential organ-dependent regulation of ROS and NADPH-recycling metabolisms. *J Plant Physiol*, 219, 71–80.

Kolbert, Z. S., Barroso, J. B., Brouquisse, R., Corpas, F. J., Gupta, K. J., Lindermayr, C. and Hancock, J. T. 2019. A forty year journey: The generation and roles of NO in plants. *Nitric Oxide*, 93, 53–70.

Lee, B. R. and Hwang, S. 2015. Over-expression of NtHb1 encoding a non-symbiotic class 1 hemoglobin of tobacco enhances a tolerance to cadmium by decreasing NO (nitric oxide) and Cd levels in *Nicotiana tabacum*. *Environ Exp Bot*, 113, 18–27.

Leshem, Y. Y., Wills, R. B. H. and Ku, V. V. V., 1998. Evidence for the function of the free radical gas-nitric oxide (NO) as an endogenous maturation and senescence regulating factor in higher plants. *Plant Physiol Biochem*, 36, 825–833.

Lin, Y. T., Li, M. Y., Cui, W. T., Lu, W. and Shen, W. B. 2012. Haem oxygenase-1 is involved in hydrogen sulfide-induced cucumber adventitious root formation. *Journal of Plant Growth Regulation*, 31, 519–528.

Lin, Y., Zhang, W., Qi, F., Cui, W., Xie, Y. and Shen, W. 2014. Hydrogen-rich water regulates cucumber adventitious root development in a heme oxygenase-1/carbon monoxide-dependent manner. *J Plant Physiol*, 171, 1–8.

Ling, T., Zhang, B., Cui, W., Wu, M., Lin, J., Zhou, W. and Shen, W. 2009. Carbon monoxide mitigates salt-induced inhibition of root growth and suppresses programmed cell death in wheat primary roots by inhibiting superoxide anion overproduction. *Plant Sci*, 177, 331–340.

Lipka, E. and Müller, S. 2014. Nitrosative stress triggers microtubule reorganization in Arabidopsis thaliana. *J Exp Bot*, 65(15), 4177–4189.

Liu, D.M., Li, J.N., Li, Z.W. and Pei, Y.X. 2020. Hydrogen sulfide inhibits ethylene-induced petiole abscission in tomato (*Solanum lycopersicum* L.). *Hortic. Res. England*, 7, 1–11.

Liu, W. Z., Kong, D. D., Gu, X. X., Gao, H. B., Wang, J. Z., Xia, M., Gao, Q., Tian, L. L., Xu, Z. H., Bao, F. and Hu, Y. 2013. Cytokinins can act as suppressors of nitric oxide in Arabidopsis. *Proceed Natl Acad Sci*, 110, 1548–1553.

Liu, Y., Xu, S., Ling, T., Xu, L. and Shen, W. 2010. Heme oxygenase/carbon monoxide system participates in regulating wheat seed germination under osmotic stress involving the nitric oxide pathway. *J Plant Physiol*, 167, 1371–1379.

Loix, C., Huybrechts, M., Vangronsveld, J., Gielen, M., Keunen, E. and Cuypers, A. 2017. Reciprocal interactions between cadmium-induced cell wall responses and oxidative stress in plants. *Front Plant Sci*, 8, 1867.

Ma, Y., Szostkiewicz, I., Korte, A., Moes, D., Yang, Y., Christmann, A. and Grill, E. 2009. Regulators of PP2C phosphatase activity function as abscisic acid sensors. *Science*, 324, 1064–1068.

Maksymiec, W. 2007. Signaling responses in plants to heavy metal stress. *Acta Physiologiae Plantarum*, 29, 177.

Mei, Y., Chen, H., Shen, W., Shen, W. and Huang L. 2017. Hydrogen peroxide is involved in hydrogen sulfide-induced lateral root formation in tomato seedlings. *BMC Plant Biol*, 17, 162.

Melo, N.K., Bianchetti, R.E., Lira, B.S., Oliveira, P.M., Zuccarelli, R., Dias, D.L., Demarco, D., Peres, L.E., Rossi, M. and Freschi, L. 2016. Nitric oxide, ethylene, and auxin cross talk mediates greening and plastid development in de-etiolating tomato seedlings. *Plant Physiol*, 170, 2278–2294.

Memon, A. R. and Schröder, P. 2009. Implications of metal accumulation mechanisms to phytoremediation. *Environ Sci Pol Res*, 16, 162–175.

Mishina, T.E., Lamb, C. and Zeier, J. 2007. Expression of a nitric oxide degrading enzyme induces a senescence programme in Arabidopsis. *Plant Cell Environ*, 30, 39–52.

Moreau, M., Lindermayr, C., Durner, J. and Klessig, D. F. 2010. NO synthesis and signaling in plants–where do we stand?. *Physiol Planta*, 138, 372–383.

Most, P. and Papenbrock, J. 2015. Possible roles of plant sulfurtransferases in detoxification of cyanide, reactive oxygen species, selected heavy metals and arsenate. *Molecules*, 20, 1410–1423.

Mostofa, M. G., Rahman, A., Ansary, M. M. U., Watanabe, A., Fujita, M. and Tran, L. S. P. 2015. Hydrogen sulfide modulates cadmium-induced physiological and biochemical responses to alleviate cadmium toxicity in rice. *Sci Rep*, 5, 14078.

Muramoto, T., Tsurui, N., Terry, M. J., Yokota, A. and Kohchi, T. 2002. Expression and biochemical properties of a ferredoxin-dependent heme oxygenase required for phytochrome chromophore synthesis. *Plant Physiol*, 130, 1958–1966.

Nanao, M. H., Vinos-Poyo, T., Brunoud, G., Thévenon, E., Mazzoleni, M., Mast, D. and Guilfoyle, T. J. 2014. Structural basis for oligomerization of auxin transcriptional regulators. *Nat Comm*, 5, 3617.

Neill, S., Barros, R., Bright, J., Desikan, R., Hancock, J., Harrison, J., Morris, P., Ribeiro, D. and Wilson, I. 2008. Nitric oxide, stomatal closure, and abiotic stress. *J Exp Bot*, 59, 165–176.

Nonogaki H. 2014. Seed dormancy and germination—emerging mechanisms and new hypotheses. *Front Plant Sci*, 5, 233.

Novikova, G.V., Mur, L.A.J., Nosov, A.V., Fomenkov, A.A., Mironov, K.S., Mamaeva, A.S., Shilov, E.S., Rakitin, V.Y. and Hall, M.A. 2017. Nitric oxide has a concentration-dependent effect on the cell cycle acting via EIN2 in Arabidopsis thaliana cultured cells. *Front. Physiol*, 8, p. 142.

Okamoto, M., Kuwahara, A., Seo, M., Kushiro, T., Asami, T., Hirai, N. and Nambara, E. 2006. CYP707A1 and CYP707A2, which encode abscisic acid 8′-hydroxylases, are indispensable for proper control of seed dormancy and germination in Arabidopsis. *Plant Physiol*, 141, 97–107.

Ou, X., Zhuang, T., Yin, W., Miao, Y., Wang, B., Zhang, Y. and Liu, B. 2015. DNA methylation changes induced in rice by exposure to high concentrations of the nitric oxide modulator, sodium nitroprusside. *Plant Mol Biol Rep*, 33, 1428–1440.

Papenbrock, J., Riemenschneider, A., Kamp, A., Schulz-Vogt, H. N. and Schmidt, A. 2007. Characterization of cysteine-degrading and H2S-releasing enzymes of higher plants-from the field to the test tube and back. *Plant Biol*, 9, 582–588.

Parani, M., Rudrabhatla, S., Myers, R., Weirich, H., Smith, B., Leaman, D. W. and Goldman, S. L. 2004. Microarray analysis of nitric oxide responsive transcripts in Arabidopsis. *Plant Biotech J*, 2, 359–366.

Park, S. Y., Fung, P., Nishimura, N., Jensen, D. R., Fujii, H., Zhao, Y. and Alfred, S. E. 2009. Abscisic acid inhibits type 2C protein phosphatases via the PYR/PYL family of START proteins. *Science*, 324, 1068–1071.

Qi, F., Xiang, Z., Kou, N., Cui, W., Xu, D., Wang, R., and Shen, W. 2017. Nitric oxide is involved in methane-induced adventitious root formation in cucumber. *Physiol Planta*, 159, 366–377.

Qiao, Z., Jing, T., Liu, Z., Zhang, L., Jin, Z., Liu, D. and Pei, Y. 2015. H 2 S acting as a downstream signaling molecule of SA regulates Cd tolerance in Arabidopsis. *Plant and Soil*, 393, 137–146.

Rentel, M.C. and Knight, M.R. 2004. Oxidative stress-induced calcium signaling in Arabidopsis. *Plant Physiol*, 135, 1471–1479.

Rodríguez-Ruiz, M., Aparicio-Chacón, M. V., Palma, J. M. and Corpas, F. J. 2019. Arsenate disrupts ion balance, sulfur and nitric oxide metabolisms in roots and leaves of pea (Pisum sativum L.) plants. *J Env Exp Bot*, 161, 143–156.

Rodríguez-Serrano, M., Romero-Puertas, M.C., Pazmiño, D.M., Testillano, P.S., Risueño, M.C., Del Río, L.A. and Sandalio, L.M. 2009. Cellular response of pea plants to cadmium toxicity: Cross talk between reactive oxygen species, nitric oxide, and calcium. *Plant Physiol*, 150, 229–243.

Samma, M.K., Zhou, H., Cui, W., Zhu, K., Zhang, J. and Shen, W. 2017. Methane alleviates copper-induced seed germination inhibition and oxidative stress in *Medicago sativa*. *Biometals*, 30, 97–111.

Sanadze, G. 1961. Absorption of molecular hydrogen by green leaves in light. *Fiziol Rast*, 8, 555–559.

Sanz, L., Albertos, P., Mateos, I., Sánchez-Vicente, I., Lechón, T., FernándezMarcos, M. and Lorenzo, O. 2015. Nitric oxide (NO) and phytohormones crosstalk during early plant development. *J Exp Bot*, 66, 2857–2868.

She, X. P. and Song, X. G. 2008. Carbon monoxide-induced stomatal closure involves generation of hydrogen peroxide in Vicia faba guard cells. *J Integr Plant Biol*, 50, 1539–1548.

Shi, H., Ye, T. and Chan, Z. 2014. Nitric oxide-activated hydrogen sulfide is essential for cadmium stress response in bermudagrass (*Cynodon dactylon* (L). Pers.). *Plant Physiol Biochem*, 74, 99–107.

Shimoda, Y., Shimoda-Sasakura, F., Kucho, K. I., Kanamori, N., Nagata, M., Suzuki, A. and Uchiumi, T. 2009. Overexpression of class 1 plant hemoglobin genes enhances symbiotic nitrogen fixation activity between *Mesorhizobium loti* and *Lotus japonicus*. *The Plant J*, 57(2), 254–263.

Shivaraj, S. M., Vats, S., Bhat, J. A., Dhakte, P., Goyal, V., Khatri, P. and Sharma, T. R. 2020. Nitric oxide and hydrogen sulfide crosstalk during heavy metal stress in plants. *Physiol Planta*, 168, 437–455.

Singh, V. P., Singh, S., Kumar, J. and Prasad, S. M. 2015. Hydrogen sulfide alleviates toxic effects of arsenate in pea seedlings through up-regulation of the ascorbate–glutathione cycle: Possible involvement of nitric oxide. *J Plant Physiol*, 181, 20–29.

Stephenson, M. and Stickland, L. H. 1931. Hydrogenase: A bacterial enzyme activating molecular hydrogen: The properties of the enzyme. *Biochem J*, 25, 205.

Stöhr, C. and Ullrich, W. R. 2002. Generation and possible roles of NO in plant roots and their apoplastic space. *J Exp Bot*, 53, 2293–2303.

Su, J., Zhang, Y., Nie, Y., Cheng, D., Wang, R., Hu, H., Chen, J., Zhang, J., Du, Y. and Shen, W. 2018. Hydrogen-induced osmotic tolerance is associated with nitric oxide-mediated proline accumulation and reestablishment of redox balance in alfalfa seedlings. *Environ Exp Bot*, 147, 249–260.

Sun, C., Liu, L., Lu, L., Jin, C. and Lin, X. 2018. Nitric oxide acts downstream of hydrogen peroxide in regulating aluminum-induced antioxidant defense that enhances aluminum resistance in wheat seedlings. *Environ Exp Bot*, 145, 95–103.

Terrón-Camero, L. C., Peláez-Vico, M. Á., Del-Val, C., Sandalio, L. M. and Romero-Puertas, M. C. 2019. Role of nitric oxide in plant responses to heavy metal stress: exogenous application versus endogenous production. *J Exp Bot*, 70, 4477–4488.

Tian, B., Zhang, Y., Jin, Z., Liu, Z. and Pei, Y. 2017. Role of hydrogen sulfide in the methyl jasmonate response to cadmium stress in foxtail millet. *Front Biosci*, 22, 530–538.

Umezawa, T., Sugiyama, N., Mizoguchi, M., Hayashi, S., Myouga, F., Yamaguchi-Shinozaki, K. and Shinozaki, K. 2009. Type 2C protein phosphatases directly regulate abscisic acid-activated protein kinases in Arabidopsis. *Proceed Nat Acad Sci*, 106, 17588–17593.

Vlad, F., Rubio, S., Rodrigues, A., Sirichandra, C., Belin, C., Robert, N. and Merlot, S. 2009. Protein phosphatases 2C regulate the activation of the Snf1-related kinase OST1 by abscisic acid in Arabidopsis. *The Plant Cell*, 21, 3170–3184.

Wang, B. L., Shi, L., Li, Y. X. and Zhang, W. H. 2010. Boron toxicity is alleviated by hydrogen sulfide in cucumber (*Cucumis sativus* L.) seedlings. *Planta*, 231, 1301–1309.

Wang, M. and Liao, W. 2016. Carbon monoxide as a signaling molecule in plants. *Front Plant Sci*, 7, 572.

Wang, X., Li, J., Liu, J., He, W. and Bi, Y. 2010. Nitric oxide increases mitochondrial respiration in a cGMP-dependent manner in the callus from Arabidopsis thaliana. *Nitric Oxide*, 23, 242–250.

Wei, B., Zhang, W., Chao, J., Zhang, T., Zhao, T., Noctor, G., Liu, Y. and Han, Y. 2017. Functional analysis of the role of hydrogen sulfide in the regulation of dark-induced leaf senescence in *Arabidopsis*. *Sci Rep*, 7, p. 2615.

Wu, Q., Su, N., Chen, Q., Shen, W., Shen, Z., Xia, Y. and Cui, J. 2015. Cadmium-induced hydrogen accumulation is involved in cadmium tolerance in Brassica campestris by reestablishment of reduced glutathione homeostasis. *PLoS ONE*, 10, 1–21.

Xie, Y., Mao, Y., Lai, D., Zhang, W., Zheng, T. and Shen, W. 2013. Roles of NIA/NR/NOA1-dependent nitric oxide production and HY1 expression in the modulation of Arabidopsis salt tolerance. *J Exp Bot*, 64, 3045–3060.

Xie, Y., Zhang, C., Lai, D., Sun, Y., Samma, M. K., Zhang, J. and Shen, W. 2014. Hydrogen sulfide delays GA-triggered programmed cell death in wheat aleurone layers by the modulation of glutathione homeostasis and heme oxygenase-1 expression. *J Plant Physiol*, 171, 53–62.

Xu, D., Cao, H., Fang, W., Pan, J., Chen, J., Zhang, J. and Shen, W. 2017. Linking hydrogen-enhanced rice aluminum tolerance with the reestablishment of GA/ABA balance and miRNA-modulated gene expression: A case study on germination. *Ecotoxicol Environ Saf*, 145, 303.

Xu, J., Wang, W., Yin, H., Liu, X., Sun, H. and Mi, Q. 2010. Exogenous nitric oxide improves antioxidative capacity and reduces auxin degradation in roots of Medicago truncatula seedlings under cadmium stress. *Plant Soil*, 326, 321–330.

Xuan, W., Zhu, F. Y., Xu, S., Huang, B. K., Ling, T. F., Qi, J. Y. and Shen, W. B. 2008. The heme oxygenase/carbon monoxide system is involved in the auxin-induced cucumber adventitious rooting process. *Plant Physiol*, 148, 881–893.

Yao, Y., Yang, Y., Li, C., Huang, D., Zhang, J., Wang, C., Li, W., Wang, N., Deng, Y. and Liao, W. 2019. Research progress on the functions of gasotransmitters in plant responses to abiotic stresses. *Plants*, 8, p.605.

Yoshida, R., Umezawa, T., Mizoguchi, T., Takahashi, S., Takahashi, F. and Shinozaki, K. 2006. The regulatory domain of SRK2E/OST1/SnRK2. 6 interacts with ABI1 and integrates abscisic acid (ABA) and osmotic stress signals controlling stomatal closure in Arabidopsis. *J Biol Chem*, 281, 5310–5318.

Zeng, J., Zhang, M. and Sun, X. 2013. Molecular hydrogen is involved in phytohormone signaling and stress responses in plants. *PLoS ONE*, 8, 1–10.

Zhang, H., and Forde, B. G. 2000. Regulation of Arabidopsis root development by nitrate availability. *J Exp Bot*, 51–59.

Zhang, H., Gannon, L., Jones, P. D., Rundle, C. A., Hassall, K. L., Gibbs, D. J. and Theodoulou, F. L. 2018. Genetic interactions between ABA signalling and the Arg/N-end rule pathway during Arabidopsis seedling establishment. *Sci Rep*, 8, 1–12.

Zhang, H., Hu, L. Y., Hu, K. D., He, Y. D., Wang, S. H. and Luo, J. P. 2008. Hydrogen sulfide promotes wheat seed germination and alleviates oxidative damage against copper stress. *J Integr Plant Biol*, 50, 1518–1529.

Zhang, H., Tang, J., Liu, X. P., Wang, Y., Yu, W., Peng, W.Y., Fang, F., Ma, D.F., Wei, Z.J. and Hu, L.Y. 2009. Hydrogen sulfide promotes root organogenesis in Ipomoea batatas, Salix matsudana and Glycine max. *J Integr Plant Biol*, 51, 1086–1094.

Zhang, Y. X., Hu, K. D., Lv, K., Li, Y. H., Hu, L. Y., Zhang, X. Q. and Zhang, H. 2015. The hydrogen sulfide donor NaHS delays programmed cell death in barley aleurone layers by acting as an antioxidant. *Oxid Med Cell Long*, 1–11.

Zhao, X., Chen, Q., Wang, Y. Shen, Z. Shen, W. and Xu, X. 2017. Hydrogen-rich water induces aluminum tolerance in maize seedlings by enhancing antioxidant capacities and nutrient homeostasis. *Ecotoxicol Environ Saf*, 144, 369–379.

Zheng, Q., Meng, Q., Wei, Y. Y. and Yang, Z. M. 2011. Alleviation of copper-induced oxidative damage in Chlamydomonas reinhardtii by carbon monoxide. *Arch Env Cont Toxicol*, 61, 220–227.

Zhou, M., Wang, W., Karapetyan, S., Mwimba, M., Marqués, J., Buchler, N. E. and Dong, X. 2015. Redox rhythm reinforces the circadian clock to gate immune response. *Nature*, 523, 472–476.

Zhu, Y., Liao, W., Wang, M., Niu, L., Xu, Q. and Jin, X. 2016. Nitric oxide is required for hydrogen gas-induced adventitious root formation in cucumber. *J Plant Physiol*, 195, 50–58.

Zilli, C. G., Santa-Cruz, D. M. and Balestrasse, K. B. 2014. Heme oxygenase-independent endogenous production of carbon monoxide by soybean plants subjected to salt stress. *Environ Exp Bot*, 102, 11–16.

11

Microbial and Plant-Assisted Bioremediation of Heavy Metal Polluted Environments

Seerat Saleem, Naveed Ul Mushtaq, Wasifa Hafiz Shah, Aadil Rasool, and Reiaz Ul Rehman
Department of Bioresources, School of Biological Sciences, University of Kashmir, Srinagar, Jammu Kashmir, India

CONTENTS

11.1 Introduction

The environment comprises of the biotic and abiotic phenomena surrounding and potentially interacting with an organism that affects its growth, development and survival (Kearney 2006). Industrialization and urbanization have led to accelerated economic growth. The people are running blindly after economic gains without thinking about the possible consequences, has led to a serious problem of environmental pollution (Liang and Yang 2019). Pollution is the presence of hazardous substances in the environment which hamper the functioning of natural processes and produce detrimental environmental and health effects (Mehndiratta et al. 2013). Anthropogenic activities as well as natural disasters such as volcanic eruptions have a great potential to harm the environment and human health, animal health, plants and trees. The various environmental problems such as air pollution, water pollution and soil pollution pose serious health threats such as cancers, asthma, birth defects and infectious diseases to the living organisms (Khan and Ghouri 2011). Mismanagement of human activities (industrial, agricultural, medical) has led to the release of xenobiotic components such as plastics and microplastics, polluting terrestrial and aquatic ecosystems worldwide (Alimba and Faggio 2019).The air we breathe in is necessary for a healthy life and the polluted air contains hazardous pollutants which may cause oxidative injury to the airways that causes inflammation, hyper-responsiveness and clinic asthma. The various pollutants that cause air pollution are particulate matter (0.1–10 μm) which deposit in the various parts of the respiratory tract, nitrogen oxides (fossil fuel combustion), ground-level ozone (formed by reaction between sunlight and nitrogen oxides), sulfur oxides, TRAP (traffic-related air pollution), carbon monoxide, heavy metals (Guarnieri and Balmes 2014). Water pollution leaves the water unfit for drinking, cooking and other uses. It is caused mostly by human activities i.e., oil spillage, sewage leakage, wastage dumping, deforestation, pesticides, herbicides, heavy metals, mining, mineral processing plants, chemicals, dumping of industrial and agricultural wastes, animal wastes. Water pollution leads to diseases such as cholera, typhoid and tuberculosis besides harming marine life to a large extent (Owa 2013). Soil is an integral part of the environment and sustains life for a wide range of organisms. Soil is greatly influenced by environmental factors and human activities. Contamination of soil

DOI: 10.1201/9781003155089-11

by various pollutants leads to the deterioration of soil, rending it unfit to grow food. These toxic compounds can cause acute toxicity, mutagenesis (genetic changes), carcinogenesis and teratogenesis (birth defects) in humans (Mishra et al. 2015). Soil pollutants are mostly xenobiotic chemicals, such as petroleum hydrocarbons, pharmaceutical products, solvents, pesticides, heavy metals from industrial activity, agricultural chemicals and improper disposal of waste. The various inorganic pollutants present in agricultural soils are nitrates, phosphates, radioactive substances, heavy metals and metalloids. These have a toxic effect on plants, microorganisms and food chain (Pirzadah et al. 2015). The various heavy metals are lead (Pb), cadmium (Cd), nickel (Ni), cobalt (Co), iron (Fe), zinc (Zn), chromium (Cr), iron (Fe), arsenic (As), silver (Ag) and the platinum group elements. Ni, Co, Cr and Cu are more toxic to plants whereas Cd, Pb, As and Hg are more toxic to animals. Metals and metalloids whose densities are >5 g/cm^3 are known as heavy metals (HMs). They can be of natural or anthropogenic origin and the excessive heavy metals in the soil result in the reduction of soil productivity and quality (Yang et al. 2018). HMs are the main pollutants in agricultural soils and their main sources are the extensive and non-standard processes of mining and smelting, sewage irrigation, sludge reuse, organic manures and fertilizer application. Sewage irrigated area is seen contaminated with Hg, Pb, Cd, As, Ni, Cr, polycyclic aromatic hydrocarbons (PAHs). Intensive livestock production generates waste materials which are high in As, Zn and Cu. These HMs, nitrates and organic pollutants are potential carcinogens (Lu et al. 2015). The various organic (that contain carbon) pollutants are hydrocarbons, PAHs, polychlorinated biphenyls (PCBs), chlorinated aromatic compounds, detergents and pesticides (Mishra et al. 2015). The most eco-friendly and cost-effective technique for the decontamination of soils is the biodegradation of the chemical contaminants by microorganisms. The bioremediation technique is the standard practice for the treatment of heavy metal contaminated soils, which employs plants and microorganisms to eradicate hazardous pollutants (Ojuederie and Babalola, 2017). Bioremediation is used for the purification of surface and subsurface soils, freshwater and marine system, groundwater and contaminated land ecosystems. Initially, bioremediation techniques were developed to treat petroleum hydrocarbon contamination in order to immobilize or transform contaminants into non-hazardous forms such as carbon dioxide, water and other non-toxic compounds. Addition of nutrients (carbon, nitrogen, phosphorus) or suitable microbial populations enhances the bioremediation process (Juwarkar et al. 2010). Nature of the pollutants, soil structure, soil pH, moisture content, hydrogeology, microbial diversity of the site, nutritional state, temperature and oxidation-reduction (redox- potential) are the various parameters that affect bioremediation (Dua et al. 2002).

11.2 Heavy Metal Polluted Soils

Group of metals with relatively high density and high atomic weight are known as HMs. These are toxic to both plants and animals and these metals can exist separately or in association with soil components such as phosphates, carbonates, free metal ions, metals attached to silicate minerals. Their availability in the soil is affected by various factors such as temperature, soil pH, organic matter, soil aeration, moisture content, water holding capacity. HMs in soil have a negative effect on the soil biological properties and the health of soil microbiome (Chibuike and Obiora 2014). The anthropogenic sources of various HMs are (1) As: pesticides, preservatives, biosolids, ore mining and smelting; (2) Cd: Pigments, paints, electroplating, phosphate fertilizers; (3) Cr: Tanneries, steel industries, fly ash; (4) Cu: Pesticides, fertilizers, biosolids, ore mining and smelting; (5) Hg: medical waste, Au-Ag mining, coal combustion; (6) Ni: Effluent, automobile batteries, kitchen appliances; (7) Pb: combustion of leaded fuel, insecticides and herbicides. On the other hand, weathering of minerals, erosion and volcanic activities, forest fires are the natural sources of heavy metals in the environment (Dixit et al. 2015). The safe limits of heavy metals in agricultural soil, as per Indian standards are 3–6, 135–270, 250–500, 300–600, 75–150 μg/g for Cd, Cu, Pb, Zn and Ni, respectively (Nagajyoti et al. 2010). Heavy metal pollution is a threat because these metal ions are non-biodegradable and are accumulated in the environment. These cannot be broken down like organic pollutants but can be transformed into less toxic forms. Heavy metals are highly toxic at higher concentrations and hence there is a need for the remediation of heavy metal polluted environments (Ayangbenro and Babalola 2017). Metal ions are mostly in stable oxidation state and they form stable bio-toxic compounds by reacting with biomolecules in bodies (Kumar and Gunasundari 2018). In plants Co, Fe, Mn, Cu, Mo, Zn and Ni are beneficial to plants in trace amounts, but they become toxic at high concentration. However, metals like As, Cd, Pb, Hg and Se are not required by plants and cause toxicity. In plants HM toxicity results in reduced productivity, drooping leaves, chlorosis, decreased nutrient uptake, metabolism disorders, weak growth, hampered seed germination (Jaiswal et al. 2018), DNA damage, protein oxidation, replacement of enzyme cofactors and transcription factors, root death, increase in reactive oxygen species (ROS), inhibition of antioxidative enzymes, senescence (Saha et al. 2017). In humans, HM toxicity causes perforation of nasal septum, respiratory cancer, central and peripheral neuropathies, dermatomes, skin cancers, proteinuria, osteomalacia, ulcers, central nervous disorders, anemia, visual defects, pneumoconiosis (Mahurpawar 2015). The conventional heavy metal remediation measures such as landfill, leaching, soil washing, chemical precipitation, electrochemical techniques, sludge filtration generate toxic waste, pollute ground water and are costly. The biologi cal methods of phytoremediation and resistant microbes are environment-friendly alternatives to physiochemical treatment (Mani and Kumar 2014).

11.3 Bioremediation

Bioremediation is an eco-friendly, non-invasive and cost-effective method used for the clean-up of contaminated sites by plants (phytoremediation) and microorganisms (microbial bioremediation) that degrade, transform or chelate the toxic compounds (Dzionek et al. 2016). Bioremediation is an ideal solution for pollution control that is ecologically sustainable and uses natural processes for the elimination of toxic contaminants (Niti et al. 2013). Microorganisms use the organic pollutants as a source of carbon and energy, enabling their destruction by cometabolism. Bioremediation carried at the place of contamination is known as *in situ* bioremediation. On the other hand, when the contaminated media is taken from its original site to a different location it is termed as *ex situ* (Dzionek et al. 2016). Intrinsic bioremediation and engineered bioremediation are the two types of *in situ* bioremediation. Intrinsic bioremediation is the natural biological attenuation of the pollutants by the indigenous microflora. The increase in metabolic activities of indigenous microbial population, by enhanced nutrient and oxygen supply is known as "Biostimulation". Engineered bioremediation involves the manipulated or stimulated biological remediation of a site, which involves introduction of specific/engineered microorganisms, that accelerate the degradation process (Bioaugmentation) (Mani and Kumar 2014). The various *in situ* bioremediation techniques which enhance the process are (i) bioventing: involves supplying air and nutrients to the contaminated soil, in order to increase the activity of the indigenous aerobic microbes, that increases sub-surface bioremediation and minimizes volatilization of pollutants; (ii) biosparging: It is used to treat aquifers contaminated with petroleum products. In this technique air is injected in the saturated zone which promotes biodegradation by upward movement of volatile compounds to the saturated zone; (iii) bioslurping: This technique uses vacuum enhanced pumping for the recovery of light non-aqueous phase liquids (LNAPLs) and other volatile, semi-volatile organic compounds and remediated capillary, unsaturated and saturated zones (Azubuike et al. 2016). Phytoremediation is an efficient *in situ* remediation technique that utilizes plants which perform rhizosphere bioremediation. The various mechanisms involved in phytoremediation are extraction, degradation, filtration, stabilization and volatilization. Bacteria and mycorrhiza forming fungi colonize these roots, which aid in degradation of the pollutants (Romantschuk et al. 2000). A rhizospheric zone developed by plant and soil microbes acts as tool for accelerating degradation (Ahemad 2012). The *ex situ* bioremediation is the excavation and pumping of the contaminated substrate. In land farming technique, the excavated soil is periodically tilled on a prepared bed. This facilitates the abiotic degradation of pollutants by microorganisms. Composting involves combining of the contaminated soil with manure or agricultural wastes, which enhances the development of a rich microbial population. Biopiles are used for the treatment of petroleum hydrocarbons contaminated surfaces. Biopile treatment is a hybrid of both land farming and composting techniques (Kensa 2011). Contaminated soil, sediment, sludge or water is also bioremediated in slurry/aqueous bioreactors through an engineered containment system. Slurry phase bioremediation involves the combination of contaminated soil with water/additives in a tank (bioreactor), which provides optimum environments for the indigenous microorganisms to degrade the pollutants. The rate of degradation in a bioreactor is greater than the *in situ* or solid-phase systems. Slurry bioreactors are used when soils have high content of clay and organic matter, recalcitrant and toxic pollutants and bioremediation needs to be accomplished in short times (Robles-González et al. 2008). The various factors affecting bioremediation process are microbial population, bioavailability of contaminants, environmental factors such as pH, temperature, water and oxygen content, nutrients, soil type. The disadvantages of *in situ* bioremediation are that it is a time-consuming process, microbial activity is affected by seasonal difference, this type of bioremediation may become uncontrollable and unmanageable sometimes and treatment additives cannot be used properly. On the other hand, the various cons of *ex situ* bioremediation are that it is not suitable for the treatment of HMs or chlorinated hydrocarbons like trichloroethylene and further processing becomes essential for clay and silt soils (Kumar and Gunasundari 2018). The bioremediation technique to be employed depends upon the nature of the pollutant, degree of pollution, cost, location and environment type (Azubuike et al. 2016). HM polluted soils are effectively treated by bioremediation. It is mostly done by *in situ* bioremediation and employing suitable microbes and plants (Chibuike and Obiora 2014).

11.4 Microbial Assisted Bioremediation of Heavy Metal Polluted Environment

The various conventional methods for the elimination of heavy metals are precipitation, ion exchange, reverse osmosis, coagulation, floatation. But these methods are expensive and lead to the production of harmful end products (Basha and Rajaganesh 2014). Microorganisms act on HMs by changing their solubility, mobility and bioavailability and hence can be used for HM bioremediation. Microorganisms use various mobilization and immobilization processes for this purpose. Mobilization involves redox reaction technique, siderophores (specific iron chelators, forming metal complexes and enhancing their solubility), complexation (metal complex formation by addition of ligand) (Pratush et al. 2018). Metals are immobilized by microbes via biosorption, bioaccumulation or extracellular precipitation (Rajendran et al. 2003). Metals at high concentration can either inhibit the microbial metabolic activities or microbes develop resistance towards high metal levels. Metal resistance in bacteria involves various defense mechanisms by which they mobilize, immobilize or transform metals which affect their uptake (Khan et al. 2009). The Czc system in *Ralstonia eutropha* (gram-negative bacteria) provides resistance against Cd, Zn and Co. The Cd resistance in *Staphylococcus, Bacillus or Listeria* is provided through Cd-efflux ATPase. Resistance traits are regulated by genes, organized in operons. These resistance mechanisms of bacteria are exploited for the remediation of contaminated soils (Ahemad 2012). Bioremediation of contaminated soil/water is mostly done by microorganisms. These microorganisms can be autochthonous i.e., naturally inhabiting the environment to be purified or microorganisms from other environments which are introduced in the contaminated environment. Various bacteria, fungi, yeast and algae are used for the bioremediation of HM polluted environment (White et al. 1997). Due to their adaptability to various environmental conditions, microbes can be used for the degradation of pollutants. Specific enzymes of microorganisms can reduce or oxidize metals (Girma 2015). Organic pollutants are mineralized into end products such as CO_2, H_2O which are required for cell growth of the microbes (Verma and Kuila 2019). Various microbial genera like *Bacillus, Enterobacter, Escherichia, Pseudomonas* help in HM remediation (Ray and Ray 2009). *Vibrio harveyi* can reportedly bioaccumulate 23.3 mg Cd^{2+}/g of dry cells; marine bacterium *Enterobacter cloacae* can reportedly chelate up to 65% of Cd, 20% Cu and 8% Co at 100 mg/L of metal concentration; *Rhodobium marinum* and *Rhodobacter sphaeroides* (purple non-sulfur bacterial isolates) have the potential of removing Cu, Zn, Cd and Pb from the environment (Das and Dash 2014). Low microbial abundance and activity in addition to lack of available nutrients and less access to the contaminants reduce the efficiency of bioremediation process. This can be overcome by the ability of microorganisms to form sessile biofilm structures. These biofilms are advantageous for bioremediation as compared to their planktonic counterparts. This is because biofilms are protected within the matrix, are able to communicate (quorum sensing), are capable of DNA sharing and nutrient recycling (Edwards and Kjellerup 2013). A biofilm is comprised either of a single bacterial species or many species of bacteria, fungi, algae and protozoa. Biofilm structures are encased within extracellular polymeric substances (EPS) which are secreted by the microbes and contain surfactants capable of solubilizing hydrophobic and recalcitrant substances (Singh et al. 2006). EPS efficiently sequesters positively charged HM ions due to its net anionic makeup. Biofilms can survive in most toxic environments and can efficiently remediate metals by using biosorption, bioaccumulation mechanisms. The presence of EPSs and biosurfactants enable the biofilms to bind and transform metals. The biosorption of various metals, such as Cd(II), Cr(VI), Pb(II), Hg(II), Ni(II), Zn(II) by different types of bacterial biomass have been reported (Mohapatra et al. 2020). *Escherichia coli* and *Staphylococcus epidermidis* biofilms were used for the removal of Cr and Zn from aqueous solution and *E.coli* was found to be more efficient in the metal removal (Quiton et al. 2018). The use of *Bacillus* biofilms (*Bacillus subtilis* and *Bacillus cereus*) was able to remove 98% of Cr(III) from chrome tanning effluent (Sundar et al. 2011). *Stenotrophomonas* spp. biofilm showed a removal efficiency of 69.9% and 78.4% for Zn and Mn, respectively, from a tannery effluent (Pani et al. 2017). The various mechanisms involved in microbial remediation of heavy metals are biosorption, bioaccumulation, bioleaching, biotransformation and biomineralization (Figure 11.1).

11.4.1 Biosorption

HMs are removed by using the biomass of bacteria, fungi, algae, which act as absorbent and adsorbent material for their removal. The renewable biomass of microorganisms is an environment-friendly bio trap for HMs. Bio traps are living or non-living organisms which can bind with toxic metals and alter their forms, facilitating their removal from the polluted water or soil (Choudhary et al. 2017). In biosorption or passive uptake, microbes (Biosorbent) bind with metal ions (sorbate) and sorb these metal ions on the binding sites of the cell wall, which leads to metal recovery. Polysaccharides, proteins and lipids present in microbial cell walls have metal-binding groups, such as carboxyl, sulfate, phosphate, amino, imidazole, sulfydryl, groups, which are involved in metal chelation. The biosorbents used are exo-polysaccharides; dead biomass

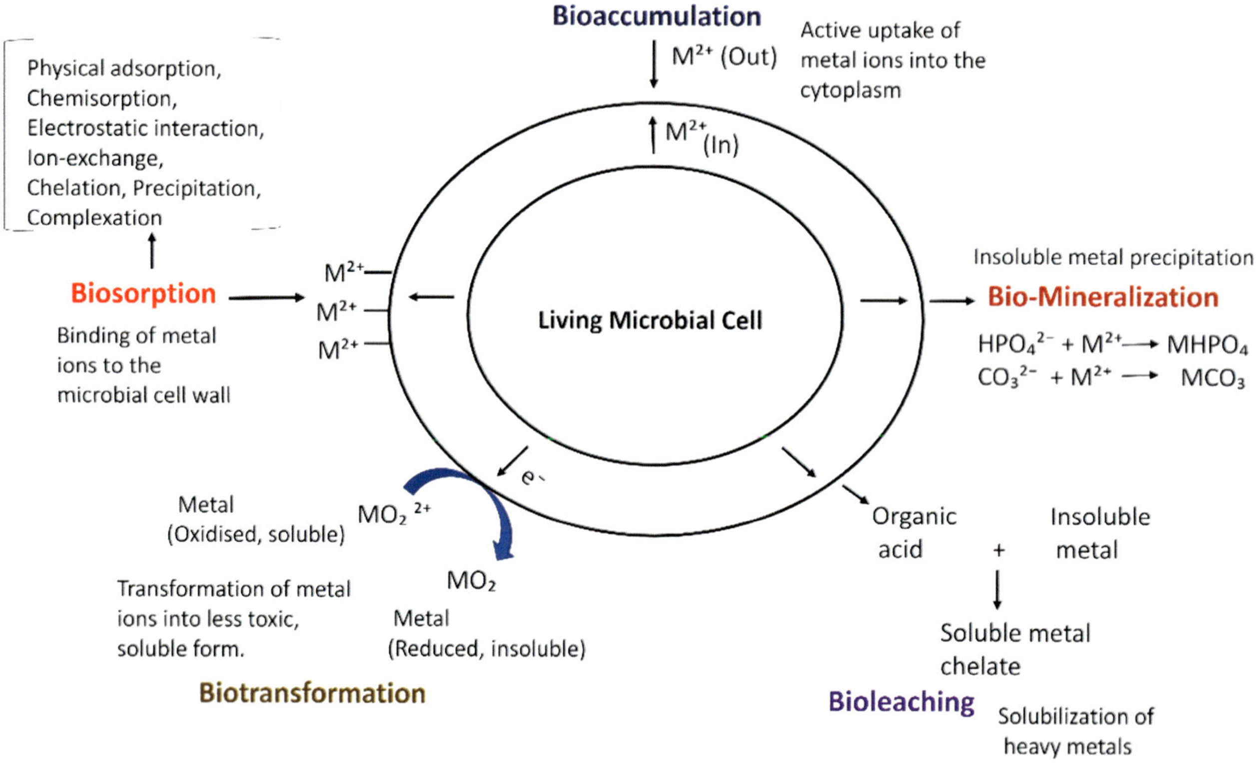

FIGURE 11.1 Microbial mechanisms used for the removal of heavy metals from contaminated sites.

and living cultures. Living cells absorb more than non-living cells. Extracellular polymeric substances of bacterial cell walls show metal-binding ability via micro-precipitation of metals, electrostatic interaction, ion exchange, surface complexation methods (Verma and Kuila 2019; Ayangbenro and Babalola 2017). The quantity of contaminants adsorbed on the sorbent's cellular surface depends upon kinetic equilibrium and composition of the cellular surface (Ayangbenro and Babalola 2017). Biosorption is a faster and reversible process. Based on metabolism, biosorption can be metabolic-dependent and metabolism-independent. On the basis of the metal location, biosorption can be extracellular accumulation/precipitation; cell surface adsorption/precipitation and intracellular accumulation (metabolism dependent, active uptake and transport of metal ions across membrane). The various metabolism independent mechanisms of biosorption are physical adsorption (Van der Wall's forces), ion exchange, complexation (complex formation between metal cations and active groups of cell surface), chemical sorption (teichoic acids). Chemisorption to a large extent is responsible for the biosorption of copper (II) (Javanbakht et al. 2014). In *pseudomonas syringae* complexation was the only mechanism which resulted in the accumulation of Cd, Zn, Cu and Hg. Environmental pH affects the metal properties and hence influences biosorption. Cheap biosorbents used for metal removal are bacteria, fungi, algae, industrial and agricultural wastes (Hansda and Kumar 2016). The biosorption capacities of some microorganisms are mentioned in Table 11.1.

Kumar et al. (2014) reported that Fungi *Aspergillus nidulans, Rhizopus arrhizus* and *Trichoderma viride* through biosorption had a maximum uptake capacity of 25.67 mg/g for Pb, 13.15 mg/g for Cd and 2.55 mg/g of Cr, respectively.

11.4.2 Bioaccumulation

Bioaccumulation is the metabolism-dependent, active uptake of HMs where metal ions pass across the cell membrane into the cytoplasm. Bioaccumulation is dependent on various

TABLE 11.1
Biosorption Capacity of Various Microorganisms

Metal	Biosorbent	pH	Biosorption Capacity (mg/g)	Reference
Aluminum(III)	*Chryseomonas luteola*	5.0	55.2	Ozdemir and Baysal (2004)
Chromium(VI)	*Bacillus licheniformis*	2.5	60.5	Zhou et al. (2007)
	Bacillus thuringiensis	2.0	83.3	Şahin and Öztürk (2005)
	Staphylococcus xylosus	1.0	143.0	Ziagova et al. (2007)
	Paecilomyces lilacinus	5.5	189.1	Sharma and Adholeya (2011)
	Spirulina sp.	5	90.91	Rezaei (2016)
Copper(II)	*Enterobacter sp. J1*	5.0	32.5	Lu et al. (2006)
	Pseudomonas putida	5.5	96.9	Uslu and Tanyol (2006)
	Aspergillus niger	5.5	17.60	Tsekova et al. (2010)
	Micrococcus luteus DE2008	6.5–7.0	408	Puyen et al. (2012)
Cadmium(II)	*Aeromonas caviae*	7.0	155.3	Loukidou et al. (2004)
	Staphylococcus xylosus	6.0	250.0	Ziagova et al. (2007)
	Pseudomonas Fluorescence	2.0	40.8	Uzel and Ozdemir (2009)
	Pseudomonas aeruginosa B237	6.0	16.89	Limcharoensuk et al. (2015)
	Aspergillus niger	5.5	69.44	Tsekova et al. (2010)
	Halomonas BVR 1sp	>8.0	12.0	Rajesh et al. (2014
Lead(II)	*Pseudomonas aeruginosa PU21*	5.5	79.5	Chang et al. (1997)
	Pseudomonas putida	5.5	270.4	Uslu and Tanyol (2006)
	Micrococcus luteus DE2008	6.5–7.0	1965	Puyen et al. (2012)
Mercury(II)	*Bacillus* sp.	6.0	7.9	Green-Ruiz (2006)
Nickel(II)	*Bacillus thuringiensis*	6.0	45.9	Öztürk (2007)
	Pseudomonas Fluorescence	2.0	12.4	Uzel and Ozdemir (2009)
	Bacillus subtilis 117S	7	155.5	Abdel-Monem et al. (2010)
Zinc(II)	*Aphanothece halophytica*	6.5	133.0	Incharoensakdi and Kitjaharn (2002)
	Streptomyces rimosus	7.5	30.0	Mameri et al. (1999)
	Tsukamurella paurometabola A155	6.0	16.75	Limcharoensuk et al. (2015)
	Streptomyces K11	6.5	48.9	Sedlakova-Kadukova et al. (2019)
	Bacillus jeotgali	7.0	222.2	Green-Ruiz et al. (2008)
Platinum(IV)	*Desulfovibrio desulfuricans*	2.0	62.5	De Vargas et al. (2004)
	Aspergillus sp.	2.0	5.49	Godlewska Żyłkiewicz et al. (2019)
Palladium(II)	*Desulfovibrio desulfuricans*	2.0	128.2	De Vargas et al. (2004)
	Aspergillus sp.	2.5–3.5	4.28	Godlewska Żyłkiewicz et al. (2019)

physiochemical and biological mechanisms. The microbes capable of bioaccumulation have tolerance towards a wide range of metals. Bioaccumulation involves sequestration, redox reactions and species transformation (Kumar and Bharadvaja 2020). In bioaccumulation, the metal ions are taken up by ATP-driven active transporter bioprecipitation (metal precipitation by excretion of sulfide or phosphate ions by bacteria) or by biotransformation. Metal ions are localized within specific organelles which undergo enzymatic detoxification and eventual are exported via efflux systems (Hansda and Kumar 2016). Biosorption is the first step of bioaccumulation. Bioaccumulation is partially reversible that needs energy and requires respiration. Organisms capable of bioaccumulation have tolerance towards high level of contaminants with higher biotransformation capabilities, changing the toxins into less harmful forms. Bacteria are seen to produce metallothioneins (metal-binding proteins) which are rich in thiol groups that increase the metal binding capacity and sequestration (Mosa et al. 2016). When compared to biosorption, bioaccumulation involves living cells and hence is cost prone. The uptake is usually low and time-consuming, moreover, the process is negatively affected by temperature and the biomass cannot be used for the next cycle. Yeasts accumulate the higher number of HMs by bioaccumulation than biosorption (Vijayaraghavan and Yun 2008). In bioaccumulation, the microbes absorb the metal ions via the same routes that supply metabolically important ions (Mg, Na) (Zabochnicka-Świątek and Krzywonos 2014). The bioaccumulation efficiency of various microorganisms is listed in Table 11.2.

11.4.3 Bioleaching

Microorganisms are used for the dissolution of metals from their mineral sources and this process is termed as bioleaching. It involves the conversion of metals into their water-soluble form which facilitates their extraction. Bioleaching is an efficient, low cost and environment-friendly method for the eradication of HMs. An example of bioleaching is the microbial oxidation of copper sulfide to copper sulfate, where the metal concentration is in aqueous phase and the remaining solids are discarded (Mishra et al. 2005). The HM elimination by bioleaching involves their solubilization and acidification. The bacterial leaching process is carried out in acidic environment (where most metal ions remain in the solution) and

TABLE 11.2

Bioaccumulation Efficiency of Various Microorganisms

Microorganism	Species	Target Metal	Initial Metal Concentration Mg L^{-1}	Uptake Efficiency	Reference
Bacteria	*Enterobacter* sp.	Hg(II)	5	100% removal	Sinha and Khare (2011)
	Sphingomonas sp.	Cu(II)	15	90% removal	Vilchez et al. (2007)
	Bacillus cereus KA18	Cr(IV)	50–250	92.25 mg g^{-1}	Akhter et al. (2017)
		Ni(II)		135 mg g^{-1}	
		Cd(II)		120 mg g^{-1}	
	Deinococcus Radiodurans	Cd(II)	100	36.86 mg g^{-1}	Jaafar et al. (2015)
	Enterobacter cloacae B1	Pb(II)	180.3	171.83 mg g^{-1}	Banerjee et al. (2015)
		Cd(II)	178.1	114.28 mg g^{-1}	
		Ni(II)	84.9	32.22 mg g^{-1}	
Fungi	*Trichoderma SP2F1*	Cu(II)	63.46	19.60 mg g^{-1}	Ting and Choong (2009)
	Aspergillus versicolor	Ni(II)	50	4.48 mg g^{-1}	Taştan et al. (2010)
		Cu(II)	50	2.08 mg g^{-1}	
	Aspergillus niger	Cu(II)	25–150	4.9–15.5 mg g^{-1}	Dursun et al. (2003)
		Pb(II)	25–250	5.3–26.2 mg g^{-1}	
		Cr(II)	25–75	5.1–6.6 mg g^{-1}	
	Beauveria bassiana	Cu(II)	30	74.13%	Gola et al. (2018)
		Ni(II)		75.00%	
		Cd(II)		63.40%	
		Zn(II)		67.80%	
		Cr(VI)		61.13%	
		Pb(II)		58.47%	
Algae	*Spirulina platensis*	Pb(II)	5–100	2.66–188.32 mg g^{-1}	Arunakumara et al. (2008)
	Chlorella spp. NKG16014	Cd(II)	5.6	39.4 mg g^{-1}	Matsunaga et al. (1999)
	Oedogonium westii	Cd, Cr, Pb	1.5	82.8; 94.0; 94.0 mg g^{-1}	Shamshad et al. (2015)
	Cladophora glomerata	Cd, Cr, Pb	1.5	78.5; 173.2; 86.4 mg g^{-1}	
	Vaucheria debaryana	Cd, Cr, Pb	1.5	74.2; 123.1; 103.3 mg g^{-1}	
	Zygnema insigne	Cd, Cr, Pb	1.5	69.8; 197.6; 81.3 mg g^{-1}	
Yeast	*Rhodotorula mucilaginosa*	Cu(II)	45.4–95.9	4.51–11.10 mg g^{-1}	Ertuğrul et al. (2009)
		Cr(IV)	49.2–129.2	6.12–26.35 mg g^{-1}	
		Ni(II)	22.3–62.2	1.76–3.58 mg g^{-1}	

mainly involves acidophilic sulfur-oxidizing bacteria, such as *Acidithiobacillus thiooxidans*, *Acidithiobacillus ferrooxidans* (use ferric ion as an alternate electron acceptor), *Thiobacillus cuprinus* belonging to the group of *Thiobacillus* (Roy and Roy 2015). Sulfur oxidizing bacteria also include thermophilic bacteria such as *Sulfolobousbrierleyi, Sulfobacillus thermosulfidoxidans, Sulfolobousambivalens.* These bacteria are chemolithoautotrophic i.e., they utilize inorganic chemicals like sulfur as energy sources. Various heterotrophic microbes carry out metal leaching by the production of organic acids (oxalic acid, citric acid and malic acid). These organic acids supply both protons and metal complexing anions. These microbes include bacterial species such as *Acetobacter, Acidophilum, Arthrobactor, Pseudomonas* and fungi (*Penicillium, Aspergillus, Fusarium*). The bioleaching process successfully removes HMs from ores, soil, industrial wastes and sewage sludges (Pathak et al. 2009). Bioleaching takes place under aerobic conditions and is done either by a direct mechanism or by an indirect mechanism. The former involves direct solubilization of metal sulfides by enzymatic oxidation stages and the latter requires the production of sulfuric acid which causes acidification of the sediments and extraction of HMs adsorbed on sediment particles. In a study conducted to evaluate the bacterial bioleaching efficiency, *A. thiooxidans* resulted in higher solubilization of metals (Cr, Cu, Zn) than *Acidithiobacillus ferrooxidans* in contaminated sediment. The solubilization of Cr, Cu, Zn was >80% and Pb was solubilized to a ratio of 63% (Akinci and Guven 2011). Indigenous sulfur-oxidizing bacteria *Acidithiobacillus thiooxidans* isolated from contaminated soil showed a solubilization efficiency of 88%, 93%, 92%, 97% for Cr, Cd, Cu and Zn, respectively. Bioleaching also resulted in undesirable loss of nitrogen, potassium and phosphorus from the soil (Kumar and Nagendran 2008). In two-step bioleaching, *Aspergillus niger* strain SY1 efficiently removed HMs from a contaminated system. The metal extraction efficiencies of Cd, Cu and Zn were 99.5, 56, 71.9 and 76.4%, respectively (Zeng et al. 2015). *Aspergillus flavus* is highly efficient in metal bioleaching and in two-step bioleaching it showed efficiencies of 16.91% for Pb, 49.66% for Cd and 65.73% for Zn and hence showing its great potential in the remediation of contaminated soil (Qayyum et al. 2019).

11.4.4 Biotransformation

Microorganisms are capable of transforming compounds through various enzymatic reactions like oxidation, reduction and hydrolysis. This process is known as biotransformation and it is the most crucial process for the remediation of metal-polluted water, soil and sediment. Bacteria and fungi are tolerant to the harmful effects of HMs and transform these from one oxidation state to another (Chaturvedi et al. 2015). Inorganic compounds exist in more than one oxidation state and it is reported that lower ionic state is more soluble and less toxic (Pratush et al. 2018). Microbes obtain energy from the oxidation of metals like Fe, S, Mn and As. Reduction of metals such as Se, Cr, U by microbes through dissimilatory reduction enables microbes to utilize these metals as terminal electron acceptor in anaerobic respiration (Juwarkar et al. 2010). Oxidation or reduction reactions result in the change of the redox state of metal ions. This redox change transforms metals into water-soluble, less toxic and precipitated form. The oxidizing agent (microorganisms) causes the metal ions to lose electrons which are accepted by electron acceptors (nitrate, sulfate and ferric oxides) (Dixit et al. 2015). Various anaerobic and aerobic bacteria reduce U(VI) to U(IV), Cr(VI) to Cr(III) which are less toxic forms. Se(VI) reduction to elemental Se^0 is used for the remediation of contaminated waters and soil. Mercuric reductase in *Bacillus* sp. reduces mercuric ion to metallic mercury, which is then released into the environment via cell membrane. Oxidation of toxic As(III) into As(V) by *Micrococcus* sp. and *Acinetobacter* sp. decreases its toxicity. Sulfur reducing bacteria have 95% removal efficiencies for Ni(II), Zn(II), Cr(VI), Cu(II) in contaminated waters. Under aerobic condition *Bacillus amyloliquefaciens* utilizes glucose for Cr(VI) reduction (Yin et al. 2019).

11.4.5 Biomineralization

Living organisms are capable of forming minerals and this is known as biomineralization. Biomineralization is used for detoxification and bio recovery. Oxides, phosphates, sulfides, silicates, carbonates, oxalates are among the various minerals precipitated by microbes which have high metal sorption capacities and redox catalysis properties (Gadd and Pan 2016). Microbial induced calcite precipitation (MICP) is a method of biomineralization where various minerals are synthesized by microorganisms from calcium carbonate via hydrolysis of urea. It includes adhesion of metals to carbonate ions for the formation of minerals (Jalilvand et al. 2020). Ureolytic bacteria secrete urease enzyme which hydrolyses urea into one mole of carbonate ion and two moles of ammonium ion. These carbonate ions then precipitate as calcium carbonate ions. Calcium carbonate minerals help in bioremediation by the biosequestration of HM cations. Various urease-producing bacterial strains showed high Cd(II) and Ni(II) removal rates i.e., 96% and 89%, respectively by microbially induced precipitation (Khadim et al. 2019). Achal et al. (2012) reported that the urease producing *Sporosarcina ginsengisoli* CR5 is highly As(III) tolerant and is a good candidate for MICP based bioremediation of As contaminated soil. Ureolytic fungi isolate *Aspergillus* sp. UF3 and *Fusarium oxysporum* UF8 showed significant calcite production and co-precipitation of radionuclide strontium carbonates (Dhami et al. 2017). *Penicillium chrysogenum* CS1 removed 98.8% of Pb from a solution by biomineralization process (Qian et al. 2017). HPO_4^{2-}, inorganic phosphate produced by *Citrobacter* sp. precipitates metals (M) as $MHPO_4$, which are retained at the cell surface (Hansda and Kumar 2016).

11.5 Phytoremediation

Inorganic (heavy metals, radionuclides) and organic pollutants like pesticides, insecticides, PAHs present in soil, water or air can be efficiently removed by the use of plants in association with soil microbes. This process is known as phytoremediation which is a cost-effective, eco-friendly *in situ* remediation technique. Plants have an immense potential of uptake of pollutants and their detoxification by various mechanisms, without affecting the top soil and preventing erosion. Phytoremediation costs 5% of the alternative cleaning methods. Fast growing and high biomass producing plants are best suited for phytoremediation (Malik et al. 2015). The term phytoremediation comprises of Greek work "phyto" meaning "plant" and Latin suffix "remedium" which denotes "restoring" (Karami and Shamsuddin 2010). HM toxicity in plants leads to oxidative stress which causes lipid peroxidation, degradation of proteins by proteases, DNA damage, downregulation of antioxidant enzymes and depletion of cellular antioxidants. Plants have evolved various physiological, biochemical and molecular mechanisms for HM tolerance. Initially, the plants attempt to reduce the metal uptake by restricting the metal ions to the apoplast, metal binding to the cell walls and cell exudates. The metal ions inside the cell go through various storage and detoxification mechanisms such sequestration by phytochelatins (PCs), metallothioneins; chelation; synthesis of metal transporters, stress-related proteins and signaling molecules. These adaptive mechanisms allow the plants to be used as agents of phytoremediation (Kalaivanan and Ganeshamurthy 2016). PCs are the thiol peptides that cause chelation of metals in the cytoplasm. Chelation is followed by sequestering of these thiol-metal complexes into the vacuoles with the aid of ATP-binding cassette transporters (Mani and Kumar 2014). Phytoremediation is a multidisciplinary approach which combines plant physiology, soil microbiology and soil chemistry. The various plant families with phytoremediation potential are Brassicaceae, Fabaceae, Euphorbiaceae, Asteraceae, Lamiaceae and Scrophulariaceae (Jadia and Fulekar 2009). High biomass, disease resistant and tolerant, non-edible plants are selected which have the potential to provide renewable energy. Also, the biomass generated during the process needs proper disposal and management (Ghosh and Singh 2005). Plants cleanse the environment by absorbing, accumulating and detoxifying contaminants polluting soil, air and water through various physical, chemical or biological processes (Jabeen et al. 2009). The aim of phytoremediation technique is the uptake of contaminants and their conversion into easily extractable form (Karami and Shamsuddin 2010). The various advantages of phytoremediation are that it is an *in situ*, solar-driven "green" technology that is easy to apply on a wide range of organic pollutants, metals and radionuclides. It is suitable for the remediation of large sites with well-studied manageable crop plants. It does not cause soil disruption and produces metal-rich plant residues which can be disposed by using comparatively inexpensive techniques. Phytoremediation is also effective in eliminating toxic secondary air and water-borne wastes. In addition to being cost-effective phytoremediation makes contaminated sites aesthetically appealing. The various limitations of phytoremediation are that it has limited application to the sites with multiple contaminants. Also, the hyperaccumulator plants are restricted in distribution. When compared to the traditional cleanup techniques, phytoremediation is a slow process (McIntyre 2003). Phytoremediation of heavy metals can be enhanced by: (1) increasing the mobility of metals in the soil. The metal mobility is increased by the use of chemicals or surfactants such as CDTA (Cyclohexane-1,2 diamine tetraacetic acid), EDTA (Ethylenediaminetetraacetic acid), DTPA (Diethylenetriaminepentaacetic acid), NTA (Nitrilotriacetic acid) and citric acid. These chemicals increase the absorption rate of metals by the plants; (2) Genetically engineered plants enhance the phytoremediation process by increased tolerance towards metals; (3) Addition of fertilizers, microbial inoculation, carbon sources, land farming techniques; (4) Using perennial crops such as *Ricinus*, *Miscanthus*, *Jatropha* and *Populous* which are efficient in bioenergy production as well as phytoremediation; (5) Wetland plants efficiently adsorb contaminants by providing more biomass (Muthusaravanan et al. 2018). The efficiency of phytoremediation can also be alleviated by lowering the pH, increasing the electrode potential (Eh), microorganisms (arbuscular mycorrhizal fungi) and plant growth-promoting rhizobacteria (PGPR) (Karami and Shamsuddin 2010). The various phytoremediation mechanisms used for the remediation of heavy metals are presented in Figure 11.2.

11.5.1 Phytoextraction/Phytoaccumulation

It is also known as phytoaccumulation, phytoabsorption or phytosequestration. This process involves the absorption of pollutants from the contaminated site by the plant roots followed by their translocation and accumulation in the aboveground biomass. Hyperaccumulator plants can accumulate metals 50–500 times greater than normal plants (Mahar et al. 2016). Metals are accumulated in the harvestable parts of the plant which are processed into ash by drying, ashing and composting. Bioharvesting of metals can be efficiently applied in mining industry (Ahmadpour et al. 2012). Brooks et al. (1977) defined the term hyperaccumulator as "the plants which could accumulate more than 1000 mg/kg Ni in their leaves". Aquatic macrophytes such as *Eichhornia crassipes* and *Centella asiatica* have an immense potential of accumulating metals. Their copper removal capacity was found to be 97.3% and 99.6%, respectively. *Cyperus roductus* is suitable for the removal of Cr for contaminated sites (Muthusaravanan et al. 2018). Using native plant species for phytoextraction is recommended because these species are less competitive and will efficiently reduce the metal concentration (Ahmadpour et al. 2012). Hyperaccumulators resist the cytotoxicity of accumulated metals due to their remarkable hyper tolerance mechanism and powerful scavenging mechanism (Alkorta et al. 2010). Tang et al. (2009) reported that *Arabis paniculate* is hyper tolerant to Pb, Zn and Cd. It could on an average accumulate 2300 mg kg^{-1} dry weight Pb, 20,800 mg kg^{-1} Zn and 434 mg kg^{-1} Cd. *Sedum alfredii* hyperaccumulated 2183 mg kg^{-1} Cd and 13,799 mg kg^{-1} Zn DW in leaves. Hence, it is an excellent Cd/Zn hyperaccumulator (Jin et al. 2009). Application of EDTA, indole-3-acetic acid (IAA) and microbial liquid (BL) to *Bryophyllum laetivirens* from municipal sludge made garden (MSMG) soil, accumulated HMs 2.1–6.8 times than the normal group. The plant hyperaccumulation of Cu, Pb, Zn, Cd, Ni ranged in 28.7–52.5, 153–429, 209–449, 3.6–10.6 and 44–111 mg kg^{-1}, respectively (Li et al. 2020). Citric acid enhanced the phytoextraction of Cd/Pb by *Ricinus communis* L., which is a bioenergy

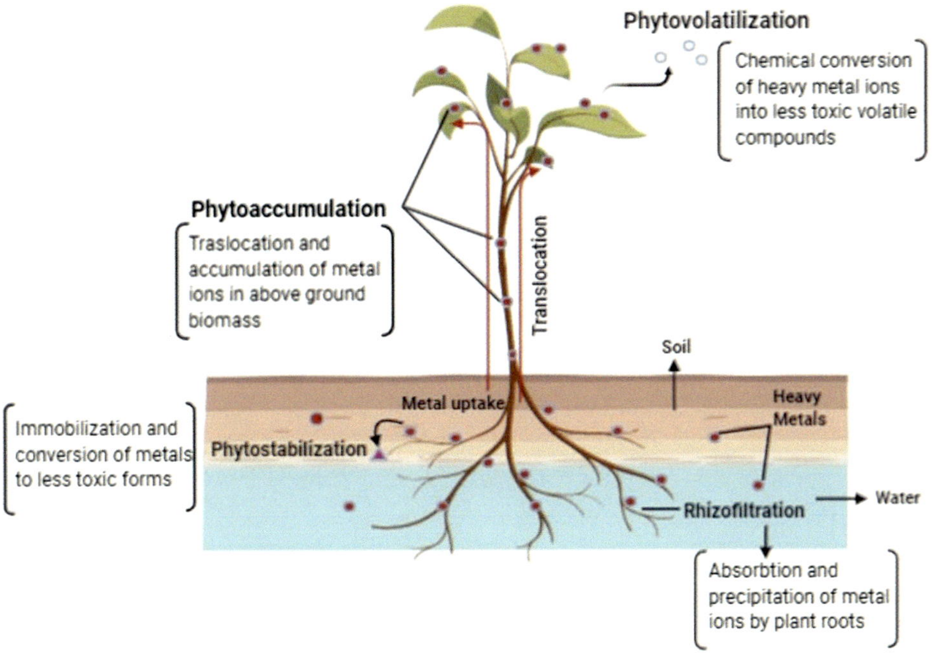

FIGURE 11.2 Mechanisms involved in phytoremediation of heavy metals.

crop (Zhang et al. 2015). Soil treated with EDTA, MGDA and GLDA chelates enhanced the metal uptake of *Zea Mays* L. Pb uptake increased from 10.6 mg plant^{-1} to 416, 398 and 416 mg plant^{-1} with 8 mmol kg^{-1} MGDA, GLDA and EDTA treatments, respectively. Likewise, Zn uptake increased from 100.9 mg plant^{-1} to 798.9, 718.9 and 530.4 mg plant^{-1} in the MGDA, GLDA and EDTA amended soils, respectively (Masoudi et al. 2020). Halophyte *Sesuvium portulacastrum* is an efficient accumulator of heavy metals from tannery affluent. The plant accumulated 22.10 mg Cd, 49.42 mg Cr, 35.10 mg Cu and 70.10 mg Zn kg^{-1} DW plant (Ayyappan et al. 2016). Bioremediation of soils by *Streptomyces* sp. increased the bioavailability of heavy metals (Zn, Cd and Pb), thereby increasing their uptake and accumulation by *Salix dasyclados* L. (Złoch et al. 2017). It was reported that *Fagopyrum tataricum* seedlings can resist, accumulate and translocate Hg. The accumulation was higher in the roots of seedlings and this capability of *F. tataricum* could be used for Hg detoxification (Pirzadah et al. 2018).

11.5.2 Phytostabilization

Phytostabilization is an in-expensive phytotechnology where the toxic metal ions are stabilized by the plants within the roots or near the rhizosphere and their uptake is limited by the crops. Phytostabilization prevents the further movement of the toxic ions by changing their speciation which leads to their deactivation and immobilization. Tolerant plants mechanically stabilize the polluted land thereby preventing bulk erosion, airborne transport and leaching of pollutants. Phytostabilization is mostly used for remediating soils, sediments and sludges (Shackira and Puthur 2019). Metal contaminants in the soil are not removed by phytostabilization but are stabilized within the root zones which reduces the risk to human health and environment. The dispersion of metal pollutants in the soil is prevented by establishing a cover crop that stabilizes the contaminated site. Phytostabilization creates an aerobic atmosphere in the root zone and the metal availability is reduced by adsorption and precipitation processes. Plants with an extended root system and large biomass are best suited for phytostabilization of large quantities of HMs. Also, the root to shoot translocation should be less (Alkorta et al. 2010). Combination of biological methods and chemicals for the phytostabilization of heavily polluted sites is termed as chemo-phytostabilization. This process involves the treatment of upper soil layers with different amendments (organic materials, liming agents, fertilizers, aluminosilicates) that help in fertilizing and adjusting the soil pH, immobilizing the metal compounds. Next, the metal tolerant plants enrich the soil with organic matter, prevent erosion, reduce the migration of contaminants and cause metal immobilization (Kucharski et al. 2005). Phytostabilization occurs through the following ways: (1) complex formation by organic acids, root exudates (sugars, polysaccharides, peptides, proteins, siderophores) – secreted by plants, these act as ligands for binding metal ions, influence pH of rhizosphere and enhance accumulation, stabilization or volatilization of contaminants; (2) precipitation within the root zone by complex formation; (3) adsorption and absorption within the roots by vacuole sequestration or cell wall binding. Metal accumulation in plant root cells is well reported. Pectin compounds present in cell walls are able to bind the metal ions. Chelation of metal ions by metal-binding ligands (PCs, metallothioneins) leads to their sequestration into the vacuole (Padmavathiamma and Li 2007). Phytostabilization traps the contaminants within the root zones and restricts their entry into the vegetative parts. *Chenopodium botrys*, *Chenopodium bijarensis*, *Chenopodium juncea*, have a high bioconcentration factor and a low translocation factor (TF) for

Mn. Hence, these plants have the potential for the phytostabilization of Mn (Cheraghi et al. 2011). High quantity of Al was restricted to the roots of *Fagopyrum* species (*F. kashmirianum* and *F. tataricum)* with TF < 1 advocating phytostabilization potential of the plants (Pirzadah et al. 2019). Galal et al. (2017) reported the TF of *vossia cuspidate* lower than 1. Also, the metals (Al, Cu, Fe, Ni) were retained from the underground parts of the plant which indicate the phytostabilization potential of the plant. Lower TF (<1) and higher accumulation of Hg (Malik et al. 2019), Pb and Al (Malik et al. 2021) in roots as compared to shoots suggested the phytostabilization capability of *Cichorium intybus* L. (Chicory) *Lupinus luteus* inoculated with PGPRs, accumulated heavy metals (Cd, Cu, Pb) in roots. PGPR improved the plant biomass and hence *Lupinus luteus* inoculated with PGPR has a phytostabilization potential (Dary et al. 2010). Pb sequestration in the root rhizosphere of *Fagopyrum tataricum* (Pirzadah et al. 2020) and *Fagopyrum kashmirianum* (Hakeem et al. 2019) was reported with TF < 1.

11.5.3 Phytovolatilization of Heavy Metals

The detoxification of heavy metal ions by their chemical conversion into less toxic volatile compounds is known as phytovolatilization. It involves the uptake of contaminants from the roots, their passage through the plant and ultimately their volatilization through the stomata in the leaves where gaseous exchange takes place. This results in the removal of harmful elements from the soil into the atmosphere (Marques et al. 2009). This technique is mostly used for organic pollutants and heavy metals like Hg and Se, which have high volatility and do not form methyl and hydride derivatives. The use of this technique is controversial as it does not remove the contaminant completely. The contaminant is transferred from the soil to the atmosphere from where it can be redeposited (Ali et al. 2013). As remediation through phytovolatilization by *Pterisvittata* resulted in 90% volatilization of As from the polluted soil. But the secondary arsenic production due to the large amount released in the atmosphere is a matter of concern. Aquatic plants (*Typha domingensis*, *Lemna obscura*, *Hydrilla verticillata Royle* and *Crinum americanum*) converted inorganic selenium into less toxic organic form which was then transpired (Gaur et al. 2014). *Arabidopsis thaliana* converts Hg^{2+} to highly volatile Hg^0. It was also reported that surrounding temperature and light intensity influences the mercury releasing ability of the leaves (Awa and Hadibarata 2020).

11.5.4 Phytofiltration

Phytofiltration is the use of aquatic plants (floating, submerged or emergent) for the removal of pollutants from streams and waste waters. The removal is mostly done by the root system of the plants (rhizofiltration) or the seedlings (blastofiltration), by the excised plant shoots (caulofiltration) (Olguín and Sánchez-Galván 2012). Among all the phytoremediation techniques, phytofiltration is specified for water treatment. The phytofiltration technique requires the plants to have a dense root system. The various factors which influence the effectiveness of phytofiltration are the physiochemical characteristics of the plants and associated microorganisms. Aquatic plant species can accumulate more than 1450-fold of heavy metal concentration in water (Rezania et al. 2016). Industrial wastewaters are metal-contaminated that pose serious threat to the environment and human health. These toxic metals can be removed by the use of hyperaccumulating plants which absorb/adsorb pollutants from the water by their roots. The biological substances of these metal tolerant plants form complexes with these metal ions at the solution-plant interface (Anawar et al. 2008). The various aquatic plants having high potential for metal removal are water hyacinths such as *Eichhornia crassipes*, *Eichhornia azurea*, *Eichhornia diversifolia*, *Eichhornia paniculate*, *Salvinia molesta*, *Salvinia herzogoi*, *Salvinia auriculata*, *Salvinia minima;* Water lettuce such as *Pistia stratiotes*, *Lemna minor (duck weed)*, *Lemna gibba*, *Azolla sp.;* submerged species like *Myriophyllum sp.*, *Potamogeton lucens*, *Ceratophyllum demersum* and emergent plants like *Limnocharis flava*, *Spartina sp.*, *Potamogeton sp.*, *Myriophyllum spicatum*, *Typha sp.*, *Scirpus sp.* (Rezania et al. 2016). Sut-Lohmann et al. 2020 reported *Monosoleum tenerum* as a suitable plant for the purpose of phytofiltration of Zn, Cu, Ni, Mn and Fe contaminated water. It is reported that aquatic plant species, such as *Ranunculus trichophyllus*, *Ranunculus peltatus subsp. saniculifolius* and *Azolla caroliniana* are highly efficient for As phytofiltration (Parmar and Singh 2015).

11.5.5 Rhizofiltration

Rhizofiltration is a phytoremediation technique where plant roots are used to absorb, concentrate and precipitate HMs from contaminated sites (Ignatius et al. 2014). Plants take up high levels of essential micro nutrients from the roots. In addition to these various non-essential metals are also absorbed by the roots. The mechanisms of mass flow and diffusion are responsible for the metal movement towards the root surface. The mass flow includes the water uptake by the plant to meet the transpiration. This results in the movement of water and the dissolved metal contaminants towards the root surface. The metal ions move into the root cells by carrier proteins also knows as transporters. The metal ions inside the roots are either immobilized in the root cells or translocated to the shoots (Yadav et al. 2011). Rhizofiltration is an effective, cost-efficient method of water treatment where plants up to 60% of their dry weight as toxic metals (Verma et al. 2016). Yadav et al. (2015) reported high removal efficiency of aquatic macrophyte *Eichhornia crappies* for Cd, Pb and Zn by rhizofiltration. Water lettuce (*Pistia stratiotes* L.) has a good rhizofiltration potential for the removal of Cd and Pb from contaminated solution (Veselý et al. 2011). Thayaparan et al. 2013 reported *Azolla pinnata* as a potential candidate for Pb rhizofiltration. The various plants with phytoremediation potential are listed in Table 11.3.

11.6 Microbial Assisted Phytoremediation of Heavy Metals

Microbes in association with plant roots (through chemotaxis and colonization) maintain rhizosphere ecology and affect the bioavailability of metals and essential nutrients in

TABLE 11.3

Plants Used in The Phytoremediation of Heavy Metals

Plants	Contaminated Area	Heavy Metals Remediated	Reference
Cajanus cajan *Pisum sativum*	Soil	As, Cd	Garg et al. (2015)
Brassica napus	Soil	Cd, Cu, Pb, Zn	Turan and Esringüu (2007)
Cicer arietinum	Soil	Pb, Cr	Dasgupta et al. (2011)
Jatropha curcas	Soil	As, Cr, Zn, Hg	Yadav et al. (2009); Marrugo-Negrete et al. (2015)
Carex riparia	Water	Ni, Zn	Ladislas et al. (2015)
Cyperus haspan	Water	Fe, Mg, Mn, Zn	Akinbile et al. (2012)
Bidens triplinervia	Soil	Pb, Zn	Bech et al. (2012)
Arabidopsis thaliana	Soil	Hg	Battke et al. (2008)
Leersia hexandra	Water	Cr	Zhang et al. (2007)
Medicago sativa	Soil	Cd	Ghnaya et al. (2015)
Lactuca sativa	Soil	As, Pb	Gunduz et al. (2012)
Bromus carinatus	Soil	Cu	Silk et al. (2006)
Verbascum virgatum	Soil	Zn	Moreira et al. (2011)
Zea mays	Soil	Cu, Zn	Tiecher et al. (2016)

the rhizospheric soil. Microbes use the root exudates as food and energy sources as they are rich in carbon. These plant–microbe associations are involved in adaptions to various metal-rich environments and hence can be utilized for the remediation of contaminated sites. Microbes improve the plant biomass, influence metal availability, aid bioaccumulation and hence enhance phytoremediation (Dotaniya et al. 2018). Plant-associated microbes directly improve phytoremediation by altering soil pH, various secretions (siderophores, organic acids, biosurfactants, polymeric substances), methylation and redox reactions, improving metal mobilization and solubilization which enhance metal uptake by the plant. Siderophores solubilize heavy metals through complexation, biosurfactants are amphiphilic molecules that develop metal complexes at the soil interface. Bacteria also change the bioavailability of metals by oxidation-reduction reactions and metal uptake through biosorption (Ullah et al. 2015). PGPR improve crop production by stimulating root growth, enhance availability of soil nutrients, improve soil fertility, provide protection against various diseases and enhance the tolerance of plants towards various stresses such as drought, salinity, heavy metal stress. The various mechanisms used by microbes to benefit the plants are the production of phytohormones, solubilization of minerals, production of ethylene reducing enzyme ACC deaminase, exopolysaccharide production and immobilization of metals (Egamberdieva et al. 2016). The PGPR includes *Serratia, Bacillus, Pseudomonas, Burkholderia, Enterobacter, Erwinia, Klebsiella, Beijerinckia, Flavobacterium and Gluconacetobacter* (Girolkar et al. 2020). *Linum usitatissimum* inoculated with arbuscular fungi showed an increase in Zn uptake. *Glomus intraradices, Glomus geosporum, Glomus mosseae* inoculation in *Pteris vittate* resulted in increased growth and As uptake. PGPR, *Bacillus megaterium* and *Azotobacter chroococcum* enhanced the growth and Pb, Zn removal in *Brassica juncea* (Yang et al. 2020). *Sphigomonads* are rhizospheric bacteria that assist metal phytoextraction by promotion of metal absorption in the root, enhancement of heavy metal transport through the stem and alleviation of metal toxicity in the plant (Waigi et al. 2017). It was reported that *Cellulosimicrobium* sp. promoted growth in alfalfa as well as increased the metal (Cr, Zn and Cu) uptake by the plant (Tirry et al. 2018). *Solanum nigrum* inoculated with *Pseudomonas* sp. Lk9 resulted in increased shoot dry biomass and higher accumulation of Cd, Zn and Cu (Chen et al. 2014). *Brassica juncea* inoculated with *Staphylococcus arlettae* strain NBRIEAG-6 showed an increased accumulation of As in the shoot and root (Srivastava et al. 2013).

11.7 Transgenic Advancement in Bioremediation

The use of nanoparticles, non-living biomass and genetically modified organisms are the recent strategies used in bioremediation of HMs. The use of genetic engineering in bioremediation is to boost the capacity of organisms to tolerate, accumulate and absorb contaminants. This involves the identification of genes from organisms involved in recovery, distribution and detoxification of metals. Transgenic *Arabidopsis thaliana* is seen to have increased bioremediation potential for As and Cd due to the increased expression of genes *AsPCS1* and *YCF1*, which are derived from garlic and baker's yeast (Gaur et al. 2014). Development of transgenic plants by transferring of genes involved in metal uptake, removal, bioaccumulation enhance the bioremediation of the contaminants. The introduction of *CUP1*gene from yeast in cauliflower resulted in 16-fold higher accumulation of Cd in the plant. Symbiotic relationship between recombinant rhizobia and *Astragalus sinicus* resulted in increased accumulation of Cd in the nodules (Azad et al. 2014). Transgenic *Brassica juncea* showed an increased accumulation of Se and Cd than the wild type. Over expression of *AtMRP7* in *Nicotiana tabacum* lead to an increased Cd tolerance and accumulation in tobacco plant. Geno-remediation (Gene-targeted bioremediation) revolves around the metal transporter genes and genes facilitating the chelate production which might enhance phytoremediation by the production of desirable hyperaccumulators (Mani and Kumar 2014). Enhancing metal/metalloid ligand production is one of the main biotechnological approaches for the engineering of plants for enhanced phytoremediation. Cysteine-rich peptides such as metallothioneins (MTs), PCs, glutathione (GSH)are used as metal-binding ligands for HM accumulation or detoxification. MT gene from pea (*Pisum sativum*), *PsMTA* when over expressed in *Arabidopsis thaliana* resulted in increased Cu^{2+} accumulation in the roots. *TaPCS1* gene from wheat encoding PC synthase, when over expressed in shrub tobacco increased the plant's tolerance and accumulation of Pb and Cd significantly (Mosa et al. 2016). Some HMs such as Hg cannot be removed by indigenous bacteria. Hence, DNA recombinant technology has a role to play. Bioremediation of heavy metals Cd, Hg, Ni, Cu, As and Fe has been successfully done by genetically engineered bacteria (Azad et al. 2014). Polyphosphate kinase and metallothionine expression in transgenic bacteria provided the bacteria with high mercury resistance and accumulation. This could enhance mercury bioremediation and also retrieve the accumulated mercury for industrial purposes (Ruiz et al. 2011). *Saccharomyces cerevisiae* transformed with metallothionein coding *Populus trichocarpa* gene (*PtMT2b*) can accumulate higher intracellular Cd and has a great potential for the bioremediation of Cd from contaminated aqueous sites (De Oliveira et al. 2020). Geva et al. (2016) reported the enhanced copper bioremediation by new strain of *Saccharomyces cerevisiae*, produced by the construction and integration of recombinant human *MT2* and *GFP-hMT2* genes into yeast cells. Transforming a *mer* operon harboring plasmid, of *Bacillus thuringiensis* into Hg resistant bacterium *Bacillus cereus* resulted in the production of transgenic bacterium *Bacillus cereus* BW-03($_p$PW-05). The transgenic bacterium was able to remove 99% of Hg supplement by volatilization and biosorption processes. Also, it resisted variations of salinity and pH (Dash and Das 2015). Genetically engineered *Ralston metallidurans* can be used for the bioremediation of Cr from industrial wastewater. Genetically modified *E. coli* SE5000 strain can accumulate Ni from an aqueous solution. *ArsR* gene expressing *E.coli* showed an enhanced bioaccumulation of As (Azad et al. 2014).

Conclusion and Future Prospects

This chapter provides a detailed account of HM pollution in the environment, pollution-related health hazards and the bioremediation of these contaminants using plants and microbes. The bioremediation of HMs is a safe and inexpensive alternative to less efficient physiochemical methods. Environmental clean-up by bioremediation uses the intrinsic biological mechanisms of microorganisms and plants, which enable them to survive in metal stress. In microorganisms, the various metal adaptive mechanisms include biosorption, bioaccumulation, precipitation, biotransformation. Phytoremediation involves the use of hyperaccumulators and the plants which in association with microbes immobilize the metal ions and transform them into less toxic forms. However, phytoremediation is not yet extensively used on a large scale and lacks commercialization. On the other hand, bioremediation is largely dependent on environmental conditions and various environmental factors determine the success rate of bioremediation. In order to enhance the bioremediation process genetic engineering is being put into use. Genetically engineered organisms have an enhanced capacity of biotransforming the toxic metals. Transgenic strategies should be used within safety limits and proper regulatory procedures. Transformation technologies such as genome editing can be used in the biotechnological improvement of bioremediation.

REFERENCES

Abdel-Monem, M. O., Al-Zubeiry, A. H. S. and Al-Gheethi, A. A. S. 2010. Biosorption of nickel by *Pseudomonas cepacia* 120S and *Bacillus subtilis* 117S. *Water Sci Technol*, 61, 2994–3007.

Achal, V., Pan, X., Fu, Q. and Zhang, D. 2012. Biomineralization based remediation of As (III) contaminated soil by *Sporosarcina ginsengisoli*. *J Hazard Mater*, 201, 178–184.

Ahemad, M. 2012. Implications of bacterial resistance against heavy metals in bioremediation: A review. *IIOABJ*, 3, 39–46.

Ahmadpour, P., Ahmadpour, F., Mahmud, T. M. M., Abdu, A., Soleimani, M. and Tayefeh, F. H. 2012. Phytoremediation of heavy metals: A green technology. *Afr J Biotechnol*, 11, 14036–14043.

Akhter, K., Ghous, T., Andleeb, S., Ejaz, S., Khan, B. A. and Ahmed, M. N. 2017. Bioaccumulation of heavy metals by metal-resistant bacteria isolated from Tagetes minuta rhizosphere, growing in soil adjoining automobile workshops. *Pak J Zool*, 49, 1841–1846.

Akinbile, C. O., Yusoff, M. S. and Zuki, A. A. 2012. Landfill leachate treatment using sub-surface flow constructed wetland by *Cyperus haspan*. *Waste Manage*, 32, 1387–1393.

Akinci, G. and Guven, D. E. 2011. Bioleaching of heavy metals contaminated sediment by pure and mixed cultures of *Acidithiobacillus* spp. *Desalination*, 268, 221–226.

Ali, H., Khan, E. and Sajad, M. A. 2013. Phytoremediation of heavy metals – concepts and applications. *Chemosphere*, 91, 869–881.

Alimba, C. G. and Faggio, C. 2019. Microplastics in the marine environment: Current trends in environmental pollution and mechanisms of toxicological profile. *Environ Toxicol Pharmacol*, 68, 61–74.

Alkorta, I., Becerril, J. M. and Garbisu, C. 2010. Phytostabilization of metal contaminated soils. *Rev Environ Health*, 25, 135–146.

Anawar, H. M., Garcia-Sanchez, A., Kul Alam, M. T. and Rahman, M. M. 2008. Phytofiltration of water polluted with arsenic and heavy metals. *Int J Environ Pollut*, 33, 292–312.

Arunakumara, K. K. I. U., Zhang, X. and Song, X. 2008. Bioaccumulation of Pb 2+ and its effects on growth, morphology and pigment contents of *Spirulina* (*Arthrospira*) *platensis*. *J Ocean Univ China*, 7, 397–403.

Awa, S. H. and Hadibarata, T. 2020. Removal of heavy metals in contaminated soil by phytoremediation mechanism: A review. *Water, Air, & Soil Pollut*, 231,1–15.

Ayangbenro, A. S. and Babalola, O. O. 2017. A new strategy for heavy metal polluted environments: A review of microbial biosorbents. *Int J Environ Res Pub Health*, 14, 94.

Ayyappan, D., Sathiyaraj, G. and Ravindran, K. C. 2016. Phytoextraction of heavy metals by *Sesuvium portulacastrum* L. a salt marsh halophyte from tannery effluent. *Int J Phytoremediation*, 18, 453–459.

Azad, M. A. K., Amin, L. and Sidik, N. M. 2014. Genetically engineered organisms for bioremediation of pollutants in contaminated sites. *Chinese Sci Bull*, 59, 703–714.

Azubuike, C. C., Chikere, C. B. and Okpokwasili, G. C. 2016. Bioremediation techniques–classification based on site of application: Principles, advantages, limitations and prospects. *World J Microbiol Biotechnol*, 32, 180.

Banerjee, G., Pandey, S., Ray, A. K. and Kumar, R. 2015. Bioremediation of heavy metals by a novel bacterial strain *Enterobacter cloaca* and its antioxidant enzyme activity, flocculant production, and protein expression in presence of lead, cadmium, and nickel. *Water, Air, & Soil Pollut*, 226, 1–9.

Basha, S. A. and Rajaganesh, K. 2014. Microbial bioremediation of heavy metals from textile industry dye effluents using isolated bacterial strains. *Int J Curr Microbiol Appl Sci*, 3, 785–794.

Battke, F., Ernst, D., Fleischmann, F. and Halbach, S. 2008. Phytoreduction and volatilization of mercury by ascorbate in *Arabidopsis thaliana*, European beech and Norway spruce. *Appl Geochem*, 23, 494–502.

Bech, J., Duran, P., Roca, N., Poma, W., Sánchez, I., Roca-Pérez, L., Boluda, R., Barcelo, J. and Poschenrieder, C. 2012. Accumulation of Pb and Zn in *Bidens triplinervia* and *Senecio* sp. spontaneous species from mine spoils in Peru and their potential use in phytoremediation. *J Geochem Explor*, 123, 109–113.

Brooks, R. R., Lee, J. and Reeves, R. D. 1977. Detection of nickeliferous rocks by analysis of herbarium species of indicator plants. *J Geochem Explor*, 7, 49–77.

Chang, J. S., Law, R. and Chang, C. C. 1997. Biosorption of lead, copper and cadmium by biomass of *Pseudomonas aeruginosa* PU21. *Water Res*, 31, 1651–1658.

Chaturvedi, A. D., Pal, D., Penta, S. and Kumar, A. 2015. Ecotoxic heavy metals transformation by bacteria and fungi in aquatic ecosystem. *World J Microbiol Biotechnol*, 31, 1595–1603.

Chen, L., Luo, S., Li, X., Wan, Y., Chen, J. and Liu, C. 2014. Interaction of Cd-hyperaccumulator *Solanum nigrum* L. and functional endophyte *Pseudomonas* sp. Lk9 on soil heavy metals uptake. *Soil Biol Biochem*, 68, 300–308.

Cheraghi, M., Lorestani, B., Khorasani, N., Yousefi, N. and Karami, M. 2011. Findings on the phytoextraction and phytostabilization of soils contaminated with heavy metals. *Biol Trace Elem Res*, 144, 1133–1141.

Chibuike, G. U. and Obiora, S. C. 2014. Heavy metal polluted soils: Effect on plants and bioremediation methods. *Appl Environ Soil Sci*, 2014, 1–12.

Choudhary, M., Kumar, R., Datta, A., Nehra, V. and Garg, N. 2017. *Bioremediation of heavy metals by microbes. Bioremediation of salt affected soils: An Indian perspective*. Springer.

Dary, M., Chamber-Pérez, M. A., Palomares, A. J. and Pajuelo, E. 2010. "In situ" phytostabilisation of heavy metal polluted soils using Lupinus luteus inoculated with metal resistant plant-growth promoting rhizobacteria. *J Hazard Mater*, 177, 323–330.

Das, S. and Dash, H. R. 2014. Microbial bioremediation: A potential tool for restoration of contaminated areas. *Microbial biodegradation and bioremediation*. Elsevier.

Dasgupta, S., Satvat, P. S. and Mahindrakar, A. B. 2011. Ability of *Cicer arietinum* (L.) for bioremoval of lead and chromium from soil. *Int J Res Eng Technol Syst*, 24, 338–341.

Dash, H. R. and Das, S. 2015. Bioremediation of inorganic mercury through volatilization and biosorption by transgenic *Bacillus cereus* BW-03 (pPW-05). *Int Biodeterior and Biodegradation*, 103, 179–185.

De Oliveira, V. H., Ullah, I., Dunwell, J. M. and Tibbett, M. 2020. Bioremediation potential of Cd by transgenic yeast expressing a metallothionein gene from *Populus trichocarpa*. *Ecotox Environ Safe*, 202, 110917.

De Vargas, I., Macaskie, L. E. and Guibal, E. 2004. Biosorption of palladium and platinum by sulfate-reducing bacteria. *J Chem Technol Biotechnol*, 79, 49–56.

Dhami, N. K., Quirin, M. E. C. and Mukherjee, A. 2017. Carbonate biomineralization and heavy metal remediation by calcifying fungi isolated from karstic caves. *Ecol Eng*, 103, 106–117.

Dixit, R., Malaviya, D., Pandiyan, K., Singh, U. B., Sahu, A., Shukla, R., Singh, P. B., Rai, J. P., Sharma, P. K., Lade, H. and Paul, D. 2015. Bioremediation of heavy metals from soil and aquatic environment: An overview of principles and criteria of fundamental processes. *Sustainability*, 7, 2189–2212.

Dotaniya, M. L., Rajendiran, S., Dotaniya, C. K., Solanki, P., Meena, V. D. Saha, J. K. and Patra, A. K. 2018. Microbial assisted phytoremediation for heavy metal contaminated soils. *Phytobiont and Ecosystem Restitution*. Springer.

Dua, M., Sethunathan, N. and Johri, A.K. 2002. Biotechnology bioremediation success and limitations. *Appl Microbiol Biotechnol*, 59, 143–152.

Dursun, A. Y., Uslu, G., Cuci, Y. and Aksu, Z. 2003. Bioaccumulation of copper (II), lead (II) and chromium (VI) by growing *Aspergillus niger*. *Process Biochem*, 38, 1647–1651.

Dzionek, A., Wojcieszyńska, D. and Guzik, U. 2016. Natural carriers in bioremediation: A review. *Electron J Biotechnol*, 19, 28–36.

Edwards, S. J. and Kjellerup, B. V. 2013. Applications of biofilms in bioremediation and biotransformation of persistent organic pollutants, pharmaceuticals/personal care products, and heavy metals. *Appl Microbiol Biotechnol*, 97, 9909–9921.

Egamberdieva, D., Abd-Allah, E. F. and da Silva, J. A. T. 2016. Microbially assisted phytoremediation of heavy metal–contaminated soils. *Plant metal interaction*. Elsevier.

Ertuğrul, S., San, N. O. and Dönmez, G. 2009. Treatment of dye (Remazol Blue) and heavy metals using yeast cells with the purpose of managing polluted textile wastewaters. *Ecol Eng*, 35, 128–134.

Gadd, G. M. and Pan, X. 2016. Biomineralization, bioremediation and biorecovery of toxic metals and radionuclides. *Geomicrobiol J*, 33, 175–178.

Galal, T. M., Gharib, F. A., Ghazi, S. M. and Mansour, K. H. 2017. Phytostabilization of heavy metals by the emergent macrophyte *Vossia cuspidata* (Roxb.) Griff.: A phytoremediation approach. *Int J Phytoremediation*, 19, 992–999.

Garg, N., Singla, P. and Bhandari, P. 2015. Metal uptake, oxidative metabolism, and mycorrhization in pigeon pea and pea under arsenic and cadmium stress. *Turk J Agric For*, 39, 234–250.

Gaur, N., Flora, G., Yadav, M. and Tiwari, A. 2014. A review with recent advancements on bioremediation-based abolition of heavy metals. *Environ Sci Process Impacts*, 16, 180–193.
Geva, P., Kahta, R., Nakonechny, F., Aronov, S. and Nisnevitch, M. 2016. Increased copper bioremediation ability of new transgenic and adapted *Saccharomyces cerevisiae* strains. *Environ Sci Pollut Res*, 23, 19613–19625.
Ghnaya, T., Mnassri, M., Ghabriche, R., Wali, M., Poschenrieder, C., Lutts, S. and Abdelly, C. 2015. Nodulation by *Sinorhizobium meliloti* originated from a mining soil alleviates Cd toxicity and increases Cd-phytoextraction in *Medicago sativa* L. *Front Plant Sci*, 6, 863.
Ghosh, M. and Singh, S. P. 2005. A review on phytoremediation of heavy metals and utilization of its by-products. *Asian J Energy Environ*, 6, 214–231.
Girma, G. 2015. Microbial bioremediation of some heavy metals in soils: An updated review. *Egypt Acad J Biol Sci*, 7, 29–45.
Girolkar, S., Thawale, P. and Juwarkar, A. 2020. Bacteria-assisted phytoremediation of heavy metals and organic pollutants: Challenges and future prospects. *Bioremediation for environmental sustainability*. Elsevier.
Godlewska Żyłkiewicz, B., Sawicka, S. and Karpińska, J. 2019. Removal of platinum and palladium from wastewater by means of biosorption on fungi *Aspergillus* sp. and yeast *Saccharomyces* sp. *Water*, 11, 1522.
Gola, D., Malik, A., Namburath, M. and Ahammad, S. Z. 2018. Removal of industrial dyes and heavy metals by *Beauveria bassiana*: FTIR, SEM, TEM and AFM investigations with Pb (II). *Environ Sci Pollut Res*, 25, 20486–20496.
Green-Ruiz, C. 2006. Mercury (II) removal from aqueous solutions by nonviable *Bacillus* sp. from a tropical estuary. *Bioresour Technol*, 97, 1907–1911.
Green-Ruiz, C., Rodriguez-Tirado, V. and Gomez-Gil, B. 2008. Cadmium and zinc removal from aqueous solutions by *Bacillus jeotgali*: pH, salinity and temperature effects. *Bioresour Technol*, 99, 3864–3870.
Guarnieri, M. and Balmes, J. R. 2014. Outdoor air pollution and asthma. *Lancet*, 383, 1581–1592.
Gunduz, S., Uygur, F. N. and Kahramanoğlu, I. 2012. Heavy metal phytoremediation potentials of *Lepidum sativum* L., *Lactuca sativa* L., *Spinacia oleracea* L. and *Raphanus sativus* L. *Her J Agric Food Sci Res*, 1, 1–5.
Hakeem, K. R., Alharby, H. F. and Rehman, R. 2019. Antioxidative defense mechanism against lead-induced phytotoxicity in *Fagopyrum kashmirianum*. *Chemosphere*, 216, 595–604.
Hansda, A. and Kumar, V. 2016. A comparative review towards potential of microbial cells for heavy metal removal with emphasis on biosorption and bioaccumulation. *World J Microbiol Biotechnol*, 32, 170.
Ignatius, A., Arunbabu, V., Neethu, J. and Ramasamy, E. V. 2014. Rhizofiltration of lead using an aromatic medicinal plant *Plectranthus amboinicus* cultured in a hydroponic nutrient film technique (NFT) system. *Environ Sci Pollut Res*, 21, 13007–13016.
Incharoensakdi, A. and Kitjaharn, P. 2002. Zinc biosorption from aqueous solution by a halotolerant cyanobacterium *Aphanothece halophytica*. *Curr Microbiol*, 45, 261–264.
Jaafar, R., Al-Sulami, A., Al-Taee, A., Aldoghachi, F. and Napes, S. 2015. Biosorption and bioaccumulation of some heavy metals by *Deinococcus radiodurans* isolated from soil in Basra governorate – Iraq. *J Biotechnol Biomater*, 5, 190.
Jabeen, R., Ahmad, A. and Iqbal, M. 2009. Phytoremediation of heavy metals: Physiological and molecular mechanisms. *Bot Rev*, 75, 339–364.
Jadia, C. D. and Fulekar, M. H. 2009. Phytoremediation of heavy metals: Recent techniques. *Afr J Biotechnol*, 8, 921–928.
Jaiswal, A., Verma, A. and Jaiswal, P. 2018. Detrimental effects of heavy metals in soil, plants, and aquatic ecosystems and in humans. *J Environ Pathol Toxicol Oncol*, 37, 183–197.
Jalilvand, N., Akhgar, A., Alikhani, H. A., Rahmani, H. A. and Rejali, F. 2020. Removal of heavy metals zinc, lead, and cadmium by biomineralization of urease-producing bacteria isolated from Iranian mine calcareous soils. *J Soil Sci Plant Nutr*, 20, 206–219.
Javanbakht, V., Alavi, S. A. and Zilouei, H. 2014. Mechanisms of heavy metal removal using microorganisms as biosorbent. *Water Sci Technol*, 69, 1775–1787.
Jin, X. F., Liu, D., Islam, E., Mahmood, Q., Yang, X. E., He, Z. L. and Stoffella, P. J. 2009. Effects of zinc on root morphology and antioxidant adaptations of cadmium-treated *Sedum alfredii* H. *J Plant Nutr*, 32, 1642–1656.
Juwarkar, A. A., Singh, S. K. and Mudhoo, A. 2010. A comprehensive overview of elements in bioremediation. *Rev Environ Sci Biotechnol*, 9, 215–288.
Kalaivanan, D. and Ganeshamurthy, A. N. 2016. Mechanisms of heavy metal toxicity in plants. *Abiotic stress physiology of horticultural crops*. Springer.
Karami, A. and Shamsuddin, Z. H. 2010. Phytoremediation of heavy metals with several efficiency enhancer methods. *Afr J Biotechnol*, 9, 3689–3698.
Kearney, M. 2006. Habitat, environment and niche: What are we modelling? *Oikos*, 115, 186–191.
Kensa, V. M. 2011. Bioremediation – an overview. *J Ind Pollut Cont*, 27, 161–168.
Khadim, H. J., Ammar, S. H. and Ebrahim, S. E. 2019. Biomineralization based remediation of cadmium and nickel contaminated wastewater by ureolytic bacteria isolated from barn horse's soil. *Environ Technol Innov*, 14, 100315.
Khan, M. A. and Ghouri, A. M. 2011. Environmental pollution: Its effects on life and its remedies. *Res World J Arts Sci Commer*, 2, 276–285.
Khan, M. S., Zaidi, A., Wani, P. A. and Oves, M. 2009. Role of plant growth promoting rhizobacteria in the remediation of metal contaminated soils. *Environ Chem Lett*, 7, 1–19.
Kucharski, R., Sas-Nowosielska, A., Małkowski, E., Japenga, J., Kuperberg, J. M., Pogrzeba, M. and Krzyżak, J. 2005. The use of indigenous plant species and calcium phosphate for the stabilization of highly metal-polluted sites in southern Poland. *Plant Soil*, 273, 291–305.
Kumar, L. and Bharadvaja, N. 2020. Microbial remediation of heavy metals. *Microbial bioremediation & biodegradation*. Springer.
Kumar, P. S. and Gunasundari, E. 2018. Bioremediation of heavy metals. *Bioremediation: Applications for environmental protection and management*. Springer.
Kumar, R., Sharma, A. K., Singh, P., Dhir, B. and Mehta, D. 2014. Potential of some fungal and bacterial species in bioremediation of heavy metals. *J Nucl Phys*, 1, 213–223.
Ladislas, S., Gerente, C., Chazarenc, F., Brisson, J. and Andres, Y. 2015. Floating treatment wetlands for heavy metal removal in highway stormwater ponds. *Ecol Eng*, 80, 85–91.

Li, F., Yang, F., Chen, Y., Jin, H., Leng, Y. and Wang, J. 2020. Chemical reagent-assisted phytoextraction of heavy metals by *Bryophyllum laetivirens* from garden soil made of sludge. *Chemosphere*, 253, 126574.

Liang, W. and Yang, M. 2019. Urbanization, economic growth and environmental pollution: Evidence from China. *Sustain Comput Infor*, 21, 1–9.

Limcharoensuk, T., Sooksawat, N., Sumarnrote, A., Awutpet, T., Kruatrachue, M., Pokethitiyook, P. and Auesukaree, C. 2015. Bioaccumulation and biosorption of Cd2+ and Zn2+ by bacteria isolated from a zinc mine in Thailand. *Ecotox Environ Safe*, 122, 322–330.

Loukidou, M. X., Karapantsios, T. D., Zouboulis, A. I. and Matis, K. A. 2004. Diffusion kinetic study of cadmium (II) biosorption by *Aeromonas caviae*. *J Chem Technol Biotechnol*, 79, 711–719.

Lu, W. B., Shi, J. J., Wang, C. H. and Chang, J. S. 2006. Biosorption of lead, copper and cadmium by an indigenous isolate *Enterobacter* sp. J1 possessing high heavy-metal resistance. *J Hazard Mater*, 134, 80–86.

Lu, Y., Song, S., Wang, R., Liu, Z., Meng, J., Sweetman, A. J., Jenkins, A., Ferrier, R. C., Luo, W. and Wang, T. 2015. Impacts of soil and water pollution on food safety and health risks in China. *Environ Int*, 77, 5–15.

Mahar, A., Wang, P., Ali, A., Awasthi, M. K., Lahori, A. H., Wang, Q., Li, R. and Zhang, Z. 2016. Challenges and opportunities in the phytoremediation of heavy metals contaminated soils: A review. *Ecotox Environ Safe*, 126, 111–121.

Mahurpawar, M. 2015. Effects of heavy metals on human health. *Int J Res Granthaalayah*, 1, 2394–3629.

Malik, B., Pirzadah, T. B., Tahir, I., Dar, T. H. and Rehman, R. 2015. Recent trends and approaches in phytoremediation. *Soil remediation and plants: Prospects and challenges*. Elsevier Publication.

Malik, B., Pirzadah, T. B., Tahir, I., Hakeem, K. R., Rather, I. A., Sabir, J. S. and Rehman, R. U. 2021. Lead and aluminium-induced oxidative stress and alteration in the activities of antioxidant enzymes in chicory plants. *Sci Hortic*, 278, 109847.

Malik, B., Pirzadah, T. B., Tahir, I. and Rehman, R. U. 2019. Growth and physiological responses in chicory towards mercury induced in vitro oxidative stress. *Plant Physiol Rep*, 24, 236–248.

Mameri, N., Boudries, N., Addour, L., Belhocine, D., Lounici, H., Grib, H. and Pauss, A. 1999. Batch zinc biosorption by a bacterial non-living *Streptomyces rimosus* biomass. *Water Res*, 33, 1347–1354.

Mani, D. and Kumar, C. 2014. Biotechnological advances in bioremediation of heavy metals contaminated ecosystems: An overview with special reference to phytoremediation. *Int J Environ Sci Technol*, 11, 843–872.

Marques, A. P., Rangel, A. O. and Castro, P. M. 2009. Remediation of heavy metal contaminated soils: Phytoremediation as a potentially promising clean-up technology. *Crit Rev Environ Sci Technol*, 39, 622–654.

Marrugo-Negrete, J., Durango-Hernández, J., Pinedo-Hernández, J., Olivero-Verbel, J. and Díez, S. 2015. Phytoremediation of mercury-contaminated soils by *Jatropha curcas*. *Chemosphere*, 127, 58–63.

Masoudi, F., Shirvani, M., Shariatmadari, H. and Sabzalian, M. R. 2020. Performance of new biodegradable chelants in enhancing phytoextraction of heavy metals from a contaminated calcareous soil. *J Environ Health Sci Eng*, 18, 655–664.

Matsunaga, T., Takeyama, H., Nakao, T. and Yamazawa, A. 1999. Screening of marine microalgae for bioremediation of cadmium-polluted seawater. *J Biotechnol*, 70, 33–38.

McIntyre, T. 2003. Phytoremediation of heavy metals from soils. *Phytoremediation*. Springer.

Mehndiratta, P., Jain. A., Srivastava, S. and Gupta, N. 2013. Environmental pollution and nanotechnology. *Environ Pollut*, 2, 49–58.

Mishra, D., Kim, D. J., Ahn, J. G. and Rhee, Y. H. 2005. Bioleaching: A microbial process of metal recovery; a review. *Met Mater Int*, 11, 249–256.

Mishra, R. K., Mohammad, N. and Roychoudhury, N. 2015. Soil pollution: Causes, effects and control. *Trop Forest Res Inst*, 3, 20–30.

Mohapatra, R. K., Behera, S. S., Patra, J. K., Thatoi, H. and Parhi, P. K. 2020. Potential application of bacterial biofilm for bioremediation of toxic heavy metals and dye-contaminated environments. *New and future developments in microbial biotechnology and bioengineering: Microbial biofilms*. Elsevier.

Moreira, H., Marques, A. P., Rangel, A. O. & Castro, P. M. 2011. Heavy metal accumulation in plant species indigenous to a contaminated Portuguese site: Prospects for phytoremediation. *Water Air Soil Pollut*, 221, 377–389.

Mosa, K. A., Saadoun, I., Kumar, K., Helmy, M. and Dhankher, O. P. 2016. Potential biotechnological strategies for the cleanup of heavy metals and metalloids. *Front Plant Sci*, 7, 303.

Muthusaravanan, S., Sivarajasekar, N., Vivek, J. S., Paramasivan, T., Naushad, M., Prakashmaran, J., Gayathri, V. and Al-Duaij, O. K. 2018. Phytoremediation of heavy metals: Mechanisms, methods and enhancements. *Environ Chem Lett*, 16, 1339–1359.

Nagajyoti, P. C., Lee, K. D. and Sreekanth, T. V. M. 2010. Heavy metals, occurrence and toxicity for plants: A review. *Environ Chem Lett*, 8, 199–216.

Naresh Kumar, R. and Nagendran, R. 2008. Changes in nutrient profile of soil subjected to bioleaching for removal of heavy metals using *Acidithiobacillus thiooxidans*. *J Hazard Mater*, 156, 102–107.

Niti, C., Sunita, S., Kamlesh, K. and Rakesh, K. 2013. Bioremediation: An emerging technology for remediation of pesticides. *Res J Chem Environ*, 17, 88–105.

Ojuederie, O. B. and Babalola, O. O. 2017. Microbial and plant-assisted bioremediation of heavy metal polluted environments: A review. *Int J Environ Res Public Health*, 14, 1504.

Olguín, E. J. and Sánchez-Galván, G. 2012. Heavy metal removal in phytofiltration and phycoremediation: The need to differentiate between bioadsorption and bioaccumulation. *N Biotechnol*, 30, 3–8.

Owa, F. D. 2013. Water pollution: Sources, effects, control and management. *Med J Soc Sci*, 4, 65–68.

Ozdemir, G. Ü. V. E. N. and Baysal, S. H. 2004. Chromium and aluminum biosorption on *Chryseomonas luteola* TEM05. *Appl Microbiol Biotechnol*, 64, 599–603.

Öztürk, A. 2007. Removal of nickel from aqueous solution by the bacterium *Bacillus thuringiensis*. *J Hazard Mater*, 147, 518–523.

Padmavathiamma, P. K. and Li, L. Y. 2007. Phytoremediation technology: Hyper-accumulation metals in plants. *Water Air Soil Pollut*, 184, 105–126.

Pani, T., Das, A. and Osborne, J. W. 2017. Bioremoval of zinc and manganese by bacterial biofilm: A bioreactor-based approach. *J Photoch Photobio B*, 175, 211–218.

Parmar, S. and Singh, V. 2015. Phytoremediation approaches for heavy metal pollution: A review. *J Plant Sci Res*, 2, 135.

Pathak, A., Dastidar, M. G. and Sreekrishnan, T. R. 2009. Bioleaching of heavy metals from sewage sludge: A review. *J Environ Manage*, 90, 2343–2353.

Pirzadah, T. B., Malik, B., Tahir, I., Hakeem, K. R., Alharby, H. F. and Rehman, R. U. 2020. Lead toxicity alters the antioxidant defense machinery and modulate the biomarkers in Tartary buckwheat plants. *Int Biodeterior Biodegradation*, 151, 104992.

Pirzadah, T. B., Malik, B., Tahir, I., Irfan, Q. M. and Rehman, R. U. 2018. Characterization of mercury-induced stress biomarkers in *Fagopyrum tataricum* plants. *Int J Phytoremediation*, 20, 225–236.

Pirzadah, T. B., Malik, B., Tahir, I., Kumar, M., Varma, A. and Rehman, R. U. 2015. Phytoremediation: An eco-friendly green technology for pollution prevention, control and remediation. *Soil remediation and plants: Prospects and challenges*. Elsevier.

Pirzadah, T. B., Malik, B., Tahir, I., Rehman, R. U., Hakeem, K. R. and Alharby, H. F. 2019. Aluminium stress modulates the osmolytes and enzyme defense system in Fagopyrum species. *Plant Physiol Biochem*, 144, 178–186.

Pratush, A., Kumar, A. and Hu, Z. 2018. Adverse effect of heavy metals (As, Pb, Hg, and Cr) on health and their bioremediation strategies: A review. *Int Microbiol*, 21, 97–106.

Puyen, Z. M., Villagrasa, E., Maldonado, J., Diestra, E., Esteve, I. and Solé, A. 2012. Biosorption of lead and copper by heavy-metal tolerant *Micrococcus luteus* DE2008. *Bioresour Technol*, 126, 233–237.

Qayyum, S., Meng, K., Pervez, S., Nawaz, F. and Peng, C. 2019. Optimization of pH, temperature and carbon source for bioleaching of heavy metals by *Aspergillus flavus* isolated from contaminated soil. *Main Group Met Chem*, 42, 1–7.

Qian, X., Fang, C., Huang, M. and Achal, V. 2017. Characterization of fungal-mediated carbonate precipitation in the biomineralization of chromate and lead from an aqueous solution and soil. *J Clean Prod*, 164, 198–208.

Quiton, K. G., Doma Jr, B., Futalan, C. M. and Wan, M. W. 2018. Removal of chromium(VI) and zinc(II) from aqueous solution using kaolin-supported bacterial biofilms of Gram-negative E. coli and Gram-positive *Staphylococcus epidermidis*. *Sustain Environ Res*, 28, 206–213.

Rajendran, P., Muthukrishnan, J. and Gunasekaran, P. 2003. Microbes in heavy metal remediation. *Indian J Exp Biol*, 41, 935–944.

Rajesh, V., Kumar, A. S. K. and Rajesh, N. 2014. Biosorption of cadmium using a novel bacterium isolated from an electronic industry effluent. *Chem Eng J*, 235, 176–185.

Ray, S. A. and Ray, M. K. 2009. Bioremediation of heavy metal toxicity-with special reference to chromium. *Al Ameen J Med Sci*, 2, 57–63.

Rezaei, H. 2016. Biosorption of chromium by using *Spirulina* sp. *Arab J Chem*, 9, 846–853.

Rezania, S., Taib, S. M., Din, M. F. M., Dahalan, F. A. and Kamyab, H. 2016. Comprehensive review on phytotechnology: Heavy metals removal by diverse aquatic plants species from wastewater. *J Hazard Mater*, 318, 587–599.

Robles-González, I. V., Fava, F. and Poggi-Varaldo, H. M. 2008. A review on slurry bioreactors for bioremediation of soils and sediments. *Microb Cell Fact*, 7, 5.

Romantschuk, M., Sarand, I., Petänen, T., Peltola, R., Jonsson-Vihanne, M., Koivula, T., Yrjala, K. and Haahtela, K. 2000. Means to improve the effect of in situ bioremediation of contaminated soil: An overview of novel approaches. *Environ Pollut*, 107, 179–185.

Roy, S. and Roy, M. 2015. Bioleaching of heavy metals by sulfur oxidizing bacteria: A review. *Int Res J Environment Sci*, 4, 75–79.

Ruiz, O. N., Alvarez, D., Gonzalez-Ruiz, G. and Torres, C. 2011. Characterization of mercury bioremediation by transgenic bacteria expressing metallothionein and polyphosphate kinase. *BMC Biotechnol*, 11, 1–8.

Saha, J. K., Selladurai, R., Coumar, M. V., Dotaniya, M. L., Kundu, S. and Patra, A. K. 2017. Agriculture, soil and environment. *Soil pollution – An emerging threat to agriculture*. Springer.

Şahin, Y. and Öztürk, A. 2005. Biosorption of chromium (VI) ions from aqueous solution by the bacterium *Bacillus thuringiensis*. *Process Biochem*, 40, 1895–1901.

Sedlakova-Kadukova, J., Kopcakova, A., Gresakova, L., Godany, A. and Pristas, P. 2019. Bioaccumulation and biosorption of zinc by a novel *Streptomyces* K11 strain isolated from highly alkaline aluminium brown mud disposal site. *Ecotox Environ Safe*, 167, 204–211.

Shackira, A. M. and Puthur, J. T. 2019. Phytostabilization of heavy metals: Understanding of principles and practices. *Plant-metal interactions*. Springer.

Shamshad, I., Khan, S., Waqas, M., Ahmad, N. and Khan, K. 2015. Removal and bioaccumulation of heavy metals from aqueous solutions using freshwater algae. *Water Sci Technol*, 71, 38–44.

Sharma, S. and Adholeya, A. 2011. Detoxification and accumulation of chromium from tannery effluent and spent chrome effluent by *Paecilomyces lilacinus* fungi. *Int Biodeterior Biodegradation*, 65, 309–317.

Silk, W. K., Bambic, D. G., O'Dell, R. E. and Green, P. G. 2006. Seasonal and spatial patterns of metals at a restored copper mine site II. Copper in riparian soils and *Bromus carinatus* shoots. *Environ Pollut*, 144, 783–789.

Singh, R., Paul, D. and Jain, R. K. 2006. Biofilms: Implications in bioremediation. *Trends Microbiol*, 14, 389–397.

Sinha, A. and Khare, S. K. 2011. Mercury bioaccumulation and simultaneous nanoparticle synthesis by *Enterobacter* sp. cells. *Bioresour Technol*, 102, 4281–4284.

Srivastava, S., Verma, P. C., Chaudhry, V., Singh, N., Abhilash, P. C., Kumar, K. V., Sharma, N. and Singh, N. 2013. Influence of inoculation of arsenic-resistant *Staphylococcus arlettae* on growth and arsenic uptake in *Brassica juncea* (L.) Czern. Var. R-46. *J Hazard Mater*, 262, 1039–1047.

Sundar, K., Sadiq, I. M., Mukherjee, A. and Chandrasekaran, N. 2011. Bioremoval of trivalent chromium using *Bacillus* biofilms through continuous flow reactor. *J Hazard Mater*, 196, 44–51.

Sut-Lohmann, M., Jonczak, J. and Raab, T. 2020. Phytofiltration of chosen metals by aquarium liverwort (*Monosoleum tenerum*). *Ecotox Environ Safe*, 188, 109844.

Tang, Y. T., Qiu, R. L., Zeng, X. W., Ying, R. R., Yu, F. M. and Zhou, X. Y. 2009. Lead, zinc, cadmium hyperaccumulation and growth stimulation in *Arabis paniculata* Franch. *Environ Exp Bot*, 66, 126–134.

Taştan, B. E., Ertuğrul, S. and Dönmez, G. 2010. Effective bioremoval of reactive dye and heavy metals by *Aspergillus versicolor*. *Bioresour Technol*, 101, 870–876.

Thayaparan, M., Iqbal, S. S., Chathuranga, P. K. D. and Iqbal, M. C. M. 2013. Rhizofiltration of Pb by *Azolla pinnata*. *Int J Environ Sci*, 3, 1811–1821.

Tiecher, T. L., Ceretta, C. A., Ferreira, P. A., Lourenzi, C. R., Tiecher, T., Girotto, E., Nicoloso, F. T., Soraini, H. H., Conti, L. D., Mimmo, T., Cesco, S. and Brunetto, G. 2016. The potential of *Zea mays* L. in remediating copper and zinc contaminated soils for grapevine production. *Geoderma*, 262, 52–61.

Ting, A. S. Y. and Choong, C. C. 2009. Bioaccumulation and biosorption efficacy of *Trichoderma isolate* SP2F1 in removing copper (Cu (II)) from aqueous solutions. *World J Microbiol Biotechnol*, 25, 1431–1437.

Tirry, N., Joutey, N. T., Sayel, H., Kouchou, A., Bahafid, W., Asri, M. and El Ghachtouli, N. 2018. Screening of plant growth promoting traits in heavy metals resistant bacteria: Prospects in phytoremediation. *J Genet Eng Biotechnol*, 16, 613–619.

Tsekova, K., Todorova, D., Dencheva, V. and Ganeva, S. 2010. Biosorption of copper (II) and cadmium (II) from aqueous solutions by free and immobilized biomass of *Aspergillus niger*. *Bioresour Technol*, 101, 1727–1731.

Turan, M. and Esringü, A. 2007. Phytoremediation based on canola (*Brassica napus* L.) and Indian mustard (*Brassica juncea* L.) planted on spiked soil by aliquot amount of Cd, Cu, Pb, and Zn. *Plant Soil Environ*, 53, 7–15.

Ullah, A., Mushtaq, H., Ali, H., Munis, M. F. H., Javed, M. T. and Chaudhary, H. J. 2015. Diazotrophs-assisted phytoremediation of heavy metals: A novel approach. *Environ Sci Pollut Res*, 22, 2505–2514.

Uslu, G. and Tanyol, M. 2006. Equilibrium and thermodynamic parameters of single and binary mixture biosorption of lead (II) and copper (II) ions onto *Pseudomonas putida*: Effect of temperature. *J Hazard Mater*, 135, 87–93.

Uzel, A. and Ozdemir, G. 2009. Metal biosorption capacity of the organic solvent tolerant *Pseudomonas fluorescens* TEM08. *Bioresour Technol*, 100, 542–548.

Verma, A., Bharagava, R. N., Kumar, V., Singh, A., Dhusia, N. and More, N. 2016. Role of macrophytes in heavy metal removal through rhizo-filtration in aquatic ecosystem. *Euro J Biotechnol Biosci*, 4, 15–20.

Verma, S. and Kuila, A. 2019. Bioremediation of heavy metals by microbial process. *Environ Technol Innov*, 14, 100369.

Veselý, T., Tlustoš, P. and Száková, J. 2011. The use of water lettuce (*Pistia stratiotes* L.) for rhizofiltration of a highly polluted solution by cadmium and lead. *Int J Phytoremediation*, 13, 859–872.

Vijayaraghavan, K. and Yun, Y. S. 2008. Bacterial biosorbents and biosorption. *Biotechnol Adv*, 26, 266–291.

Vilchez, R., Pozo, C., Gómez, M. A., Rodelas, B. and González-López, J. 2007. Dominance of *sphingomonads* in a copper-exposed biofilm community for groundwater treatment. *Microbiology*, 153, 325–337.

Waigi, M. G., Sun, K. and Gao, Y. 2017. *Sphingomonads* in microbe-assisted phytoremediation: Tackling soil pollution. *Trends Biotechnol*, 35, 883–899.

White, C., Sayer, J. A. and Gadd, G. M. 1997. Microbial solubilization and immobilization of toxic metals: Key biogeochemical processes for treatment of contamination. *FEMS Microbiol Rev*, 20, 503–516.

Yadav, A. K., Pathak, B. and Fulekar, M. H. 2015. Rhizofiltration of heavy metals (cadmium, lead and zinc) from fly ash leachates using water hyacinth (*Eichhornia crassipes*). *Int J Environ*, 4, 179–196.

Yadav, B. K., Siebel, M. A. and van Bruggen, J. J. 2011. Rhizofiltration of a heavy metal (lead) containing wastewater using the wetland plant *Carex pendula*. *CLEAN–Soil, Air, Water*, 39, 467–474.

Yadav, S. K., Juwarkar, A. A., Kumar, G. P., Thawale, P. R., Singh, S. K. and Chakrabarti, T. 2009. Bioaccumulation and phyto-translocation of arsenic, chromium and zinc by *Jatropha curcas* L.: Impact of dairy sludge and biofertilizer. *Bioresour Technol*, 100, 4616–4622.

Yang, Q., Li, Z., Lu, X., Duan, Q., Huang, L. and Bi, J. 2018. A review of soil heavy metal pollution from industrial and agricultural regions in China: Pollution and risk assessment. *Sci Total Environ*, 642, 690–700.

Yang, Y., Liu, Y., Li, Z., Wang, Z., Li, C. and Wei, H. 2020. Significance of soil microbe in microbial-assisted phytoremediation: An effective way to enhance phytoremediation of contaminated soil. *Int J Environ Sci Technol*, 17, 2477–2484.

Yin, K., Wang, Q., Lv, M. and Chen, L. 2019. Microorganism remediation strategies towards heavy metals. *Chem Eng J*, 360, 1553–1563.

Zabochnicka-Świątek, M. and Krzywonos, M. 2014. Potentials of biosorption and bioaccumulation processes for heavy metal removal. *Pol J Environ Stu*, 23, 551–561.

Zeng, X., Wei, S., Sun, L., Jacques, D. A., Tang, J., Lian, M., Ji, Z., Wang J., Zhu, J. and Xu, Z. 2015. Bioleaching of heavy metals from contaminated sediments by the *Aspergillus niger* strain SY1. *J Soils Sediment*, 15, 1029–1038.

Zhang, H., Chen, X., He, C., Liang, X., Oh, K., Liu, X. and Lei, Y. 2015. Use of energy crop (*Ricinus communis* L.) for phytoextraction of heavy metals assisted with citric acid. *Int J Phytoremediation*, 17, 632–639.

Zhang, X. H., Liu, J., Huang, H. T., Chen, J., Zhu, Y. N. and Wang, D. Q. 2007. Chromium accumulation by the hyperaccumulator plant *Leersia hexandra* Swartz. *Chemosphere*, 67, 1138–1143.

Zhou, M., Liu, Y., Zeng, G., Li, X., Xu, W. and Fan, T. 2007. Kinetic and equilibrium studies of Cr (VI) biosorption by dead *Bacillus licheniformis* biomass. *World J Microbiol Biotechnol*, 23, 43–48.

Ziagova, M., Dimitriadis, G., Aslanidou, D., Papaioannou, X., Tzannetaki, E. L. and Liakopoulou-Kyriakides, M. 2007. Comparative study of Cd (II) and Cr (VI) biosorption on *Staphylococcus xylosus* and *Pseudomonas* sp. in single and binary mixtures. *Bioresour Technology*, 98, 2859–2865.

Złoch, M., Kowalkowski, T., Tyburski, J. and Hrynkiewicz, K. 2017. Modeling of phytoextraction efficiency of microbially stimulated *Salix dasyclados* L. in the soils with different speciation of heavy metals. *Int J Phytoremediation*, 19, 1150–1164.

12

Role of Plant-Associated Microbes in Phytoremediation of Heavy Metal Polluted Soils

Manoj Parihar
ICAR-Vivekananda Parvatiya Krishi Anusandhan Sansthan (VPKAS), Almora, Uttarakhand, India

Amitava Rakshit
Department of Soil Science & Agricultural Chemistry, Institute of Agricultural Science, Banaras Hindu University, Varanasi, UP, India

Manoj Kumar Chitara
Department of Plant Pathology, College of Agriculture, GBPUAT, Pantnagar, Uttarakhand, India

Hanuman Singh Jatav
S.K.N. Agriculture University, Jobner, Rajasthan, India

Vishnu D. Rajput
Academy of Biology and Biotechnology, Southern Federal University, Stachki, Rostov-on-Don, Russia

Ashish Kumar Singh
ICAR-Vivekananda Parvatiya Krishi Anusandhan Sansthan (VPKAS), Almora, Uttarakhand, India

Kiran Rana
Department of Agronomy, Institute of Agricultural Science, Banaras Hindu University, Varanasi, UP, India

Surendra Singh Jatav
Department of Soil Science & Agricultural Chemistry, Institute of Agricultural Science, Banaras Hindu University, Varanasi, UP, India

Mohsina Anjum
Department of Soil Science & Agricultural Chemistry, University of Agricultural Sciences, Bangalore, India

Tatiana Minkina
Academy of Biology and Biotechnology, Southern Federal University, Rostov-on-Don, Russia

Utkarsh Kumar
ICAR-Vivekananda Parvatiya Krishi Anusandhan Sansthan (VPKAS), Almora, Uttarakhand, India

CONTENTS

DOI: 10.1201/9781003155089-12

12.1 Introduction

Environmental pollution is gaining attention in last few years due to rapidly expanding of urbanization, industrialization and unscientific agriculture practices which poses a great risk on plant growth, human health and ecosystem sustainability (Chibuike and Obiora 2014; Ojuederie and Babalola 2017). Among these pollutants, metal contaminates in our soil and water body with increasing concentration renders them unsuitable for living organism for further use. In the long term, they alter the functionality and viability of soil by reducing their buffering, filtering and transforming capacity as a result of poor physical, chemical and biological quality. Although some of the heavy metal plays a vital role in various biochemical and physiological reaction and are important for normal functioning of the biological system while metals such as As (arsenic), Pb (lead) and Cd (cadmium) are not essential for plant and human growth (Kavamura and Esposito 2010; Li et al. 2013). These essential heavy metal require in very minute quantity and sometimes term as micronutrient i.e. Zn Cu, Fe, Mn and Ni and their concentration above certain threshold level are toxic and hampered enzymatic process, damage to protein, nucleic acids, lipids, DNA and produce reactive oxygen species (ROS) to counteract the oxidative damage (Jadia and Fulekar 2009; Chibuike and Obiora 2014).

The remediation of heavy metal polluted soils is of prime importance for environmental safety and sustainability. Various approaches such as physical and chemical methods including excavation, thermal treatments, solidification, soil washing and pyrometallurgical separation are available but have limited use and unsuitable for developing nations due to greater involvement of cost, input and destruction to soil (Schnoor 1997; Mulligan et al. 2001; Dermont et al. 2008). In such circumstances, phytoremediation as in situ and passive technology can be a promising approach to decontaminate the polluted sites contains low to medium concentration of heavy metals (Bharti et al. 2014; Ojuederie and Babalola 2017; Rakshit et al. 2017; Sikdar and Kundu 2018). The major advantage of phytoremediation is their large applicability and ecofriendly nature provides a huge number of additional benefits to the environment. However, similar to other technology, phytoremediation also has some limitation and require a longer period with greater dependability on plant species, contaminant type and quantity, soil and climate type. In addition to this, biological methods are generally applicable to clean up the contaminate present in surface soil. For successful phytoremediation approach, it is essential to remove heavy metals from contaminated sites in greater amount by the plants which can be achieved through greater resistance and larger biomass producing plant capability. Regarding this, plant-microbe interaction could be imperative under stress condition and survival of both the partners. In this mutual agreement, plant obtains greater nutrient, water and favorable condition which is extremely essential for better plant growth and successfully phytoremediation approach. In this chapter, we provide an overview of plant-microbe association under metal stress condition and how they could be implemented effectively in phytoremediation program.

12.2 Rhizobiome in Heavy Metal Contaminated Soil

Soil contains a plethora of the microorganisms which are an intrinsic part of the soil biological constituents. The rhizobiome is divided into three key separate but interacting surface: (1) rhizosphere – microbial community colonized in the narrow soil zone near root, (2) rhizoplane – the microbes community colonized on the root surface as well as the firmly attached soil particles on the roots, and (3) endosphere – the root tissues colonized by certain rhizospheric microbes (Edwards et al. 2015), commonly found in the hyperaccumulator plants. The rhizospheric microbiome contains a diverse group of bacteria predominantly firmicutes, proteobacteria, actinobacteria belonging to genus Acetobacter, Achromobacter, Arthrobacter, Azospirillum, Azotobacter, Bacillus, Clostridium, Enterobacter, Flavobacterium, Frankia, Pseudomonas, Staphylococcus, Streptomyces, Rhizobium (Pires et al. 2017; Bashan et al. 2008). The rhizospheric microbes play a key role in soil contaminated with heavy metal by alleviating metal stress by accelerating the secreting root exudates. The root exudates enhance the capability of the host plant to withstand heavy metal contaminated soil by accelerating the detoxification process (absorption accumulation and fixation of heavy metal) (Seshadri et al. 2015). Endophytic bacteria such as rhizobium help to remediate the heavy metal contaminated soil (Jing et al. 2007; Checcucci et al. 2017). Apart from the bacteria, these soil also contains fungi (Ascomycota and Basidiomycota) particularly the arbuscular mycorrhiza fungi present the close association with the hyperaccumulator plants helps to remediate the heavy metal contaminated soils (Sepheri et al. 2012). Interaction of rhizobiome in heavy metals contaminated soil is depicted in Figure 12.1.

12.3 Various Phytoremediation Approaches

Phytoremediation defines as the use of plants and microbes for the removal or remediation of various contaminants from soil, sediments and water. The term phytoremediation originated from the Greek word *phyto* and Latin word *remedium* and was first used in the 1980s for the management of degraded or pollutes soil using plants (Neil 2007). According to Wang et al. (2012) phytoremediation refer as use of plants and associated microbes, soil amendments and agronomic approaches to managing the various environmental contaminates. The remediation of polluted sites via phytoremediation is categorized in phytoextraction, phytostabilization, rhizofiltration, rhizodegradation, phytodegradation and phytovolatilization (Chibuike and Obiora 2014; Ansari et al. 2014).

Phytoextraction plants uptake toxic metals from contaminated sites and accumulate in above-ground plant parts such as stem, leaves, and inflorescence, which further need to be stored for safe disposal (Zhao and McGrath 2009; Ali et al. 2013; Sytar et al. 2016). For successful extraction of heavy metals, plants should have some characteristics such as high biomass production, vigorous root growth, tolerance to metal

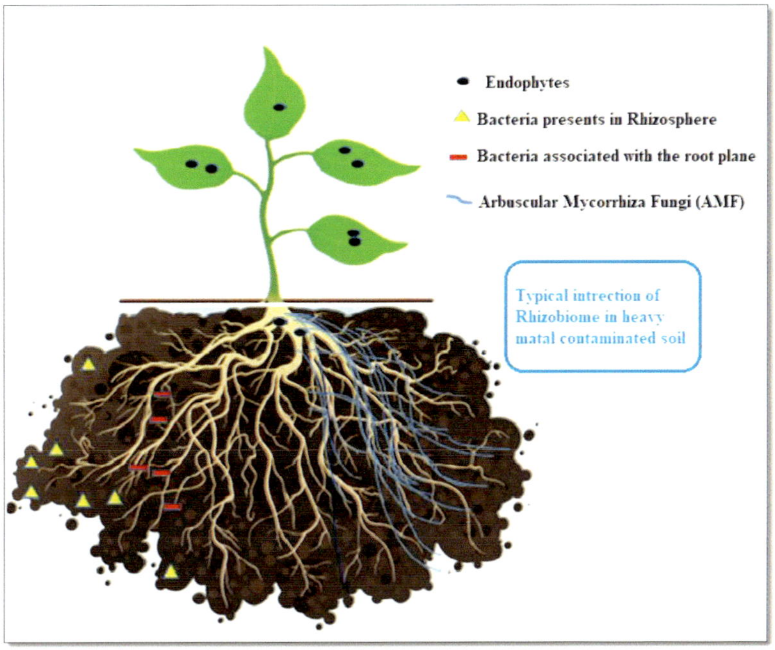

FIGURE 12.1 Rhizobiome in heavy metal contaminated soil.

stress environment, high bioconcentration factor (BCF) and biological absorption coefficient (BAC) (Vamerali et al. 2010). Generally, plants having high tolerance ability to metal stress restrict the entry of these metals to their aerial parts. However, some plant species called as hyperaccumulators (Table 12.1), can accumulate high concentration of specific metals ion (≥0.1% of plant dry weight) in their aerial parts without showing any toxic symptoms (Garbisu and Alkorta 2001).

Some plant species of Brassicaceae, Poaceae, and Euphorbiaceae family are found suitable for this purpose. Rhizofiltration employs under the aquatic condition where plant with high metal tolerant capacity absorbs and accumulates pollutant in roots or other submerge organ and restrict contaminates to moving them deeper in water (Kvesitadze et al. 2006; Mukhopadhyay and Maiti 2010). In rhizofiltration, plants grow hydroponically and uptake the heavy metals either intracellularly by making compartmentalization and accumulates them in the vacuole or extracellularly by precipitation or surface adsorption (Salt et al. 1995; Phieler et al. 2013). By using this technique removal of radio nuclides like U, ^{137}Cs, and ^{90}Sr is also possible and sunflower breeds found promising for this approach (Dushenkov et al. 1995, 1997).

However, rhizofiltration as a cheap and efficient approach is more suitable where a large volume of water contaminates is present with a low concentration of pollutants. Phytostabilization involves the immobilization and stabilization of metals pollutant in soil (Singh 2012) by sorption, complexation and precipitation and reduce the further risk of environmental degradation (Mendez and Maier 2007). Plants with an extensive root system and large aboveground biomass are more suitable for phytostabilization because of reduced leaching and minimize soil erosion (Kvesitadze et al. 2006; Ali et al. 2013). Phytostabilization reduce the hazard of the pollutant without influencing their concentration in soil (Arthur et al. 2005). Selection of suitable plants for phytostabilization may require high tolerance to metals stress, vigorous growth to establish dense ground cover, high BCF, lower root to shoot translocation, high transpiration rate, short-living and easy to cultivate (Kucharski et al. 2005; Yoon et al. 2006).

Phytodegradation includes the use of microbes and plants to decompose the organic pollutants into its basic constitute or transforming further by secreting degrading enzymes and product (Garbisu and Alkorta 2001). Rhizodegradation refers to the decomposition of contaminates mostly organics in plant rhizosphere with the help microbial activity (Arthur et al. 2005; Ali et al. 2013). Microbial degradation of organic contaminates facilitated by both extracellular and intracellular enzymes and also provides energy and carbon to microbes (Vallero 2010).

Phytovolatilization described as the uptake of pollutants and dispersed them into the atmosphere through leaf surface (Moreno et al. 2005) at comparatively low concentration (Mueller et al. 1999). Metals such as Se, As, Hg and compounds like tetrachloroethane, trichloromethane, tetrachloromethane is reported to be remediated by using phytovolatilization (Raskin et al. 1997; Susarla et al. 2002; Ali et al. 2013). This technique is successfully reported for mercury, where Hg^{2+} is converted into less toxic Hg^{0} by inserting bacterial reductase genes for Hg into Arabidopsis thaliana L. and tobacco (*Nicotiana tabacum* L.) (Heaton et al. 1998; Bizily et al. 1999). Despite lower soil erosion and site disturbance associated with this technique, precipitation of volatile compound present in the atmosphere and reposition back into the soil system make this approach most controversial (Henry 2000).

TABLE 12.1

List of Various Metal Hyperaccumulators Plant Species for Phytoextraction and Phytostabilization

Metal	Plant Species	Plant Tissue	Metal Accumulation	Location	References
As	*Corrigiola telephiifolia*	Aboveground part (2110 μg g^{-1}),	AP/S (6.8), R/S ratio (4), AP/R (1.6)	Bustarviejo, Madrid	García-Salgado et al. (2012)
Cd	*Arabidopsis halleri* spp. *Germmifera*	Shoot	Plant extracted (22.87 ± 9.21) % of total Cd concentration from the soil	Tome City, Japan	Zhang et al. (2017)
As	*Pteris vittata* (PV)	Aboveground tissues	43–71% of the As extruded out of the fronds of PV	N/A	Datta et al. (2017)
Cu, Zn, As and Cd	*Eleocharis acicularis*	Shoots	BCFs for Cu, Zn, As, Cd and Pb as large as 28.0, 8.0, 7.4, 11.8 and 1.7, respectively	Mine drainage site in southwestern Japan	Sakakibara et al. (2011)
Ni	*Isatis pinnatiloba*	Aboveground tissue	1441 mg Ni kg^{-1} biomass	Serpentine soil Turkish	Altinözlü et al. (2012)
Cd and Pb	*Ricinus communis* L.	Plant shoot	Citric acid ↑ Cd and Pb contents in plant shoots by about 78% and 18–45%, respectively	Pot exp. China	Zhang et al. (2015)
Ni	*Alyssum murale*	Plant biomass	4730 to 20,100 mg kg^{-1}	SE Europe	Bani et al. (2010)
As	*P. cretica* cv Mayii	Leaves and stems	Higher As conc. (~50 times) was obtained in the stems and leaves compared to the roots	NC, USA	Baldwin and Butcher (2007)
Pb	*Noccaea caerulescens*	Pb accumulated mainly in the roots (0.16–0.23 wt% dry mass)	High levels of Pb^{2+} tolerance and accumulate 1700–2300 mg kg^{-1} DW	Queensland, Australia	Dinh et al. (2018)
Cu, Cd, Ni	*Cannabis sativa* L.	Aboveground biomass	Plant accumulate Cu (1530), Cd (151) and Ni (123 mg kg^{-1})	Rawalpindi, Pakistan	Ahmad et al. (2016)
Pb	*Tagetes minuta* L. and *Bidens pilosa* L.	Leaves	*Tagetes minuta* L. and *Bidens pilosa* L. accumulated 380.5 and 100.6 μg Pb g^{-1} DW, respectively	Córdoba City, Argentina	Salazar and Pignata (2014)
Hg and Cd	*Azolla pinnata*	Bioaccumulation	Hg content in *A. pinnata* was 310, 580, and 667 mg kg^{-1} while Cd(II) content was 402.2, 650, and 740 mg kg^{-1} for the 0.5, 1.0, and 3.0 mg L^{-1} each metal conc., respectively	Singrauli, India,	Rai (2008)
Cr	*A. filiculoides, A. microphylla, A. pinnata*	Whole plant	Accumulate 5000–15,000 mg Cr kg^{-1} tissue and BCF ranged between 243 and 4617 for three species	IARI, New Delhi	Arora et al. (2006)
Cr	*Brassica napus*	Whole plant	Compost and *B. licheniformis* strain enhanced *B. napus* accumulation of Cr in pot experiment	AltaMurgia Park, Italy	Brunetti et al. (2012)
Cd	*Rorippa globosa*	Leaves	Cd accumulated in the leaves was up to 218.9 μg Cd g^{-1} dry weight (DW)	Shenyang Station, China	Sun et al. (2010)

12.4 Rhizospheric Microbes under Metal Stress

Heavy metals (HM) have been extensively exploited due to its wide application in industrial, pharmaceutical, domestic and agricultural use (synthesis of pesticide). Heavy application has HM leads to contamination of food chain at each trophic level such as accumulation in soil, water, plants and animals. Indiscriminate use of heavy metals has raised a serious concern over ecological as well as public health globally. HM stress has found to affect organisms on cellular as well molecular level by affecting cell membrane to DNA damage and carcinogenesis (Chang et al. 1996; Beyersmann and Hartwig 2008). To combat the problems of heavy metals, phytoremediation could be an efficient solution. Soil microbes living in the vicinity of root so-called rhizospheric microbes (eukaryotes and prokaryotes) play a very prominent role in bio-remediation under contaminated soil (Khan et al. 2000) and popularly known as rhizoremediation (Kuiper et al. 2004).

Heavy metal contaminated soils are predominantly composed of aerobic bacterial genera like *Bacillus*, *Pseudomonas*, *Arthrobacter, Azotobacter, Klebsiella, Enterobacter* and *Alcaligenes* that can tolerate HM contamination (Pires et al. 2017; Kloepper et al. 1989; Glick 1995). Rhizobacteria like *Thalspi caerulescens* have remarkable potential in accumulation of zinc (Zn) from the soil and can be actively used to reduce non-labile zinc contamination in soil (Kamaludeen and Ramasamy 2008). Use of *Pseudomonas putida* found to reduce Cd phytotoxicity in plants and enhanced accumulation of Cd from the soil system (Huang and Tao 2005). Similarly, *P. maltophilia* reduces the mobility and toxicity of Cr^{6+}, Hg, Pb and Cd (Blake et al. 1993). *Escherichia coli* and *Moraxella* sp. found to accumulate 25-fold cadmium and mercury from contaminated soil (Bae et al. 2000; Bae et al. 2003). HM stress severely affect the nodule formation and nitrogenase activities of microbes but Rhizobia are an important PGP in the rhizosphere region that significantly influence their nodulation ability and nitrogenase enzyme activity under HM stress environment. Although some HM tolerant strain of rhizobia has been reported successfully to accomplish N fixation process (Mishra et al. 2017) and their detoxifying capacity is also well known to improve the condition of metal-polluted sites (Checcucci et al. 2017).

Fungi belonging to division Basidiomycota, Ascomycota and Arbuscular mycorrhizae group have been found to colonize in HM polluted soils (Narendrula-Kotha and Nkongolo 2017; Khan et al. 2000). AM fungi and some rhizospheric microbes have the ability to binding HM present in the external environment to the cell surface and mobilize them into the cell (Ehrlich 1997). There is a certain nematode parasite of marine fish such as *Echinocephalus* sp. and *Ascaris* sp. which have been found as natural bioremediatory to bioaccumulate HM in their body (Azmat et al. 2008). Polycyclic aromatic hydrocarbon (PAH) mixture has been found to cause a significant change in dominance and assemblage of nematode species that prove to use nematode as a potential organism for bioindicator for coastal environment monitoring against PAH contamination (Louati et al. 2015). These rhizospheric microbes transformed the availability of HM by various mechanism but more specifically it will depend on soil characteristics, HM concentration and their nature, plant species, microbial dynamics and diversity (Kong and Glick 2017; Mishra et al. 2017).

12.5 Mechanism Followed by Microbes to Alleviate Metal Stress

Generally, HM tolerating microbes tackle the problem toxicity by enhancing plant growth and altering physicochemical properties of soil. Heavy metal tolerant (HMT) microbes detoxify heavy metals through acidification, complexation, chelation, redox reaction and precipitation. Merdy et al. (2009) suggested that acidic soil with a pH below 6, highly favors the absorption of HM by microbes. Decomposition of organic matters by rhizospheric microbes releases organic acids like oxalic, gluconic, acetic and malic acids which decreases soil pH and sequesters soil metal ions (Turnau and Kottke 2005; Ullah et al. 2015; Gube 2016). A similar experiment conducted by Seneviratne et al. (2017) suggested that rhizospheric organism like bacteria and fungi releases organic acid as a chelating agent against HM. Citric and oxalic acid secreted by *Beauveria caledonica* helps in solubilizing of Cu, Cd, Pb and Zn metals (Fomina et al. 2005). Mycelial structure of mycorrhizae has been found to penetrates deeply into the soil aggregates and helps in chelating HM of contaminated soils (Gadd et al. 2014).

Roots are known to exudates certain organic compounds which play an important role in forming metal complexes to mobilize metals and providing energy sources to the microbial communities. Root exudates are highly rich in amino acids, organic acids and phytochelatin like compound which helps in binding HM compounds. Release of H^+ and certain enzymes from root exudates is known to help in acidifying and transferring electron in the rhizosphere which helps in enhancing the metal bioavailability to the root (Ma et al. 2016).

The alteration in redox reactions by rhizospheric microbes (using OM c-Cyts, Pcc or MtrABC like systems) have been proven to have a role in detoxification of HM toxicity in soil (Amstaetter et al. 2009; Shi et al. 2016). There are various instances in which HM also have been oxidized by various enzymes i.e. Cu effluxes are enhanced by multicopper oxidases (CueO, CuiD, CopR), Chromate reductase reducing Cr^{6+} to Cr^{3+} (Chatterjee et al. 2009). Similarly As-oxidizing bacteria (*Bacillus* spp. and *Geobacillus* sp.) have been confirmed in the transformation of As^{3+} to lesser toxic As^{5+} (Majumder et al. 2013). In another mechanism, bioaccumulation mediated by HMT-microbes largely known to be responsible for HM uptake and detoxification. Bioaccumulation happens in two ways: (1) passive absorption through biosorption by non-living biomasses in soil and (2) active absorption mediated by living cells and depends on energy and metabolism for metal transport (Gutierrez-Corona et al. 2016).

Plant cell-wall and functional groups like -OH, -SH and -COOH possess an affinity to HM that helps to bind metals during biosorption. Metal-binding also carried out by PC and MT (metallothioneins) like metal-binding peptide secreted by rhizospheric microbes and plants (Miransari 2011). There are various studies which suggest the role of MT in HM toxicity alleviation. MT biosynthesis increased in response to Pb toxicity after exposure of *Bacillus cereus*, *Providencia vermicola strain* (Murthy et al. 2011; Sharma et al. 2017). Once HM reaches to the cell system it gets sequestered or compartmentalized into different cellular organelles like Zn, Cd, Cu in vacuoles of mycorrhizae (*Glomus intraradices*) (Gonzalez-Guerrero et al. 2008).

Rhizospheric microbes also are known to excrete EPS (extracellular polymeric substances) like glycoproteins, polysaccharides, lipids and different kind of soluble peptides which possess sufficient amount of anions to recover as well removal of metals through biosorption (Ayangbenro and Babalola 2017). EPS excreted by rhizospheric microbes serve as biofilm which provides a protective sheath to the microbial cell against HM and this sheath also helps in transformation of toxic metal ions into nontoxic forms (Gupta and Diwan 2017).

Heavy metal tolerating plant growth-promoting (HMT-PGP) rhizobacteria helps in detoxification of HM toxicity by enhancing the availability of metals to plants and microbial communities. They release an iron-chelating compound i.e. Siderophore, under Fe stress in the soil to enhance the uptake of iron. It acts as an iron solubilizing agent to facilitate the uptake of iron by the plants. Siderophores also have been known to farm heavy metal complexes in soil and enhancing the bioavailability of HM to the plants (Glick and Bashan 1997; Rajkumar et al. 2010). Siderophore produced by *Pseudomonas aeruginosa* possess ability to bind 16 different metals including Ag^+, Al^{3+}, Cd^{2+}, Co^{2+}, Cr^{2+}, Cu^{2+}, Eu^{3+}, Ga^{3+}, Hg^{2+}, Mn^{2+}, Ni^{2+}, Pb^{2+}, Sn^{2+}, Tb^{3+}, Tl^+, and Zn^{2+} (Braud et al. 2009). Siderophore production also has been observed to be produced by ectomycorrhizal fungi like *Scleroderma verrucosum*, *Suillus luteus*, and *Rhizopogon luteolus* under iron-deficient soil (Goodell et al. 1997; Machuca et al. 2007). Exact mechanism lying behind siderophore-mediated detoxification of soil is still left an excellent area to explore. Therefore, it is suggested that siderophore producing rhizomicrobes are providing an excellent opportunity to remediate the HM contaminated soils.

In another mechanism, biosurfactants as an effective complexing agent form a complex with heavy metals in the soil system and desorb it from soil matrix and increase metal detoxification. Biosurfactant released by microbes possesses high surface and emulsifying activities with hydrophobic tail and a hydrophilic head. These biosurfactant has been categorized into fatty acids, lipids, lipopeptides and glycolipids. Due to the potential properties of anionic nature, biodegradability, low toxicity and high surface activity, they are used in HM contaminated soil. Biosurfactant activities in *Bacillus subtilis*, *Pseudomonas aeruginosa*, and *Tomlopsis bornhicola* have been well proved to detoxify HM (Mulligan et al. 1999).

12.6 Plant and Microbes Interaction under Metal Stress

Plant rhizospheric interaction with soil microbes is imperative to alleviate the heavy metal stress (Figure 12.2) by employing various mechanisms includes the production of plant growth hormones (Leyval et al. 1997), osmoprotectants (Upadhyay and Singh 2015), greater nutrient solubilization and immobilization of toxic metals (Ma et al. 2013).

These beneficial microbes either enhance the overall growth and biomass production of plant or improve the metal tolerance by altering their bioavailability (Figure 12.3) which otherwise achieved by using chemical amendments (Figure 12.3).

Microbes living in root rhizosphere synthesize and secrete secondary metabolites due to abundant supply of roots exudates which enhances root biomass and provides a greater surface area to accelerate metal absorption and ultimately their phytoremediation ability (Egamberdieva et al. 2014; Egamberdieva et al. 2016). Besides, root exudates secreted by plants i.e. organic acids, amino acids, and phytochelatins (PC) changes the bioavailability of heavy metals and also provide nutrients and energy to rhizospheric microbes which in turn improve overall plant growth and development under stress condition (Mishra et al. 2017). Metal chelating agents such as phytosiderophores released by plant roots can be further complemented by soil microbes and improve the metal tolerance in plants (Raskin 1997; Vamerali et al. 2010). For example, Sheng et al. (2008) and Khan et al. (2015) found that the inoculation of Pb-resistant strain *Pseudomonas fluorescens* in *B. napus* and Cr-resistant *Microbacterium arborescens* HU33 in ryegrass, increases the root growth by 21–35% and 39%, respectively, under these heavy metal stress environment. Under stress condition, the plant produces a higher amount of ethylene which restricted root growth (Barnawal et al. 2012) but some metal resistant bacterial strain *P. myrsinacearum* RC6b, *A. xylosoxidans*, *Pseudomonas koreensis* AGB-1, *Acinetobacter* sp.

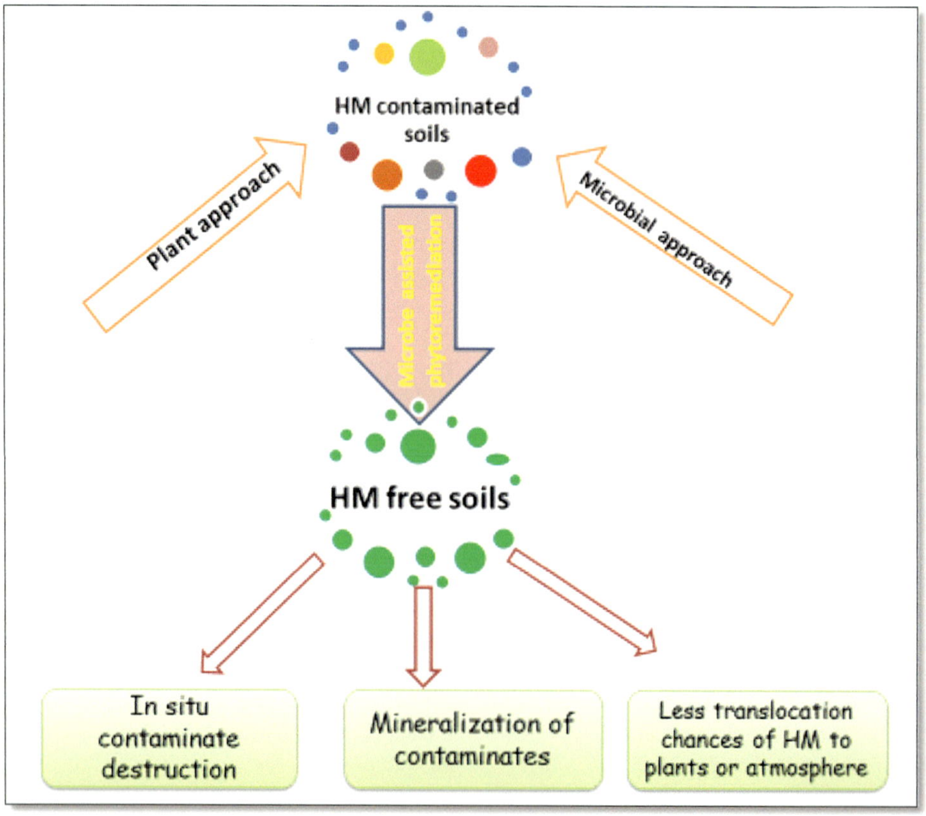

FIGURE 12.2 Microbe assisted phytoremediation.

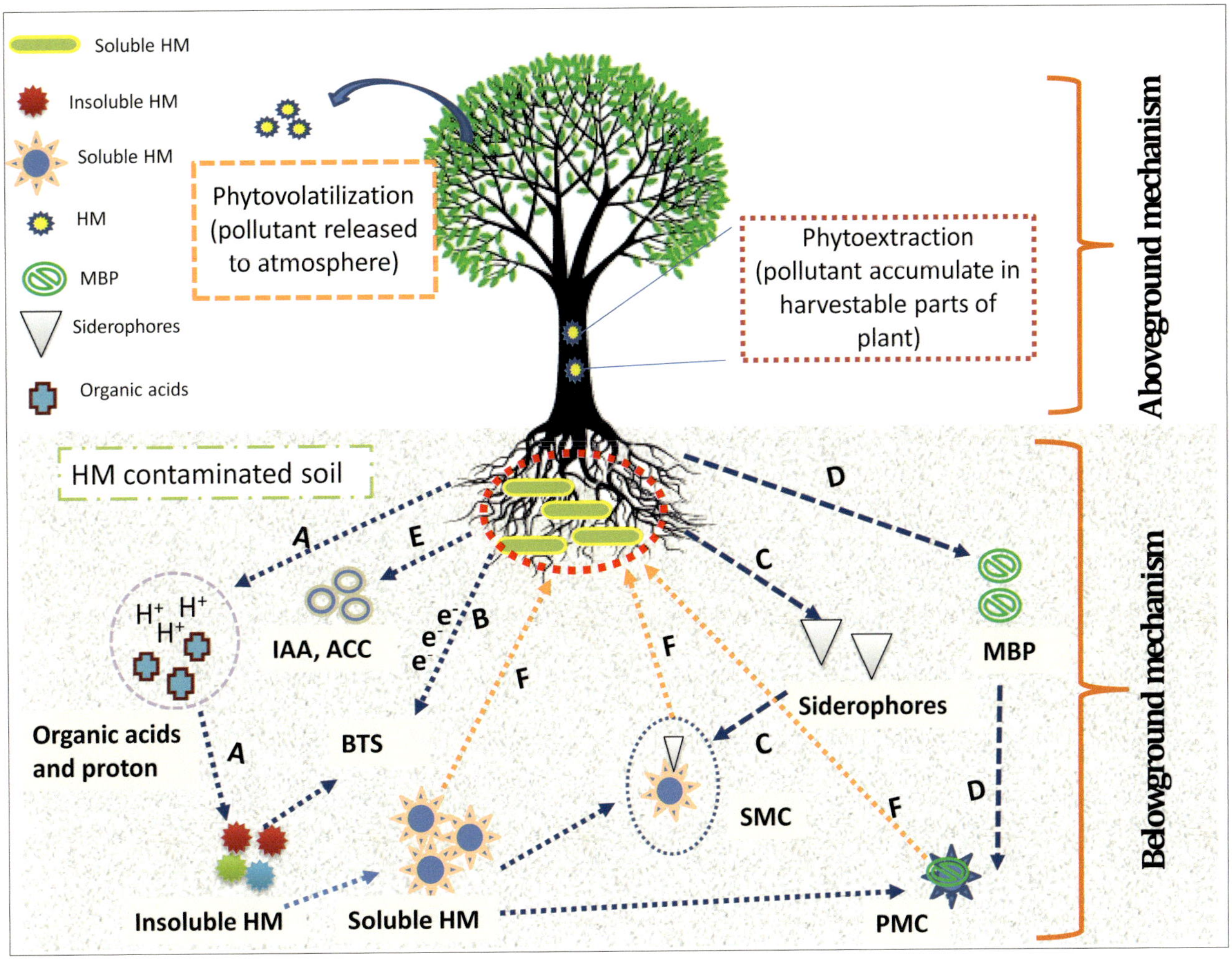

FIGURE 12.3 Plants and rhizospheric microbes mediated phytoremediation mechanism which describe: (A) solubilization of insoluble heavy metal compound by organic acids; (B) biotransformation of HM by redox reaction; (C) formation of siderophore metal complex; (D) metal binding protein complex; (E) plant growth promotion using IAA, ACC deaminase, etc., released form microbes or plant. All the processes (A–E) transform the metal availability and improve their translocation in plant (F) or stabilize them in soil.

and *Bacillus* sp. reported to produce ACC deaminase activity which breaks down the ethylene precursor ACC and use it as a source of nitrogen (Ma et al. 2009, 2013; Babu et al. 2015). Role of plant associated microbes in phytoremediation of heavy metal polluted soils is depicted in Table 12.2.

For successful phytoremediation, greater bioavailability of metal ions and their subsequent absorption and accumulation in plant biomass is critically important and for the same, the interaction of plant and microbes contribute significantly by mobilizing metal ions due to their enzymatic activity (Egamberdieva et al. 2016). Jiang et al. (2008) reported that *Burkholderia* sp. strain J62 is capable to solubilize the resistant fraction of Pb and Cd metals while Saravanan et al. (2007) found *Gluconacetobacter diazotrophicus*, an endophytic diazotroph to improve the availability of various insoluble Zn compounds by producing 5-ketogluconic acid. Moreover, the release of proton and enzymes along with root exudates acidify the pH of rhizosphere and enhance the bioavailability of heavy metals (Ma et al. 2016). In another mechanism, Saif and Khan (2018) have reported lower metal stress in chickpea plants grown under Cr- and Ni-contaminated soils when inoculating with *Pseudomonas aeruginosa* due to lower activity of proline and antioxidant enzyme activity such as APX, CAT, GR and SOD. Generally, plants exhibited higher production of antioxidant enzyme activity under stress condition to manage oxidate damage performed by reactive oxygen species (ROS) (Das and Roychoudhury 2014). In addition to this, Kuramshina et al. (2018) found that inoculation of *Bacillus subtilis* to *Sinapis alba* L. reduces the manifestation of oxidative stress under a higher level of Cd and Ni and also intensify the phytoextraction of these metal ion due to entophytic ability of *Bacillus subtilis*. Moreover, immobilization of toxic metals by microbes using intracellular sequestration, precipitation and adsorption mechanism render them unavailable to plant and improve their growth and development (Gadd 2004; Joner and Leyval 2000; Fein et al. 2001; Zaidi and Musarrat 2004). Another important plant-microbe association known as arbuscular mycorrhiza (AM) fungi colonizing 80 to 90% of higher land plants (Schüßler et al. 2001; Parihar et al. 2019a) is essential for sustainable agriculture production under various biotic and abiotic stress (Evelin et al. 2009; Kohler et al. 2009; Parihar et al. 2019b). AM fungal assisted plant sustenance under metal stress is well documented (Audet and Charest 2007;

TABLE 12.2

Role of Plant Associated Microbes in Phytoremediation of Heavy Metal Polluted Soils

Plant Type	Microbes Involved	Metal	Mechanism Followed by Microbes/Plant	Effect on Plant Performance	Reference
Alfalfa	*Cellulosimicrobium* sp.	Cr(VI)	Production of IAA (12.24 to 96.14 μg mL^{-1}), siderophores (+) and P solubilization (2.66)	Bacterium ↑ plant growth by 68% and 28% for shoot and root control plants, respectively, and also uptake of Cr, Zn and Cu in root and shoot	Tirry et al. (2018)
Brassica juncea	*B. subtilis* SJ-101	Ni	Production of IAA, solubilizing inorganic phosphate	↑ Plant growth attributes of *B. juncea* grown in Ni-amended soil and accumulated 0.147% Ni (with SJ-101) and 0.094% Ni (without SJ-101) on dry mass basis	Zaidi et al. (2006)
Cicuta virosa L.	*Pseudomonas putida* *Rhodopseudomonas* sp.	Zn	Metal-chelating compound, IAA production	Microbe inoculated seedlings had significantly greater numbers of leaves, numbers of stems, and fresh and dry masses of roots than the control seedlings ($P<0.05$)	Nagata et al. (2015)
Cajanus cajan	*Glomus mosseae*	Cd and Pb	Biosorption	AM fungi inoculation ↑ the number of nodules, their dry weights, leghemoglobin content nitrogenase activity while ↓ metal uptake compared with the control	Garg and Agrawal (2011)
Solanum nigrum	*Pseudomonas* sp. Lk9	Cd, Zn and Cu	Production of IAA, siderophores, ACCD and solubilization of P	Shoot dry biomass by 14% and the total of Cd by 46.6%, Zn by 16.4% and Cu by 16.0% accumulated in aerial parts, compared to those of non-inoculated control. *S. nigrum* and Lk9 significantly ↑Soil microbial biomass C by 39.2% and acid phosphatase activity by 28.6% compared to only *S. nigrum*	Chen et al. (2014)
Pelargonium graveolens	*P. monteilii* PsF84, *P. plecoglossicida* PsF610	Cr	Production of IAA, siderophores and solubilization of P	Isolate PsF84 ↑ the dry biomass of shoot, root, essential oil yield and chlorophyll by 44, 48, 43, 31%, respectively, over control. The corresponding ↑ with the isolate PsF610 were 38%, 40%, 39% and 28%, respectively	Dharni et al. (2014)
Noccaea caerulescens, Thlaspi perfoliatum	*Microbacterium* sp. NCr-8, *Arthrobacter* sp. NCr-1, *Bacillus* sp. NCr-5, *Bacillus* sp. NCr-9 and *Kocuria* sp. NCr-3	Ni	Production of IAA, siderophores and ACCD	Inoculated plants showed increased shoot biomass, root length and root-to-shoot Ni translocation compared to control	Visioli et al. (2014)
Amaranthus hypochondriacus, A. mangostanus, Solanum nigrum and *Z. mays*	*Rahnella* sp. JN27	Cd	Production of IAA, siderophores, ACCD and solubilization of P	Enhanced plant growth and Cd uptake compared with control	Yuan et al. (2014)
Brassica juncea	*Pseudomonas brassicacearum* and *Rhizobium leguminosarum*	Zn	Bacteria-induced metal chelation	*R. leguminosarum* alone and in combination with *P. brassicacearum* enhanced plant growth and Zn phytoextraction	Adediran et al. (2015)
Robinia pseudoacacia and legume herbs	*Rhizophagus intraradices*	Pb	AMF inoculation enhances Pb uptake and accumulation in the root system	AMF influences plant biomass, plant photosynthesis, macronutrient acquisition in legume tree co-cultured with legume herbs	Yang et al. (2016)

Miransari 2011). AM fungi employ various mechanisms such as the production of chelating agent and phytohormones, interaction with the plant plasma membrane and cell wall, improved uptake of nutrient, interaction with other soil microbes and effects on plant roots (Garg and Pandey 2015; Miransari 2017). AM fungi modulate the heavy metal absorption by either enhancing their uptake under low concentration or restricted uptake in the toxic concentration of metal (Frey et al. 2000). In a study, Shabani and Sabzalian (2016) reported that AM fungi alleviate Ni stress in tall fescue by hampering Ni transport for roots to the aboveground biomass. Similarly, Allah et al. (2015) found that mycorrhizal fungi improve the plant growth, chlorophyll content and membrane stability and mitigate adverse effects of cadmium stress.

Conclusion

In this chapter, we have critically examined the possible threat of heavy metal pollution on the environment and their remediation using plant-microbe association with various mechanism followed by them under stress condition. Microbes improve the overall plant efficiency by improving nutrient availability, tolerance to metal toxicity, induced systematic resistance, production of growth hormone and antioxidant enzymes which resulted in greater phytoextraction and accumulation of metals in plant harvestable biomass. The interaction effect between various microbes along with plants species is needed to be studied extensively under metal stress condition to identify their suitable strain or species for successful phytoremediation approach. Application of microbial consortium and assessment of introduced and indigenous microorganism in the rhizospheric region is important and needed to investigate more precisely. Further technological advancement as a biotechnological tool to develop transgenic plants and genetically modified soil microbes with complete biosafety regulation should be taken into consideration. Moreover, to understand the feasibility of phytoremediation and to make a practical recommendation, it is necessary to replicate our earlier findings in field conditions.

REFERENCES

Adediran, G. A., Ngwenya, B.T., Mosselmans, J. F. W., Heal, K.V., and Harvie, B. A. 2015. Mechanisms behind bacteria induced plant growth promotion and Zn accumulation in *Brassica juncea*. *J Hazard Mater* 283, 490–499.

Ahmad, R., Tehsin, Z., Malik, S. T., Asad, S. A., Shahzad, M., Bilal, M., Shah, M. M., and Khan, S. A. 2016. Phytoremediation potential of hemp (*Cannabis sativa* L.): identification and characterization of heavy metals responsive genes. *Clean Soil Air Water* 44, 195–201.

Ali, H., Khan, E., and Sajad, M. A. 2013. Phytoremediation of heavy metals–concepts and applications. *Chemosphere* 91, 869–881.

Allah, E. F., Abeer, H., Alqarawi, A. A., and Hend, A. A. 2015. Alleviation of adverse impact of cadmium stress in sunflower (*Helianthus annuus* L.) by arbuscular mycorrhizal fungi. *Pak J Bot* 47, 785–795.

Altinözlü, H., Karagöz, A., Polat, T., and Ünver, İ. 2012. Nickel hyperaccumulation by natural plants in Turkish serpentine soils. *Turkish J Bot* 36, 269–280.

Amstaetter, K., Borch, T., Larese-Casanova, P., and Kappler, A. 2009. Redox transformation of arsenic by Fe (II)-activated goethite (α-FeOOH). *Environ Sci Technol* 44, 102–108.

Ansari, A. A., Gill, S. S., Gill, R., Lanza, G.R., and Lee, N. 2014. *Phytoremediation: management of environmental contaminants*. vol. 1. Springer.

Arora, A., Saxena, S., and Sharma, D. K.2006. Tolerance and phytoaccumulation of chromium by three Azolla species. *World J Microb Biot* 22, 97–100.

Arthur, E. L., Rice, P. J., Rice, P. J., Anderson, T. A., Baladi, S.M., Henderson, K. L., and Coats, J. R. 2005. Phytoremediation—an overview. *Critical Rev Plant Sci*, 109–122.

Audet, P., and Charest, C. 2007. Dynamics of arbuscular mycorrhizal symbiosis in heavy metal phytoremediation: meta-analytical and conceptual perspectives. *Environ Pollut* 147, 609–614.

Ayangbenro, A., and Babalola, O. 2017. A new strategy for heavy metal polluted environments: a review of microbial biosorbents. *Int J Environ Res Public Health* 14, 94.

Azmat, R., Fayyaz, S., Kazi, N., Mahmood, S.J., and Uddin, F. 2008. Natural bioremediation of heavy metals through nematode parasite of fish. *Biotechnol* 7, 139–143.

Babu, A. G., Shea, P. J., Sudhakar, D., Jung, I. B., and Oh, B. T. 2015. Potential use of *Pseudomonas koreensis* AGB-1 in association with Miscanthus sinensis to remediate heavy metal(loid)-contaminated mining site soil. *J Environ Manage* 151, 160–166.

Bae, W., Chen, W., Mulchandani, A., and Mehra, R. 2000. Enhanced bioaccumulation of heavy metals by bacterial cells displaying synthetic phytochelatins. *Biotechnol Bioeng* 70, 518–523.

Bae, W., Wu, C. H., Kostal, J., Mulchandani, A., and Chen, W. 2003. Enhanced mercury biosorption by bacterial cells with surface-displaced MeR. *Appl Environ Microbiol* 69, 3176–3180.

Baldwin, P. R., and Butcher, D. J. 2007. Phytoremediation of arsenic by two hyperaccumulators in a hydroponic environment. *Microchem J* 85, 297–300.

Bani, A., Pavlova, D., Echevarria, G., Mullaj, A., Reeves, R. D., Morel, J. L., and Sulce, S. 2010. Nickel hyperaccumulation by the species of Alyssum and Thlaspi (Brassicaceae) from the ultramafic soils of the Balkans. *Bot Serb* 34, 3–14.

Barnawal, D., Bharti, N., Maji, D., Chanotiya, C. S., and Kalra, A. 2012. 1-Aminocyclopropane-1-carboxylic acid (ACC) deaminase-containing rhizobacteria protect Ocimum sanctum plants during waterlogging stress via reduced ethylene generation. *Plant Physiol Biochem* 58, 227–235.

Bashan, Y., Puente, M. E., De-Bashan, L. E., and Hernandez, J. P. 2008. Environmental uses of plant growth-promoting bacteria. In: Barka, E.A., Clément, C. (Eds.) *Plant microbe interactions* 661, 69–93.

Beyersmann, D., and Hartwig, A. 2008. Carcinogenic metal compounds: recent insight into molecular and cellular mechanisms. *Arch Toxicol* 82, 493–512.

Bharti, R. P., Vastava, A. S., Soni, N., Tiwari, A., More, S., and Choudhary, J. R. 2014. Phytoremediation of heavy metal toxicity and role of soil in rhizobacteria. *Int J Sci Res* 4, 1–5.

Bizily, S. P., Rugh, C. L., Summers, A. O., and Meagher, R. B. 1999. Phytoremediation of methylmercury pollution: Mer B expression in Arabidopsis thaliana confers resistance to organomercurials. *Proc Natl Acad Sci USA* 96, 6808–6813.

Blake, R. C., Choate, D. M., Bardhan, S., Revis, N., Barton, L. L., and Zocco, T. G. 1993. Chemical transformation of toxic metals by a Pseudomonas strain from a toxic waste site. *Environ Toxicol Chem* 12, 1365–1376

Braud, A., Hannauer, M., Mislin, G. L., and Schalk, I. J. 2009. The *Pseudomonas aeruginosa* pyochelin-iron uptake pathway and its metal specificity. *J Bacteriol*191, 3517–3525.

Brunetti, G., Farrag, K., Soler-Rovira, P., Ferrara, M., Nigro, F., and Senesi, N. 2012. The effect of compost and Bacillus licheniformis on the phytoextraction of Cr, Cu, Pb and Zn by three brassicaceae species from contaminated soils in the Apulia region, Southern Italy. *Geoderma* 170, 322–330.

Chang, L.W., Magos, L., and Suzuki, T. 1996. *Toxicology of metals*. Boca Raton, FL: CRC Press.

Chatterjee, S., Sau, G. B., and Mukherjee, S. K. 2009. Plant growth promotion by a hexavalent chromium reducing bacterial strain, *Cellulosimicrobium cellulans* KUCr3. *World J Microbiol Biotechnol* 25, 1829–1836.

Checcucci, A., Bazzicalupo, M., and Mengoni, A. 2017. Exploiting nitrogen-fixing rhizobial symbionts genetic resources for improving phytoremediation of contaminated soils. In: *Enhancing cleanup of environmental pollutants*. Springer.

Chen, L., Luo, S., Li, X., Wan, Y., Chen, J., and Liu, C. 2014. Interaction of Cd-hyperaccumulator *Solanum nigrum* L. and functional endophyte Pseudomonas sp. Lk9 on soil heavy metals uptake. *Soil Biol Biochem* 68, 300–308.

Chibuike, G. U., and Obiora, S. C. 2014. Heavy metal polluted soils: effect on plants and bioremediation methods. *Appl Environ Soil Sci.* 752708.

Das, K., and Roychoudhury, A. 2014. Reactive oxygen species (ROS) and response of antioxidants as ROS-scavengers during environmental stress in plants. *Front Environ Sci* 2, 53.

Datta, R., Das, P., Tappero, R., Punamiya, P., Elzinga, E., Sahi, S., Feng, H., Kiiskila, J., and Sarkar, D. 2017. Evidence for exocellular arsenic in fronds of *Pteris vittata*. *Sci Rep* 7, 2839.

Dermont, G., Bergeron, M., Mercier, G., and Richer-Laflèche, M. 2008. Soil washing for metal removal: a review of physical/chemical technologies and field applications. *J Hazard Mater* 152, 1–31.

Dharni, S., Srivastava, A. K., Samad, A., and Patra, D. D. 2014. Impact of plant growth promoting Pseudomonas monteilii PsF84 and Pseudomonas plecoglossicida PsF610 on metal uptake and production of secondary metabolite (monoterpenes) by rose-scented geranium (Pelargonium graveolens cv. bourbon) grown on tannery sludge amended soil. *Chemosphere* 117, 433–439.

Dinh, N., Van Der Ent, A., Mulligan, D. R., and Nguyen, A.V. 2018. Zinc and lead accumulation characteristics and in vivo distribution of Zn^{2+} in the hyperaccumulator *Noccaea caerulescens* elucidated with fluorescent probes and laser confocal microscopy. *Environ Exp Bot* 147, 1–12.

Dushenkov, S., Kumar, P. B. A. N., Motto, H., and Raskin, I. 1995. Rhizofiltration: the use of plants to remove heavy metals from aqueous streams. *Env Sci Technol* 29, 1239–1245.

Dushenkov, S., Vasudev, D., Kapulnik, Y., Gleba, D., Fleisher, D., Ting, K. C., and Ensley, B. 1997. Removal of uranium from water using terrestrial plants. *Environ Sci Technol* 31, 3468–3474.

Edwards, J., Johnson, C., Santos-Medelli, C., Lurie, E., Podishetty, N. K., Bhatnagar, S., *et al.* 2015. Structure, variation, and assembly of the root-associated microbiomes of rice. *Proc Natl Acad Sci USA* 112, 911–920.

Egamberdieva, D., Abd-Allah, E. F., and da Silva, J. A. T. 2016. Microbially assisted phytoremediation of heavy metal-contaminated soils. *Plant metal interaction*. Elsevier.

Egamberdieva, D., Botir, H., Hashem, A., and Abd-Allah, E. F. 2014. Characterization of salt tolerant Enterobacter hormaechei strain associated with tomato root grown in arid saline soil. *J Pure Appl Microbiol* 8, 4231–4239.

Ehrlich, H. L. 1997. Microbes and metals. *Appl Microbiol Biotechnol* 48, 687–692.

Evelin, H., Kapoor, R., and Giri, B. 2009. Arbuscular mycorrhizal fungi in alleviation of salt stress: a review. *Ann Bot* 104, 1263–1280.

Fein, J. B., Martin, A. M., and Wightman, P. G. 2001. Metal adsorption onto bacterial surfaces: development of a predictive approach. *Geochim Cosmochim Acta* 65, 4267–4273.

Fomina, M., Hillier, S., Charnock, J. M., Melville, K., Alexander, I. J., and Gadd, G. M. 2005. Role of oxalic acid overexcretion in transformations of toxic metal minerals by Beauveria caledonica. *Appl Environ Microbiol* 71, 371–381.

Frey, B., Zierold, K., and Brunner, I. 2000. Extracellular complexation of Cd in the Hartig net and cytosolic Zn sequestration in the fungal mantle of Picea abies–Hebeloma crustuliniforme ectomycorrhizas. *Plant Cell Environ* 23, 1257–1265.

Gadd, G. M. 2004. Microbial influence on metal mobility and application for bioremediation. *Geoderma* 122, 109–119.

Gadd, G. M., Bahri-Esfahani, J., Li, Q., Rhee, Y. J., Wei, Z., Fomina, M., and Liang, X. 2014. Oxalate production by fungi: significance in geomycology, biodeterioration and bioremediation. *Fungal Biol Rev* 28, 36–55.

Garbisu, C., and Alkorta, I. 2001. Phytoextraction: a cost-effective plant-based technology for the removal of metals from the environment. *Bioresour Technol* 77, 229–236.

García-Salgado, S., García-Casillas, D., Quijano-Nieto, M. A., and Bonilla-Simón, M. M. 2012. Arsenic and heavy metal uptake and accumulation in native plant species from soils polluted by mining activities. *Water Air Soil Pollut* 223, 559–572.

Garg, N., and Aggarwal, N. 2011. Effects of interactions between cadmium and lead on growth, nitrogen fixation, phytochelatin, and glutathione production in mycorrhizal *Cajanus cajan* (L.) Millsp. *J Plant Growth Regul* 30, 286–300.

Garg, N., and Pandey, R. 2015. Effectiveness of native and exotic arbuscular mycorrhizal fungi on nutrient uptake and ion homeostasis in salt-stressed *Cajanus cajan* L. (Millsp.) genotypes. *Mycorrhiza* 25, 165–180.

Glick, B. R. 1995. The enhancement of plant growth by free-living bacteria. *Can J Microbiol* 41, 109–117.

Glick, B. R., and Bashan, Y. 1997. Genetic manipulation of plant growth-promoting bacteria to enhance biocontrol of phytopathogens. *Biotechnol Adv* 15, 353–378.

Gonzalez-Guerrero, M., Melville, L. H., Ferrol, N., Lott, J. N., Azcon-Aguilar, C., and Peterson, R. L. 2008. Ultrastructural localization of heavy metals in the extraradical mycelium and spores of the arbuscular mycorrhizal fungus Glomus intraradices. *Can J Microbiol* 54, 103–110.

Goodell, B., Jellison, J., Liu, J., Daniel, G., Paszczynski, A., Fekete, F., Krishnamurthy, S., Jun, L., and Xu, G. 1997. Low molecular weight chelators and phenolic compounds isolated from wood decay fungi and their role in the fungal biodegradation of wood. *J Biotechnol* 53, 133–162.

Gube, M. 2016. 4 Fungal molecular response to heavy metal stress. *Biochemistry molecular biology*. Springer.

Gupta, P., and Diwan, B. 2017. Bacterial exopolysaccharide mediated heavy metal removal: a review on biosynthesis, mechanism and remediation strategies. *Biotechnol Rep* 13, 58–71.

Gutierrez-Corona, J. F., Romo-Rodriguez, P., Santos-Escobar, F., Espino-Saldana, A. E., and Hernandez-Escoto, H. 2016. Microbial interactions with chromium: basic biological processes and applications in environmental biotechnology. *World J Microbiol Biotechnol* 32, 191.

Heaton, A. C., Rugh, C. L., Wang, N. J., and Meagher, R. B. 1998. Phytoremediation of mercury-and methylmercury-polluted soils using genetically engineered plants. *J Soil Contam*7, 497–509.

Henry, J. R. 2000. In: An overview of phytoremediation of lead and mercury. NNEMS Report, Washington, DC, pp. 3–9.

Henry, J. R. 2000. Overview of the phytoremediation of lead and mercury. In: *Overview of the phytoremediation of lead and mercury*. EPA.

Huang, Y., and Tao, S. 2005. The role of arbuscular mycorrhiza on change of heavy metal speciation in rhizosphere of maize in wastewater irrigated agriculture soil. *J Environ Sci* 17, 276–280.

Jadia, C. D., and Fulekar, M. H. 2009. Phytoremediation of heavy metals: recent techniques. *Afr J Biotechnol* 8, 6.

Jiang, C. Y., Sheng, X. F., Qian, M., and Wang, Q. Y. 2008. Isolation and characterization of a heavy metal-resistant Burkholderia sp. from heavy metal-contaminated paddy field soil and its potential in promoting plant growth and heavy metal accumulation in metal-polluted soil. *Chemosphere* 72, 157–164.

Jing, Y-D., Zhen-Li, H., and Yang, X-e. 2007. Role of soil rhizobacteria in phytoremediation of heavy metal contaminated soils. *J Zhejiang Univ Sci B*, 8, 192–207.

Joner, E. J., and Leyval, C. 2000. Bioavailability of heavy metals in the mycorrhizosphere. In: *Trace elements in the rhizosphere*. CRC Press.

Kamaludeen, S. P. B., and Ramasamy, K. 2008. Rhizoremediation of metals: harnessing microbial communities. *Indian J Microbiol* 48, 80–88.

Kavamura, V. N., and Esposito, E. 2010. Biotechnological strategies applied to the decontamination of soils polluted with heavy metals. *Biotechnol Adv* 28, 61–69.

Khan, A. G., Kuek, C., Chaudhry, T. M., Khoo, C. S., and Hayes, W. J. 2000. Role of plants, mycorrhizae and phytochelators in heavy metal contaminated land remediation. *Chemosphere* 41, 197–207.

Khan, M. U., Sessitsch, A., Harris, M., Fatima, K., Imran, A., Arslan, M., Shabir, G., Khan, Q. M., and Afzal, M. 2015. Cr-resistant rhizo-and endophytic bacteria associated with Prosopis juliflora and their potential as phytoremediation enhancing agents in metal-degraded soils. *Front Plant Sci* 5, 755.

Kloepper, J. W., Lifshitz, R., and Zablotowicz, R. M. 1989. Free-living bacterial inocula for enhancing crop productivity. *Trends Biotechnol* 7, 39–44.

Kohler, J., Hernandez, J. A., Caravaca, F., and Roldán, A. 2009. Induction of antioxidant enzymes is involved in the greater effectiveness of a PGPR versus AM fungi with respect to increasing the tolerance of lettuce to severe salt stress. *Environ Exp Bot* 65, 245–252.

Kong, Z., and Glick, B. R. 2017. The role of plant growth-promoting bacteria in metal phytoremediation. *Advances in microbial physiology*, Elsevier.

Kucharski, R., Sas-Nowosielska, A., Małkowski, E., Japenga, J., Kuperberg, J. M., Pogrzeba, M., and Krzyżak, J. 2005. The use of indigenous plant species and calcium phosphate for the stabilization of highly metal-polluted sites in southern Poland. *Plant Soil* 273, 291–305.

Kuiper, I., Lagendijk, E. L., Bloemberg, G.V., and Lugtenberg, B. J. 2004. Rhizoremediation: a beneficial plant-microbe interaction. *Mol Plant Microbe Interact* 17, 6–15.

Kuramshina, Z. M., Smirnova, Y. V., and Khairullin, R. M. 2018. Cadmium and nickel toxicity for Sinapis alba plants inoculated with endophytic strains of *Bacillus subtilis*. *Russ J Plant Physiol* 65, 269–277.

Kvesitadze, G., Khatisashvili, G., Sadunishvili, T., and Ramsden, J. J. 2006. *Biochemical mechanisms of detoxification in higher plants: basis of phytoremediation*. Springer Science & Business Media.

Leyval, C., Turnau, K., and Haselwandter, K. 1997. Effect of heavy metal pollution on mycorrhizal colonization and function: physiological, ecological and applied aspects. *Mycorrhiza* 7, 139–153.

Li, X., Liu, L., Wang, Y., Luo, G., Chen, X., Yang, X., Hall, M. H., Guo, R., Wang, H., Cui, J., and He, X. 2013. Heavy metal contamination of urban soil in an old industrial city (Shenyang) in Northeast China. *Geoderma* 192, 50–58.

Louati, H., Said, O. B., Soltani, A., Cravo-Laureau, C., Duran, R., Aissa, P., Mahmoudi, E., and Pringault, O. 2015. Responses of a free-living benthic marine nematode community to bioremediation of a PAH mixture. *Environ Sci Pollut Res* 22, 15307–15318.

Ma, Y., Oliveira, R. S., Freitas, H., and Zhang, C. 2016. Biochemical and molecular mechanisms of plant-microbe-metal interactions: relevance for phytoremediation. *Front Plant Sci* 7, 918.

Ma, Y., Rajkumar, M., and Freitas, H. 2009. Inoculation of plant growth promoting bacterium Achromobacter xylosoxidans strain Ax10 for the improvement of copper phytoextraction by *Brassica juncea*. *J Environ Manage* 90, 831–837.

Ma, Y., Rajkumar, M., Luo, Y., and Freitas, H. 2013. Phytoextraction of heavy metal polluted soils using Sedum plumbizincicola inoculated with metal mobilizing *Phyllobacterium myrsinacearum* RC6b. *Chemosphere* 93, 1386–1392.

Machuca, A., Pereira, G., Aguiar, A., and Milagres, A. M. F. 2007. Metal-chelating compounds produced by ectomycorrhizal fungi collected from pine plantations. *Lett Appl Microbiol* 44, 7–12.

Majumder, A., Bhattacharyya, K., Bhattacharyya, S., and Kole, S. C. 2013. Arsenic-tolerant, arsenite-oxidising bacterial strains in the contaminated soils of West Bengal, India. *Science Total Environ* 463, 1006–1014.

Mendez, M. O., and Maier, R. M. 2007. Phytostabilization of mine tailings in arid and semiarid environments-an emerging remediation technology. *Environ Health Perspect* 116, 278–283.

Merdy, P., Gharbi, L. T., and Lucas, Y. 2009. Pb, Cu and Cr interactions with soil: sorption experiments and modelling. *Colloids Surf A: Physicochem Eng Asp* 347, 192–199.

Miransari, M. 2011. Hyperaccumulators, arbuscular mycorrhizal fungi and stress of heavy metals. *Biotechno Adv* 29, 645–653.

Miransari, M. 2017. Arbuscular mycorrhizal fungi and heavy metal tolerance in plants. *Arbuscular mycorrhizas and stress tolerance of plants*. Springer.

Mishra, J., Singh, R., and Arora, N. K. 2017. Alleviation of heavy metal stress in plants and remediation of soil by rhizosphere microorganisms. *Front Microbiol* 8, 1706.

Moreno, F. N., Anderson, C. W., Stewart, R. B., and Robinson, B. H. 2005. Mercury volatilisation and phytoextraction from base-metal mine tailings. *Environ Pollut*136,341–352.

Mueller, B., Rock, S., Gowswami, D., and Ensley, D. 1999. Phytoremediation decision tree. *Prepared by Interstate Technology and Regulatory Cooperation Work Group*, pp.1–36. https://clu-in.org/download/partner/phytotree.pdf

Mukhopadhyay, S., and Maiti, S. K. 2010. Phytoremediation of metal enriched mine waste: a review. *Global J Environ Res* 4, 135–150.

Mulligan, C. N., Yong, R. N., and Gibbs, B. F. 1999. Removal of heavy metals from contaminated soil and sediments using the biosurfactant surfactin. *J Soil Contam* 8, 231–254.

Mulligan, C. N., Yong, R. N., and Gibbs, B. F. 2001. Remediation technologies for metal-contaminated soils and groundwater: an evaluation. *Eng Geol* 60, 193–207.

Murthy, S., Bali, G., and Sarangi, S. K. 2011. Effect of lead on metallothionein concentration in lead resistant bacteria Bacillus cereus isolated from industrial effluent. *Afr J Biotechnol* 10, 15966–15972.

Nagata, S., Yamaji, K., Nomura, N., and Ishimoto, H. 2015. Root endophytes enhance stress-tolerance of *Cicuta virosa* L. growing in a mining pond of eastern Japan. *Plant Species Biol* 30, 116–125.

Narendrula-Kotha, R., and Nkongolo, K. K. 2017. Microbial response to soil liming of damaged ecosystems revealed by pyrosequencing and phospholipid fatty acid analyses. *PloS One* 12, 1.

Neil, W. 2007. *Phytoremediation: methods and reviews*. Humana Press, Totowa, pp. V–VII.

Ojuederie, O., and Babalola, O. 2017. Microbial and plant-assisted bioremediation of heavy metal polluted environments: a review. *Int J Environ Res Public Health*, 14, 1504.

Parihar, M., Meena, V. S., Mishra, P. K., Rakshit, A., Choudhary, M., Yadav, R. P., Rana, K., and Bisht, J. K. 2019a. Arbuscular mycorrhiza: a viable strategy for soil nutrient loss reduction. *Arch Microbiol* 201(6), 723–735.

Parihar, M., Rakshit, A., Singh, H. B., and Rana, K. 2019b. Diversity of arbuscular mycorrhizal fungi in alkaline soils of hot sub humid eco-region of Middle Gangetic Plains of India. *Acta Agric Scand, Section B-Soil & Plant Science*, 1–12.

Phieler, R., Voit, A., and Kothe, E. 2013. Microbially supported phytoremediation of heavy metal contaminated soils: strategies and applications. *Geobiotechnology I*. Springer.

Pires, C., Franco, A. R., Pereira, S. I., Henriques, I., Correia, A., Magan, N., and Castro, P. M. 2017. Metalloid-contaminated soils as a source of culturable heterotrophic aerobic bacteria for remediation applications. *Geomicrobiol J* 34, 760–768.

Rai, P. K. 2008. Phytoremediation of Hg and Cd from industrial effluents using an aquatic free floating macrophyte *Azolla pinnata*. *Int J Phytoremediat* 10, 430–439.

Rajkumar, M., Ae, N., Prasad, M. N. V., and Freitas, H. 2010. Potential of siderophore-producing bacteria for improving heavy metal phytoextraction. *Trends Biotechnol* 28, 142–149.

Rakshit, A., Pal, S., Parihar, M., and Singh, H. B. 2017. Bioremediation of soils contaminated with Ni and Cd: an overview. *Adaptive soil management: From theory to practices*. Springer.

Raskin, I., Smith, R. D., and Salt, D. E. 1997. Phytoremediation of metals: using plants to remove pollutants from the environment. *Curr Opin Biotech* 8, 221–226.

Saif, S., and Khan, M. S. 2018. Assessment of toxic impact of metals on proline, antioxidant enzymes, and biological characteristics of Pseudomonas aeruginosa inoculated *Cicer arietinum* grown in chromium and nickel-stressed sandy clay loam soils. *Enviro Monit Assess* 190, 290.

Sakakibara, M., Ohmori, Y., Ha, N. T. H., Sano, S., and Sera, K. 2011. Phytoremediation of heavy metal contaminated water and sediment by *Eleocharis acicularis*. *Clean Soil Air Water* 39, 735–741.

Salazar, M. J., and Pignata, M. L. 2014. Lead accumulation in plants grown in polluted soils. Screening of native species for phytoremediation. *J Geochem Explor* 137, 29–36.

Salt, D. E., Blaylock, M., Kumar, N. P., Dushenkov, V., Ensley, B. D., Chet, I., and Raskin, I. 1995. Phytoremediation: a novel strategy for the removal of toxic metals from the environment using plants. *Biotechnol* 13, 468.

Saravanan, V. S., Osborne, J., Madhaiyan, M., Mathew, L., Chung, J., Ahn, K., and Sa, T. 2007. Zinc metal solubilization by Gluconacetobacter diazotrophicus and induction of pleomorphic cells. *J Microbiol Biotechnol* 17, 1477–1482.

Schnoor, J. L. 1997. Phytoremediation: Technology evaluation report (TE-98-01). Groundwater Remediation Technologies Analysis Center.

Schüßler, A., Schwarzott, D., and Walker, C. 2001. A new fungal phylum, the Glomeromycota: phylogeny and evolution. *Mycol Res* 105, 1413–1421.

Seneviratne, M., Seneviratne, G., Madawala, H. M. S. P., and Vithanage, M. 2017. *Role of rhizospheric microbes in heavy metal uptake by plants*. Agro-Environmental Sustainability Springer.

Sepheri, M., Habib, K., and Mehdi, Z. 2012. Fungi and their role in phytoremediation of heavy metal-contaminated soils. *Fungi as bioremediators*. Springer.

Seshadri, B., Bolan, N. S., and Naidu, R. 2015. Rhizosphere-induced heavy metal (loid) transformation in relation to bioavailability and remediation. *Soil Sci Plant Nutr* 15, 524–548.

Shabani, L., and Sabzalian, M. R. 2016. Arbuscular mycorrhiza affects nickel translocation and expression of ABC transporter and metallothionein genes in Festuca arundinacea. *Mycorrhiza* 26, 67–76.

Sharma, J., Shamim, K., Dubey, S. K., and Meena, R. M. 2017. Metallothionein assisted periplasmic lead sequestration as lead sulfite by Providencia vermicola strain SJ2A. *Sci Total Environ* 579, 359–365.

Sheng, X. F., Xia, J. J., Jiang, C. Y., He, L. Y., and Qian, M. 2008. Characterization of heavy metal-resistant endophytic bacteria from rape (*Brassica napus*) roots and their potential in promoting the growth and lead accumulation of rape. *Environ Pollut* 156, 1164–1170.

Shi, L., Dong, H., Reguera, G., Beyenal, H., Lu, A., Liu, J., *et al.* 2016. Extracellular electron transfers mechanisms between microorganisms and minerals. *Nat Rev Microbiol* 14, 651–662.

Sikdar, S., and Kundu, M. 2018. A review on detection and abatement of heavy metals. *Chem Bio Eng Rev* 5, 18–29.

Singh, S. 2012. Phytoremediation: a sustainable alternative for environmental challenges. *Int J Gr Herb Chem* 1, 133–139.

Sun, R., Jin, C., and Zhou, Q. 2010. Characteristics of cadmium accumulation and tolerance in Rorippa globosa (Turcz.) Thell., a species with some characteristics of cadmium hyperaccumulation. *Plant Growth Regul* 61, 67–74.

Susarla, S., Medina, V. F., and McCutcheon, S. C. 2002. Phytoremediation: an ecological solution to organic chemical contamination. *Ecol Eng* 18, 647–658.

Sytar, O., Brestic, M., Taran, N., and Zivcak, M. 2016. Plants used for biomonitoring and phytoremediation of trace elements in soil and water. *Plant metal interaction*. Elsevier.

Tirry, N., Joutey, N. T., Sayel, H., Kouchou, A., Bahafid, W., Asri, M., and El Ghachtouli, N. 2018. Screening of plant growth promoting traits in heavy metals resistant bacteria: prospects in phytoremediation. *J Genetic Eng Biotechnol* 16, 613–619.

Turnau, K., and Kottke, I. 2005. Fungal activity as determined by microscale methods with special emphasis on interactions with heavy metals. In: Dighton, J. and White, J. F. (Eds.) *The fungal community*. Boca Raton, CRC Press, pp. 287–305.

Ullah, A., Heng, S., Munis, M. F. H., Fahad, S., and Yang, X. 2015. Phytoremediation of heavy metals assisted by plant growth promoting (PGP) bacteria: a review. *Environ Exp Bot* 117, 28–40.

Upadhyay, S. K., and Singh, D. P. 2015. Effect of salt-tolerant plant growth-promoting rhizobacteria on wheat plants and soil health in a saline environment. *Plant Biol* 17, 288–293.

Vallero, D. A. 2010. Applied microbial ecology: bioremediation. *Environ Biotechnol*. pp 325–400. Academic, San Diego. http://dx.doi.org/10.1016/B978-0-12-375089-1.10007-8

Vamerali, T., Bandiera, M., and Mosca, G. 2010. Field crops for phytoremediation of metal-contaminated land. A review. *Environ Chem Lett* 8, 1–17.

Visioli, G., D'Egidio, S., Vamerali, T., Mattarozzi, M., Sanangelantoni, A. M. 2014. Culturable endophytic bacteria enhance Ni translocation in the hyperaccumulator *Noccaea caerulescens*. *Chemosphere* 117, 538–544.

Wang, H. B., Xie, F., Yao, Y. Z., Zhao, B., Xiao, Q. Q., Pan, Y. H., and Wang, H. J. 2012. The effects of arsenic and induced-phytoextraction methods on photosynthesis in Pteris species with different arsenic-accumulating abilities. *Environ Exp Bot* 75, 298–306.

Yang, Y., Liang, Y., Han, X., Chiu, T. Y., Ghosh, A., Chen, H., and Tang, M. 2016. The roles of arbuscular mycorrhizal fungi (AMF) in phytoremediation and tree-herb interactions in Pb contaminated soil. *Sci Rep* 6, 20469.

Yoon, J., Cao, X., Zhou, Q., and Ma, L. Q. 2006. Accumulation of Pb, Cu, and Zn in native plants growing on a contaminated Florida site. *Sci Total Environ* 368, 456–464.

Yuan, M., He, H., Xiao, L., Zhong, T., Liu, H., Li, S., Deng, P., Ye, Z., and Jing, Y. 2014. Enhancement of Cd phytoextraction by two Amaranthus species with *endophytic Rahnella sp.* JN27. *Chemosphere* 103, 99–104.

Zaidi, S., and Musarrat, J. 2004. Characterization and nickel sorption kinetics of a new metal hyper-accumulator *Bacillus sp.* *J Environ Sci Health A* 39, 681–691.

Zaidi, S., Usmani, S., Singh, B. R., Musarrat, J. 2006. Significance of *Bacillus subtilis* strain SJ-101 as a bioinoculant for concurrent plant growth promotion and nickel accumulation in *Brassica juncea*. *Chemosphere* 64, 991–997.

Zhang, H., Chen, X., He, C., Liang, X., Oh, K., Liu, X., Lei, Y. 2015. Use of energy crop (*Ricinus communis* L.) for phytoextraction of heavy metals assisted with citric acid. *Int J Phytoremediat* 17, 632–639.

Zhang, Z., Wen, X., Huang, Y., Inoue, C. Liang, Y. 2017. Higher accumulation capacity of cadmium than zinc by Arabidopsis halleri ssp. Germmifera in the field using different sowing strategies. *Plant Soil* 418, 165–176.

Zhao, F. J., and McGrath, S. P. 2009. Biofortification and phytoremediation. *Curr Opin Plant Biol* 12, 373–380.

13

Strategies to Reduce Heavy Metal Contamination in Soil-Plant System

Muhammad Bilal Shakoor
College of Earth and Environmental Sciences, University of the Punjab, Lahore, Pakistan

Usman Iftikhar
Department of Environmental Sciences and Engineering, Government College University Faisalabad, Faisalabad, Pakistan

Sajid Rashid Ahmad and Sana Ashraf
College of Earth and Environmental Sciences, University of the Punjab, Lahore, Pakistan

Mujahid Farid
Department of Environmental Sciences, University of Gujrat, Hafiz Hayat Campus, Gujrat, Pakistan

Tanzeela Kokab
College of Earth and Environmental Sciences, University of the Punjab, Lahore, Pakistan

CONTENTS

13.1 Introduction

Extensive disposal of contaminants in the agricultural fields has put considerable pressure on all compartments of the environment (Grimm et al. 2008; Alloway 2013). Soil contamination with the toxic heavy metals is major problem for food safety and human health. Approximately 5 million locations of soil contamination in about 500 million ha of area have been identified where heavy metals are present more than permissible levels (Liu et al. 2015). The main causes of heavy metals entry into soil and water bodies are human practices such as warfare and military relevant products, fossil fuel, pesticides, electronics, smelting, mining, solid/liquid waste addition and contaminated water for irrigation. For instance, the coal has a variety of heavy metals for instance nickel (Ni), lead (Pb), cobalt (Co), cadmium (Cd), copper (Cu), zinc (Zn), chromium (Cr) and mercury (Hg) and the concentration of these heavy metals spans from 0.1 to 18 mg kg^{-1}. The abovementioned toxic elements are released in air or soil in flue gas emission, bottom ash, vapors and fly ash during combustion of petroleum products

DOI: 10.1201/9781003155089-13

(Nalbandian 2012). Improper dumping of mines related wastes, construction and industrial wastes in agricultural soil mainly results in serious contamination of soils. Application of phosphorus (P) rich-chemicals, biosolids, Cu-containing pesticides/insecticides, manure application and edible crop irrigation with domestic wastewater are important sources from which toxic heavy metals enter into the crop fields (Bolan et al. 2014).

Contrary to organic contaminants, the heavy metals in soil are persistent, irreversible and resistant to degradation. The pollution of heavy metals in the soils can cause increased dietary exposure via the soil–plant–food chain transfers. A number of adverse impacts were caused by Cd on the cellular compounds largely by inducing oxidant–antioxidant disturbance in plants. Moreover, Cd is a known carcinogen, caused myocardial infarction, peripheral artery illness, diabetic nephropathy, hypertension and most importantly itai–itai disorder in Japan (Ghosh and Indra 2018). Lead is not considered to be an essential metal for humans infect higher Pb consumption can result in damaging effects on skeletal, nervous, endocrine, enzymatic, circulatory and immune systems of the human body (Kankia and Abdulhamid 2014). Hence, the elimination of heavy metals from soils is not just essential to protect pollution from the sources as well as to improve the treatment potential of contaminated crop fields. For many years, different in-situ and ex-situ techniques are being used to restrict, remediate and restore the contaminated crop fields for example surface capping, extraction, soil flushing, electrokinetic, vitrification, solidification and phytoremediation. Different methods have been applied for heavy metals treatment for example thermal remediation, solidification/stabilization, chemical leaching, phytoremediation for Hg, Zn, Cd, Pb, respectively (Hseu et al. 2014; Al-Wabel et al. 2014; Babu et al. 2013; Makino et al. 2007).

The remediation techniques are usually categorized into five types such as chemical, physical, electrical, biological and thermal treatments. They can be also be classified into three divisions: based on containment, transport and transformation methods. Generally, soil remediation techniques follow varying mechanisms and have different pros and cons. These soil treatment techniques differ substantially with respect to efficiency and cost-effectiveness in when applied in the field (Khalid et al. 2017). Major factors affecting the choice and applicability of the soil remediation methods include long-term persistence, cost, large-scale accessibility, excess of selected heavy metal level, acceptance, physicochemical characteristics (volume, mobility, toxicity) and type of media (heavy metals or organic pollutants). Here we aimed to review the existing soil remediation methods for heavy metals polluted agricultural fields considering their technical processes, merits and demerits, working principles, potential applicability and application status. Data from this book chapter is anticipated to help in choosing suitable remediation methods for elimination of heavy metals pollution from urban and agricultural fields with respect to specific scenarios.

13.2 Sources of Heavy Metal Exposure of Polluted Soils

Heavy metals and metalloids are originated from the natural soil environment (described as lithosphere source) and through many human activities. Mainly metals and metalloids are present in such a form that they are not directly available for uptake by plants. Naturally, by the process of weathering of parent material, heavy metal (loid)s are present at trace levels (1000 mg kg^{-1}) in the soil environment, and these are toxic substances (Pierzynski et al. 2000; Kabata-Pendias and Pendias 2001). However, these are more bioavailable to plant uptake when produced by anthropogenically (Lamb et al. 2009).

Soils are more prone to different groups of heavy metals pollutants, which entirely rely on different kinds of industry sectors. These toxic metals are behaved like pollutants in soils due to these factors (1) the rate of their production is more increased by anthropogenic sources as compared to natural sources, (2) they become hazardous by exposing directly to the environment due to their transport activities, (3) the quantity of heavy metals is more in the scraped product in comparison to the receiving soils, (4) the heavy metals are bioavailable in soils when they present in their chemical form (i.e., species). Various countries have established different regulations regarding the concentration of heavy metals in the soil environment to minimize the risk of contamination and ensure the safety of natural resources (Table 13.1).

TABLE 13.1

Permissible Concentration (mg kg^{-1}) of Various Metals in the Soils of Different Countries

Heavy Metal	Canada	America	European Union	Australia	Japan	Taiwan, China	Mainland, China
Ni	≤50	≤72	≤100	≤60		≤200	≤40–≤60 depending on pH
As	≤12	≤0.11	≤50	≤20	In soil solution, ≤0.01 mg L^{-1} and in paddy soil <15 mg kg^{-1}	≤60	≤30/40–≤20/25 depending on pH
Se				≤1	In soil solution, ≤0.01 mg L^{-1}		
Zn	≤200	≤1100	≤250	≤200		≤600	≤200–≤300 depending on pH
Cu	≤63	≤270	≤150	≤100	In paddy soil, <125 mg kg^{-1}	≤200	≤50–≤100 depending on pH
Hg	≤6.6	≤1	≤2	≤1	≤0.5 μg L^{-1} for total Hg	≤2	≤0.30–≤1.0 depending on pH
Cr	≤64	≤11	≤200	≤50	In soil solution, ≤0.05 mg L^{-1} as Cr(VI)	≤250	≤250/150–≤350/250 depending on pH
Cd	≤1.4	≤0.48	≤10	≤3	In soil solution, ≤0.01 mg L^{-1} and in grown rice <0.4 mg kg^{-1}	≤5	≤0.30–≤0.60 depending on pH
Pb	≤70	≤200	≤200	≤300	≤0.01 mg L^{-1} in soil solution	≤500	≤250–≤350 depending on pH

13.3 In Situ and Ex Situ Remediation Methods

13.3.1 In Situ Methods

During the treatment by in situ methods extraction and transfer of polluted soil to off-site waste disposal are not required, thus minimizing land degradation, decreasing sensitivity to staff and the local population to pollutants, and significantly reducing treatment costs (Figure 13.1 and Table 13.2). Nevertheless, actual field variables, including the permeability of the soil, temperature, pollution depth, and possible deep chemical leachate, need to be carefully considered (Olexsey and Parker 2006).

13.3.1.1 Surface Capping

The polluted site is coated with a waterproof coating to form a secure, safe surface. This method of remediation used in this situation is not actually "soil remediation" since no steps are taken to reduce heavy metals accumulation in the soil. While there is effectiveness, there is a chance of exposure to polluted soil either by physical contact or unintentional swallowing. The surface cap is an impermeable shield to the absorption of soil pollutants into groundwater, preventing surface water pollution from touching the surface. However, the soil loses its natural functions when it is capped, thereby affecting the plant growth. The place can be reused as either a football ground or a parking place (NJDEP 2014). A capping system that complies with the environmental aspects and appropriate treatment goals should be selected. Various capping materials may be used for single-layer caps or multilayer structures, containing asphalt, clay, high-density polyethene, and concrete. With sufficient structural strength and dynamic integrity, the surface layer can stretch 60–90 cm to the pollution site limit (Rumer and Ryan 1995). Dikes, slopes and ditches are built to irrigation and divert runoff from surface caps. Re-vegetation of multilayer cap systems can be accomplished by the use of topsoil upon the impermeable cap layer. This method is suitable for high concentrations of toxins in soils (e.g., $I_{geo} > 3$). This technique is only favourable for small areas. Without such a cap covering, the building of the surface becomes impossible (Ohio EPA 2000). At the remediation site and hydrogeology of the site, the seasonal and depth changes must be considered during the treatment by remediation. If the "sliding risk" is measured, the cap must be set on a sloped floor. When remediation is performed around residential neighbourhoods, beaches, or walkways, there has to be a concern for the neighbour's approval. Surface capping provides the fastest and most efficient way to minimize damage from soil contamination.

13.3.1.2 Soil Flushing

In this technique, extraction fluid is usually pass through the contaminated soil to eliminate the heavy metals. This fluid is processed and disposed of, ensuring it is reused or recycled. This technology can be applied to coarse-textured soils that are particularly permeable (CLU-IN 2017). It requires adding

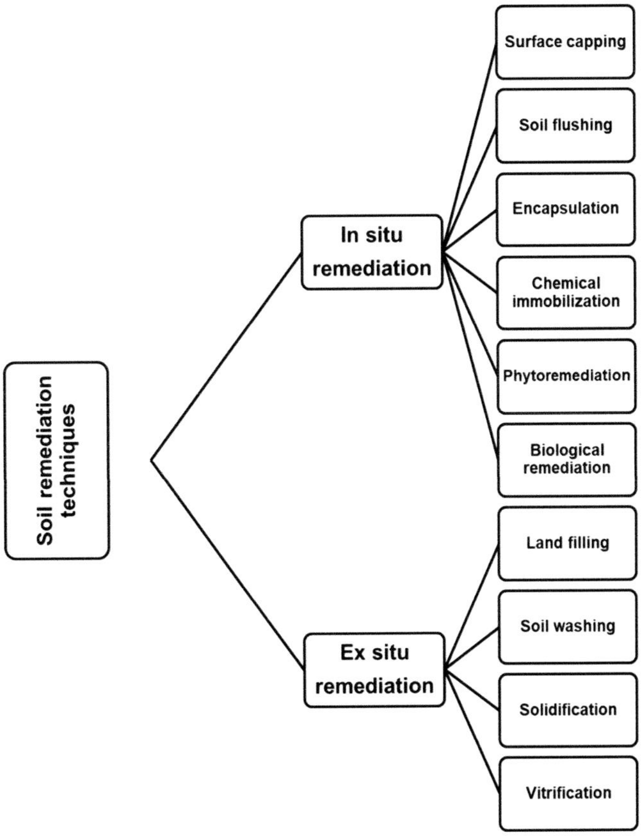

FIGURE 13.1 Soil remediation methods used for the treatment of heavy metals.

TABLE 13.2

Summary of Remediation Methods Applied for the Treatment of Heavy Metals Contaminated Soils

Method	Applicability	Mechanisms	Benefits	Drawbacks	Current Status
Surface-capping	High contamination load and in-situ application	Heavy metals are physical contained	Less cost, most secure, simple to fix	For small locations, specific geological area, land cultivation loss	Extensively applied
Encapsulation	In-situ method for highly polluted site	Isolation and physical containment	Installation is swift and highly secure	Small-scale less deep contaminated areas, expensive, crops are not able to grow	Treatment of mixed and radioactive waste
Electrokinetics	In-situ high to moderate contamination	Electricity is used to remove pollutants	Minimum soil eruption, contaminant removal	Time taking, less efficacy, suitable for soils with fine texture having not much permeability	Under progress with pilot scale illustrations
Soil flushing	It is an in-situ method, suitable for high to moderate pollution and for coarse soil	Solutions of different chemicals are used to remove the pollutants	Easy installation, less soil disturbance, requires cost-effective, contaminant elimination	Suitable for soils with coarse-textured, high potential for groundwater contamination	Used for mixed waste but limited applicability
Immobilization/ stabilization	In-situ method for highly pollutant loads	Physiochemical transformation of the pollutants by deactivating them	Promptly effective Easy to implement and requires less cost	For specific metals, provide temporary efficacy and pollutants retained in soil	Not approved legally, and removal is momentary
Phytoremediation	In-situ method with contamination treatment (low to moderate)	Plants are used to remove pollutants	Accepted by public, cost is less, fit for large areas, easy to implement	For shallow pollution, time taking, less effective, metals-specific application	Developing with pilot illustrations
Biological remediation	Do	Microorganisms are used to treat the pollutants	Less costly, simple to execute, minimum disturbance of soil	Less efficient, applied as a supplementary method major treatment methods	Not widely applied for heavy metals removal, only for organic pollutants
Landfilling	Ex-situ, high contamination	Physical transformation of the pollutants by deactivating them	Safe to use, quick cleanup	Cost effective, need land for dumping	Applicable broadly
Washing of soil	Moderate to high contamination level and applied as ex-situ method	Use chemicals and mechanically means for the removal of pollutants	Quick results, highly effective	Soil is disturbed largely	In use on regular basis
Vitrification	Both in situ and ex situ applications, high contamination	Pollutants are deactivated by thermal vitrification	High efficacy	Cost effective, only for small sites, environmental features of soil is lost	In use on regular basis
Solidification	Do	Physical solidification of soil	Can be executed swiftly, and efficacy is higher	Cost effective, only for small sites, environmental features of soil is lost	In use on regular basis

an extraction solvent to the soil environment. In order to efficiently remove heavy metals from the soil, an extraction fluid with a specific formula must be created. A variety of agents have been used to chelate heavy metals, and EDTA is the most efficient. Using batch tests, it was observed that removing metalloids from a reinforced loamy sand (pH 6.1, organic matter (OM) concentration 8.7%) at 0.01 M and 1:25 soil/solution proportion. EDTA has succeeded across tartaric and citric acid. The solutions have differing degrees of metal mobilization efficiencies depending on the metal species (Cu, Ni, Zn, Cd, Pb) being mobilized. The concurrent extraction of the metal fractions revealed that EDTA mobilized more non-residual percentages of metals than tartaric and citric acid (Wuana et al. 2010).

Soil flushing requires a technical and expensive installation for storage well or subsurface draining, one can also extract the flushing elutriate through groundwater if the aquifer is shallow. In soils with high OM and high buffering ability and clay content, metal extraction its ability is low (Shammas 2009).

The effectiveness of the procedure varies depending on the soil, its thin layers, and its orientation. The projected cost of flushing the soil was between \$20 and \$104 million gallons per square mile, which rises as porousness of soil reduces, and the aquifer extends (Iturbe et al. 2004; FRTR 2012). The procedure for extracting contaminants from toxic areas has been more widely performed with surfactant solutions than other techniques.

13.3.1.3 Encapsulation

This method also referred to as liner, barrier wall, or cutoff wall process, is a rehabilitative approach parallel to surface capping. Contaminated soil is trapped in a correctly constructed physical obstacle device having less permeability covers, enveloping underwater walls and, in rare cases, shield floors. Polluted areas are segregated, and toxins are sealed, preventing the off-site distribution of chemicals and bio-exposure of chemicals on-site (Khan et al. 2004; Meuser 2013). The low permeability covers, typically plastic cloth sheets or clay layers, reduce the infiltration of surface water and thereby avoid pollutants from escaping through groundwater. Underground impermeable contraction is the source of toxins from horizontal movement to adjacent areas by possible diffusion and subsurface flow. The critical problem in encapsulation is the design of underground horizontal impermeable barriers at pollution sites. Various installation systems have been applied, like slurry walls, sheet pile walls, thin walls, and walls for injection (Meuser 2013). To create trenches, the trench must first be excavated in the field using a ditch cutter or a bucket in slurry wall technique. After the asphalt is mixed with bentonite clay, it is poured into the trench. Thin HDPE board, brick-like pile of stone, and glass can be used in place of slurry wall materials. The wall could be made up of several parts, but the walls must be integrated to ensure the integrity of the wall. In a wall structure, a strong beam is vibrated through the earth through contact with a vibrating layer. Clay-cement-water slurry is then poured into the trench to create a wall of 0.15 meter. Alternatives include precast concrete plates or wooden sheets and aluminum/steel beams that can be put used. The holes are cast from the ground upwards using a rotary steel casing. The casing has the hole outline ground in and is then filled with small quantities of the desired cement and sprayed with the remainder of the cement floating in water to create the desired wall (Bradl and Xenidis 2005) with low hydraulic conductivity, low shrinkage, ample strength and long-term reliability (Meuser 2013).

Encapsulation might be restricted to significant pollution zones. Technology is usually favoured over other cost-effective approaches to control sites where asbestos, heavy petroleum hydrocarbons, polycyclic aromatic hydrocarbons (PAHs), radionuclides, and blended waste are present (Meuser 2013). It is particularly useful for areas where vertical barriers are designed to prevent water pollution.

13.3.1.4 Chemical Immobilization

It is often referred to as in-situ stabilization/solidification (S/S), immobilizing or capturing contaminants in polluted soils by introducing chemical material into the soil to solidifying or transforming transient toxins fragments (i.e., soluble and interchangeable structures) to crystallites (Tajudin et al. 2016).

Solidification/stabilization was intended to alter waste's physical properties to make it less undesirable to humans and the environment (Sherwood and Qualls 2001). Stabilization helps with the treatment of waste by converting the pollutant to a less oxidized and dangerous state. The leftover material solidified in the solidification stage. The precise chemical makeup of the polluted area and the amount of water present would influence S/S technologies. The outcomes of these variables will contribute to the failure of the S/S operation, either by hindering the material's binding with the bonding of the waste, delaying the mixture environment, reducing the matrix stiffness, or reducing the strength of the solidified area (USEPA 1990). It sends pollutants to low hazardous substances and increases the material's strength and structural stability. The reagents are cheap and conveniently obtained, thus being useful for a variety of pollutants. This approach can be used to work with multiple forms of soils with high output rates. The only appropriate equipment needed for it is readily accessible and convenient.

However, this approach also does not kill or extract toxins from the soil. Volatiles and particulate matter are produced as a result of the treatment process. The other concern is long-term performance, which remains unknown. In a study by Al-Wabel et al. (2014), biochar values of 0.0%, 1.0%, 3.0%, and 5.0% with two levels of soil moisture (75 and 100% of field potential, FC) were investigated on the abundance and immobilization of Pb, Cd, Fe, Zn, Cu and Mn to the plants of maize. The shoot heavy metal concentrations decreased in maize plants with stabilizing soils by the addition of biochar (except for Fe, at 75% FC) in response to growing the application rates, with a peak decline of 53.2% for Cd, 28.8% for Zn, 51.3% for Mn and 60.9% for Cu, and at levels of soil moisture of 75% FC and 100% FC, respectively. This research indicates that biochar can be a useful amendment medium to reduce phytotoxicity and increase heavy metals' stability.

13.3.1.5 Phytoremediation

Phytoremediation utilizes plants to absorb pollutants. There are three major types: phytovolatilization, phytostabilization and phytoextraction (Shen and Chen 2000). The phytoremediation methodology, while depending on sun-powered energy-driven innovation, provides an aesthetic approach. It is usually used to treat different pollutants with low production of secondary wastes in the environment. It is a cheap and effective process with a low concentration of pollutants in large polluted areas. The upper layer of soil is viable for the cultivation of different crops to enhance agriculture activity. The process of phytoremediation needs more than one growing season for proper treatment of heavy metals, this is the major drawback of this process, and It is therefore restricted to soils just under three meters from groundwater (Faisal et al. 2004).

For phytoremediation treatment, hyperaccumulator plants are recommended (Memon et al. 2001; Memon and Schröder 2009). Van der Ent et al. 2013 indicated that in the dry foliage with crops present in their ecosystems, there are some amounts of heavy metals and metalloids, such as 300 mg kg^{-1} for Cu, Co and Cr: 10,000 mg kg^{-1} for Mn, 3000 mg kg^{-1} for Zn: 100 mg kg^{-1} for Se, Cd and Tl: 1000 mg kg^{-1} for Pb, Ni. Typically, shoot metal accumulation by hyperaccumulator plant is 100-fold higher than non-accumulator plants without impacting crop yield (Lasat 2002; Chaney et al. 2007). A plant (*Pteris vittate*) has been shown to have the hyperaccumulation potential of As (Ma et al. 2001) with root adsorption. This can take up and accumulate about 1442 and 7526 mg kg^{-1} As in its parts from polluted areas.

13.3.1.6 Biological Remediation

It is a mechanism that utilizes microbes to eliminate toxins using simple degradation (characteristic bio-degradation) or by adding microscopic agents, nutrient substitutes, electron supplies, or electron receptors (updated bioremediation) to the bioremediation process (EPA 2001). Microbes cannot degrade the toxic heavy metals, but they can alter action and change by modifying their physicochemical characteristics. The bioremediation process consists of precipitation, extracellular complexation, intracellular uptake, and redox reactions. Leaching of microbes is an essential and reasonable advancement for extracting valuable metals from the concentrates of minerals. Microbial leaching technology can potentially remediate the sites of mining, mineral industries site, treat waste sludge, and rehabilitate soils containing heavy metals (Bosecker 2001). This method is environmentally friendly and is generally 60–70% cheap than other various advancements. Bioremediation might be limited to certain factors, such as temperatures, oxygen levels, and pH. Usually, it takes more time in the treatment process, and it also depends on the type of microbes for the treatment of specific metal (FAQs 2012).

It was estimated that the oxidation of iron must be supported by microbes of *Leptothrix ochracea* and *Gallionella ferruginea* (Katsoyiannis and Zouboulis 2004). The researchers have done some lab tests where these microbes and iron oxide were put in a medium, providing a suitable environment for the As adsorption. As a result, the sorption of As(III) cannot done onto iron oxides. These microbes oxidized As(III) to As(V) in order to help the binding on Fe(III); thus, even at substantial concentrations of As (200 mg L^{-1}), the total removal efficiency of As improved to 95%.

13.3.2 Ex-Situ Remediation Techniques

In this type of soil remediation, polluted soil is extracted from the natural site (Figure 13.1 and Table 13.2). The excavated soil is then transported to the area where it has to be treated. The treated soil then shifted to a suitable site for disposal. Ex situ remediation is expensive than in situ remediation treatment. High cost is needed to excavate, transport and dispose-off the soil to clean the sites. However, improved and efficient treatment of soil can be done in a very small period of time.

13.3.2.1 Landfilling

It is very common and most simple way for remediation of soil which is also called "dig and haul". In this treatment process the soil from the contaminated area is removed and sifted to a site where it can be buried in the landfill area (Table 13.2). The inner structure of landfill is based on a particular engineering. The usage, construction and design of the safe landfill are according to legal laws and regulations. The main components include; water-resistant liner layer, pipes for leachate & wall inclusion. The important components among them are; the liner layers of clay and geotextile/plastic film. The leachate collection system is designed to avoid the percolation and leakage of leachate into groundwater.

It is the most-simple method of treating the harmful waste locations and commonly used in the United States. The total expense in the United States, from the hazardous waste site to landfill area and disposal site, ranged from USD300 to 500 ton^{-1} (FRTR 2012).

13.3.2.2 Soil Washing

It is the physiochemical process to eliminate the toxic heavy metals from the polluted soil. In this process the soil from the contaminated area is extracted and washed with special formulated solutions. The process is initiated with crushing of soil which is followed by screening process to remove the coarse particles, such as pieces of wood, plastic and stones. There are also magnetic particles in the soil, which can be removed with magnets. The soil (<5 mm) which has been screened is then subjected to mechanical agitation to mix it with formulated solutions. The screened soil is sieved to remove uneven particles of gravel (>0.05 mm) from clay and silt fraction (>0.05 mm). As, coarser pieces are not much contaminated so moved back to the previous site after washing process. While recovery of clay and silt involves the settling, rinsing and transporting them to the parent site. The wastewater generated while rinsing and in washing solutions is recyclable, reusable and can be shifted to the treatment plant for final dumping. The sludge produced during WWT is stabilized and solidified before its burial.

Soil washing solutions have tendency to change the soil acidity, complexation, redox potential and ionic capacity thus enhancing the movement of heavy metals. The perfect ideal solutions should be less toxic and biodegradable. Also, the mobility and solubility of heavy metal pollutants should be enhanced by washing solutions. However, these solutions don't make much interface with ingredients of soil. To make efficient washing solutions different chemicals were tested which include; polyglutamic acid, ethylene-diaminetetraacetic acid, diethylene-triamine pentaacetic acid, carbonates and bicarbonates, NaOH, $CaCl_2$, $FeCl_3$, NH_4Cl, $C_2H_4O_2.H_3N$, O4S2-2, isopropyl alcohol, HCl, H_2SO_4, HNO_3, H_3PO_4, HCOOH, CH_3COOH, oxal $(COOH)_2$, $C_6H_8O_7$ (Alghanmi et al. 2015; Zhu et al. 2015; Bilgin and Tulun 2016; Yang et al. 2017). As, the metal species varies so does the efficiency, which is greatly affected by the soil texture, organic components and soil pH (FRTR 2012). Some solutions has most high effectiveness which include; EDTA, HCl, subcritical water. The formulated solutions work under specific conditions such as agitation time (0.5–5 h), the soil/solution ratio (20–50 g L^{-1}) and effective concentration of the washing solution (1 molar hydrochloric acid or 0.2 molar EDTA). Sequential washing shall be made with various solutions if the sole washing does not remove complex pollutants efficiently.

In mineral processing various instruments such as centrifuges, hydro-cyclones, trommels and screens can also be used during soil washing process. In 1980s, U.S. and European countries like Sweden introduced the mobile soil washing system to lessen the charge of soil transportation and to clean the pollutant site and device was employed at the field area (FRTR 2012). Since 1995, U.S., Canada, Australia, Europe and Korea carried out pilot scale and field scale projects with treatment capacity of 0.2 to 10 ton per hour (USEPA 2013). This treatment process requires less time and high cost. The cost in U. S. in year 2011 ranges for large scale as USD70 per cubic meter to USD183 per cubic meter of soil for small scale (FRTR 2012).

13.3.2.3 Solidification

This treatment method is also known as micro-encapsulation. In this method the soil having metal contaminants is extracted from the site and shifted to another site for treatment. The screening is done to remove coarser particle. And with the help of extruder binding materials are mixed. Water proof solid structure is formed on mixing the binding substance, which ceases the metal pollutants. Binding substances can be replaced with stabilizing agents to stop the movement of contaminants (FRTR 2012).

The implementation of this technique is USD120–220 per cubic meter of soil (USEPA 2016). It is highly expensive but less time taking. The waste dispensation rate is 1000 kg per hour. Due to formation of solid structures its major drawback is the volume of the waste become double and it requires further disposal process. The material is replaceable over the site with frequent examination according to the regulations of "Comprehensive Environmental Response, Compensation, and Liability Act" of 1980 (CERCLA).

13.3.2.4 Vitrification

Since 1980, the development and testing was made on this technique. This is a thermal technique which requires intensive heat >1500°C to convert the waste into glasslike solid. The soil is converted into lava form to make it glass after lowering temperature. Organic contaminants will be smashed while the toxic heavy metals are specially encapsulated within the matrix of glass. And the resulting material will have the properties like; chemically inert, durable, resistant to leaching, strong. This method is categorized into three types on the basis of source of energy: (1) electrical vitrification (heat is generated by inserting the graphite electrodes at high voltage), (2) plasma vitrification (electrical discharge induced gas plasma is used to obtain high temperature), and (3) thermal vitrification (rotary retort with polluted soil is heated externally with the natural gas and microwave radiation) (Khan et al. 2004). The major drawback of this method is that, soil can't support agriculture. It is applied as in situ and ex situ. In case of ex situ thermal and plasma vitrification and in case of in-situ electrical and plasma vitrification are used. Ex situ is harmful due to exposure to radioactive and dispersive chemicals, although it is easy to handle.

In the ex situ method, refractory lined rotating container is filled with contaminated soil. The container has furnace containing electric arc, plasma torches, microwave radiations, gas burners and gas emitters. Glassy solid is formed when the processed soil is converted into liquefied dross that try to flow outside. it is formed at temperature > 1100°C. While gaseous effluents are collected separately for further treatment (Bradl and Xenidis 2005).

Vitrification has some limitations as it can't be applied to all soils such as, with high moisture (10%), maximum organic matter content (7%), soils having volatile or flammable organics. Moreover, 2–3% monovalent basic cations such as Na and K must be present. The vitrified waste is durable and strong but it will start to degrade due to weathering of waste while storing. Vitrification has been nominated as "best demonstrated available technology" by USEPA for heavy metal waste and radioactive waste (Meegoda et al. 2003). U.S. energy department has introduced the portable vitrification system to carry out field operations. This system consists of off-gas treatment unit, process-control components and electrical distribution unit. The United States is carrying out four in-situ projects on full-scale and small. This ex-situ method could be carried out by using previously present metal-processing equipment (CPEO 2016).

13.4 Choice of Suitable Remediation Method

Abovementioned methods for treatment of heavy metals-rich soils have several benefits and drawbacks. There are several factors which will limit the application of specific technique. These include; site geography, characteristics of pollutants, objective of the treatment, time requirement, finance, cost efficacy, execution, willingness and acceptance by general public (USEPA 2012). Each factor is important while applying a specific treatment technique.

Sometimes two or more treatment methods are required at specific step and at specific site during the remediation process. For instance; at highly polluted sites chemical stabilization becomes necessary to lessen the toxic effects and bioavailability of various metals in the soil. This will permit the plants to settle and initiate phytoremediation. This will lead to gradual reinstate the functions of the ecosystem in the polluted soil.

Conclusion

Soil is considered an important component of terrestrial ecosystem which plays a prominent role dead biomass management. The emission of heavy metals from various sources and subsequent accumulation in soils has significantly deteriorated the quality of soils. The major sources (rapid urbanization and industrialization) of soil pollution with heavy metals are grave concern due to its potential effect on plants, animals and human health. Various advanced methods are commonly used which could be regarded as the best available methods for removal of heavy metals from soils thereby increasing the plant growth.

The aim of remediation is usually to minimize the level of pollution load at a contaminated site in order to avoid the further destruction of ecosystem, reduce/prevent human exposure and other living organisms to toxic chemicals. The treatment method at a particular location differs based on the characteristics of hazardous waste, contamination depth, soil type and natural conditions prevailing at the contaminated soil location. The benchmarks for choosing a remediation method include: (i) short-term or long-term efficiency at fulfilling the remediation aims, (ii) significant reduction the volume of chemicals, (iii) decrease in contaminants toxicity and (iv) cost-effectiveness.

Different other aspects also influence selection of treatment method such as land disposal of toxic waste is banned under current environmental laws. Since a range of methods can usually control pollution at contaminated soil location thus choice of a suitable method or methods might be a challenging task.

The traditional in situ or ex situ methods for clean up or restoration of soils with heavy metals pollution have some disadvantages with respect to changes in soil characteristics, native microflora, cost, and problem of secondary pollution production. In comparison, plant-based phytoremediation method is a superior option in order to resolve the issue. Being eco-friendly and less costly this method could serve the purpose without requiring the costly apparatus and intensive are of heavy metal polluted soils. However, longer time duration, need of background knowledge of contaminated site and pollutant characteristics limits the scope of this method. Research is in progress to find native plants with high biomass producing ability to treat soils with heavy metal pollution. The potential application of a remediation method is considered to be project-specific, affected by various factors as discussed above. Treatability investigations helps in selecting the best viable method and must be carried out before implementation at large-scale.

REFERENCES

Alghanmi, S. I., A. F Al Sulami, T. A. El-Zayat, B. G. Alhogbi, M. A. Salam. 2015. Acid leaching of heavy metals from contaminated soil collected from Jeddah, Saudi Arabia: kinetic and thermodynamics studies. *Int Soil Water Conser Res* 3, 196–208.

Alloway, B. J. 2013. Heavy metals in soils: Trace metals and metalloids in soils and their bioavailability, *Environmental pollution* (Vol. 22, pp. 50–102). Whiteknights, UK: Springer.

Al-Wabel, M. I., A. R. A. Usman, A. H, El-Naggar, A. A. Aly, H. M. Ibrahim, A. Elmaghraby, A. et al. 2014. Conocarpus biochar as a soil amendment for reducing heavy metal availability and uptake by maize plants. *Saudi J Biolog Sci* 22, 503–511.

Babu, A. G., J. D. Kim, B. T. Oh. 2013. Enhancement of heavy metal phytoremediation by Alnus firma with endo-phytic Bacillus thuringiensis GDB-1. *J Hazard Mat* 250–251, 477–483.

Bilgin, M., S. Tulun. 2016. Heavy metals (Cu, Cd and Zn) contaminated soil removal by EDTA and FeCl3. *Global NEST J* 18, 98–107.

Bolan, N., A. Kunhikrishnan, R. Thangarajan, J. Kumpiene, J. Park, T. Makino, et al. 2014. Remediation of heavy metal(loid)s contaminated soils—To mobilize or to immobilize? *J Hazard Mat* 266, 141–166.

Bosecker, K. 2001. Microbial leaching in environmental clean-up programs. *Hydrometallurgy* 59(2–3), 245–248.

Bradl, H., Xenidis, A. 2005. Remediation techniques. In: Bradl, H.B. (Ed.), *Heavy Metals in the Environment: Origin, Interaction and Remediation*. Academic Press, Cambridge, MA, pp. 165–261.

Chaney, R. L., J. S. Angle, C. L. Broadhurst, C. A. Peters, R. V. Tappero, D. L. Sparks. 2007. Improved understanding of hyperaccumulation yields commercial phytoextraction and phytomining technologies. *J Environ Qual* 36, 1429–1443.

CLU-IN. 2017. *In-situ flushing: Overview. Contaminated site clean-up information.* U.S. Environmental Protection Agency, Washington, DC. https://clu-in.org/techfocus/default.focus/sec/In_Situ_Flushing/cat/Overview/ (accessed 31 January 2018).

CPEO. 2016. *Stabilization/solidification—vitrification.* Center for Public Environmental Oversight, Mountain View, CA. http://www.cpeo.org/techtree/ttdescript/ssvit.htm (accessed 1 February 2018).

EPA. 2001. Use of bioremediation at superfund sites. Solid waste and emergency response (5102G). U.S. Environmental Protection Agency Office of Solid Waste and Emergency Response Technology Innovation Office, Washington, DC, 20460.

FAQs. 2012. Bioremedial cleaning products for a cleaner, Greener Planet. Environmental Solution, Inc. http://www.totalbiosolution.com/index.php/freqently-asked-guestions.

FRTR. 2012. Remediation technologies screening matrix and reference guide, version 4.0. Federal Remediation Technologies Roundtable, Washington, DC.

Ghosh, K., N. Indra. 2018. Cadmium treatment induces echinocytosis, DNA damage, inflammation, and apoptosis in cardiac tissue of albino Wistar rats. *Environ. Toxicol Pharmacol* 59, 43–52.

Grimm, N. B., D. Foster, P. Groffman, J. M. Grove, C. S. Hopkinson, K. J. Nadelhoffer, et al. 2008. The changing land-scape: Ecosystem responses to urbanization and pollution across climatic and societal gradients. *Front Ecol and Environ* 6(5), 264–272.

Hseu, Z. Y., Y. T. Huang, H. C. His. 2014. Effects of remediation train sequence on decontamination of heavy metal-contaminated soil containing mercury. *J Air and Waste Manag Assoc* 64(9), 1013–1020.

Iturbe, R., C. Flores, C. Chavez, A. Ramirez, L. G. Torres. 2004. In situ flushing of contaminated soils from a refinery: Organic compounds and metal removals. *Remediat J* 14, 141–152.

Kabata-Pendias, A., H. Pendias. 2001. Trace metals in soils and plants. CRC Press, Boca Raton, FL (Ed. 3). Visit the CRC Press. www.crcpress.com. (Accessed on 15 January 2021)

Kankia, H. I., Y. Abdulhamid. 2014. Determination of accumulated heavy metals in benthic invertebrates found in Ajiwa Dam, Katsina State, Northern Nigeria. *Arch Appl Sci Res* 6(6), 80–87.

Katsoyiannis, I. A., A. I. Zouboulis. 2004. Application of biological processes for the removal of arsenic from groundwaters. *Water Res* 38, 17–26.

Khalid, S., M. Shahid, N. K. Niazi, B. Murtaza, I. Bibi, C. Dumat. 2017. A comparison of technologies for remediation of heavy metal contaminated soils. *J Geochem Explor* 182, 247–268.

Khan, F. I., T. Husain, R. Hejazi. 2004. An overview and analysis of site remediation technologies. *J Environ Manag* 71, 95–122.

Lamb, D. T., H. Ming, M. Megharaj, R. Naidu. 2009. Heavy metal (Cu, Zn Cd and Pb) partitioning and bioaccessibility in uncontaminated and long-term contaminated soils. *J Hazard Mat* 171, 1150–1158.

Lasat, M. M. 2002. Phytoextraction of toxic metals: A review of biological mechanisms. *J Environ Qual* 31, 109–120.

Liu, W., D. Wei, J. Mi, Y. Shen, B. Cui, C. Han. 2015. Immobilization of Cu(II) and Zn(II) in simulated polluted soil using sulfurizing agent. *Chem Eng J* 277, 312–317.

Ma, L. Q., K. M. Komar, C. Tu, W. Zhang, Y. Cai, E. D. Kennelley. 2001. A fern that hyperaccumulates arsenic. *Nature* 409, 579.

Makino, T., T. Kamiya, H. Takano, T. Itou, N. Sekiya, K. Sasaki, et al. 2007. Remediation of cadmium-contaminated paddy soils by washing with calcium chloride—Verification of on-site washing. *Environ Pollut* 147, 112–119.

Meegoda, J., A. Ezeldin, H. Fang, H. Inyang. 2003. Waste immobilization technologies. *Pract Period Hazard, Toxic, Radioact Waste Manag* 7, 46–58.

Memon, A. R., D. Aktoprakligil, A. Ozdemir, A. Vertii. 2001. Heavy metal accumulation and detoxification mechanism in plants. *Turk J Bot* 25, 111–121.

Memon, A. R., P. I. Schröder. 2009. Implications of metal accumulation mechanisms to phytoremediation. *Environ Sci Pollut Res* 16, 162–175.

Meuser, H. 2013. Soil remediation and rehabilitation: Treatment of contaminated and disturbed land. Springer, Dordrecht, The Netherlands.

Nalbandian, H. 2012. Trace element emissions from coal. IEA Clean Coal Center, London, UK.

NJDEP. 2014. Technical guidance on the capping of sites undergoing remediation. New Jersey Department of Environmental Protection, Trenton, NJ.

Ohio EPA. 2000. Final Covers for Hazardous Waste Surface Impoundments, Waste Piles and Landfills. Ohio Environmental Protection Agency, Columbus, OH.

Olexsey, R.A., Parker, R.A. 2006. Current and future in situ treatment techniques for the remediation of hazardous substances in soil, sediments, and groundwater. In: Twardowska, I., Allen, H.E., Häggblom, M.M., Stefaniak, S. (Eds.), Soil and Water Pollution Monitoring, Protection and Remediation. NATO Science Series vol. 69. Springer, Dordrecht, The Netherlands, pp. 211–219.

Pierzynski, G. M., J. T. Sims, G. F. Vance. 2000. Soils and environmental quality (2nd ed., p. 584). CRC Press, London.

Rumer, R. R., M. E. Ryan. 1995. Barrier containment technologies for environmental remediation applications. Wiley, New York, NY.

Shammas, N. 2009. Management and removal of heavy metals from contaminated soil. In: Wang, L. K., Chen, J. P., Hung, Y., Shammas, N. K. (Eds.), Heavy metals in the environment. Taylor & Francis, London, UK, pp. 381–429.

Shen, Z. G., H. M. Chen. 2000. Bioremediation of heavy metal polluted soils. *Rural Eco-Environment* 16, 39–44.

Sherwood, L. J., R. G. Qualls. 2001. Stability of phosphorus within a wetland soil following ferric chloride treatment to control eutrophication. *Environ Sci Technol* 35, 4126–4131.

Tajudin, S. A. A., M. A. M. Azmi, A. T. A. Nabila. 2016. Stabilization/solidification remediation method for contaminated soil: A review. *IOP Conf. Ser.: Mater Sci Eng* 136, 012043.

USEPA. 1990. Interference mechanisms in waste stabilization/solidification processes. Tech. Rep. EPA/540/A5-89/004, United States Environmental Protection Agency, Office of Research and Development, Cincinnati, OH, USA.

USEPA. 2012. In situ flushing site profiles. U.S. Environmental Protection Agency, Washington, DC. https://clu-in.org/products/isf/usersearch/list.cfm (accessed 31 January 2018).

USEPA. 2013. Soil washing application. U.S. Environmental Protection Agency, Washington, DC. https://clu-in.org/techfocus/default.focus/sec/soil_washing/cat/application/ (accessed 1 February 2018).

USEPA. 2016. Solidification. U.S. Environmental Protection Agency, Washington, DC. https://clu-in.org/techfocus/default.focus/sec/Solidification/cat/Overview/ (accessed 31 January 2018).

Van der Ent, A., A. J. M. Baker, R. D. Reeves, A. J. Pollard, H. Schat. 2013. Hyperaccumulators of metal and metalloid trace elements: Facts and fiction. *Plant and Soil* 362, 319–334.

Wuana, R.A., F. E. Okieimen, J. A. Imborvungu. 2010. Removal of heavy metals from a contaminated soil using organic chelating acids. *Int J Environ Sci Technol* 7, 485–496.

Yang, Z., C. Dong, C. Chen, Y. Sheu, C. Kao. 2017. Using polyglutamic acid as soil-washing agent to remediate heavy metal-contaminated soils. *Environ Sci Pollut Res Int.* https://doi.org/10.1007/s11356-017-9235-7.

Zhu, G., Q. Guo, J. Yang, H. Zhang, R. Wei, C. Wang. M. Peters, X. Zhou, J. Yang. 2015. Washing out heavy metals from contaminated soils from an iron and steel smelting site. *Front Environ Sci Eng* 9, 634–641.

14

Nano-Biorem: A New Concept toward Remedial Study

Arpita Tripathi
Biological Product Laboratory, Department of Botany, University of Allahabad, Prayagraj, UP, India

Ravikant Singh
Department of Biotechnology, Swami Vivekanand University, Sagar, MP, India

Saket Jha
Biological Product Laboratory, Department of Botany, University of Allahabad, Prayagraj, UP, India

Anand Pandey
Department of Plant Sciences, Avvaiyar Government College for Women, University of Puducherry, Karaikal, India

Anupam Dikshit
Biological Product Laboratory, Department of Botany, University of Allahabad, Prayagraj, UP, India

CONTENTS

14.1 Introduction

The toxic pollutants are increasing at alarming levels in the environment which are deteriorating the quality of environment, disturbing the ecosystem, and have adverse impact on the human health. According to the reports of the Outlook on the Global Agenda 2015, rising pollution in developing countries is the sixth most significant global trend, and in Asia, it is third (World Economic Forum; Singh et al. 2020). Many workers proposed several in situ and ex situ technologies for large-scale clean-up of contaminated sites (Singh et al. 2020). However, application of such methodologies at wider level would be affected due to certain limiting factors such as high operational and maintenance cost, high energy requirements, destructive methodologies, time constraints, etc. (Zelmanov and Semiat 2008).

Nanotechnology plays a crucial role nowadays in medicine, textiles, pharmaceutics, electronics optics, cosmetics, and

DOI: 10.1201/9781003155089-14

sports etc., Application of Nanotechnology in environmental remediation was also observed in last few years. Recent investigations and works on environmental remediation using nanotechnology exhibit the potentials of nanotechnology (Tratnyek and Johnson 2006; Mueller and Nowack 2010; Singh and Misra 2014; Singh and Misra 2016; Patil et al. 2016).

Since, sustainable remediation efficiently reduces the concentrations of pollutants to risk-based levels as well as minimizing the ancillary environmental impacts such as greenhouse gas emissions, waste generation and natural resource consumption, among others, so it got a peer attention (Reddy and Adams 2010). However, bioremediation requires long treatment time and it may not be effective if high contaminant concentrations that are toxic to microorganisms exist. The chapter reviews the current application of nanoparticles for remediation of contaminated soils, potential toxicity of nanoparticles, interaction of nanoparticles with soil constituents, challenges of using nanoparticles for site remediation, and finally potential opportunities to integrate nanoparticles with bioremediation to overcome each technology's limitations and develop a new and effective remediation technology are presented (Cecchin et al. 2017).

Recently multiple technologies have been incorporated in a single system as a complete solution for the decontamination and restoration of sites economically. Bioremediation is one of the best methods to combat the issue of contamination in an economic and eco-friendly manner. According to Saxena et al. (2019), use of microbes in bioremediation process to remediate the pollutants present in water and soil is one of the essential phenomenon. More hazardous substances are converted in less toxic or non-toxic substances in a bioremediation treatment using microbes. Due to high selectivity, specificity, cost and energy efficiency, minimal requirements etc., this technique gain an appropriate attention and advantage. However, bioremediation takes a longer duration for degradation of toxic compounds, so it is clear that this treatment process has some limitations too.

Another limitation of application bioremediation is that the sites which are severely contaminated with highly toxic and hazardous pollutants does not show proper results (Azubuike et al. 2016). As every method has pros and cons, the integration of different remediation methods could be a better solution to tackle remediation problems. Use of nanotechnology for bioremediation received a lot of attention in the past few years due to its untold advantages. Here we discuss briefly about nanotechnology and variety of nanostructured materials for the treatment of organic and inorganic contaminants from environment followed by detailed description of nano-bioremediation technique, its application processes, and methods (Singh et al. 2020).

14.2 Nanotechnology

The understanding and control of matter at dimension between 1–100 nanometers with novel application is known as nanotechnology. This definition is given by US National Nanotechnology Initiative (NNI). Properties like reactivity, magnetism, stability, and optical characteristics of any nanoparticle depends on the distinctive size, shape, and structure of the nanoparticles. Due to these properties of nanoparticles, they are widely used in different fields like drug delivery, textiles, cosmetics, water purification, food packaging, and several other industrial uses. Recently, use of nanotechnology due to its specific properties became more and more as injection of nanoparticles to the site of action, flexibility for in situ and ex situ application (Singh et al. 2020).

14.2.1 Distinctive Features of NPs

Nanoparticles and nanomaterials have unique size-dependent physical and chemical properties, so during last decade both got a significant attention. Due to small size nanoparticles exhibit a number of special properties relative to bulk material (Yadav et al. 2017). They are small enough to sequester their electrons and produce quantum effects. Due to these unique properties many nanoparticles such as gold are widely used in various fields such as photonics, catalysis, electronics and biomedicine etc. Radioactive wastes from nuclear power plants and nuclear weapon production, such as uranium, can be bioremediate using nanoparticles. Uranium contaminated wastewaters can be cleanup by cells and S-layer proteins of *Bacillus sphaericus* (Duran et al. 2007). Biological systems has unique ability to be self-organized and to synthesize molecules that have highly selective properties (Yadav et al. 2017).

14.2.2 Biomolecular Production of NPs

High cost of physical and chemical process of biosynthesis of nanoparticles is one of the main concerns in this context. Redox reaction is the main reaction involved in the biosynthesis of nanoparticles. For reduction of metal compounds into their respective nanoparticles, reducing properties of the microbial enzymes or the plant phytochemicals are important factors. The production of nanomaterials is currently estimated to be in the millions of tons worldwide and is expected to increase dramatically in the near future (BCC Research 2007). Not only vascular plants, but also microorganisms such as bacteria, yeasts, algae, fungi and actinomycetes can be used for biosynthesis of nanoparticles (Sastry et al. 2003).

14.2.2.1 Plant Mediated NPs Production

Green synthesis of nanoparticles by plants is gaining importance now-a-days because of single step biosynthesis process, absence of toxicants and occurrence of natural capping agents (Gurunathan et al. 2009). Nowadays many plants are investigated for the synthesis of nanoparticles. Reduction of metals by water soluble phytochemicals takes less time as compared to the fungi and bacteria. Therefore, plants are the solution to resolve the issue of nanoparticle synthesis other than bacteria and fungi. It is evident from compiled information that effect of nanoparticles varies from plant to plant and depends on their mode of application, size and concentrations (Siddiqui et al. 2015). Previous works and reviews reveals that the research on nanoparticles, essentiality for plants, is in the early stages; more rigorous study is needed to understand physiological, biochemical, and molecular mechanisms of plants in reference to nanoparticles and further work is required to explore the mode of action of NPs, their interaction with biomolecules, and their impact on the regulation of gene expressions in plants (Yadav et al. 2017).

14.2.2.2 Bacterial Mediated NPs Production

Bacteria have ability to mobilize the metals and some of them can reduce metal ions to precipitate metals at nanometer scale. Bacteria are one of the most potent biological agents for the synthesis of nanoparticles like gold, silver, platinum, palladium, titanium, titanium dioxide, magnetite, cadmium sulfide, hence considered as "bio-factory". The use of bacteria as a source of enzymes which will catalyze specific reactions resulting in inorganic nanoparticles may be a new rational biosynthesis strategy and use of enzymes, microbial enzymes, vitamins, polysaccharides, biodegradable polymers, microorganisms, and biological systems for synthesis of nanoparticles (Iravani 2014).

Nanoparticles can be synthesized in pure state which are free from other cellular proteins by extracellular secretion of enzymes and then purified further by filtering. The special metal binding abilities of the bacterial cells and S-layers make them useful for technical applications in bioremediation and nanotechnology (Yadav et al. 2017). The large-scale synthesis of nanoparticles using bacteria is appealing because it doesn't need any hazardous, toxic and expensive chemical materials for synthesis and stabilization processes (Iravani 2014).

14.2.2.3 Yeast and Fungi Mediated NPs Production

Fungi are an excellent source of various extracellular enzymes which influence nanoparticle synthesis. They have been widely used for the biosynthesis of nanoparticles and therefore the mechanistic aspects governing the nanoparticle formation have also been documented for a couple of them. In comparison to bacteria, fungi produce higher amount of nanoparticles due to greater volume of those proteins which directly translate to nanoparticle formation (Mohanpuria et al. 2008). When plant extracts are used for the generation of nanoparticles, then it is clearly observed that the rate of formation of NPs are greater than that of microbiological methods as they generate nanoparticles at a much slower rate.

Fungi have a foothold over other biological systems due to wide diversity, easy culture methods, reducing time and increasing cost-effectiveness. This successively provides an eco-friendly approach for nanoparticle synthesis. Genetic engineering techniques can be employed to improve the particle properties in near future (Saxena et al. 2014). In synthesis of various enzymes and rapid climb with the utilization of straight forward nutrients, yeast strains possess certain benefits over bacteria and therefore the synthesis of metallic nanoparticles employing the yeast is being considered.

14.3 Contaminated Sites Clean-Up via Nanoparticles

Due to their small size (usually less than 100 nm), high surface area and unique chemical characteristics, nanoparticles have been extensively studied and are being implemented with increasing frequency (Tosco et al. 2014; Thome et al. 2015) explained that the morphology, particle size distribution, specific surface area, surface charge and crystallographic characterization are the important characteristics that help understand the behavior of the nanoparticles. Ramamurthy and Eglal (2014) described iron nanoparticles in the form of beads with smooth surface. The application of nanoparticles for remediation of recalcitrant organic compounds such as tetrachloroethylene (TCE) and polyvinyl biphenyls (PCBs) has proven to be a fast and effective alternative to slow degradation of these compounds via microbial processes. Among many nano-materials used, the nano-scale zero-valent iron, also simply known as nano-iron (nZVFe), is found to be practical to inject into subsurface environments at the contaminated sites due to its low toxicity and cost of production (Cecchin et al. 2017). nZVFe has received the attention of numerous scientific investigations, accounting for over 90% of the work done in the area (Yan et al. 2013). In general, nZVFe is shown to be an effective technology if sufficient nZVFe mobility in the porous medium can be induced to achieve adequate distribution in the contaminated soil zones (Figure 14.1). In recent years, the applications of

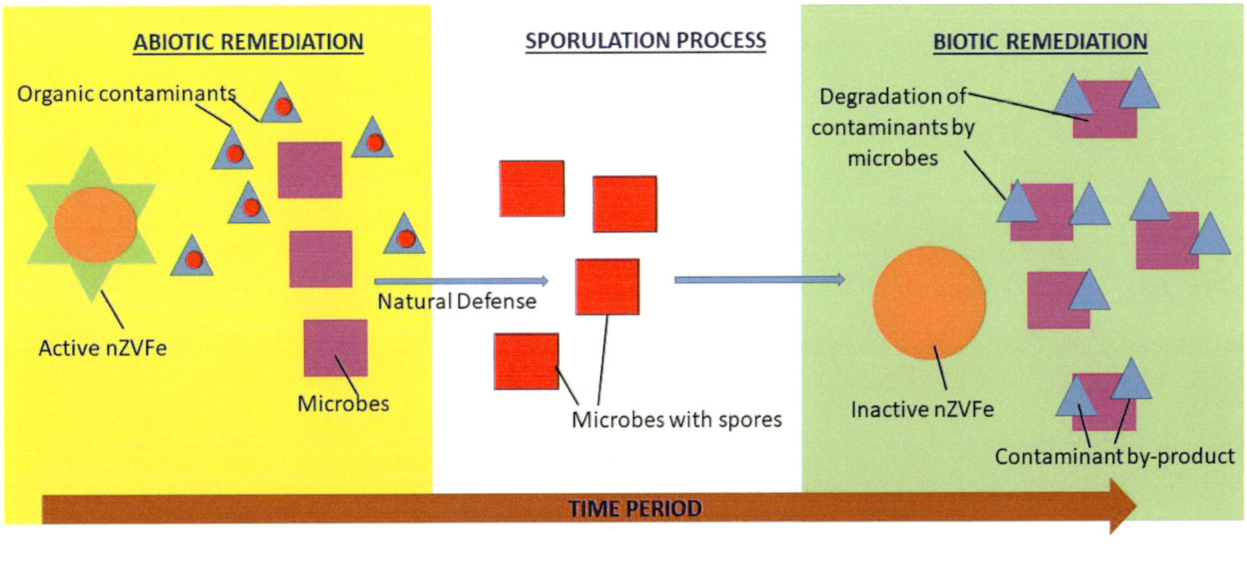

FIGURE 14.1 Graphical representation of nZVFe, microbes and organic contaminants during different phases of nano-biorem.

nZVFe increased significantly, becoming one of the most used in-situ methods for the remediation of contaminated sites with toxic compounds. Several studies investigated its reactivity, transport and application in pilot-scale studies and field-scale applications (Liu and Zhang 2010). Zero-valent iron (ZVI), in the form of iron filings, have been used extensively as a reactive material for the construction of permeable reactive barriers (PRB) (Gillham and O'Hannesin 1994). Gillham had been identified with the use of ZVI in PRBs over the years, such as rapid loss of reactivity of ZVI, excessive gas formation, synthesizes the precipitates near barriers, places where the contamination was close to the bedrock, creates problems in digging and need to recycle the reactive material used in the system. Zhang et al. (1998) presented a study in which they used for the first time nano-scale zero-valent iron (nZVFe) particles as a way to enhance the reductive dechlorination of TCE (Figure 14.1). The use of iron as a reducing agent of contaminants has been explored by O'Hannesin and Gillham (1998), and reported that it can be uses for the treatment of water contaminated by persistent lethal contaminants (chlorinated aliphatic hydrocarbons). The first applications of zero-valent iron in field trials, was using this as a fill material in permeable reactive barriers (Zanetti and Fiore 2005). Figure 14.1 shows a schematic model of such Nano-bioremediation processes, evaluating existing effects on soil microbiota.

14.4 Interaction of NPs with Edaphic Microbes

The interaction of nanoparticles with soils (edaphic factors) and their local microbes under natural condition is always a difficult to analyzation as well as the short time periods of monitoring. Thus, common to visualizes the interaction of NPs with soil and its microbes by using oxido-reduction potential such as using the nano-zero valent Iron (nZVFe) which exhibit same mechanism when dissolved in ground water as well as possesses electrical conductivity and concentration of iron in the system (Fe0, Fe2þ, Fe3þ) (Kirschling and Tilton 2010; Tosco et al. 2012; Cecchin et al. 2017).

The availability of edaphic microbes depends on edaphic characteristic such as soil texture and chemical constituents and thence, remediates the contaminants present on soils. The use of nZVFe stabilized with organic polymers showed good growth promoters as well as exhibits maximum biodegradation of toxic contaminants (Xiu et al. 2010; Comba et al. 2011). The co-formation of contaminants with Fe^{+2} or Fe^{+3} after the injection of nano-ferro suspension can chemically altered the nanoparticle surfaces under surrounding environmental conditions (Tosco et al. 2012; Yan et al. 2013). These compounds can directly influence the local flora and may interfere with long term immobilization of inorganic contaminants such as chromium (VI) and uranium (IV) (Cecchin et al. 2017).

14.5 Nano-Bioremediation (Nano-Biorem)

The removal or degradation of contaminants from environment by using microbes or plants with association of nanotechnology is called as "Nano-Biorem" or nano-bioremediation (Yadav et al. 2017). The concept "Nano-Biorem" is now emerging technique for the remediation of toxic contaminants/compounds present in the environment (Yadav et al. 2017). The use this novel concept Nano-Biorem is eco-friendly and applied easily for the removal of contaminants (Singh and Walker 2006).

The basic category of bioremediation is use of microbes, plants or their derived products (enzymatic approaches). The inclusion of nanotechnology enhances the remedial activity especially on plants based remediation of several heavy metals and other organic contaminants such as atrazine, chlorpyrifos, etc., which are easily degraded by using nano-zero valent ions (Ghormade et al. 2011). The use of enzyme based associated with nanoparticles can also be used in phytoremedial processes (Singh 2010). For example, several complex organic compounds, such as long-chain hydrocarbons and organochlorines, are particularly resistant to microbial and plant degradation (Table 14.1).

14.5.1 Application of Nano-Biorem

14.5.1.1 With Nanoparticles

Despite of many available technologies, single technology for remediation of contamination cannot proved to be an efficient and sustainable means (Nemecek et al. 2016; Cecchin et al. 2017). As a result, development and combined use of technologies in a sequential manner can resolve the loopholes of earlier available techniques (Dinesh et al. 2012; Koenig et al. 2016). Aiming the combination of biotic and abiotic degradation, Le et al. (2015) conducted a study for the degradation of a solution containing Aroclor 1248 (PCB) using nZVFe (1000 mg/L) and subsequent using biodegradation with *Burkholderia xenovorans*. A study was conducted by Bokare et al. (2010) in a contaminated site for the integration of bioremediation practice and reductive process through nanoparticles along with 5 g/L of triclosan. Considering the high persistence and low biodegradability of the compound, it carried out a dechlorination process in anaerobic environment with 2,3,7,8-tetrachlorodibenzo-p-dioxin through the nanoparticles of Pd/Fe, yielding 100% of dechlorination in only 10 h of exposure. After dechlorination, the byproduct obtained (dibenzo-p-dioxin), not amenable to remediation by the nanoparticles was totally degraded by microbial metabolism (Figure 14.1). Singh et al. (2013) conducted a study evaluating the integrated remediation of soil contaminated with Lindane. For the implementation of the study, the researchers evaluated the behavior of *Sphingomonas* sp. when subjected to different concentrations of CMC-Pd/Fe in nano-scale form (Figure 14.2, Table 14.1).

14.5.1.2 With Other Nanomaterial

There are various reasons for various NMs to be utilized in bioremediation; for instance, when the substance is delivered to nano scale, area per unit mass of a substance increases; hence, a bigger amount of the substance can inherit contact with surrounding materials and this affects the reactivity (Figure 14.2, Table 14.1). In the field of bioremediation, application of different NMs based on various reasons for example, NMs show quantum effect; therefore, less activation energy is required to make the chemical reactions feasible (Rizwan

TABLE 14.1

Remediation of Environmental Contaminants Using Nano-Biorem

S. No.	Nanoparticles	Bio-Agent (s)	Contaminant	Remarks	Ref(s)
1	Fe_3O_4 NP	*Sphingomonas* sp.	Carbazole	Degraded higher carbazole. Best suitable in recycling and degradation of organic wastes.	Wang et al. (2007)
				Easily separable from microbes by using magnets and can be reusable for degradayon of carbazole.	Li et al. (2013)
		Rhodococcus rhodochrous	Chlorophenols	Efficient degradation of chlorophenols and easily reusable.	Hou et al. (2016)
	Magnetic Fe_3O_4 NP	*Pseudomonas delafieldii*	Dibenzothiophene	Showed greater biodesulfurization of dibenzothiophene and can be easily reusable.	Shan et al. (2005)
2	Pd and Fe	Laccase derives from *Trametes versicolor*	Triclosan	Degraded by product converted to non-toxic compounds by laccase.	Bokare et al. (2010)
		Sphingomonas wittichii	1,3,7,8-tetrachlorodibenzo-p-dioxin (2,3,7,8-TeCDD)	It degraded the toxic dioxin isomer and convert to usable by product.	Bokare et al. (2010)
		Burkholderia xenovorans	Polychlorinated biphenyl (PCB) Aroclor 1248	Efficiently dechlorinated the bi-, tri-, tetra-, penta-, hexa-chlorinated biphenyls into biodegradable intermediates which further degraded by *B. xenovorans*	Le et al. (2015)
3	Bio-Pd NP	*C. pasteurianum*	Cr(VI)	It directly moves to the cell membrane and cytosol where it easily catalyzed the Cr(VI) and produces the hydrogen gas.	Chidambaram et al. (2010)
4	nZVFe-C-A beads	*Bacillus subtilis, E. coli, Acinetobacter junii*	Cr(VI)	It entrapped calcium alginate beads and easily removed Cr(VI)	Ravikumar et al. (2016)
	nZVFe	*Paracoccus* sp.	Nitrate	Low dose enhanced denitrification along with low microbial toxicity but in higher doses reduce the denitrification rate.	Liu et al. (2014)
		Dehalococcoides spp.	TCE	It stimulates the metabolic activity of methanogens as well as activates the dechlorinating bacterial activity for removal of TCE converted into ethane as by-products.	Xiu et al. (2010b)
		Sphingomonas sp.	PBDEs	Effectively degraded the PBDEs through reductive debromination followed by biological oxidation.	Kim et al. (2012)
	Bimetallic iron-based NPs	*Nicotiana tabaccum* L.	Hexabromocyclododecane (HBCD)	Remediated the HBCD from polluted soil.	Le et al. (2019)
	Polyvinylpyrrolidone (PVP) coated iron oxide NPs	*Halomonas* sp.	Pb and Cd	It helps in removal of heavy metals from both water and soil.	Cao et al. (2020)
5	Carbon nanotubes (CNTs)	*Shewanella oneidensis*	Cr(VI)	It efficiently removed the Cr(VI) contaminants.	Yan et al. (2013)
6	nZVPd	*Shewanella oneidensis*	PCBs	Effectively dechlorinated the PCBs producing less toxic by-products.	Windt et al. (2005)

et al. 2014). For the toxic material detection surface plasma on resonance phenomenon that is exhibited by NPs can be used. Various metallic and nonmetallic NMs of different shape and size can be applicable for environmental cleanup. For instance, it can use a variety of single metal NPs, bimetallic NPs, carbon base NMs, and so forth, for the reason that (i) NPs can diffuse or penetrate into a contamination zone where micro particles cannot reach and (ii) they have higher reactivity to redox-amenable contaminants (Rizwan et al. 2014).

It is observed that oxide-coated Fe^0 can form weak and outer-sphere complexes with contaminants such as carbon-tetrachloride (CT). Oxide coating increases the reactivity and hence, through electron transfer. Methane, carbon monoxide, or format, whereas benzoquinone, by tri-chloro-ethene are the different categories of CT and other chlorinated aliphatic hydrocarbons can be categorized into chemicals that have lower toxic values in batch experiments and field assessments (Nurmi et al. 2005). For the degradation of pentachlorophenol (PCP) through a photo electro catalytic reaction TiO_2 nanotubes can prove to be useful (Quan et al. 2005).

For reductive dichlorination single metal NPs can be useful. Palladium, Pd(0) NPs are often deposited on the cell wall and

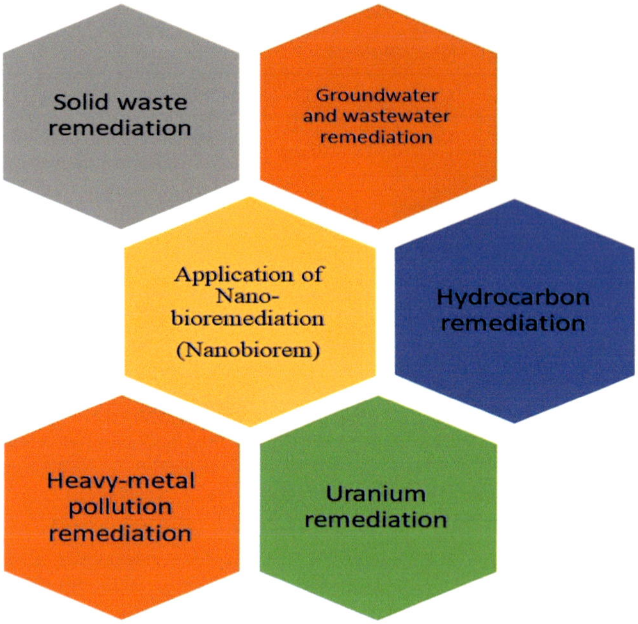

FIGURE 14.2 Application of nano-biorem.

inside the cytoplasm of *Shewanella oneidensis* and should be charged with H^{*} radicals by adding different substrates like hydrogen, acetate, and format as electron donors during a bio-reductive assay containing Pd(II).

NPs are often further used to immobilize microbial cells which can degrade or bio recover specific chemicals. By applying an external magnetic field to these microbial cells (such as *Pseudomonas delafieldii*), these magnetic NP coated cells concentrated at a specific location on the reactor wall, separated from the bulk solution, and recycled for the management of the identical substrate.

These incorporate applications for solid waste, ground and wastewater, petroleum and its relevant products (hydrocarbon), soil remediation, uranium remediation and heavy metal contamination remediation. The capability of NMs to dwindle pollution production is in progress and could potentially drive the most revolutionary changes in the field of environment field in the upcoming years (Rizwan et al. 2014).

14.5.1.3 With Nanocrystals and CNTs

The extraordinary properties of carbon-based nano-materials such as nano crystals and carbon nano-tubes (CNTs) facilitate novel technologies to make out and crack a vast range of environmental applications: sorbents, high flux membranes, depth filters, antimicrobial agents, environmental sensors, renewable energy technologies and prevention strategies for pollution (Guo et al. 2012). NMs like single-walled carbon nanotubes (SWCNTs) multi-walled carbon nanotubes (MWCNTs) and hybrid carbon nanotubes (HCNTs) have been evaluated as the elimination of ethyl-benzene from aqueous solution. The stable amount removed by SWCNTs was elevated as compared to that of MWCNTs and the SWCNTs was better for ethyl-benzene sorption as compared to HCNTs and MWCNTs (Figure 14.2, Table 14.1).

Isotherms study indicates that the ethyl-benzene sorption by SWCNTs was provided best fit by BET isotherm expression. For that reason, it could be used for prevention of ethyl-benzene borne diseases and for environmental pollution remediation (Bina et al. 2012). In recent times, cyclo-dextrins (CD) and CNT(s) have been incorporated in water treatment and pollutant monitoring purpose.

CNTs that are an environmental friendly adsorbent was found to be immobilized by calcium alginate (CNTs/CA) have been prepared and their property of copper adsorption has been worked upon by the use of equilibrium studies. Experimental results showed that even at a lower pH of 2.1 CNTs/CA has high copper removal efficiency and reaches 69.9%. The adsorption potential of copper CNTs/CA can reach 67.9 mg/g at a copper equilibrium concentration of 5 mg/L (Li et al. 2010).

14.5.1.4 With Single Enzyme NPs

In bioremediation, enzymes (specific and effective proteins) plays a role of biocatalysts. Enzymes found to be less efficient because they lose their activity due to oxidation that results in a reduced amount of stability and a shorter lifetime. In order to increase the stability, longevity, and reusability of the enzymes joining them to magnetic iron NPs could be useful. Separation of enzymes from reactants or products by applying a magnetic field could become convenient if they get attached to the magnetic iron NPs (Figure 14.2).

In order to accomplish this purpose, trypsin and peroxides (catabolic enzymes), have been applied to uniform core shell magnetic nanoparticles (MNPs). Based on the scientific study it indicates that the time period and enzyme activity increase significantly from a few hours to weeks and are more stable, efficient and economical values were recorded for that of MNP enzyme conjugates. The shielding of enzymes by that of MNPs prevents them from becoming oxidized those results in increased life time as based on the scientific notes. Because of

TABLE 14.2
Removal of Pollutants Mediated by Nanoparticles and Bio-Based Treatments

Types of Nano-Materials	Organisms Used	Chemical Presentation	Pollutants	Ref(s)
Nano-crystals	Enzymatic degradation by bacterial overexpressed organophosphorus hydrolase	ZnS	P-nitrophenol and acid orange 7	Torres-Martinez et al. (2001)
Nanoparticles	*Sphingomonas* sp.	Nanoscale nZVFe, Ti, Mn, Ag, Au	Decarbonated diphenyl ether, chlorinated hydrocarbons, pathogens	Kim et al. (2012)
Nano-powders	Soil microbes	FeO nano-powder	Azo dye direct red 23	Kos et al. (2014)
Nano-membranes	A biological extract of *Cynomorium coccineum* L.	Thin film composite polyamide	Cyanide compounds	Sebeia et al. (2019)
Nano-composites	*Arthrobacter globiformis*	Microbial immobilized nanocellulose composites	Herbicide	Liu et al. (2018)
Nano-tubes	Enzyme organophosphate hydrolase-MWNT paper	Unzipped carbon nanotube (CNT), single-walled CNT and multi- walled CNT	Organophosphates and heavy metals	Mechrez et al. (2014); Fosso-Kankeu et al. (2014)
Nano-sponge	Two organo-clay (Dellite 67G and Dellite 43B)	Cyclodextrin- based, highly cross-linked polymers	Triclopyr (3,5,6-Trichloro-2-pyridinyloxyacetic acid)	Baglieri et al. (2013)

the MNPs, NP-enzyme conjugates can become more productive because their high magnetization can be separated efficiently magnetically (Qiang et al. 2007; Rizwan et al. 2014).

14.5.1.5 With Novel Polymeric NPs

14.5.1.5.1 Remediation of Organic Contaminants

The solubilization rate and mobility of hydrophobic organic contaminants, such as polycyclic aromatic hydrocarbons (PAHs) has been limited by their sorption in soil. In order to enhance the bioavailability of phenanthrene (PHEN), polyethylene-glycol modified urethane acrylate (PMUA) precursor chain has been developed. The mineralization rate of PHEN crystal in water was found to be increased with PMUA NPs. The use of PMUA particles combined with bacterial inoculants found more accessible and valuable for in situ bioremediation (Rizwan et al. 2014), described in Table 14.2.

14.5.1.5.2 Remediation of Soil Contaminants

Poly-nuclear aromatic hydrocarbons (PAHs) that are hydrophobic organic groundwater contaminants), are difficult to remove after they sorb strongly to soils. For the soil remediation contaminated with PAHs, Amphiphilic polyurethane (APU) NP has been synthesized that to be used to resolve this issue. The chains that can be emulsified and cross-linked in water are made up of those particles that incorporate polyurethane acrylate anionomer (UAA) or poly (ethylene-glycol), modified urethane acrylate (PMUA) precursors. The colloidal size of the resulting particles (17–97 nm) as measured by dynamic light scattering. PAH desorption can be enhanced by APU particles and transportation is comparable to that of surfactant micelles, but contrary the surface-active components of micelles, the individual cross-linked precursor chains in APU particles cannot be sorbed to soil surface. In the aqueous phase, in spite of the concentration, the APU particle shows stability.

Engineered APU particles needs to be worked upon to achieve desired properties and scientific studies show that the APU particles with hydrophobic interior regions that confer a high affinity for phenanthrene (PHEN) and hydrophilic surfaces that encourage particle mobility in soil can be designed. During the synthesis of chain, an alteration in the size of the hydrophobic segment used can control APU particles affinity for contaminants such as PHEN. The charge density or the size of the pendent water-soluble chains that reside on the particle surface determines the mobility of colloidal APU suspensions in soil. For a variety of contaminant type and soil conditions different NPs can be produced by altering the properties related to the particle and thus enhance its optimization (Tungittiplakorn et al. 2004).

14.5.1.5.3 Remediation of Heavy Metal Contaminants

Globally, environmental pollution has become a problematic situation particularly with heavy metal and over last two decades a special attention has been paid to minimize the pollution sources and to remediate the polluted soil and water resources (Vazquez-Nunez et al. 2020). In order to find the native accumulator plants a field study was conducted in a dried waste pool of a lead mine. In dried waste pool by the application of flame absorption atomic method concentration of heavy metal had been determined both in the soil and the plants. The concentration all the toxic metals (Cu, Zn, Pb, and Ni) were found to be higher than it is found in the natural soil (Vazquez-Nunez et al. 2020).

Heavy metal accumulation was reported in six dominant vegetation, namely, *Gundelia tournefortii, Centaurea virgata, Reseda lutea, Scariola orientalis, Eleagnum angustifolia*, and *Noaea mucronata* (Vazquez-Nunez et al. 2020). *Noaea mucronata*is the best Pb accumulator and also a good accumulator for Zn, Cu, and Ni, but in case of Fe its *Reseda lutea* and for Cd its *Marrubium vulgare* as per the result based study (Vazquez-Nunez et al. 2020). The bioaccumulation ability of nanoparticles prepared from *N. mucronata* was evaluated in experimental water containers. During three days of bioremediation the experiment shows that heavy metals concentration decreases many-fold (Mohsenzadeh and Chehregani 2012; Vazquez-Nunez et al. 2020).

14.6 Present and Future Scenario of Application of Nano-Biorem

Nanotechnology application is not restricted only to environmental biotechnology but innovations from other fields of nanotechnology may appear to move at a faster pace, as in the field of medical nanotechnology. Nanostructure functionalization with bio-molecules may be a neighborhood worth exploring (Bolisetty and Mezzenga 2016).

This work has focused attention on the importance of discovering a contemptible source of biomolecules as a key to this sort of development. Moreover, several advantages being possessed by natural proteins. For instance, their production via well-known recombinant technologies in a commercial manner and could encompass 20 variant amino acids in order to present a large combinatorial capacity to cooperate with several other molecules, also for the creation of new catalytic surfaces and structures (Ljubetič et al. 2017).

For NPs functionalization to provide eco-friendly methodologies biotechnology can provide a way to it. Recently the bacteria *Komagataeibacter sucrofermentans* introduced in the biological toolbox to production of novel cellulose-like polymers functionalized with custom moieties (Gao et al. 2019). These bacteria are cultured with traditional bioreactor methodologies and fed with glucose monomers ornamented with the specified chemical modifications leading to their biological incorporation within the polymer. This approach detour the utilization of complex solvents, stoichiometry, and the production of byproducts that are environmentally dangerous. The optimization of this bio-system by classic and next-generation biotechnological modifications holds a great possibility to make things easier for the synthesis of a bulk quantity of cellulose-based NMs having a number of applications (Sharma et al. 2019). Recent developments within the field of RNA-based fungicides indicate that it's a feasible technology for substituting traditional biochemical fungicides. Double-strand RNAs, designed to hybridize with vital mRNAs of fungal pathogens, are sprayed on leaves or fruits and induce expression silencing within the pathogen (Wang et al. 2016).

However, the short mean-life of naked RNAs within the environment is an obstacle to beat (Wang et al. 2016). Traditional clay nano-sheets were tested as protectants of double-strand RNA and probed to reinforce the mean-life of the bio-molecules and extend the biocidal actions against the fungal pathogen (Mitter et al. 2017). Although this technology was created for the protection of aerial plant organs, it's recently been applied in root protection. Chariou et al. (2019) studied a number of nano-scale encased architectures ranging from traditional silica particles to recombinant or plant-based virus capsules. The proof-of-concept was administered with a nematicide and therefore the biologically derived capsules proved superior in soil penetration and cargo release. These bio-nanoparticles (biodegradable) do not generate organic pollution that proves to be an eco-friendly approach.

In the field of environmental biotechnology by the means of DNA hybridization technique, the 3D construction of DNA structures constitutes an edge research area that is still needs to be explore (Li et al. 2017). These architectures provide a new toolbox for functionalization, as illustrated by the work of DNA nano robots that can be loaded with intractable molecules (other DNAs/RNAs or proteins) that serve as cargo, fasteners or springs. Li et al. (2018) found that the planning of DNA sheets that interact with the protein thrombin that becomes compartmentalized, when other DNA molecules seals the nano-sheet acting as fasteners.

A nanotube that is a DNA origami here, will solitary unlock at nucleolin (tumor-produced protein), and release the thrombin cargo that induces tumor coagulation and necrosis. This development exposed the view of programmable and complex mechanics that biomolecular-based bio-robots could accomplish. Though this is a biomedicine-based development, but it presents an exceptional paradigm of bio-molecular interactions that can further be extended to environmental applications. Recently, Li et al. (2017), were reviewed biotechnological options to succeed in the economic feasibility of DNA/RNA production like chip synthesis, recombinant bacteria and present bacteria ready to export RNAs. An additional example of sophisticated indexation of nanotechnology to get access of remote information where a chemically coded choice diagram with inorganic molecules to identify and irretrievably inform the presence of analytes in the air, resembling soot, ammonia and tri-methylamine (Koman et al. 2018).

Inorganic sensors can also code their self-precipitation through photo-induction and ion changes in their environment (Brandl et al. 2015). In order to desalinate water molecular biotechnology could provide an efficient way to resolve the issue. The incorporation of the bacterial protein aquaporin-Z into polymeric membranes and demonstrated that it could exclude salt and yield purified water (Kumar et al. 2007). Inclusion of modern protein modeling algorithms and biology techniques, this approach was used to supply porin proteins that shows improved exclusion activities for water solutes(organic and inorganic) (Chowdhury et al. 2018). The development of O_2 biosensors by the use of the technological use of oxygen-sensitive proteins present in humans and plants is an emerging field originally thought to be for in vivo applications, they will be adapted to organize functional NMs ready to stoichiometrically answer O_2 levels (Licausi and Giuntoli 2021).

These concepts, explored with proteins, have also been demonstrated with DNA molecules ready to exclude complex analytes like proteins and open a good area of sensing and purification opportunities (Diederichs et al. 2019). Recently, Ryu et al. (2019) and ÁAlvarez et al. (2018) reviewed the sector of trans-membrane proteins incorporated into membranes and their coupling to different transductors and applications in gas monitoring, pesticide detection, microarrays, and energy harvesting within the future, enzymes could also be incorporated within the above-discussed arrangements to watch more complex pollutants or their combinations.

This concept is used in healthcare monitoring where biosensors have been fixed in a plethora of materials like patches, temporary tattoos or wrists, among others (Kim et al. 2019). Notably, the economic market of diabetes management directs these transformations. Though, the security of workers doing dangerous tasks in radioactive, potentially toxic, or enclosed areas can benefit from the development of real-time monitoring bio-nanotechnologies.

Conclusions

For decontamination of soil, air or water present nanotechnologies could be used but cost-effective production means needs to be focused. Nano-biorem might enormously contribute to sustainability because it offers environmental advantages and is cheap when compared to other technologies; even more the range of applications of NMs, coupled with biological treatments, proved to be highly efficient for the degradation of contaminants that makes a new way to counter environmental challenges more effectively. The use of nanomaterials in the field of nano-biorem (nano-bioremediation) initially reduces the contaminants to biodegradation levels and then promotes it to reach the risk-based levels. Therefore, it provides the proficient and effective solution for the remediation of contamination sites. Attributable to its potentiality, its application will increase in nearby future that will play a pivotal role in sustainable development.

Acknowledgment

Authors are thankful to UGC for providing fund.

REFERENCES

ÁAlvarez, P. J. J., Chan, C. K., Elimelech, M., Halas, N. J., and Villagrán, D. 2018. Emerging opportunities for nanotechnology to enhance water security. Nat. Nanotechnol. 2018, 13:, 634.

Azubuike, C. C., Chikere, C. B., and Okpokwasili, G. C. 2016. Bioremediation techniques–classification based on site of application: Pprinciples, advantages, limitations and prospects. World J Microbiol. Biotechnol. 32(11): 180.

Baglieri, A., Nègre, M., Trotta, F., Bracco, P., and Gennari, M. 2013. Organo-clays and nanosponges for acquifer bioremediation: aAdsorption and degradation of triclopyr. J. Environ. Sci. Health Part B Pestic. Food Contam. Agric. Wastes. 48: 784–792.

BCC Research., 2007. Market Research Reports aAnd Technical Publications., Wellesley, MA.

Bina, B., Pourzamani, H., Rashidi, A., and Amin, M. M. 2012. Ethylbenzene removal by carbon nanotubes from aqueous solution. J. Environ. Public Health. 2012: 1–8.

Bokare, V., Murugesan, K., Kim, Y. M., Jeon, J. R., Kim, E. J., and Chang, Y. S. 2010. Degradation of tri-closan by an integrated nano-bio redox process. Bioresour. Technol. 101(16): 6354–6360.

Bokare, V., Murugesan, K., Kim, Y. M., Jeon, J. R., Kim, E. J., and Chang, Y. S. 2010. Degradation of triclosan by an integrated nano-bio redox process. Bioresour. Technol. 101: 6354–6360.

Bolisetty, S., and Mezzenga, R. 2016. Amyloid-carbon hybrid membranes for universal water purification. Nat. Nanotechnol. 11: 365.

Brandl, F., Bertrand, N., Lima, E. M., and Langer, R. 2015. Nanoparticles with photoinduced precipitation for the extraction of pollutants from water and soil. Nat. Commun. 6: 1–10.

Cao, X., Alabresm, A., Chen, Y. P., Decho, A. W., and Lead, J. 2020. Improved metal remediation using a combined bacterial and nanoscience approach. Sci. Total Environ. 704: 135378.

Cecchin, I., Reddy, K.R., Thomé, A., Tessaro, E.F. and Schnaid, F., 2017. Nanobioremediation: Integration of nanoparticles and bioremediation for sustainable remediation of chlorinated organic contaminants in soils. Int. Biodeteriorat. Biodegrad. 119:419–428.

Chariou, P. L., Dogan, A. B., Welsh, A. G., Saidel, G. M., Baskaran, H., and Steinmetz, N. F. 2019. Soil mobility of synthetic and virus-based model nanopesticides. Nat. Nanotechnol. 2019, 14: 712–718.

Chidambaram, D., Hennebel, T., Taghavi, S., Mast, J., Boon, N., Verstraete, W., van der Lelie, D., and Fitts, J. P. 2010. Concomitant microbial generation of palladium nanoparticles and hydrogen to immobilize chromate. Environ. Sci. Technol. 44(19): 7635–7640.

Chowdhury, R., Ren, T., Shankla, M., Decker, K., Grisewood, M., Prabhakar, J., Baker, C., Golbeck, J. H., Aksimentiev, A., Kumar, M., et al. 2018. Pore dDesigner for tuning solute selectivity in a robust and highly permeable outer membrane pore. Nat. Commun. 9: 1–10.

Comba, S., Molfetta, A. D., and Sethi, R., 2011. A comparison between field applications of nano-, micro-, and millimetric zero-valent iron for the remediation of contaminated aquifers. Water Air Soil Pollut. 215: 595–607.

Diederichs, T., Pugh, G., Dorey, A., Xing, Y., Burns, J. R., Hung Nguyen, Q., Tornow, M., Tampé, R., and Howorka, S. 2019. Synthetic protein-conductive membrane nanopores built with DNA. Nat. Commun. 10: 1–11.

Dinesh, R., Anandaraj, M., Srinivasan, V., and Hamza, S. 2012. Engineered nanoparticles in the soil and their potential implications to microbial activity. Geoderma. 173--174:, 19–27.

Duran, N., Marcato, P. D., De Souza, G. I. H., Alves, O. L., and Esposito, E. 2007. J. Biomed. Nanotechnol. 3: 203–208.

Fosso-Kankeu, E., Mulaba-Bafubiandi, A. F., and Mishra, A. K. 2014. Prospects for iImmobilization of mMicrobial sSorbents on cCarbon nNanotubes for bBiosorption: bBioremediation of hHeavy mMetals pPolluted wWater. In Mishra, A. K. (ed.), Application of Nanotechnology in Water Research, 1st ed.; Mishra, A.K., Ed.; John Wiley & Sons, Inc.,: Hoboken, NJ, USA. Vol. 1,: pp. 213–235.

Gao, M., Li, J., Bao, Z., Hu, M., Nian, R., Feng, D., An, D., Li, X., Xian, M., and Zhang, H. 2019. A natural in situ fabrication method of functional bacterial cellulose using a microorganism. Nat. Commun. 10: 1–10.

Ghormade, V., Deshpande, M.V. and Paknikar, K.M., 2011. Perspectives for nano-biotechnology enabled protection and nutrition of plants. Biotechnol. Adv. 29(6): 792–803.

Gillham, R. W., and O'Hannesin, S. F. 1994. Enhanced degradation of halogenated aliphatics by zero-valent iron. Ground Water. 32: 958–967.

Guo, R., Guo, X., Yu, D., and Hu, J. 2012. Application research in water treatment of PAMAM dendrimer. Chem. Industry Eng. Prog. 31: 671–675.

Gurunathan, S., Kalishwaralal, K., Vaidyanathan, R., Venkataraman, D., Pandian, S. R., Muniyandi, J., Hariharan, N., and Eom, S. H. 2009. Colloid. Surface. B. 74: 328–335.

Hou, J., Liu, F., Wu, N., Ju, J., and Yu, B. 2016. Efficient biodegradation of chlorophenols in aqueous phase by magnetically immobilized aniline-degrading *Rhodococcus rhodochrous* strain. J. Nanobiotechnol. 14: 5.

Iravani, S., 2014. Bacteria in nanoparticle synthesis: current status and future prospects. International scholarly research notices. 2014: 1–18 http://dx.doi.org/10.1155/2014/359316

Kim, J., Campbell, A. S., de Ávila, B. E. F., Wang, J. 2019. Wearable biosensors for healthcare monitoring. Nat. Biotechnol. 37: 389–406.

Kim, Y. M., Murugesan, K., Chang, Y. Y., Kim, E. J. and Chang, Y.S. 2012. Degradation of polybrominated diphenyl ethers by a sequential treatment with nanoscale zero valent iron and aerobic biodegradation. J. Chem. Technol. Biotechnol. 240: 525–532.

Kim, Y. M.,; Murugesan, K.,; Chang, Y. Y.,; Kim, E. J.;, and Chang, Y. S. 2012. Degradation of polybrominated diphenyl ethers by a sequential treatment with nanoscale zero valent iron and aerobic biodegradation. J. Chem. Technol. Biotechnol. 2012, 240:, 525–532.

Kirschling, T. L., and Tilton, R. D. 2010. Impact of nanoscale zero valent iron on geochemistry and microbial populations in trichloroethylene contaminated aquifer materials. Environ. Sci. Technol. 44: 3474–3480.

Koenig, J. C., Boparai, H. K., Lee, M. J., O'Carroll, D. M., Barnes, R. J., and Manefield, M. J. 2016. Particles and enzymes: combining nanoscale zero valent iron and organochlorine respiring bacteria for the detoxification of chloro-ethane mixtures. J. Hazard. Mater. 308: 106–112.

Koman, V. B., Liu, P., Kozawa, D., Liu, A. T., Cottrill, A. L., Son, Y., Lebron, J. A., and Strano, M. S. 2018. Colloidal nanoelectronic state machines based on 2D materials for aerosolizable electronics. Nat. Nanotechnol. 13: 819–827.

Kos, L., Sójka-Ledakowicz, J., Michalska, K., and Perkowski, J. 2014. Decomposition of azo dye C.I. Direct Yellow 86 by the Ffenton process in the presence of nanoparticles of iron oxides. Fibres Text. East. Eur. 5: 114–120.

Kumar, M., Grzelakowski, M., Zilles, J., Clark, M., and Meier, W. 2007. Highly permeable polymeric membranes based on the incorporation of the functional water channel protein aquaporin Z. Proc. Natl. Acad. Sci. USA. 104: 20719–20724.

Le, T. T., Nguyen, K. H., Jeon, J. R., Francis, A. J., and Chang, Y. S. 2015. Nano/bio treatment of polychlorinated biphenyls with evaluation of comparative toxicity. J Hazard Mater. 287: 335–341.

Le, T. T., Nguyen, K. H., Jeon, J. R., Francis, A.J., and Chang, Y. S. 2015. Nano/bio treatment of polychlorinated biphenyls with evaluation of comparative toxicity. J. Hazard. Mater. 287: 335–341.

Le, T. T.,; Yoon, H.,; Son, M. H.,; Kang, Y. G.,; and Chang, Y. S. 2019. Treatability of hexabromocyclododecane using Pd/Fe nanoparticles in the soil-plant system: eEff_ects of humic acids. Sci. Total Environ. 2019, 689:, 444–450.

Li, J., Green, A. A., Yan, H., and Fan, C. 2017. Engineering nucleic acid structures for programmable molecular circuitry and intracellular biocomputation. Nat. Chem. 9: 1056.

Li, S., Jiang, Q., Liu, S., Zhang, Y., Tian, Y., Song, C., Wang, J., Zou, Y., Anderson, G. J., Han, J. Y., et al. 2018. A DNA nano-robot functions as a cancer therapeutic in response to a molecular trigger in vivo. Nat. Biotechnol. 36: 258.

Li, Y., Du, X., Wu, C., Liu, X., Wang, X., and Xu, P. 2013. An efficient magnetically modified microbial cell bio-composite for carbazole biodegradation. Nanoscale Res Lett. 8(1): 522.

Li, Y., Liu, F., and Xia, B., et al. 2010. Removal of copper from aqueous solution by carbon nanotube/calcium alginate composites. Journal of Hazardous Materials. 177 (1--3): 876–880.

Licausi, F., and Giuntoli, B. 2021. Synthetic biology of hypoxia. New Phytologist. 229(1): 50–56.

Liu, J., Morales-Narváez, E., Vicent, T., Merkoçi, A., and Zhong, G. H. 2018. Microorganism-decorated nanocellulose for efficient diuron removal. Chem. Eng. J. 354: 1083–1091.

Liu, Y., Li, S., Chen, Z., Megharaj, M., and Naidu, R. (2014.) Influence of zero-valent iron nanoparticles on nitrate removal by *Paracoccus* sp. Chemosphere. 108: 426–432.

Liu, Z. G., and Zhang, F. S. 2010. Nano-zero-valent iron contained porous carbons developed from waste biomass for the adsorption and dechlorination of PCBs. Bioresour. Technol. 101:, 2562–2564.

Ljubetič, A., Lapenta, F., Gradišar, H., Drobnak, I., Aupič, J., Strmšek, Ž., Lainšček, D., Hafner-Bratkovič, I., Majerle, A., Krivec, N. and Benčina, M., 2017. Design of coiled-coil protein-origami cages that self-assemble in vitro and in vivo. Nature biotechnology, 35(11):1094–1101. https://doi.org/10.1038/nbt.3994

Ljubetič˘c, A., Lapenta, F., Gradišar, H., Drobnak, I., Aupič˘c, J., Strmšek, Ž., Lainšč˘cek, D., Hafner-Bratkovič˘c, I., Majerle, A., and Krivec, N., et al. 2017. Design of coiled-coil protein-origami cages that self-assemble *in vitro* and *in vivo*. Nat. Biotechnol. 35: 1094–1101.

Siddiqui M.H., Al-Whaibi M.H., Firoz M., Al-Khaishany M.Y. (2015) Role of Nanoparticles in Plants. In: Siddiqui M., Al-Whaibi M., Mohammad F. (eds) Nanotechnology and Plant Sciences. Springer, Cham. https://doi.org/10.1007/978-3-319-14502-0_2

Mechrez, G., Krepker, M. A., Harel, Y., Lellouche, J. P., and Segal, E. 2014. Bio-catalytic carbon nanotube paper: aA "one-pot" route for fabrication of enzyme-immobilized membranes for organophosphate bioremediation. J. Mater. Chem. B. 2: 915–922.

Mitter, N., Worrall, E. A., Robinson, K. E., Li, P., Jain, R. G., Taochy, C., Fletcher, S. J., Carroll, B. J., Lu, G. Q., Xu, Z. P. 2017. Clay nanosheets for topical delivery of RNAi for sustained protection against plant viruses. Nat. Plants. 3: 1–10.

Mohanpuria, P., Rana, N.K. and Yadav, S.K., 2008. Biosynthesis of nanoparticles: technological concepts and future applications. J. Nanopart. Res. 10(3): 507–517.

Mohsenzadeh F., and Chehregani Rad, A. 2012. "Bioremediation of heavy metal pollution by nano-particles of *Noaeamucronata*. International Journal of Bioscience, Biochemistry and Bioinformaticsm. 2: 85–89.

Mueller, N. C., and Nowack, B. 2010. Nanoparticles for remediation: Ssolving big problems with little particles. Elements. 6(6): 395–400.

Nemecek, J., Pokorný, P., Lhotský, O., Knytl, V., Najmanova, P., Steinova, J., Cerník, M., Filipova, A., Filip, J., and Cajthaml, T. 2016. Combined nano-biotechnology for in-situ remediation of mixed contamination of groundwater by hexavalent chromium and chlorinated solvents. Sci. Total Environ. 563–564:, 822–834.

Nurmi, J. T., Tratnyek, P. G., and Sarathy V., et al. 2005. Characterization and properties of metallic iron nanoparticles: spectroscopy, electrochemistry, and kinetics. Environ Sci Technol.Environmental Science and Technology. 39 (5): 1221–1230.

Nurmi, J. T., Tratnyek, P. G., Sarathy, V., Baer, D. R., Amonette, J. E., Pecher, K., Wang, C., Linehan, J. C., Matson, D. W., Penn, R. L., and Driessen, M. D. 2005. Characterization and properties of metallic iron nanoparticles: Sspectroscopy, electrochemistry, and kinetics. Environ. Sci. Technol. 39: 1221–1230.

O'Hannesin, S. F., and Gillham, R. W. 1998. Long-term performance of an in situ 'iron wall' for remediation of VOCs. Ground Water. 36: 164–170.

Patil, S. S., Shedbalkar, U. U, Truskewycz, A., Chopade, B. A., and Ball, A. S. 2016. Nanoparticles for environmental clean-up: Aa review of potential risks and emerging solutions. Environ Technol. Innov. 5: 10–21.

Qiang, Y., Sharma, A., Paszczynski, A., and Meyer, D. 2007. Conjugates of magnetic nanoparticle-enzyme for bioremediation. In: Proceedings of the 2007 NSTI Nanotechnology Conference and Trade Show. 4: 656–659.

Quan, X., Yang, S., Ruan, X., and Zhao, H. 2005. "Preparation of titania nanotubes and their environmental applications as electrode,". Environmental Science and Technology. 39 (10): 3770–3775.

Ramamurthy, A. S., and Eglal, M. M. 2014. Degradation of TCE by TEOS coated nZVI in the presence of Cu(II) for groundwater remediation. J. Nanomater. 2014:, 1e9.

Ravikumar, K. V. G., Kumar, D., Kumar, G., Mrudula, P., Natarajan, C., and Mukherjee, A. 2016. Enhanced Cr (VI) removal by nano-zero-valent iron-immobilized alginate beads in the presence of a biofilm in a continuous-flow reactor. Ind. Eng. Chem. Res. 55(20): 5973–5982.

Reddy, K. R., and Adams, J. A. 2010. Sustainable Remediation of Contaminated Sites. Momentum Press, New York, NY.

Rizwan, Md., Singh, M., Mitra, C. K., and Morve, R. K. 2014. Ecofriendly Application of Nanomaterials: Nanobioremediation. J. Nanopart. 2014: 1–7. https://doi.org/10.1155/2014/431787

Ryu, H., Fuwad, A., Yoon, S., Jang, H., Lee, J. C., Kim, S. M., and Jeon, T. J. 2019. Biomimetic membranes with transmembrane proteins: State of the art in transmembrane protein applications. Int. J. Mol. Sci. 20: 1437.

Sastry, M., Ahmad, A., Khan, M. I., and Kumar, R., 2003. Biosynthesis of metal nanoparticles using fungi and actinomycete. Curr. Sci. 85 (2): 162–170.

Saxena, G., Purchase, D., Mulla, S. I., Saratale, G. D., and Bharagava, R. N. 2019. Phytoremediation of heavy metal-contaminated sites: Eco-environmental concerns, field studies, sustainability issues, and future prospects. Rev Environ Contam. Toxicol. https://doi.org/10.1007/398_2019_24.

Saxena, J., Sharma, M. M., Gupta, S., and Singh, A. 2014. Emerging role of fungi in nanoparticles synthesis and their applications. World J. Pharmacy Pharmaceut. Sci. 3 (9): 1586–1613.

Sebeia, N., Jabli, M., Ghith, A., and Saleh, T. A. 2019. Eco-friendly synthesis of *Cynomorium coccineum* extract for controlled production of copper nanoparticles for sorption of methylene blue dye. Arab. J. Chem. 2019, 13: 4263–4274.

Shan, G., Xing, J., Zhang, H., and Liu, H. 2005. Bio-desulfurization of dibenzothiophene by microbial cells coated with magnetite nanoparticles. Appl. Environ. Microbiol. 71(8): 4497–4502.

Shan, G., Xing, J., Zhang, H., and Liu, H. 2005. Bio-desulfurization of dibenzothiophene by microbial cells coated with magnetite nanoparticles. Applied and Environmental Microbiology. 71 (8): 4497–4502.

Sharma, A., Thakur, M., Bhattacharya, M., Mandal, T., and Goswami, S. 2019. Commercial application of cellulose nano-composites – A review. Biotechnol. Rep. 2019, 21: 00316.

Singh, B. K. 2010. Exploring microbial diversity for biotechnology: the way forward. Trend Biotechnol. 28 (3): 111–116.

Singh, B. K., and Walker, A. 2006. Microbial degradation of organophosphorus compounds. FEMS Microbiol. Rev. 30 (3): 428–471.

Singh, R., and Misra, V. (eds.). 2014. Application of zero-valent iron nanoparticles for environmental cleanup. In: Tiwari A and Syväjärvi M (eds) Advanced Materials For Agriculture, Food, And Environmental Safety. Scrivener Publishing LLC, Willey, pp. 385–420. https://doi.org/10.1002/9781118773857.ch14

Singh, R., and Misra, V. 2016. Stabilization of zero-valent iron nanoparticles: Rrole of polymers and surfactants. Handbook of Nnanoparticles. Springer International Publishing, Cham., pp. 985–1007.

Singh, R., Behera, M., and Kumar, S. 2020. Nano-bioremediation: An Innovative Remediation Technology for treatment and management of Contaminated Sites. In Bhargava, R. N., and Saxena, G. (eds.), Bioremediation of Industrial Waste for Environmental Safety. Springer Nature, Singapore Pte Ltd.2020. pp. 165–182.

Singh, R., Manickam, N., Mudiam, M. K. R., Murthy, R. C., and Misra, V. 2013. An integrated (nano-bio) technique for degradation of -HCH contaminated soil. J. Hazard. Mater. 258: 35–41.

Thome, A., Reddy, K. R., Reginatto, C., and Cecchin, I. 2015. Review of nanotechnology for soil and groundwater remediation: Brazilian perspectives. Water, Air Soil Pollut. 226:, 1–20.

Torres-Martínez, C.L., Kho, R., Mian, O. I., and Mehra, R. K. 2001. Efficient photocatalytic degradation of environmental pollutants with mass-produced ZnS nanocrystals. J. Colloid Interface Sci. 240(2): 525–532.

Tosco, T., Coisson, M., Xue, D., and Sethi, R. 2012. Zero-valent iron nanoparticles for groundwater remediation: surface and magnetic properties, colloidal stability, and perspectives for field application. In: Chiolerio, A. A. P. (eEd.), Nanoparticles Featuring Electromagnetic Properties: Ffrom Science to Engineering. Research Signpost, Kerala., pp. 201–223.

Tosco, T., Papini, M. P., Viggi, C. C., and Sethi, R. 2014. Nanoscale zerovalent iron particles for groundwater remediation: Aa review. J. Clean. Prod. 1: 10–21.

Tratnyek, P. G., and Johnson, R. L. 2006. Nanotechnologies for environmental cleanup. Nano Today. 1 (2): 44–48.

Tungittiplakorn, W., Lion, L. W., Cohen, C., and Kim, J. Y. 2004. Engineered polymeric nanoparticles for soil remediation. Environmental Science and Technology. 38(5): 1605–1610.

Vazquez-Nunez, E., Molina-Guerrero, C. E., Pena-Castro, J. M., Fernandez-Luqueno, F., and Rosa-Alverez, M. G. de la. 2020. Use of nNanotechnology for the bBioremediation of cContaminants: aA review. Processes. 8 (826): 1–17. http://dx.doi.org/10.3390/pr8070826.

Wang, M., Weiberg, A., Lin, F. M., Thomma, B. P. H. J., Huang, H. D., and Jin, H. 2016. Bidirectional cross-kingdom RNAi and fungal uptake of external RNAs confer plant protection. Nat. Plants. 2: 1–10.

Wang, X., Gai, Z., Yu, B., Feng, J., Xu, C., Yuan, Y., Lin, Z., and Xu, P. 2007. Degradation of carbazole by microbial cells immobilized in magnetic gellan gum gel beads. Appl. Environ. Microbiol. 73 (20): 6421–6428.

Windt, W. D., Aelterman, P., and Verstraete, W. 2005. Bioreductive deposition of palladium (0) nanoparticles on *Shewanellaoneidensis* with catalytic activity towards reductive dechlorination of polychlorinated biphenyls. Environ. Microbiol. 7(3): 314–325.

Xiu, Z. M., Gregory, K. B., Lowry, G. V., and Alvarez, P.J. J. 2010. Effect of bare and coated nano-scale zero-valent iron on *tceA* and *vcrA* gene expression in *Dehalococcoides* spp. Environ. Sci. Technol. 44: 7647–7651.

Xiu, Z. M., Jin, Z. H., Li, T. L., Mahendra, S., Lowry, G. V., Alvarez, P. J. (2010b.) Effects of nano-scale zerovalent iron particles on a mixed culture dechlorinating trichloroethylene. Bioresour Technol. 101 (4): 1141–1146.

Yadav, K. K., Singh, J. K., Gupta, N., and Kumar, V. 2017. A review of nNano-bioremediation technologies for eEnvironmental cleanup: A novel bBiological aApproach. Journal of Materials and Environmental Sciences. 8 (2): 740–757.

Yan, F. F., Wu, C., Cheng, Y. Y., He, Y. R., Li, W. W., and Yu, and H. Q. 2013. Carbon nanotubes promote Cr (VI) reduction by alginate-immobilized *Shewanellaoneidensis* MR-1. Biochem Eng J. 77: 183–189.

Yan, W., Lien, H. L., Koel, B. E., and Zhang, W. X. 2013. Iron nanoparticles for environmental clean-up: recent developments and future outlook. Environ. Sci. Process. Impacts 15:, 63–77.

Yan, W., Lien, H.L., Koel, B.E., and Zhang, W. X., 2013. Iron nanoparticles for environmental clean-up: recent developments and future outlook. Environ. Sci. Process. Impacts 15: 63–77.

Zanetti, M. C., and Fiore, S. 2005. Evaluation of mutual connections between zero-valent iron reactivity and groundwater composition in trichloroethylene degradation. Ann. Chim. 95: 779–789.

Zelmanov, G., and Semiat, R. 2008. Iron (3) oxide-based nanoparticles as catalysts in advanced organic aqueous oxidation. Water Reseasrch. 42(1–2): 492–498.

Zhang, W., Wang, C., and Lien, H. 1998. Treatment of chlorinated organic contaminants with nanoscale bimetallic particles. Catal. Today. 40: 387–395.

15

Cadmium Uptake, Toxicity, and Tolerance in Plants

Aditi Shreeya Bali
Dyal Singh College, Karnal, India

Gagan Preet Singh Sidhu
Centre for Applied Biology in Environment Sciences, Kurukshetra University, Kurukshetra, India

CONTENTS

15.1 Introduction

Both geogenic and anthropogenic activities contribute towards the addition of toxic heavy metals in the soils that might act as the root cause of heavy metal entry inside the food chain (Sidhu et al. 2016; Sidhu, 2016; Bali et al. 2020). Heavy metals are the non-biodegradable, persistent inorganic contaminants and are reported to induce toxicity in the living organisms (Sidhu et al. 2018). Among the heavy metals, cadmium (Cd) is reported to be one of the toxic heavy metal (Sidhu et al. 2017; Rizwan et al. 2018). Agency for Toxic Substances and Disease Registry (ATSDR) in the year 2012 has issued a report on ˜30,000 ton Cd production annually throughout the world (ATSDR 2012). ATSDR has listed Cd to the seventh position on the basis of its toxicity and potential exposure to humans (ATSDR 2019). Cd is extremely mobile and noxious in the soil-plant interface. Cd entry in the agriculture soils poses high risk to the human health. Excessive utilization of chemical fertilizers, sewage water use for irrigation purposes, sludge application in agriculture, metal working industries, atmospheric depositions have contributed its accumulation in soils (Dharma-Wardana, 2018). The agricultural soils of Asian countries like India and China are currently facing the extreme problems related to Cd contamination (Mori et al. 2016, Kumar et al. 2019).

Cd in plants has no physiological significance and is taken up by the plant roots due to its high mobility. Due to high Cd mobility, its entry in the food chain is quite common. In the mid-20th century, Japan has faced the serious issue related to the outbreak of itai-itai disease due to the long-term consumption of Cd-contaminated rice (Horiguchi et al. 1994). The toxic ionic form of Cd present in soil is Cd^{2+} and is taken up by the plant roots with the help of zinc/iron regulated transporters (Asgher et al. 2015). According to Sidhu et al. (2017), Cd in plants affect growth, induce changes in the nutrient uptake, pose pessimistic effects on the biosynthesis of chlorophyll, thus hamper photosynthesis in plants. Studies related to the decrease in the efficacy of carbon assimilation pathways are correlated with Cd instigated degradation of chlorophyll-protein complexes, chlorophyll structure and pigments associated with photosynthesis (Parmar et al. 2013). Cd excess in the growth medium is known to cause negative effects on the activities of vital mitochondrial enzymes like malate dehydrogenase, isocitrate dehydrogenase and succinate dehydrogenase (Bansal et al. 2002). Additionally, Cd stress in plants restricts the entry of vital nutrients that in turn induce nutritional deficiency.

Cd is also known to promote oxidative stress by causing oxidation of proteins, lipids and nucleic acids, induce variations in metabolism of carbohydrates and antioxidant defense

DOI: 10.1201/9781003155089-15

response in plants (Ahmad et al. 2016; Sidhu et al. 2017). Cd excess inactivates antioxidant enzymes like superoxide dismutase (SOD), catalase (CAT), ascorbate peroxidase (APX), glutathione peroxidase (GPX) and glutathione reductase (GR) accountable for detoxifying the cells. On the parallel, Cd is known to disturb metabolic pathways in plants by the enhanced generation and accumulation of oxygen free radicals and reactive oxygen species (ROS) that further promote peroxidation of lipid membranes. Accumulation of ROS and oxidation of lipid membranes pose catastrophic damage to the biomolecules and ROS might also act as signaling molecules that hinder the plant responses to the adaptation and modulates cell death (Guo, J. et al. 2019).

Plants have developed various enzymatic and non-enzymatic antioxidant mechanisms that enabled them to withstand and cope stressful conditions instigated due to high ROS levels. The critical enzymatic mechanisms involve the antioxidant enzymes that impart a key role in antioxidant defense system. Antioxidant enzyme like SOD acts as a frontline barrier for oxidative stress and disintegrates superoxide ion to H_2O_2 (Sidhu et al. 2016). CAT, APX and GPX are involved in catalysing H_2O_2 into H_2O and O_2, donating electrons to the guaiacol (Sidhu et al., 2018). Non-enzymatic antioxidants include phenols, proline, carotenoids, phytochelatins, metallothioneins, vitamin C and glutathione. Among the non-enzymatic antioxidants, proline is a solute that acts as a chelant to form Cd-chelant complex, hence protect the plants from oxidative damage. Phytochelatins, metallothioneins and glutathione are well known to detoxify Cd stress in plants. In light of the available studies, exploiting various enzymatic and non-enzymatic antioxidants in mitigating Cd toxicity at molecular level is a sustainable adaptable approach opted by the plants. This chapter is divided into two parts: (i) Cd uptake, transport and the effect of its toxicity on the morphological, physiological and biochemical attributes of the plants and (ii) tolerance strategies adopted by plants to combat Cd toxicity.

15.2 Cd in the Environment

Cd is a bluish-grey metal having atomic number 48, atomic weight 112.41 and density 8.65 g/cm^3 (Bali et al. 2020). It has electronic configuration of [Kr], $4d^{10}$, $5s^2$ and belongs to group 12 of the periodic table. According to Asgher et al. (2015), Cd is insoluble in water and has soft, malleable and ductile characters. Cd exist in two oxidation states (+1 and +2), however +2 oxidation state is the frequent occurring oxidation state. Cd in solution occurs in divalent oxidation state (Cd^{2+}). Though pH of the solution effect its mobility due to the solubility of organic matter, ionic pair formation and metal hydrolysis. The divalent ionic state of Cd is not redox sensitive, but it retain itself to redox sensitive conditions due to the formation of complexes with $CdHS^+$, and in the form of precipitates like galena, sphalerite, chalcopyrite that contain Cd as an impurity. Cadmium oxide (CdO) is an amorphous form and is formed due to the burning of Cd in the atmosphere.

Both anthropogenic and natural sources contribute towards Cd contamination of soil/water ecosystems. Natural sources include weathering of rocks, volcanoes, hydrothermal vents, biogenic materials, atmospheric depositions and sea spray (ATSDR 2012). The predominant source of Cd in the environment is weathering of rocks followed by forest fires, volcanic eruptions and meteoric dust (Kubier et al. 2019). Anthropogenic Cd sources include production of phosphate fertilizers, cement manufacturing, fossil fuel combustion, electroplating, road dust, incineration of municipal wastes and use of sewage sludge in agriculture (ATSDR 2012). According to Bigalke et al. (2017), the common reason for Cd release by anthropogenic activities is the excessive use of phosphate fertilizers in agriculture. Cd is present as an impurity in phosphate fertilizers that is liable to alter the soil chemistry. In European soils, phosphate fertilizers contain an average of 36 mg Cd kg^{-1} P_2O_5 (Six and Smolders 2014), 77 mg Cd kg^{-1} P_2O_5 in Eastern Mediterranean countries (Azzi et al. 2017). Therefore, utilization of excessive fertilizers in the agriculture add Cd to the soil, hence it can enter the food chain and pose great threat to the health of living biota.

15.3 Cd Uptake in Plants

Secretion of secondary metabolites increases Cd solubility in the soil (Bali et al. 2020). Soil acidification spike Cd bioavailability to the plants. Cd enters the plants as divalent ion (Cd^{2+}) by two stage process: apoplastic adsorption and symplastic uptake. In plants, apoplastic uptake involves the movement of water within the cell in the intercellular spaces along the cell walls. While in symplastic pathway, water moves across the cytoplasm and plasmodesmata. The plasma membrane poses resistance to the flow of water in the symplastic pathway.

The initial step of metal uptake does not require energy and is a passive pathway. The cationic Cd species (Cd^{2+}) gets accumulated in the root zone from where its uptake is dependent upon the soil pH and cell wall exchange properties i.e. the release of anionic carboxyl group. Increase in pH deprotonate carboxyl group and enhance electrostatic interaction between positively charged Cd^{2+} and negatively charged carboxylic group at the root apoplast. The second step i.e. symplastic movement is a slow metabolic step and requires energy. The transport of Cd through the root cell membrane is the initial step of its transport and uptake by the plants. Besides, several protein transporters are indulged in Cd uptake by the roots. It primarily depends upon the concentration of Cd in the soil, plant species and the quantity of micronutrients present in the soil. The mechanisms related to Cd transport across the plant membranes are still lacking. Therefore, complete knowledge in Cd transport is important to device the tolerance strategies adopted by the plants either to alleviate Cd from the external medium, or to limit Cd uptake by the plants.

Cd enter inside the roots as Cd^{2+} mainly via three transporters (i) natural resistance associated macrophage proteins (NRAMP) like OsNRAMP1, OsMRAMP5 and AtNRAMP6; (ii) zinc/iron-regulated transporter-like proteins (ZIP) transporters like AtIRT1, TcZIP4 transporters and (iii) low-affinity calcium transporters (TaLCT1) (Corguinha et al. 2015). ZIP family of transporters have been identified in fungi, bacteria, plants and animals and are highly involved in the uptake of metal ions in plants. The crucial feature of ZIP family of transporters lies in the potential of transport heavy metals from the extracellular space and from the lumen of the organelles to the cytoplasm. IRT1 is the first discovered member of ZIP family of transporters that was identified first in *Arabidopsis thaliana* for the uptake of Fe^{2+}, Mn^{2+}, Zn^{2+}, Cu^{2+}, Ni^{2+}, Co^{2+} and Cd^{2+}. It has been noticed that the transgenic plants in which AtIRT1 gene has been upregulated showed high Cd uptake and acclimatization than wild type mutants under the conditions of iron deficiency from the soil (Connolly et al. 2002). In a study conducted by Lin et al. (2009), the upregulation of IRT3 gene was higher in *Arabidopsis halleri* compared to *Arabidopsis thaliana* that implies a positive correlation with Cd/Zn accumulation in *Arabidopsis halleri*. Expression of AtIRT3 in *Arabidopsis thaliana* showed an increment in Zn accumulation in shoots and iron accumulation in roots of transgenic lines, but no uptake and transport of Cd in *Arabidopsis* plants was observed. Adding to it, AtIRT3 expression in *Arabidopsis halleri* was reported to respond to Zn stress but no upregulation was found in case of Cd, Na and Cu (Talke et al. 2006). According to Bughio et al. (2002) under iron deficient conditions, OsIRT1 a homologous to AtIRT1 is highly expressed in the rice roots that resulted in the increased Cd uptake in the rice roots. A comparative study related to the genome-wide upregulation of ZIP proteins in *Oryza sativa* and *Arabidopsis thaliana* revealed the distinct pattern of the expression of ZIP transporters in these plants under Cd toxicity. Root ZIP transporters were expressed in *Arabidopsis* while shoot ZIP transporters were expressed in rice plants. In the yeast cells three ZIP transporters AtIRT1, OsZIP1 and OsZIP3 were expressed that grant enhanced Cd uptake and transport in plants (Zheng et al. 2018). AtIRT1, a plasma membrane protein has high affinity for divalent heavy metals like Cd^{2+} and mediates the uptake and accumulation of Cd by the plant roots (Nishida et al. 2011; Huang et al. 2020). AtIRT1 absorbs metal from the soil and is located on the outer root layer. The work conducted by Zhang, X. et al. (2020) revealed the expression of VsRIT1 contribute towards Cd uptake by the roots of Zhangye Mawan variety of *Vicia faba*. Furthermore, root tips of the Zhangye Mawan variety have showed higher Cd influx in the roots when exposed to different Cd treatments. In another study, the expression of ZIP genes (NtZIP1, NtZIP2, NtZIP4 and NtIRT1) in the apical, middle and basal root parts resulted in the uptake and acclimatization of Cd and Zn in the roots and shoots of *Nicotiana tabacum* plants (Palusińska et al. 2020). Two plasma membrane-localized protein transporters OsZIP5 and OsZIP9 overlap in the root epidermis and act synergistically to influx Cd and Zn inside the roots of rice plants (Tan et al. 2020). Additionally, overexpression of OsZIP9 remarkably enhanced Zn/Cd concentrations in the shoots and rice grains.

Yellow stripe-like (YSL) are the oligopeptide family transporters and are involved in the metal transport through the formation of metal-nicotianamine complexes through the cell membrane. The plants in the rhizosphere secrete organic acids as root exudates that form complex with the Cd ions from where it is taken up by the plant roots with the assistance of YSL family of transporter proteins. ZmYSL is the first member of YSL gene and is isolated from *Zea mays* is indulged in the transport of Cd-phytosiderophore complex in the plants at a relative slow rate. Cd excess in the growth medium upregulated SnYSL3 gene in *Solanum nigrum* (Feng et al. 2017). The heterologous upregulation of SnYSL in the yeast revealed the transport of nicotianamine complex having Cd and other metals like Zn and Fe. The authors in their research suggested the potential role of the expressed SnYSL transporter protein in forming metal-nicotianamine complex, which mediates Cd transport along with other metals. Three YSL protein transporters (TcYSL3, TcYSL5 and TcYSL7) have been expressed in *Thlaspi caurulescens*, a Cd, Ni and Zn hyperaccumulator (Gendre et al. 2007). It is noteworthy that the exposure of these three heavy metals did not upregulated TcYSL genes. Expression of TcYSL5 and TcYSL7 gene transporters in yeast showed no Cd, Zn and Ni-nicotianamine influx. However, the expression of TcYSL3 in yeast showed Ni and Fe uptake through the formation of Ni- and Fe–nicotianamine complex but no such complex is formed for Cd.

Natural resistance associated macrophage proteins (NRAMP) are the integral membrane protein transporters involved in iron transport in plants. Upregulation of HvNramp5 iron transporter in the plasma membrane has been involved in Cd influx. The expression of HvNramp5 was found to be higher in the root tips compared to the basal root parts. The authors concluded that HvNramp5 transporter facilitates the uptake of Cd and Mn but not Fe in the roots of the barley plants (Wu et al. 2016). Under Cd stress, SaNramp6 transporters expressed in transgenic *Arabidopsis thaliana* plants exhibited high Cd acclimatization as demonstrated with the increased Cd influx, high Cd concentration and translocation factors (Chen et al. 2017). Likewise, Sui et al. (2018) studied the functionality expression of Nramp5 transporters in enhancing Cd uptake by the roots of rice, wheat and maize plants. The hydroponic tests revealed Cd uptake in rice range from 2.5-8.1 and was 1.1–3.6 times that of wheat and maize, respectively. The maximum Cd influx velocity in rice was 6.5 and was ~2.2 times more than wheat and maize. The plasma membrane localised OsNRAMP5 was absolutely expressed 4–5 times in the roots of rice compared to TaNRAMP5D in wheat or *Zm*NRAMP5 in maize (Sui et al. 2018). This showed that NRAMP transporter genes are highly involved in Cd uptake and translocation in plants. OsNRAMP1 is highly involved in cellular Cd uptake and the variation in Cd acclimatisation in different crop plants is probably due to the variation in the expression levels of OsNRAMP1 transporters. It has also been found that OsNRAMP5 is highly involved in Cd and Mn influx into the roots from the external solution (Sasaki et al. 2012). Contrarily, OsNRAMP3 played no role in Cd and Fe uptake in plant roots but its expression in the vascular nodes enhanced Mn uptake in the roots of the plants (Yamaji et al., 2013). To the best of our knowledge, OsNRAMP4 transporter re rarely involved in the uptake and transport of divalent metal cations (Cd, Zn, Mn, etc.) but actively participate in the uptake of trivalent metal cations (Al) by the plant roots (Xia et al. 2011). Moreover, Cd exposure expressed a tonoplast protein transporter NcNRAMP3 in the roots of *Noccaea caerulescens* that enable the plants to uptake and transport Cd, and Zn. The role of NcNRAMP3 in Cd influx across the endodermal plasma membrane imparts a critical role in Cd influx in the stele and its root to shoot transport (Wei et al. 2009). The expression of NcNRAMP3 contributes towards the hyperaccumulation of Cd in the tissue parts of *Noccaea caerulescens*.

Once Cd enters inside the root cells, it can be excluded out to the external medium by the specific transporters (Kim et al. 2007), stored in the vacuoles of the root cells (Ueno et al. 2010) or can be retained in the vasculature for its long distance transport mediated by HMA2 and HMA4 (Wong and Cobbett 2009). HMA3 participate in retaining ~895% Cd inside the vacuoles while some transporters are involved in Cd efflux out of the roots to prevent Cd toxicity and promote plant tolerance towards Cd stress. Pleiotropic drug resistance (PDR) gene AtPDR8 expressed in *Arabidopsis* is found to be localized at the plasma membrane and root epidermal cells provide Cd resistance in plants by efflux Cd ions out of the plasma membrane of root epidermal cells. The transgenic plants of *Arabidopsis thaliana* overexpressed AtPDR8 gene were found to be more tolerant towards Cd exposure. Cd concentration in transgenic plants was decreased in AtABCG36 overexpressed plants, thus suggesting the role of AtABCG36 transporters in effluxing Cd from roots to the external medium. This relation was further confirmed with the presence and upregulation of AtABCG36 especially in the root epidermal cells (Theodoulou, 2000).

In the root cortex, number of important processes occurs at specific sites and the cells are indulged in the Cd loading into the root xylem. Prior to Cd loading to the xylem vessels, Cd entry inside the stele induces its apoplastic movement. Cd loading is the vital part of Cd transport, its acquisition, translocation and redistribution in various plant tissues is a function of Cd perpetuate inside the roots and loading into the xylem. Apoplastic barriers, vacuolar quenching and formation of Cd complex with phytochelatins are responsible for holding Cd inside the root cells (Mendoza-Cózatl et al. 2011). Protein family heavy metal P1B-ATPase is involved in loading Cd inside the roots. Courbot et al. (2007) in their study demonstrated the colonization of a quantitative trait locus for Cd tolerance with a gene encoding a heavy metal ATPase like HMA4. The knockdown of AhHMA4 in *A. halleri* had resulted an increase in Cd loading in the xylem but decremented the concentration of Cd acclimatized in aerial tissues, thus suggested high expression of transcript level is required for the effective hyperaccumulation of heavy metals in hyperaccumulators (Hanikenne et al. 2008). Furthermore, it has been reported by Mills et al. (2005) that the reduced expression of AtHMA4 has caused an increase in sensitivity of *Arabidopsis thaliana* exposed to Cd stress. Adding to it, the upregulation of HMA4 protein transporter in the yeast cells proliferated well under Cd toxicity and promoted Cd efflux out of the cells through the plasma membrane (Korenkov et al. 2009). Moreover, members of ABC transporter family like AtMRP3 are also involved in Cd transport across different plant tissues (Verbruggen et al. 2009).

In the stem nodes, the xylem-to-phloem transport system is activated that allows Cd uptake and translocation in plants. This movement of Cd is regulated by OsNRAMP5 protein like transporters (Yoneyama et al. 2015). In rice plants, the downregulation of OsNRAMP5 hindered Cd uptake and root-to-shoot Cd translocation. This further had induced a remarkable reduction in Cd accumulation in rice grains (Yoneyama et al. 2015). The same condition has been correlated by Rodda et al. (2011) who demonstrated Cd uptake by plant roots, its loading to the xylem, and its acclimatization at the shoot base is followed by Cd entry inside the phloem and the xylem-to-phloem transport contributes towards Cd entry inside the leaves and grains. According to Yadav (2010), the long distance Cd transport is facilitated by the formation of Cd complex with phytochelatins, glutathione present in the phloem sap. Through, the phloem channel Cd- phytochelatins and Cd-glutathione complexes are actively transported. The sequestration of Cd in the subcellular and extracellular compartment of the leaf tissues enables the plants to withstand and tolerate Cd toxicity (Sidhu et al. 2019). Besides, the physiological processes in metal accumulating plants like transpiration facilitate the plants in metal transport across the plant tissues by the formation of metal-low molecular weight complex.

15.4 Cd Toxicity in Plants

Cd has no functional role and is a toxic metal for plants. Cd excess induce pessimistic effects on plants and disturb the physiological and biochemical status of the plants by disrupting opening and closing of stomata, nutrient and water uptake, hinders nitrogen and carbohydrate metabolism affect photosynthesis and respiration processes in plants (Sidhu et al. 2017; Bali et al. 2020). According to Wang et al. (2014), growth of rice plants was extensively affected and reduced due to excessive Cd concentration in the growth medium. Additionally, plant exposure to Cd ranging from 0.2 to 1.0 mM imposes negative effect on the germination potential of rice seeds (Ahsan et al. 2007). Cd toxicity in plants alter the expression of various enzymes involved in Calvin cycle, retard the activity of PSI and PS II and promote the dissociation of ribulose-1,5-biphosphate carboxylase oxygenase enzyme (RuBisCO). Cd excess directly affects the pigment concentration in plants and cause deleterious effects on the chlorophyll biosynthesis. Wang et al. (2014) reported a decrease in the photosynthesis and photosynthetic gas exchange attributes in rice plants grown under Cd stress. Prolonged and high Cd exposure in plants stimulates the generation and accumulation of reactive oxygen species (ROS), cause oxidation of lipids and denaturation of proteins (Plata et al. 2014). Furthermore, Cd abundance in the growth medium induce changes in cell cycle, affects the expression of genes and cause programmed cell death in plants (Sidhu et al. 2017).

In addition to the toxic Cd effects on the morphological, physiological and biochemical attributes, Cd is also known to cause alterations in the plant genetic material. Cd excess in the plants binds to the proteins and DNA, cause damage to the nucleic acids, proteins, promote chromosomal aberrations, disrupt cell division, influence the production of stress signaling molecules and stress related genes (Gill and Tuteja 2011; Gill 2014). Long term Cd exposure in plants may cause modification of genetic material and induces genomic variations. Sun et al. (2015) in hydroponic experiments revealed the variation in genetic expression of two barley genotypes grown under 5 µM Cd for 15 days. The authors demonstrated the overexpression of 812 genes having 382 upregulated and 131 downregulated in W6nk2 genotype a low Cd grain accumulation genotype, while in Zhenong genotype having high Cd accumulation in the grains 303 genes were upregulated and 106 were downregulated. In another study, 100 µM and 150 µM Cd exposure in sorghum plants increased the expression of 18 proteins while 15 proteins were downregulated. The proteins were mostly expressed in the mitochondria, chloroplast membranes, ribosomes, nucleus and cytoplasm and impart a key role in the plant metabolism, transcription, translation and transport mechanisms in plants exposed to Cd stress (Roy et al. 2016). Different Cd levels (50, 100, 150 and 200 µM) in *Brassica juncea* increment the expression of mRNA along with phytochelatin synthase gene (Shanmugaraj et al. 2013). Alterations in gene expression were also found in proteins and chloroplasts. The plant defense mechanisms in response to Cd toxicity can be easily understood by studying the involvement of molecular mechanisms adopted by plants during Cd distress. The expression of ZmMPK3-1 and ZmMPK6-1 genes during Cd stress in the roots of *Zea mays* activated the activities of antioxidant enzymes like peroxidase to combat Cd instigated oxidative stress in the plants (Yue et al. 2016). This activation of specific genes in response to Cd excess enabled the plants to withstand Cd toxicity. According to Castro et al. (2015), exposure of 32 mg l^{-1} Cd to *Theobroma cacao* seedlings induced the expression of genes related to the biosynthesis of class III peroxidase ˜26% in leaves and ˜57% in the roots. Besides, Cd excess in *Zea mays* pose pessimistic effects on the expression of psbA, psaB and rbcL genes associated with the photosynthesis (Franic, 2018). The findings from the different researchers suggested the catastrophic effect posed by Cd at molecular level that range from seed germination to the fruit development in plants.

15.5 Cd Tolerance Mechanisms Adopted by Plants

As mentioned above, Cd acclimatization pose toxic and catastrophic effects on the growth, physiological, biochemical and molecular levels in plants. There are two detoxification strategies adopted by the plants to nullify the toxic effects of Cd on the plants: (i) stress avoidance strategy and (ii) tolerance strategy. These adaptive mechanisms to withstand Cd stress can work together in association either to limit Cd entry and its acclimatization in the plant tissues or to neutral Cd toxicity inside the plants. Plants have evolved a well-developed immune system in the form of enzymatic and non-enzymatic antioxidant response to counter Cd induced toxicity in plants (Figure 15.1). In addition, various signaling pathways in the

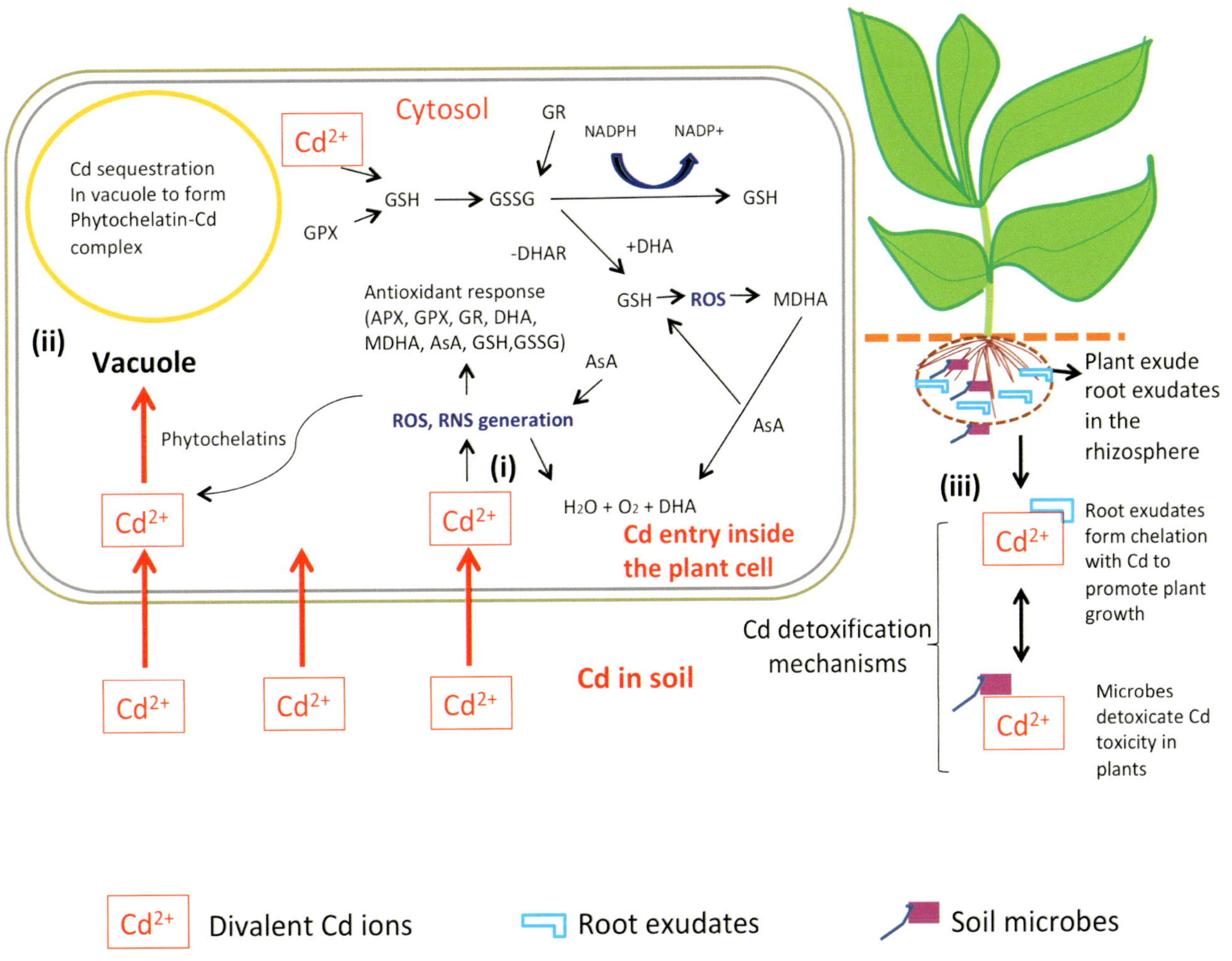

FIGURE 15.1 Cadmium detoxification inside the plant cell. (i) Cd entry inside the plant cell leads to the generation of reactive oxygen species (ROS) by the overproduction of hydrogen peroxide, hydroxyl ion and superoxide ion, etc. that trigger the activity of antioxidant enzymes, thiol and non-thiolic compounds like GPX, guaiacol peroxidase; GR, glutathione reductase; DHA, dehydroascorbate; MDHA, monodehydroascorbate; AsA, ascorbic acid, GSH, reduced glutathione; GSSG, oxidized glutathione. (ii) Cd is quenched inside the vacuole as Cd-phytochelatin complex. (iii) Secretion of root exudates form complex with Cd and microorganisms assist the plants to detoxicate Cd instigated toxicity in plants.

form of phytohormone synthesis and ROS generation are activated in plants during Cd exposure. Restoration of photosynthetic pigments, synthesis of phytochelatins and antioxidants enhance plant tolerance towards increased Cd concentrations in the growth medium. Cd sequestration in root cell wall and leaf vacuole, chelation, induction of antioxidant enzymes and proteins enable the plants counter-act oxidative stress caused due to Cd distress.

15.5.1 Plant Hormones

Plant hormones regulate the growth and development of plants and also serve as a signal for Cd toxicity (Pan et al. 2020). Salicylic acid, auxin, cytokinin, gibberellins, ethylene, abscisic acid jasmonic acid, brassinosteroids (BR), and strigolactones (SL) are the plant hormones that actively participate in providing tolerance response in plants towards Cd distress (Amir et al. 2019). Under Cd exposure, variety of hormonal effects has been revealed in plants (Table 15.1). Plant hormones regulates cell signaling under Cd excess, trigger the activation of membrane-localized transporter proteins that directly/indirectly provide stress tolerance in plants (Marques et al. 2019). Exposure of *Koelreuteria paniculata* seedlings to varied Cd treatments (50, 150, 250 and 500 µM) has shown variable effect on the activation of different phytohormones (Yang et al. 2018). Jasmonic acid and indole – 3 acetic acid level decreased at highest Cd concentration, while abscisic acid level increased on exposure to varied Cd treatments in *Koelreuteria paniculata* seedlings (Yang et al. 2018). Likewise, the level of abscisic acid in Cd treated (40 µM) *Glycine max* plants increased in the first 24 hours, while a decrement in abscisic acid level was observed after 140 hours (Pérez Chaca et al. 2014). Additionally, Pérez Chaca et al. (2014) demonstrated the role of abscisic acid and jasmonic

TABLE 15.1

Phytohormones Modulate Cd Tolerance in Plants

Plant	Phytohormones	Cd Concentration/ Growth Medium	Time Span	Effect on Plant Traits	References
Brassica napus	Abscisic acid	0, 50, 100, 200, 400 µM; Soil	6 weeks	Increased plant fresh weight with decreased malondialdehyde level in the leaves and antioxidant response.	Meng et al. (2009)
Solanum lycopersicum	Brassinosteroids	3, 6, 9, 12 mg kg^{-1}; Soil	90 days	Foliar spray of the brassinosteroids enhanced the photosynthesis and maintained balance on the production of oxygen free radicals with increased antioxidant response.	Hayat et al. (2012)
Vicia faba	Jasmonic acid	150 mg l^{-1}; hydroponics	60 days	Jasmonic acid application dramatically increased SOD, CAT, APX and GR activities with a parallel decrease in the levels of H_2O_2 and MDA levels.	Ahmad et al. (2017)
Nymphaea tetragona	Salicylic acid	150 mg l^{-1}; hydroponics	9 days	Salicylic acid addition enhanced the levels of photosynthetic pigments, glutathione and phytochelatins in plant tissues to combat Cd stress.	Gu et al. (2018)
Arabidopsis thaliana	Gibberellic acid	50 µM; petri dish	12 days	Gibberellic acid decreased the expression of Cd uptake gene IRT1 in Arabidopsis plants, thus mediated Cd tolerance in plants against Cd distress.	Zhu et al. (2012)
Triticum aestivum	Ethylene	200 mg kg^{-1}; Soil	30 days	Sulphur and selenium mediated the biosynthesis of ethylene that assists to reverse Cd induced oxidative stress by modulating proline and glutathione metabolism in the plant tissues.	Khan et al. (2015)
Kandelia obovata	Abscisic acid	110 µM; nutrient solution	7 days	Abscisic acid concentration increased in the root epidermis and exodermis that increment the expression of KoSODs in root tissue to counter Cd toxicity.	Pan et al. (2020)
Solanum melongena	Cytokinin	3, 9 mg kg^{-1}; nutrient solution	4 days	Enhanced the activities of antioxidant enzymes with increased pigment concentrations to tolerate Cd induced oxidative stress in plants.	Singh and Prasad (2016)
Solanum lycopersicum	28-homobrassinoloid	3, 9 mg kg^{-1}; hydroponics	15 days	28-homobrassinoloid alleviated Cd stress by regulating photosynthesis and nitrogen metabolism with increased protein content, carbohydrate content and enhanced activity of nitrogen assimilation enzymes.	Singh and Prasad (2017)
Trigonella foenum-graecum	Indole-3-acetic acid	1, 3, 6, 9, 12, 15 mg kg^{-1}; Soil	15 days	Indole-3-acetic acid increased the activities of antioxidant enzymes and maintain redox homeostasis, thus decreased the accumulation of ROS in the plant tissues generated due to Cd excess.	Bashri and Prasad (2016)

acid in promoting plant growth in *Glycine* plants grown under Cd stress. It has also been noticed that pre-soaking the crop plant seeds in phytohormones rich medium confer the plants to alleviate Cd toxicity. Study conducted by Munzuroglu and Zengin (2006) revealed the positive effect of pre-soaked barley seeds in gibberellic acid on the coleoptiles growth. Furthermore, Cd treated maize plants showed an increase in the growth of coleoptiles segments when the seeds were pretreated with salicylic acid and indole-3-acetic acid (Karcz and Kurtyka 2007).

The exogenous application of salicylic acid in plants imparts crucial role in large number of physiological plant processes. Studies have demonstrated the effect of pretreated salicylic acid on the incremented growth, biomass and photosynthetic activity of plants under Cd stress (Semida et al. 2015; Roychoudhury et al. 2016). According to Semida et al. (2015), pre-soaked *Phaseolus vulgaris* seeds with salicylic acid increased the antioxidant activities to combat Cd toxicity. In *Zea mays*, the supplementation of salicylic acid remarkably reduced the acclimatization of Cd in the roots (Gondor et al. 2016). Pre-treated barley seedlings with 600 μM reduced the toxic effects of Cd on the germination and early seedling growth with the parallel increment in the activity of hydrolytic enzyme (Kalai et al. 2016). A similar study conducted by Espanany et al. (2015) enhanced the rate of germination, root-shoot length and biomass of fenugreek seeds pretreated with 100 mg l^{-1} salicylic acid.

15.5.2 Cd Quenching

Cd quenching is an adaptive mechanism adopted by the plants to withstand Cd toxicity and to tolerate severe environmental factors. Plant roots are the primary targets for Cd stress, where the process of Cd detoxification is effective (Bali et al. 2020). Cd entry inside the plants and its root to shoot transport can be reduced by the effective exclusion mechanisms like Cd acquisition by the microbes and by forming chelation with root exudates released by the plants (Bali et al. 2020). Plants have the inherent potential to limit Cd entry and its mobility inside the plant tissues by the binding of Cd to the cell wall and plasma membrane, hence reduce its transport to the aerial parts by the formation of complex in the rhizosphere. Studies conducted by Fu et al. (2011) and Xin et al. (2018) in *Phytolacca americana* and *Triarrhena sacchariflora* revealed Cd binding in the cell wall of the plants, thus protect them from Cd toxicity. The enhanced plant tolerance is associated with the presence of pectin and suberin as cell wall associated glycolipids (Zhi-bin et al. 2016) or due to the presence of carboxylic and hydroxyl groups (Huguet et al. 2012) that assist Cd binding on the plant cell wall.

Once Cd cross the root cell wall, cell vacuoles as central compartments are involved in quenching Cd. Sequestration of Cd inside the vacuole reduce Cd concentration in the cytoplasm, hence decrease Cd toxicity in plants. Cd sequestration is controlled by the transporter proteins present on the plasma membrane like ZIP transporters that assist the transport of micro-nutrients like Zn, Fe, Cu and Mn and toxic metals like Cd. Cd exposure (50 μM) for 4 days in *Arabidopsis thaliana* upregulated the expression of transporter genes like HMA3, MTP3, IRT2, NAS1 and NAS2, that modulate the sequestration of Cd in the roots and limit its translocation to aerial plant parts (Wu et al. 2016). Recently, hydroponic experiments conducted by Faheem et al. (2020) illustrated the effect of Cd (10 and 15 μM l^{-1}) on the molecular status and Cd accumulation in aerial tissues of two rice cultivars (Heizhan 43 and Yinni 801). Results revealed the upregulation of protein transporters induced Cd sequestration in roots and its translocation to aerial plant parts. Additionally, expression of OsHMA2 and OsHMA3 in the roots of Heizhan 43 cultivar induces the vacuolar sequestration of Cd and suggested Cd translocation from roots to the grains (Faheem et al. 2020). Furthermore, *Iris lactea* plants on Cd exposure expressed IHMA2 protein transporter in its roots that regulated root to shoot Cd transport via xylem (Guo, Q. et al. 2019).

15.5.3 Antioxidant Defense Response in Plants

Antioxidant defense strategy is a vital detoxification mechanism adopted by plants to cope Cd toxicity. The antioxidant defense response in plants include both enzymatic (catalase, superoxide dismutase, guaicol peroxidase, ascorbate peroxidase, glutathione reductase) and non-enzymatic antioxidants (phenolic compounds, amino acids, a-tocopherols, alkaloids and glutathione). Cd excess in the growth medium induce the production and accumulation of ROS, induce oxidative stress in the plant tissues but this defensive plant system enable the plants to counter oxidative stress by converting the toxic compounds into less toxic forms and assist the plants to maintain cellular redox homeostasis (Sidhu et al. 2017).

Various genes have been modulated on exposure to Cd in plants that alters the antioxidant machinery. However, studies have demonstrated an altered expression of genes in plants grown under Cd stressed conditions. According to Alaraidh et al. (2018), number of primers was detected to estimate the transcription profile of antioxidant enzyme encoding genes in fenugreek plants exposed to different concentration of Cd, Pb and Cr (2.5–10 mg l^{-1}) for seven days. Antioxidant enzyme like POD and CAT were expressed poorly under Cd treatment, while a fairly moderate expression was observed in case of ascorbate peroxidase. Nevertheless, Han et al. (2015) observed an upregulation of TaEXPB23 gene that enhanced the activity of GR and DHAR in transgenic tobacco plants. Expression of gene like ZmVTE4 that encodes $^{\gamma}$-tocopherol-methyltransferase incremented the concentration of $^{\gamma}$-tocopherol and α-tocopherol in transgenic tobacco pants that in turn increased the tolerance potential of the plants towards Cd stress (Zhu et al. 2019). In another study, upregulation of glutaredoxin obtained from the leaves of chickpea enhanced the growth, antioxidant enzyme activity (APX, GPX, GR, SOD and CAT) with a parallel increment in the tolerance potential of *Arabidopsis thaliana* plants grown under 500 μM Cd treatment (Kumar et al. 2020).

15.5.4 Plant-Microbe Interactions

Microbial community in the soil contaminated with Cd impart a pivotal role in increasing plant growth along with conferring plant tolerance towards increased Cd concentrations in the growth medium (Bali et al. 2020). Wu et al. (2016)

found a remarkable increase in the pH, electrical conductivity, organic matter concentration and the amount of bioavailable Cd content in the soil with the inoculation of arbuscular mycorrhizal fungi. Likewise, inoculation of *Funneliformis caledonium* and *Glomus versiforme* in Cd-contaminated soil enhanced the accumulation of phosphorus and plant biomass with a parallel decrease in Cd concentration in root and shoot part of cucumber plants (Hu et al. 2016). The concentration of non-enzymatic antioxidants like proline, phenols, etc. increased significantly to counter Cd induced oxidative stress generated by the accumulation of MDA and H_2O_2 content in tomato seedlings (Hashem et al. 2016).

Plant growth promoting rhizobacteria (PGPR) are also known to grant plant tolerance to increased Cd toxicity. PGPR like strain CBMB20 from *Methylobacterium oryzae* and strain CBMB40 from *Burkholderia* sp. obtained from rice are known to increase the growth and biomass of tomato plants grown under Cd and Ni stress. Furthermore, these PGPR strains reduced the metal uptake, accumulation and their translocation to the aerial parts of the tomato plants. In a similar study conducted by Guo and Chi (2014) revealed the increment in the growth, biomass, nutrient level and chlorophyll content of the soybean plants when grown under Cd stress. Furthermore, this symbiotic association decrement the concentration of Cd in the shoots of the plants with a parallel increase in the concentration of iron as nutrient in the plant tissues. Contrarily, the microorganisms like PGPR are also known to enhance Cd distress in plants. A study conducted by Sharma and Archana (2016) revealed the important mechanisms for solubilization of mineral nutrients, exudation of siderophores, organic acids, phytohormones, formation of Cd-chelate complex and Cd immobilization for promoting Cd uptake and its acclimatization in the plant tissues. Thus, before opting microorganism-based techniques for remediating Cd-contaminated ecosystems, it is vital to take precautionary measures to select and screen a plant species for the specific experiments like phytoextraction, phytostabilization, etc.

15.5.5 Proline Biosynthesis

Proline is an amino acid, synthesized in the chloroplasts and cytoplasm of the plant cell. It is not a vital amino acid for the plant metabolism but is known to impart a critical role during plant-metal interactions. Proline is known to form Cd-proline complex, a relatively less toxic compound, enhance the concentration of carbon and nitrogen levels in plants and also known to ameliorate plant antioxidant defense system (Zouari et al.2016). Recent studies have demonstrated high proline levels in plant tissues on exposure to high Cd stress in plants. The concentration of proline in root-shoot part of *Olea europaea* increased dramatically when treated with high Cd regimes (Zouari et al. 2016). It has also been suggested that the exogenous addition of proline enhanced internal production of proline in root and shoot part of *Olea europaea* plants under Cd stress with a parallel increase in the levels of photosynthetic pigment concentration and micro nutrient levels like Ca, K and Mg, while a remarkable decrement in the concentration of oxidative stress markers like relative electrolyte leakage, LPO, and H_2O_2. In another study, Rady et al. (2019) studied the interplay role of proline and silicon in combating combined stress induced by 100 mM NaCl + 1.0 mM Cd^{2+} in *Phaseolus vul*garis plants. The authors conveyed that combined addition of proline and silicon maintain cell water balance, restore nutrient concentration, and improved the activities of both enzymatic and non-enzymatic antioxidant enzymes to promote Cd stress tolerance in plants (Rady et al. 2019). In hydroponic experiments, Szalai et al. (2020) reported that 50 μM Cd treatment enhanced the synthesis of abscisic acid in the leaves of wild mutants of wheat plants that instigate metabolism of proline and polyamines in roots. The metabolic shift from the synthesis of proline and polyamines to phytochelatins was found to influence high Cd tolerance in the plants (Szalai et al. 2020).

15.5.6 Phytochelatin Biosynthesis

Synthesis of phytochelatins is the effective detoxification strategy adopted by plants to combat Cd toxicity. Glutathione along with the enzyme like phytochelatin synthase act as a precursor for phytochelatin synthesis in plants (Peco et al. 2020). Phytochelatins are the peptide compounds especially (c-Glu-Cys)n-Gly, while other phytochelatin compounds include (c-Glu-Cys)n-Glu, (c-Glu-Cys)n-Ser and (c-Glu-Cys)n-c-Ala (Ahmad et al. 2019; Hendrix et al. 2017). Phytochelatins are known to form chelation with heavy metals to form relatively less toxic compounds. Phytochelatins form complex with Cd (phytochelatin-Cd complex) and contribute towards Cd transport from cytosol into the vacuole in the inactive state (Huang et al. 2019).

High Cd regimes impart the production of phytochelatins to enhance plant tolerance towards Cd distress. The chelation of Cd with phytochelatins enables the plants to sequester Cd in the inactive state in the vacuole. Exogenous supplementation of glutathione enhanced Cd acclimatization in the roots of *Brassica campestris* exposed to 20 μM Cd treatment (Huang et al. 2019). It has been observed that glutathione addition induce the synthesis of phytochelatins that form complex with Cd, sequester Cd inside the root vacuole and limit its translocation to the aerial plant parts. Jiang et al. (2020) reported the physiological and transcriptomic response of *Dendrobium officinale* towards increasing Cd concentrations in the growth medium. Cd stress (14 mg l^{-1}) remarkably increased the antioxidant defense strategy in the plants. Furthermore, 2469 genes were expressed after 15-day treatment. The pathways for sulphate glutathione metabolism and phenylpropanoid metabolism were identified to be crucial for Cd stress tolerance (Jiang et al. 2020).

In another study, Bhuyan et al. (2020) revealed vanillic acid modulate enhanced antioxidative response, glyoxylase enzyme activity and phytochelatin synthesis in roots of rice plants on 72 hour Cd exposure at 2.0 mM Cd treatment. The enhanced antioxidant response and phytochelatin production enabled the rice plants to tolerate and withstand Cd stress. Shah et al. (2020) postulated the expression of ATPase genes (CsHA2, CsHA3, CsHA4, CsHA8, and CsHA9) in *Cucumis sativus* seedlings on addition of a metabolite 2-hydroxymelatonin grown under 50 mg kg^{-1} Cd treatment. The authors noticed the potential utility of 2-hydroxymelatonin to enhance nutritional capacity by modulating gene expression and phytochelatin synthesis in *Cucumis sativus* plants grown in Cd-contaminated soils.

15.5.7 Organic Amendments

Various types of organic amendments in the rhizosphere are known to limit Cd solubility in the soil and its uptake and mobility in the plant tissues. Humic acid in soil form complex with Cd and has the potency to reduce Cd bioavailability in the soil. Addition of humic acid increases the nutrient level of the soils along with the enhanced soil pH, cation exchange capacity but decrease Cd bioavailability. Haghighi et al. (2013) reported a linear increase in plant biomass and decrement in the Cd bioavailability and antioxidant response (SOD and POD) in lettuce leaves when treated with 100 and 1000 mg l^{-1} humic acid. Song et al. (2020) demonstrated that the foliar application of humic acid (1.0 g l^{-1}) on the leaves of *Hybrid Pennisetum* grown under different Cd treatments increased the pigment concentration and net photosynthetic rate of the plants. Besides, the malondialdehyde content in the tissues of the plants decreased by ˜4.4–12.2% compared to Cd only treated plants. These findings suggest that the foliar application of humic acid mitigate Cd induced toxicity in plants, hence reduce the risk of soil Cd contamination. In a similar study Ozfidan-Konakci et al. (2018) hydroponically amended Cd treated wheat seedlings with 750 and 1500 mg l^{-1} humic acid for 7 days. It has been noticed that 750 mg l^{-1} humic acid treatment triggered the activities of monodehydroascorbate reductase (MDHAR) and dehydroascorbate reductase (DHAR), and dehydroascorbate (DHA) content. Application of humic acid in Cd treated plants decreased the levels of MDA and H_2O_2 in wheat leaves, thus assist the plants to tolerate Cd stress by regulating water status, photosynthetic apparatus and antioxidant metabolism (Ozfidan-Konakci et al. 2018). Likewise, Zhang, H. et al. (2020) postulated the reduction in Cd uptake in rice plants on the treatment of 4 and 8 g kg^{-1} humic acid in agricultural soil.

15.5.8 Biochars

Biochars are the organic material derived during the burning of organic material in the limited supply of oxygen. Recently, different biochar have gained the interest of the scientific world in having potential capacity to immobilize Cd effectively in the contaminated soils. Pak choi plants grown in Cd contaminated soils showed toxicity with reduced biomass and negative effect on the pigment concentration (Khan et al., 2017). Soil amendment with different biochar (tomato green waste and chicken manure) increased plant biomass and nutrient bioavailability. Both the biochar were found to be efficient in immobilizing Cd in the soil by 35–54% and 26–43% and decrement Cd acclimatization in shoots of pak choi plant by 64–85% and 55–80%, respectively. The effective immobilization of Cd by the different biochar enables the plants to tolerate and withstand Cd toxicity. Naeem et al. (2020) studied the effect of acid treated wheat straw biochar on the physiology, grain production, Cd acclimatisation and its tolerance in *Chenopodium quinoa* plants. Results revealed that the biochar amended soils increased the plant biomass, pigment concentration and other gas exchange related attributes in plants. Furthermore, oxidative stress markers like LPO and H_2O_2 content in roots and leaves of the quinoa plants decreased dramatically in biochar amended soils compared to the control. This signified that acid treated wheat straw biochar is effective in regulating Cd instigated oxidative distress by decrementing Cd acclimatisation and a meaty antioxidant response. Farooq et al. (2020) studied the impact of the application of combined intrinsic zinc concentration (49 mg kg^{-1}) and biochar (20 g kg^{-1}) on the wheat productivity. The results revealed that the combined application of amendments increased the grain yield (9.51%), and decreased Cd concentration in wheat seeds by 95.6% and in straw by 7.1%. Additionally, the combined application of intrinsic zinc and biochar enabled the wheat plants to tolerate high Cd levels. Likewise, Zhu et al. (2020) noticed the combined application of biochar (3%) and biofertilizer (1.5%) decreased Cd concentration in the tissues of *Gossypium hirsutum* grown under varied Cd treatments (1, 2, or 4 mg Cd kg^{-1}). Moreover, the combined application of amendments decreased the electrolyte leakage rate and MDA content but dramatically increased the antioxidant response in cotton plants. Similar observations on the biochar mediated Cd tolerance in plants was observed by Rajendran et al. (2019) and ur Rehman et al. (2020) in *Oryza sativa*; by Rizwan et al. (2019) in *Zea mays* and by Moradi et al. (2019) in *Crocus sativus*.

Conclusions

Cd is a toxic heavy metal and is responsible to cause reduction in growth, biomass, physiological and biochemical alterations in plants. Cd excess in the growth medium induces toxicity and promotes the generation and acclimatisation of oxygen free radicals that further reduces the antioxidant metabolism in plants. High Cd uptake by the plants poses severe damage to DNA and instigates dreadful changes in nuclear and chloroplast gene expression. During the past few decades, number of studies regarding the toxic effects of Cd on the morphological, physiological and molecular traits of numerous plant species has been conducted. Though, a gap in the knowledge still persist in the scientific world related to the mechanisms and strategies adopted by plants to withstand and tolerate Cd induced toxicity. Therefore, future research should aim at the following studies to resolve the unsolved issues of Cd stress tolerance in plants: (i) environmental sound amendments/techniques should be adopted to reduce and limit Cd toxicity in plants; (ii) combined application of different amendments with potential efficacy to minimize Cd toxicity in plants needs to be identified; (iii) pilot scale studies and field related research is required to validate the effects of soil characteristics, temperature and rhizosphere microbes on plant response and the molecular mechanisms involved in Cd detoxification; (iv) molecular tools need to be adopted to identify and screen genes responsible to reduce Cd uptake and its accumulation in plants; (v) further studies using phytohormones, secondary metabolites, microorganisms and other biodegradable organic amendments for reducing Cd uptake and for enhancing plant tolerance towards Cd needs to be conducted in future.

REFERENCES

Agency for Toxic Substances and Disease Registry (ATSDR), 2012. CERCLA priority list of hazardous substances.

Agency for Toxic Substances and Disease Registry (ATSDR), 2019. CERCLA priority list of hazardous substances.

Ahmad, J., Ali, A. A., Baig, M. A., Iqbal, M., Haq, I., and Qureshi, M. I. 2019. Role of phytochelatins in cadmium stress tolerance in plants. *Cadmium Toxicity and Tolerance in Plants*. Academic Press.

Ahmad, P., Abd Allah, E. F., Hashem, A., Sarwat, M., and Gucel, S. 2016. Exogenous application of selenium mitigates cadmium toxicity in *Brassica juncea* L. (Czern & Cross) by up-regulating antioxidative system and secondary metabolites. *J Plant Growth Regul*, *35*, 936–950.

Ahmad, P., Alyemeni, M. N., Wijaya, L., Alam, P., Ahanger, M. A., and Alamri, S. A. 2017. Jasmonic acid alleviates negative impacts of cadmium stress by modifying osmolytes and antioxidants in faba bean (*Vicia faba* L.). *Arch Agro Soil Sci*, 63, 1889–1899.

Ahsan N, Lee S. H., Lee D. G., Lee H., Lee S. W., Bahk J. D., and Lee B. H. 2007. Physiological and protein profiles alternation of germinating rice seedlings exposed to acute cadmium toxicity. *C R Biol* 330, 735–746.

Alaraidh, I. A., Alsahli, A. A., and Abdel Razik, E. S. 2018. Alteration of antioxidant gene expression in response to heavy metal stress in *Trigonella foenum-graecum* L. *S. Afr. J. Bot*, 115, 90–93.

Amir, R., Munir, F., Kubra, G., Iqbal, T., and Khan, M. 2019. Plant signaling molecules and cadmium stress tolerance. In: Hasanuzzaman, M., Prasad, M.N.V., and Nahar, K. (eds.).*Cadmium Tolerance in Plants*. Academic Press.

Asgher, M., Khan, M. I. R., Anjum, N. A., and Khan, N. A. 2015. Minimising toxicity of cadmium in plants-role of plant growth regulators. *Protoplasma*, 252, 399–413.

Azzi, V., Kazpard, V., Lartiges, B., Kobeissi, A., Kanso, A., and El Samrani, A. G. 2017. Trace metals in phosphate fertilizers used in Eastern Mediterranean countries. *CLEAN–Soil Air Water*, 45, 1–8.

Bali, A. S., Sidhu, G. P. S., and Kumar, V. 2020. Root exudates ameliorate cadmium tolerance in plants: A review. *Environ Chem Lett*, 18,1243–1275.

Bansal, P., Sharma, P., and Goyal, V. 2002. Impact of lead and cadmium on enzyme of citric acid cycle in germinating pea seeds. *Biol Plant*, *45*, 125–127.

Bashri, G., and Prasad, S. M. 2016. Exogenous IAA differentially affects growth, oxidative stress and antioxidants system in Cd stressed *Trigonella foenum-graecum* L. seedlings: Toxicity alleviation by up-regulation of ascorbate-glutathione cycle. *Ecotox Environ Saf*, 132, 329–338.

Bhuyan, M. H. M., Parvin, K., Mohsin, S. M., Mahmud, J. A., Hasanuzzaman, M., and Fujita, M. 2020. Modulation of cadmium tolerance in rice: Insight into vanillic acid-induced upregulation of antioxidant defense and glyoxalase systems. *Plants*, 9(2), 188.

Bigalke, M., Ulrich, A., Rehmus, A., and Keller, A. 2017. Accumulation of cadmium and uranium in arable soils in Switzerland. *Environ Pollut*, 221, 85–93.

Bughio, N., Yamaguchi, H., Nishizawa, N. K., Nakanishi, H., and Mori, S. 2002. Cloning an iron regulated metal transporter from rice. *J Exp Bot*, 53, 1677–1682.

Castro, A. V., de Almeida, A. A. F., Pirovani, C. P., Reis, G. S. M., Almeida, N. M., Mangabeira, P. A. O. 2015. Morphological, biochemical, molecular and ultrastructural changes induced by Cd toxicity in seedlings of *Theobroma cacao* L. *Ecotoxicol Environ Saf*, 115, 174–186.

Chen, S., Han, X., Fang, J., Lu, Z., Qiu, W., Liu, M., Zhuo, R. 2017. *Sedum alfredii* SaNramp6 metal transporter contributes to cadmium accumulation in transgenic *Arabidopsis thaliana*. *Sci Rep*, 7, 1–13.

Connolly, E. L., Fett, J. P., and Guerinot, M. L. 2002. Expression of the IRT1 metal transporter is controlled by metals at the levels of transcript and protein accumulation. *Plant Cell*, 14, 1347–1357.

Corguinha, A. P. B., de Souza, G. A., Gonçalves, V. C., Carvalho, C. D. A., de Lima, W. E. A., Martins, F. A. D., Yamanaka, C. H., Francisco, E. A. B., and Guilherme, L. R. G. 2015. Assessing arsenic, cadmium, and lead contents in major crops in Brazil for food safety purposes. *J Food Compos Anal*, 37, 143–150.

Courbot, M., Willems, G., Motte, P., Arvidsson, S., Roosens, N., Saumitou-Laprade, P., and Verbruggen, N. 2007. A major quantitative trait locus for cadmium tolerance in *Arabidopsis halleri* colocalizes with HMA4, a gene encoding a heavy metal ATPase. *Plant Physiol*, 144, 1052–1065.

Dharma-Wardana M. W. C. 2018. Fertilizer usage and cadmium in soils, crops and food. *Environ Geochem Health*, 40: 2739–2759.

Espanany, A., Fallah, S., and Tadayyon, A. 2015. The effect of halopriming and salicylic acid on the germination of Fenugreek (*Trigonella foenum-graecum*) under different cadmium concentrations. *Not Sci Biol*, 7, 322–329.

Faheem, M. A., Sehar, S., Chen, G., Chen, Z. H., Jilani, G., Chaudhry, A. N., and Shamsi, I. H. 2020. Cadmium-zinc cross-talk delineates toxicity tolerance in rice via differential genes expression and physiological/ultrastructural adjustments. *Ecotox Environ Saf*, 190, 110076.

Farooq, M., Ullah, A., Usman, M., and Siddique, K. H. 2020. Application of zinc and biochar help to mitigate cadmium stress in bread wheat raised from seeds with high intrinsic zinc. *Chemosphere*, 260, 127652.

Feng, S., Tan, J., Zhang, Y., Liang, S., Xiang, S., Wang, H., and Chai, T. 2017. Isolation and characterization of a novel cadmium-regulated yellow stripe-like transporter (SnYSL3) in *Solanum nigrum*. *Plant Cell Rep*, 36, 281–296.

Franic, M. 2018. *Effects of Cadmium on Photosynthetic Parameters in Different Maize Genotypes*. University of Josip Juraj Strossmayer in Osijek.

Fu, X., Dou, C., Chen, Y., Chen, X., Shi, J., Yu, M., and Xu, J. 2011. Subcellular distribution and chemical forms of cadmium in *Phytolacca americana* L. *J Hazard Mater*, 186, 103–107.

Gendre, D., Czernic, P., Conéjéro, G., Pianelli, K., Briat, J. F., Lebrun, M., and Mari, S. 2007. TcYSL3, a member of the YSL gene family from the hyper-accumulator *Thlaspi caerulescens*, encodes a nicotianamine-Ni/Fe transporter. *Plant J*, 49, 1–15.

Gill, M. 2014. Heavy metal stress in plants: A review. *Int J Adv Res*, 2, 1043–1055.

Gill, S. S., and Tuteja, N. 2011. Cadmium stress tolerance in crop plants: Probing the role of sulfur. *Plant Sig Behav*, 6, 215–222.

Gondor, O. K., Pál, M., Darkó, É., Janda, T., and Szalai, G. 2016. Salicylic acid and sodium salicylate alleviate cadmium toxicity to different extents in maize (*Zea mays* L.). *PLoS One*, 11, e0160157.

Gu, C. S., Yang, Y. H., Shao, Y. F., Wu, K. W., and Liu, Z. L. 2018. The effects of exogenous salicylic acid on alleviating cadmium toxicity in *Nymphaea tetragona* Georgi. *South Afr J Bot*, 114, 267–271.

Guo, J., and Chi, J. 2014. Effect of Cd-tolerant plant growth-promoting rhizobium on plant growth and Cd uptake by Lolium multiflorum Lam. and Glycine max (L.) Merr. in Cd-contaminated soil. *Plant Soil*, 375(1-2), 205–214.

Guo, J., Qin, S., Rengel, Z., Gao, W., Nie, Z., Liu, H. and Zhao, P. 2019. Cadmium stress increases antioxidant enzyme activities and decreases endogenous hormone concentrations more in Cd-tolerant than Cd-sensitive wheat varieties. *Ecotoxicol Environ Saf,172*, 380–387.

Guo, Q., Tian, X., Mao, P., and Meng, L. 2019. Functional characterization of IlHMA2, a P1B2-ATPase in Iris lactea response to Cd. *Environ. Exp. Bot*, 157, 131–139.

Haghighi, M., Kafi, M., and Khoshgoftarmanesh, A. 2013. Effect of humic acid application on cadmium accumulation by lettuce leaves. *J. Plant Nutr*, 36(10), 1521–1532.

Han, Y., Chen, Y., Yin, S., Zhang, M., and Wang, W. 2015. Overexpression of TaEXPB23, a wheat expansin gene, improves oxidative stress tolerance in transgenic tobacco plants. *J. Plant Physiol*, 173, 62–71.

Hanikenne, M., Talke, I. N., Haydon, M. J., Lanz, C., Nolte, A., Motte, P., and Krämer, U. 2008. Evolution of metal hyperaccumulation required cis-regulatory changes and triplication of HMA4. *Nature*, 453, 391–395.

Hashem, A., Abd_Allah, E. F., Alqarawi, A. A., Al Huqail, A. A., Egamberdieva, D., and Wirth, S. (2016). Alleviation of cadmium stress in Solanum lycopersicum L. by arbuscular mycorrhizal fungi via induction of acquired systemic tolerance. *Saudi J. Biol. Sci*, 23(2), 272–281.

Hayat, S., Alyemeni, M. N., and Hasan, S. A. 2012. Foliar spray of brassinosteroid enhances yield and quality of *Solanum lycopersicum* under cadmium stress. *Saudi J Biol Sci*, 19, 325–335.

Hendrix, S., Schröder, P., Keunen, E., Huber, C., and Cuypers, A. (2017). Molecular and cellular aspects of contaminant toxicity in plants: The importance of sulphur and associated signalling pathways. *Advances in Botanical Research*. Academic Press.

Horiguchi H., Teranishi H., Niiya K., Aoshima K., Katoh T., Sakuragawa N., and Kasuya M. 1994. Hypoproduction of erythropoietin contributes to anemia in chronic cadmium intoxication: Clinical study on Itai-itai disease in Japan. *Arch Toxicol*, 68, 632–636.

Hu, J., Tsang, W., Wu, F., Wu, S., Wang, J., Lin, X., and Wong, M. H. 2016. Arbuscular mycorrhizal fungi optimize the acquisition and translocation of Cd and P by cucumber (Cucumis sativus L.) plant cultivated on a Cd-contaminated soil. *J. Soils. Sediments*, 16(9), 2195–2202.

Huang, X., Duan, S., Wu, Q., Yu, M., and Shabala, S. 2020. Reducing cadmium accumulation in plants: Structure–function relations and tissue-specific operation of transporters in the spotlight. *Plants*, *9*, 223.

Huang, Y., Zhu, Z., Wu, X., Liu, Z., Zou, J., Chen, Y., Su, N., and Cui, J. 2019. Lower cadmium accumulation and higher antioxidative capacity in edible parts of Brassica campestris L. seedlings applied with glutathione under cadmium toxicity. *Environ. Sci. Pollut. Res*, 26(13), 13235–13245.

Huguet, S., Bert, V., Laboudigue, A., Barthès, V., Isaure, M. P., Llorens, I., and Sarret, G. 2012. Cd speciation and localization in the hyperaccumulator *Arabidopsis halleri*. *Environ Exp Bot*, 82, 54–65.

Jiang, W., Wu, Z., Wang, T., Mantri, N., Huang, H., Li, H., Tao, Z., and Guo, Q. 2020. Physiological and transcriptomic analyses of cadmium stress response in Dendrobium officinale seedling. *Plant Physiol. Biochem*, 148, 152–165.

Kalai, T., Bouthour, D., Manai, J., Bettaieb Ben Kaab, L., and Gouia, H. 2016. Salicylic acid alleviates the toxicity of cadmium on seedling growth, amylases and phosphatases activity in germinating barley seeds. *Arch Agron Soil Sci*, 62, 892–904.

Karcz, W., and Kurtyka, R. 2007. Effect of cadmium on growth, proton extrusion and membrane potential in maize coleoptile segments. *Biol Plant*, 51, 713.

Khan, K. Y., Ali, B., Cui, X., Feng, Y., Yang, X., and Stoffella, P. J. (2017). Impact of different feedstocks derived biochar amendment with cadmium low uptake affinity cultivar of pak choi (Brassica rapa ssb. chinensis L.) on phytoavoidation of Cd to reduce potential dietary toxicity. *Ecotoxicol. Environ. Saf*, 141, 129–138.

Khan, M. I. R., Nazir, F., Asgher, M., Per, T. S., and Khan, N. A. 2015. Selenium and sulfur influence ethylene formation and alleviate cadmium-induced oxidative stress by improving proline and glutathione production in wheat. *J Plant Physiol*, 173, 9–18.

Kim, D. Y., Bovet, L., Maeshima, M., Martinoia, E., and Lee, Y. 2007. The ABC transporter AtPDR8 is a cadmium extrusion pump conferring heavy metal resistance. *Plant J*, 50, 207–218.

Korenkov, V., King, B., Hirschi, K., and Wagner, G. J. 2009. Root-selective expression of AtCAX4 and AtCAX2 results in reduced lamina cadmium in field-grown *Nicotiana tabacum* L. *Plant Biotechnol J*, 7, 219–226.

Kubier, A., Wilkin, R. T., and Pichler, T. 2019. Cadmium in soils and groundwater: A review. *Appl Geochem*, *108*, 104388.

Kumar, A., Dubey, A. K., Kumar, V., Ansari, M. A., Narayan, S., Meenakshi, Kumar, S., Pandey, V., Shirke, P. A., Pande, V., and Sanyal, I. 2020. Over-expression of chickpea glutaredoxin (CaGrx) provides tolerance to heavy metals by reducing metal accumulation and improved physiological and antioxidant defence system. *Ecotoxicol. Environ. Saf*, 192, 110252.

Kumar, V., Sharma, A., Kaur, P., Sidhu, G. P. S., Bali, A. S., Bhardwaj, R., and Cerda, A. 2019. Pollution assessment of heavy metals in soils of India and ecological risk assessment: A state-of-the-art. *Chemosphere*, *216*, 449–462.

Lin, Y. F., Liang, H. M., Yang, S. Y., Boch, A., Clemens, S., Chen, C. C., and Yeh, K. C. 2009. *Arabidopsis* IRT3 is a zinc-regulated and plasma membrane localized zinc/iron transporter. *New Phytol*, *182*, 392–404.

Marques, D. N., Carvalho, M. E. A., Piotto, F. A., Batagin-Piotto, K. D., Nogueira, M. L., Gaziola, S. A., and Azevedo, R. A. 2019. Antioxidant defense response in plants to cadmium stress. In: Hasanuzzaman, M., Prasad, M.N.V., and Nahar, K. (eds.). *Cadmium Tolerance in Plants* (pp. 423–461). Academic Press.

Mendoza-Cózatl, D. G., Jobe, T. O., Hauser, F., and Schroeder, J. I. 2011. Long-distance transport, vacuolar sequestration, tolerance, and transcriptional responses induced by cadmium and arsenic. *Curr Opin Plant Biol*, 14, 554–562.

Meng, H., Hua, S., Shamsi, I. H., Jilani, G., Li, Y., and Jiang, L. 2009. Cadmium-induced stress on the seed germination and seedling growth of *Brassica napus* L., and its alleviation through exogenous plant growth regulators. *Plant Growth Regul*, 58, 47–59.

Mills, R. F., Francini, A., da Rocha, P. S. F., Baccarini, P. J., Aylett, M., Krijger, G. C., and Williams, L. E. 2005. The plant P1B-type ATPase AtHMA4 transports Zn and Cd and plays a role in detoxification of transition metals supplied at elevated levels. *FEBS letters*, 579(3), 783–791.

Moradi, R., Pourghasemian, N., and Naghizadeh, M. 2019. Effect of beeswax waste biochar on growth, physiology and cadmium uptake in saffron. *J. Clean. Prod*, 229, 1251–1261.

Mori M., Kotaki K., Gunji F., Kubo N., Kobayashi S., Ito T., and Itabashi H. 2016. Suppression of cadmium uptake in rice using fermented bark as a soil amendment. *Chemosphere*, 148, 487–494.

Munzuroglu, O., and Zengin, F. K. 2006. Effect of cadmium on germination, coleoptile and root growth of barley seeds in the presence of gibberellic acid and kinetin. *J Environ Biol*, 27, 671–677.

Naeem, M. A., Shabbir, A., Amjad, M., Abbas, G., Imran, M., Murtaza, B., Tahir, M., and Ahmad, A. 2020. Acid treated biochar enhances cadmium tolerance by restricting its uptake and improving physio-chemical attributes in quinoa (Chenopodium quinoa Willd.). *Ecotoxicol. Environ. Saf*, 191, 110218.

Nishida, S., Tsuzuki, C., Kato, A., Aisu, A., Yoshida, J., and Mizuno, T. 2011. AtIRT1, the primary iron uptake transporter in the root, mediates excess nickel accumulation in Arabidopsis thaliana. *Plant Cell Physiol*, 52, 1433–1442.

Ozfidan-Konakci, C., Yildiztugay, E., Bahtiyar, M., and Kucukoduk, M. 2018. The humic acid-induced changes in the water status, chlorophyll fluorescence and antioxidant defense systems of wheat leaves with cadmium stress. *Ecotoxicol. Environ. Saf*, 155, 66–75.

Palusińska, M., Barabasz, A., Kozak, K., Papierniak, A., Maślińska, K., and Antosiewicz, D. M. 2020. Zn/Cd status-dependent accumulation of Zn and Cd in root parts in tobacco is accompanied by specific expression of ZIP genes. *Plant Biol*, 20, 37.

Pan, C., Lu, H., Liu, J., Yu, J., Wang, Q., Li, J., and Yan, C. 2020. SODs involved in the hormone mediated regulation of H2O2 content in *Kandelia obovata* root tissues under cadmium stress. *Environ Pollut*, 256, 113272.

Parmar, P., Kumari, N., and Sharma, V. 2013. Structural and functional alterations in photosynthetic apparatus of plants under cadmium stress. *Bot Stud*, 54, 45.

Peco, J. D., Campos, J. A., Romero-Puertas, M. C., Olmedilla, A., Higueras, P., and Sandalio, L. M. 2020. Characterization of mechanisms involved in tolerance and accumulation of Cd in Biscutella auriculata L. *Ecotoxicol Environ Saf*, 201, 110784.

Pérez Chaca, M. V., Vigliocco, A., Reinoso, H., Molina, A., Abdala, G., Zirulnik, F., and Pedranzani, H. 2014. Effects of cadmium stress on growth, anatomy and hormone contents in *Glycine max* (L.) Merr. *Acta Physiol Plant*, 36, 2815–2826.

Plata, S., J., Meyssen, D., Cuypers, A., Escobar, C., and Hernández, L. E. 2014. Glutathione is a key antioxidant metabolite to cope with mercury and cadmium stress. *Plant Soil*, 377, 369–381.

Rady, M. M., Elrys, A. S., El-Maati, M. F. A., and Desoky, E. S. M. 2019. Interplaying roles of silicon and proline effectively improve salt and cadmium stress tolerance in Phaseolus vulgaris plant. *Plant Physiol. Biochem*, 139, 558–568.

Rajendran, M., Shi, L., Wu, C., Li, W., An, W., Liu, Z., and Xue, S. 2019. Effect of sulfur and sulfur-iron modified biochar on cadmium availability and transfer in the soil–rice system. *Chemosphere*, 222, 314–322.

Rizwan, M., Ali, S., ur Rehman, M. Z., Adrees, M., Arshad, M., Qayyum, M. F., Ali, L., Hussain, A., Chatha, S. A. S., and Imran, M. 2019. Alleviation of cadmium accumulation in maize (Zea mays L.) by foliar spray of zinc oxide nanoparticles and biochar to contaminated soil. *Environ. Pollut*, 248, 358–367.

Rizwan, M., Ali, S., ur Rehman, M. Z., Rinklebe, J., Tsang, D. C., Bashir, A., and Ok, Y. S. 2018. Cadmium phytoremediation potential of Brassica crop species: a review. *Sci Total Environ*, 631, 1175–1191.

Rodda, M. S., Li, G., and Reid, R. J. 2011. The timing of grain Cd accumulation in rice plants: The relative importance of remobilisation within the plant and root Cd uptake post-flowering. *Plant Soil*, 347, 105–114.

Roy, S. K., Cho, S. W., Kwon, S. J., Kamal, A. H. M., Kim, S. W., Oh, M. W., Lee, M. S., Chung, K. Y., Xin, Z., and Woo, S. H. 2016. Morpho-physiological and proteome level responses to cadmium stress in sorghum. *PLoS One*, 11(2), e0150431.

Roychoudhury, A., Ghosh, S., Paul, S., Mazumdar, S., Das, G., and Das, S. 2016. Pre-treatment of seeds with salicylic acid attenuates cadmium chloride-induced oxidative damages in the seedlings of mungbean (*Vigna radiata* L. Wilczek). *Acta Physiol Plant*, 38, 11.

Sasaki, A., Yamaji, N., Yokosho, K., and Ma, J. F. 2012. Nramp5 is a major transporter responsible for manganese and cadmium uptake in rice. *Plant Cell*, 24, 2155–2167.

Semida, W. M., Mostafa, M. R., Taia, A. A. E. M., Saad, M. H., and Magdi, T. A. 2015. Alleviation of cadmium toxicity in common bean (*Phaseolus vulgaris* L.) plants by the exogenous application of salicylic acid. *J Hortic Sci Biotech*, 90, 83–91.

Shah, A. A., Ahmed, S., Ali, A., and Yasin, N. A. 2020. 2-Hydroxymelatonin mitigates cadmium stress in cucumis sativus seedlings: Modulation of antioxidant enzymes and polyamines. *Chemosphere*, 243, 125308.

Shanmugaraj, B. M., Chandra, H. M., Srinivasan, B., and Ramalingam, S. 2013. Cadmium induced physio-biochemical and molecular response in *Brassica juncea*. *Int J Phytoremediation*, 15, 206–218.

Sharma, R. K., and Archana, G. 2016. Cadmium minimization in food crops by cadmium resistant plant growth promoting rhizobacteria. *Appl Soil Ecol*, 107, 66–78.

Sidhu, G. P. S. 2016. Heavy metal toxicity in soils: Sources, remediation technologies and challenges. *Adv Plants Agric Res*, *5*, 445–446.

Sidhu, G. P. S., Bali, A. S., and Bhardwaj, R. 2019. Use of fungi in mitigating cadmium toxicity in plants. In: Hasanuzzaman, M., Prasad, M.N.V., and Fujita, M. (eds.). *Cadmium Toxicity and Tolerance in Plants*. (pp. 397–426). Academic Press.

Sidhu, G. P. S., Bali, A. S., Singh, H. P., Batish, D. R., and Kohli, R. K. 2018. Ethylenediamine disuccinic acid enhanced phytoextraction of nickel from contaminated soils using *Coronopus didymus* (L.) Sm. *Chemosphere*, 205, 234–243.

Sidhu, G. P. S., Singh, H. P., Batish, D. R., and Kohli, R. K. 2016. Effect of lead on oxidative status, antioxidative response and metal accumulation in *Coronopus didymus*. *Plant Physiol Bioch*, 105, 290–296.

Sidhu, G. P. S., Singh, H. P., Batish, D. R., and Kohli, R. K. 2017. Tolerance and hyperaccumulation of cadmium by a wild, unpalatable herb *Coronopus didymus* (L.) Sm.(Brassicaceae). *Ecotoxicol Environ Saf*, 135, 209–215.

Singh, S., and Prasad, S. M. 2016. Kinetin ameliorates cadmium induced toxicity on growth, pigments and photosynthesis by regulating antioxidant potential in tomato seedlings. *Int J Sci Eng Appl Sci*, 2, 1–10.

Singh, S., and Prasad, S. M. 2017. Effects of 28-homobrassinoloid on key physiological attributes of *Solanum lycopersicum* seedlings under cadmium stress: Photosynthesis and nitrogen metabolism. *Plant Growth Regul*, 82, 161–173.

Six, L., and Smolders, E. 2014. Future trends in soil cadmium concentration under current cadmium fluxes to European agricultural soils. *Sci Total Environ*, 485, 319–328.

Song, X., Chen, M., Chen, W., Jiang, H., and Yue, X. 2020. Foliar application of humic acid decreased hazard of cadmium toxicity on the growth of Hybrid Pennisetum. *Acta Physiol. Plant*, 42(8), 1–11.

Sui, F. Q., Chang, J. D., Tang, Z., Liu, W. J., Huang, X. Y., and Zhao, F. J. 2018. Nramp5 expression and functionality likely explain higher cadmium uptake in rice than in wheat and maize. *Plant Soil*, 433, 377–389.

Sun, H., Chen, Z. H., Chen, F., Xie, L., Zhang, G., Vincze, E., and Wu, F. 2015. DNA microarray revealed and RNAi plants confirmed key genes conferring low Cd accumulation in barley grains. *BMC Plant Biol*, 15, 1–17.

Szalai, G., Tajti, J., Hamow, K. Á., Ildikó, D., Khalil, R., Vanková, R., Dobrev, P., Misheva, S. P., Janda, T., and Pál, M. 2020. Molecular background of cadmium tolerance in Rht dwarf wheat mutant is related to a metabolic shift from proline and polyamine to phytochelatin synthesis. *Environ. Sci Pollut Res*, 27, 1–13.

Talke, I. N., Hanikenne, M., and Krämer, U. 2006. Zinc-dependent global transcriptional control, transcriptional deregulation, and higher gene copy number for genes in metal homeostasis of the hyperaccumulator *Arabidopsis halleri*. *Plant Physiol*, 142, 148–167.

Tan, L., Qu, M., Zhu, Y., Peng, C., Wang, J., Gao, D., and Chen, C. 2020. Zinc transporter5 and zinc transporter9 function synergistically in zinc/cadmium uptake. *Plant Physiol*, 183, 1235–1249.

Theodoulou, F. L. 2000. Plant ABC transporters. *BBA*, 1465, 79–103.

Ueno, D., Yamaji, N., Kono, I., Huang, C. F., Ando, T., Yano, M., and Ma, J. F. 2010. Gene limiting cadmium accumulation in rice. *PNAS*, 107, 16500–16505.

ur Rehman, M. Z., Batool, Z., Ayub, M. A., Hussaini, K. M., Murtaza, G., Usman, M., Naeem, A., Khalid, H., Rizwan, M., and Ali, S. (2020). Effect of acidified biochar on bioaccumulation of cadmium (Cd) and rice growth in contaminated soil. *Environ Technol Innov*, 19, 101015.

Verbruggen, N., Hermans, C., and Schat, H. 2009. Mechanisms to cope with arsenic or cadmium excess in plants. *Curr Opin Plant Biol*, 12, 364–372.

Wang Y., Jiang X., Li K., Wu M., Zhang R., Zhang L., and Chen G. 2014. Photosynthetic responses of *Oryza sativa* L. seedlings to cadmium stress: physiological, biochemical and ultrastructural analyses. *Biometals*, 27, 389–401.

Wei, W., Chai, T., Zhang, Y., Han, L., Xu, J., and Guan, Z. 2009. The *Thlaspi caerulescens* NRAMP homologue TcNRAMP3 is capable of divalent cation transport. *Mol Biotechnol*, 41, 15–21.

Wong, C. K. E., and Cobbett, C. S. 2009. HMA P-type ATPases are the major mechanism for root-to-shoot Cd translocation in *Arabidopsis thaliana*. *New Phytol*, 181, 71–78.

Wu, D., Yamaji, N., Yamane, M., Kashino-Fujii, M., Sato, K., and Ma, J. F. 2016. The HvNramp5 transporter mediates uptake of cadmium and manganese, but not iron. *Plant Physiol*, 172, 1899–1910.

Xia, J., Yamaji, N., and Ma, J. F. 2011. Further characterization of an aluminum influx transporter in rice. *Plant Signal Behav*, 6, 160–163.

Xin, J. P., Zhang, Y., and Tian, R. N. 2018. Tolerance mechanism of *Triarrhena sacchariflora* (Maxim.) Nakai. seedlings to lead and cadmium: Translocation, subcellular distribution, chemical forms and variations in leaf ultrastructure. *Ecotox Environ Saf*, 165, 611–621.

Yadav, S. K. 2010. Heavy metals toxicity in plants: An overview on the role of glutathione and phytochelatins in heavy metal stress tolerance of plants. *S Afr J Bot*, 76, 167–179.

Yamaji, N., Sasaki, A., Xia, J. X., Yokosho, K., and Ma, J. F. 2013. A node-based switch for preferential distribution of manganese in rice. *Nat Commun*, 4, 1–11.

Yang, L. P., Zhu, J., Wang, P., Zeng, J., Tan, R., Yang, Y. Z., and Liu, Z. M. 2018. Effect of Cd on growth, physiological response, Cd subcellular distribution and chemical forms of *Koelreuteria paniculata*. *Ecotox Environ Saf*, 160, 10–18.

Yoneyama, T., Ishikawa, S., and Fujimaki, S. 2015. Route and regulation of zinc, cadmium, and iron transport in rice plants (*Oryza sativa* L.) during vegetative growth and grain filling: Metal transporters, metal speciation, grain Cd reduction and Zn and Fe biofortification. *Int J Mol Sci*, 16, 19111–19129.

Yue, R., Lu, C., Qi, J., Han, X., Yan, S., Guo, S., Liu, L., Fu, X., Chen, N., Yin, H., Chi, H., and Tie, S. 2016. Transcriptome analysis of cadmium-treated roots in maize (*Zea mays* L.). *Front Plant Sci*, 7, 1–11.

Zhang, H., Xie, S., Bao, Z., Tian, H., Carranza, E. J. M., Xiang, W., Yao, L., and Zhang, H. 2020. Underlying dynamics and effects of humic acid on selenium and cadmium uptake in rice seedlings. *J. Soils Sediments*, 20(1), 109–121.

Zhang, X., Li, X., Tang, L., Peng, Y., Qian, M., Guo, Y., and Xia, Y. 2020. The root iron transporter 1 governs cadmium uptake in *Vicia sativa* roots. *J Hazard Mater*, 398, 122873.

Zheng, X., Chen, L., and Li, X. 2018. *Arabidopsis* and rice showed a distinct pattern in ZIPs genes expression profile in response to Cd stress. *Bot Stud*, 59, 22.

Zhi-bin, L., He, J., Polle, A., and Rennenberg, H. 2016. Heavy metal accumulation and signal transduction in herbaceous and woody plants: Paving the way for enhancing phytoremediation efficiency. *Biotech Advan*, 34, 1131–1148.

Zhu, Q., Zhang, J., Yu, H., Li, L., Chen, X., Jiang, M., and Tan, M. 2019. Maize Cd-tolerant ZmVTE4 encoding c-tocopherol-methyl-transferase alleviated Cd-toxicity through its product a-tocopherol. *Environ Exp Bot*, 158, 171–179.

Zhu, X. F., Jiang, T., Wang, Z. W., Lei, G. J., Shi, Y. Z., Li, G. X., and Zheng, S. J. 2012. Gibberellic acid alleviates cadmium toxicity by reducing nitric oxide accumulation and expression of IRT1 in *Arabidopsis thaliana*. *J Hazard Mater*, 239, 302–307.

Zhu, Y., Wang, H., Lv, X., Zhang, Y., and Wang, W. 2020. Effects of biochar and biofertilizer on cadmium-contaminated cotton growth and the antioxidative defense system. *Sci. Rep*, 10(1), 1–12.

Zouari, M., Ben Ahmed, C., Elloumi, N., Bellassoued, K., Delmail, D., Labrousse, P., Ben Abdallah, F., and Ben Rouina, B. 2016. Impact of proline application on cadmium accumulation, mineral nutrition and enzymatic antioxidant defense system of Olea europaea L. cv Chemlali exposed to cadmium stress. *Ecotoxicol Environ Saf*, 128, 195–205.

16

Ecotoxicological Impacts of Arsenic on Plants and Its Remediation Approaches

Mohamed Gomaa and Mona F.A. Dawood
Department of Botany & Microbiology, Faculty of Science, Assiut University, Assiut, Egypt

CONTENTS

16.1 Introduction

Arsenic (As) is the twentieth abundant metalloid on the Earth's crust. Arsenic shares properties with both metals and nonmetals; thereby it is classified as a metalloid. As is listed to be one of the most toxic materials that could be poisonous to humans based on the Agency for Toxic Substances and Disease Registry list of 2017. Approximately 200 million people in around 70 countries have been exposed to this metalloid (Sodhi et al. 2019; Abedi and Mojiri 2020). In this respect, the International Agency of Research on Cancer (IARC) classified As as a class I carcinogen that causes acute and chronic toxicity depending on the type of exposure (Martinez et al. 2011). The World Health Organization (WHO) and the U.S. Environmental Protection Agency (EPA) list a threshold of 10 µg/L of arsenic in drinking water. Unfortunately, water sources in Bangladesh, India, China, Argentina, Ghana, Chile, Vietnam, Canada, Laos, Mexico, the United States and several other countries have been reported to contain the toxicant in levels that exceed this threshold (Basu et al. 2014; Hettick et al. 2015). However, As is a toxic metalloid and is considered as a considerable global groundwater pollutant, affecting water sources in East and South Asia and in South American countries (Kobya et al. 2020).

As enters agricultural lands and the ecosystem naturally from rocks, As-enriched minerals, forest fires, volcanoes as well as anthropogenic activities such as mining, herbicides, phosphate fertilizers, smelting, industrial processes, coal combustion, and timber preservatives (Rahaman et al. 2020). Arsenic is ubiquitously present in Earth's crust; however, numerous human interventions such as rapid industrialization use of As-based pesticides, insecticides and discharge of industrial wastes in water bodies leads to cumulative increase in As in the environment and has become a global challenge (Mishra et al. 2021). Weathering of Himalayan rocks containing sulfide minerals is the primary source of As deposition into rivers. Further, As bearing minerals when undergone oxidation lead to the activation of iron (Fe) hydroxides, oxy-hydroxides, and oxides, thus contributing to As in soil. Besides this, the microbial leaching is another process that liberates the As in soil from Fe oxides (Mishra et al. 2021). Further, several man-made actions including excessive use of As-based chemicals,

DOI: 10.1201/9781003155089-16

insecticides, pesticides and fertilizers are responsible for the elevated level of As in the environment (Sharma et al. 2014). Above-mentioned natural and anthropogenic activities raised the As content in ground and drinking water as it is in range of 10–50 µg L^{-1} which is toxic to various life forms (WHO 2001). The average amount of As in agricultural fields that receive As-comprising pesticides and defoliants ranges from 5 mg/kg to 2553 mg/kg (Abedi and Mojiri 2020).

Thus, it is important to declare the uptake of As by plants, its toxic effects on plants, thus we can declare the methods of its bioremediation especially using ecofriendly methods as phytobial remediation methods. All these issues will be illustrated in the present chapter.

16.2 Uptake of Inorganic and Organic Arsenic

Inorganic forms of As as AsIII and AsV, is more toxic than As-organic forms as dimethylarsinic acid (DMA, $(CH_3)_2AsOOH$) and monomethylarsonic acid (MMA, $CH_3AsO(OH)_2$) (Islam et al. 2017). Inorganic As is the dominant form of As in soil and groundwater. Under aerobic conditions in soil, AsV dominates, whereas, in submerged conditions, AsIII is the dominant species (Abedin et al. 2002). Distinguishing between species is important due to the differences in properties and mechanisms of toxicity (Amaral et al. 2014). As V may generally enter the roots of rice crops through phosphate transporters (PHTs), primarily PHT1 (phosphate transporterl)-type transporters (Luan et al. 2018) during the regulation of inorganic phosphorous (Mishra et al. 2017). In rice (*Oryza sativa*), 13 PHTs from the PHT1 family in rice have been detected to modulate Pi transport and uptake (Ye et al. 2015) through the highly efficient silicon (Si) uptake pathway. On the other hand, AsIII may take their way to plants by the nodulin 26-like intrinsic (NIPs) aquaporin channels accompanied by silicic acid and ammonia (Mishra et al. 2017). NIP proteins are amongst intrinsic proteins which comprise the family of major membrane channel proteins (Pommerrenig et al. 2015). NIPs are classified into three main groups, namely, NIP-I, NIP-II and NIP-III, with regard to the common substrate selectivity and consistency of amino acid composition (Mitani et al. 2008). The NIP-II group (such as OsNIP3;1, AtNIP5;1, AtNIP6;1 and ZmNIP3;1) has been recorded for the boron uptake and transport in several plants (Pommerrenig et al. 2015). NIP-III group members (such as OsNIP2;1, OsNIP2;2, HvNIP2;1, HvNIP2;2 and CmNIP2;1) are revealed to be vital for the effectual uptake and translocation of Si (Pommerrenig et al. 2019). A family of 10 NIP proteins is found in rice. Bienert et al. (2008) reported that NIPs in rice as OsNIP2;1 (Lsi1) and OsNIP3;2 (Lsi2) that facilitate AsIII influx into the root cells of rice.

Organic methylated As species (MMA and DMA) might be occur in soil due to microbial activity or past utilization of methylated As compounds, cacodylic acid or sodium salt of MMA and DMA as pesticides (Mishra et al. 2017). Microorganisms in soil may convert As species from AsV to AsIII and further to MMA and DMA. Suriyagoda et al. (2018) reported that DMA and MMA might be influxed by the silicic acid transporter Lsi1. Plant roots are capable of absorbing DMA and MMA, but the rate of uptake is lower than those of inorganic As species and decrease with increasing numbers of methyl groups (Zhao et al. 2013). MMA(V) is partially reduced to trivalent MMA(III) in rice roots, but only MMA(V) is translocated to shoots (Guillod-Magnin et al. 2018). DMA(V) is mobile in plants and may easily transfer from root to shoot. Muehe et al. (2019) explained that methylated As is taken up gradually into rice relative to inorganic species but that it is freely translocated to grains.

16.3 Factors Affecting Arsenic Uptake by Plants

Many conditions viz., soil pH, soil texture, organic matter (OM) and sulfide concentrations, may affect As uptake by plants (Hossain et al. 2015). Soil texture may influence As mobility owing to differences in charges on the soil surface, which controls the adsorption and desorption procedures in soil. Soils with high amounts of clay have a higher As retention potential than coarse-textured soils. Bakhat et al. (2019) reported that As uptake and concentrations are higher in plants grown in loamy sand than in plants grown in silty clay loam soils. Furthermore, As is 5-times more toxic in sand and loam than in clay soil, and its available species is an important factor associated with phytotoxicity (Quazi et al. 2011).

16.3.1 Soil pH

Tu and Ma (2003) reported that pH is a prime factor affecting As uptake. As the pH of soil increase, the mobilization of As in soil increases. Generally, an increase in soil pH results in a release of anions from their exchange positions, along with AsV and AsIII (Moreno-Jiménez et al. 2015). Tu and Ma (2003) reported that redox potential and pH influence As species. For instance, at pH lower than 6.9, $H_2AsO_4^-$ becomes the primary species, and at a high pH, $HAsO_4^{2-}$ is dominant. High soil pH (generally pH 8.5) increases the negative surface charges, such as hydroxyl ions, thereby facilitating the desorption of As from Fe oxides and the resulting mobilization of As in the root area; these conditions, in turn, enhance As accumulation in plants (Mitra et al. 2017).

16.3.2 Soil Organic Matter

OM in soil may influence the mobility and bioavailability of As over redox reactions, anions (phosphate, DOC and silicate), As–OM complexation, and competitive adsorption (Syu et al. 2019). Theoretically, OM insolubilizes As via binding As with phenolic OH, carboxylate and sulfhydryl groups with/without ternary complexes (Suda and Makino 2016). Norton et al. (2013) stated that OM is crucial in the mobilization of As from paddy fields because microbes utilizing OM consume oxygen that account for a reduction in redox potential causing As dissolution from FeOOH. The use of OM to As-polluted soils may exert different effects on the As accumulation and growth of rice plants (Syu et al. 2019). For instance, biochar may improve As reduction and release in flooded paddy soils (Qiao et al. 2018), but augmenting farmyard manure to soils with high amounts of As leads to a reduction of plant growth (Norton et al. 2013). Norton et al. (2013) stated that OM may also play two other roles in As availability in soils: by desorbing As species from soil surface exchange sites and complexing As species with dissolved organic matter (DOM). Qiao et al. (2018) found that lactate and biochar can enhance As reduction and release in flooded paddy soils through stimulation of the active arsenate-respiring bacteria Geobacter species, which probably increases the bioavailability of As to rice plants. Jia et al. (2013) observed that the application of biogas slurry significantly increased the accumulation of As in rice plants, especially for methylated As species. However, Rahaman et al. (2011) reported that OM amendments could decrease the uptake of As by rice plants.

16.3.3 Minerals Availability (Nitrogen, Phosphorous and Sulfate) in Soil

The procedure of Fe redox cycling may be influenced by N cycling. The coupled NO3 reduction and Fe(II) oxidation can diminish As in paddy environments (Kumarathilaka et al. 2018). Phosphate and arsenate are analogues and may compete for the same sorption sites on soil particles and rice absorbs AsV via phosphate transporters (Cao et al. 2017). The adding of phosphate commonly has two consequences: (i) raised downward move of As resulting in increased leaching from the topsoil and (ii) enhanced accessibility of As in the soil solution. AsV also acts as a phosphate analogue with respect to transport across the root plasma membrane (Abedin et al. 2002). Pigna et al. (2010) reported that As toxicity become more toxic when As pollution coexists with low P-availability. Sulfur (S) is an element that interacts toughly with As, particularly under reducing conditions; the reduced forms of S can make a binding with As(III) (Boye et al. 2017). Srivastava et al. (2016) stated that S is important for plant growth as it regulates As tolerance via combination of As with S-containing ligands (glutathione and phytochelatins). Zhang et al. (2011) reported a reduction in the translocating As from roots to shoots when rice plants pretreated with high sulfate. Furthermore, Anwar et al. (2018) illustrated that Fe and Mn-rich products, such as goethite, nontronite, pyrolusite, ferruginous smectites, and birnessite, absorb large amounts of As(V), thus As mobility may be low. A coating of Fe hydroxides/oxides identified as iron plaque, is produced on the roots of aquatic plant species. Iron plaque is produced due to the oxidation of roots by releasing oxygen and oxidants into the rhizosphere (Liu et al. 2005; Ultra et al. 2009). Iron plaque also restrains the uptake of As by plants, possibly due to its adsorption or co-precipitation procedures (Lee et al. 2013). Liu et al. (2005) expressed that Mn and Fe plaque can reduce the uptake of As in rice seedlings.

16.3.4 Arsenic Speciation

Another important factor in the interaction between plants and arsenic is speciation. Xu et al. (2014) stated that the species of arsenic present makes a noticeable effect on uptake and translocation. AsO_4^{3-}/H3AsO_4 is generally more easily uptake by roots than other inorganic species (Kumar et al. 2014). This property, along with the affinity of AsO_4^{3-}/H_3AsO_4 to substitute phosphate in reactions, has translated into the use of phosphate fertilizers to prevent arsenic accumulation in crops such as tobacco (Zvobgo et al. 2015). Once it is inside the plant it has been determined that AsO_4^{3-}/H_3AsO_4 is readily reduced to AsO_3^{3-}/H_3AsO_3 (Finnegan and Chen 2012). The concentrations of AsO_4^{3-}/H_3AsO_4 and AsO_3^{3-}/H_3AsO_3 tend to fluctuate in plant systems but stable ratio over time (Amaral et al. 2014).

16.4 Phytotoxicity of Arsenic on Plants

Most of the arsenic and heavy metals present in plants come from the cultivation soil (Rodriguez-Iruretagoiena et al. 2015), which in most cases is initially contaminated via irrigation water and fertilizers (Chakraborty et al. 2014). The biosafety of using products irrigated by arsenic- or metal-contaminated water is a worldwide health concern. Even plants that contain only trace levels of a contaminant can still contribute to the total risk of exposure (Zhu et al. 2014). Plants grown in areas of high arsenic contamination have been shown to exhibit slightly reduced biomass (Sultana et al. 2015), but the problem of contamination is broader than just the size of the plant (Hettick et al. 2015). Mishra et al. (2014) stated that vegetables with higher water content accumulate more arsenic. In conformity, Hüvely et al. (2014) found that lettuce showed a high concentration of arsenic in the roots. Crops such as maize, tobacco, lotus roots, spinach, bitter gourd, Chinese jujubes, arum, radish, and most grains have exhibited high arsenic accumulation (Bhattacharyya et al. 2003; Schmidt 2014; Herbert et al. 2014; Pena-Fernandez et al. 2014; Islam et al. 2012; Chen et al. 2013; Zhu et al. 2014; Chakraborty et al. 2014). Rice is an inherently efficient accumulator of arsenic and constitutes a major exposure route for arsenic's inorganic species (Ma et al. 2014). The high sensitivity of rice to contamination is attributed to its nodulin 26-like intrinsic aquaporin channels, which allow arsenic to interfere with amino acid mechanisms and affect the nutritional value of the grain (Kumar et al. 2014). This makes growing rice problematic worldwide. It becomes especially difficult in flooded paddy fields, which facilitates the mobilization of AsO_3^{3-}/H_3AsO_3 and enhances its bioavailability in the soil (Chakraborty et al. 2014). Rice is so efficient an accumulator that Sultana et al. (2015) even incorporated rice in experimental phytoremediation techniques. On the other hand, analysis of tomato roots grown in contaminated soil showed a high concentration of arsenic uptake (Mishra et al. 2014), but another study found very little accumulation of arsenic or any metals in the fruit portion of the plant (Rodriguez-Iruretagoiena et al. 2015). This means that tomato roots serve as an effective protection mechanism and might make it an effective crop for areas where contamination is a concern. Extreme arsenic concentrations have still been found to decrease root length and fruit size in tomatoes (Calvo et al. 2006).

Numerous physiological and biochemical processes are changed due to As toxicity which relies on its content, bioavailability, speciation, translocation and uptake in plants. Major changes due to As stress will be discussed in the following sections.

16.4.1 Seed Germination, Plant Growth and Yield Attributes

As is extremely phytotoxic to plants as it diminishes plant growth and crop yield (Islam et al. 2017). Shakoor et al. (2019) studied the effect of AsIII and AsV on some rice varieties. They found that grains germination was slightly affected at 0.5 and 1 mgL^{-1}, and a decrease of 10% in germination was detected at 2 mg L^{-1} but was 20% at 0.5 mgL^{-1} of AsV. Thus, AsV was found to be more toxic than AsIII. The decreased plant height and total dry weight were obtained due to the effect of As on *Oryza sativa* (Singh et al. 2015a), *Brassica napus* (Farooq et al. 2016) and *Triticum aestivum* (Sharma et al. 2017). As toxicity can be due to the accumulation of arsenate in the root and the reduction of phosphorus translocation and consequently phosphorus deficiency (Gupta and Ahmad 2014).

Arsenic significantly reduces the seed germination, morphological changes like leaf number, root and shoot length and photosynthesis which ultimately reduces the yield and fruit production (Cozzolino et al. 2010; Srivastava et al. 2011; Singh et al. 2018, 2020). As reduced the vegetative and root growth in tomato plants (Miteva 2002). Shaibur and Kawai (2009) reported that the shoot and root dry mass of sorghum was repressed by higher As levels. Moreover, a decrease in plants biomass with increasing As concentration in irrigation water has also been reported by Pigna et al. (2008). Khan and Gupta (2018) reported that roots of plants are the major parts to get exposure and accumulation of As that may affect their proliferation and elongation. This could be due to translocation in plants, As accumulation decreases from root to aboveground parts (Pigna et al. 2010).

Leão et al. (2016) stated that As toxicity caused a reduction of relative growth rate and tolerance index of *Azolla caroliniana*. This was associated with increase the arsenic uptake by *A. caroliniana* with intensified concentration of As, reaching 386.1 µg g^{-1} DW at the highest dose. Also, Farooq et al. (2016) reported a decrement of growth rate is a typical plant response to As exposure. Nabi et al. (2019) found that the soil-applied As proved toxicity for the growth of plants at the highest concentration as it causes a reduction in plant height and fresh weight of *Mentha arvensis*. Ahmad et al. (2020a) stated that As stress negatively affect the lengths and weights of shoot and roots of faba bean plants. Mousavi et al. (2020) reported that As treatment at 50 µM had a negative effect on the height of rice varieties T. hashemi and Fajr by 22.8 and 13.9%, respectively than the controls. Furthermore, As treatment significantly decreased the total dry weight of both rice varieties in a concentration-dependent manner. The highest decrease observed in T. hashemi and Fajr varieties at 50 µM As by 22.2 and 32%, respectively compared to untreated plants. Alsahli et al. (2020) denoted that As reduced the growth parameters in Pea (*Pisum sativum* L.) plants in terms of lengths, fresh and dry weight of shoots and roots. In addition, Shoot tolerance index and root tolerance index also decreased to 44.67 and 35.16, respectively. As-induced reduction in plants growth may occur due to disturbance in cellular processes at molecular, biochemical and physiological levels in plants (Khalid et al. 2017; Rafiq et al. 2017, 2018; Abbas et al. 2018).

Arsenic has been reported to cause 'straight head' disease which is a physiological disorder of rice in Bangladesh, characterized by sterility of the florets/spikelets leading to reduced grain yield (Rahman et al. 2008). Mousavi et al. (2020) reported that tiller, grains panicle^{-1}, filled grain percentage, and 1000-grain weight significantly decreased in two rice varieties under As treatments and the highest reduction was observed under high As concentration.

16.4.2 Photosynthetic Pigments

Increased As accumulation has been reported to decrease chlorophyll and photosynthesis rate, thereby reducing rice growth and yield (Rahman et al. 2007). Rahman et al. (2007) showed that the highest correlation was found between chlorophyll content and rice yield under As toxicity, indicating the importance of chlorophyll content in rice yield under As toxicity. A negative effect of As stress has been demonstrated on photosynthesis and chlorophyll metabolism, leading to impaired biosynthesis and accelerated degradation of the pigments (Abbas et al. 2018; Singh et al. 2016). A considerable reduction in chlorophyll content of the pea plants by As stress which is analogous to some earlier studies on *Zea mays* (Emamverdian et al. 2015) and *Trifolium pratense* (Hasanuzzaman et al. 2017). Anjum et al. (2011) reported that a remarkable reduction in chlorophyll pigment synthesis under As stress results from the shortage of adaptive adjustments of photosystems-I and II. Previous studies reported that As-induced reduction in carotenoid pigment in different plant species such as mungbean (Srivastava et al. 2017) and chickpea (Dwivedi et al. 2012). Also, Nabi et al. (2019) reported that the photosynthetic parameters including the total content of chlorophyll and carotenoids, chlorophyll fluorescence (Fv/Fm) were decreased by application of As to soil for Mint (*Mentha arvensis* L.). Similar reductions of chlorophyll a, b, total chlorophyll, and carotenoids contents in *Vicia faba* plants treated with 10 and 20 µM were reported by Ahmad et al. (2020a). Alsahli et al. (2020) reported that arsenic stress reduced total chlorophyll, carotenoid, transpiration rate (E), net photosynthetic rate (Pn) and stomatal conductance (gs) in pea plants. The reduction in Pn can be attributed to the damage caused by As in both photochemical and biochemical steps of the photosynthesis process (Abbas et al. 2018).

In other studies, disturbances in ATP, chlorophylls, and photosynthesis have been reported in As-stressed plants (Singh et al. 2020). As damages the chloroplast membrane and dismantles the photosynthetic process and consequently causes significant decrease in pigment content, rate of CO_2 fixation as well as functional activity of photosystem II (PSII) (Stoeva and Bineva 2003). Other physiological parameters such as efficiency of PS II, rate of transpiration, gas exchange and photo-protective compounds as carotenoids were found to be decreased under As stress (Milivojevic et al. 2006). However, the change in the fluorescence parameters is not significantly observed in the primary leaves of soybean indicating less effect of AsV on photosynthetic electron transport. Further, AsV uncouples photophosphorylation that leads to decrease in ATP synthesis associated with inhibition in the germination of *Zea mays* (Kaya et al. 2020). Mousavi et al. (2020) observed reduction of chlorophyll a and chlorophyll b contents as well as chlorophyll fluorescence yield (Fv/Fm) under As treatment for two rice varieties. Kaya et al. (2020) denoted that As toxicity significantly decreased chlorophyll *a* and *b* contents as well as photosystem II efficiency (Fv/Fm) in maize plants in comparison with unstressed plants. Arsenic stress leads to a concomitant oxidative stress mainly by suppressing electron transfer during the photosynthesis process (Zhang et al. 2019). These studies greatly revealed the damaging impacts of arsenic on photosynthetic pigments and their attributes, thus maintaining the pigment content is a distinguishable criterion of As-tolerance.

16.4.3 Metabolic Disorders and Hormonal Status of Plants under Arsenic Toxicity

Arsenic interferes with the plant metabolic processes and targets growth, causes oxidative stress often leads to cell death (Singh et al. 2015b, c). Arsenic in high concentrations is known to impede key plant metabolic phenomena (Gupta et al. 2017; Mitra et al. 2017). The phytotoxic effects of As result in physiological changes in plants, including reduced chlorophyll, transpiration, and photosynthetic capacity; chlorosis, withered and yellow leaves; and reduced reproductive capacity and yield (Liu et al. 2014). Arsenic stress leads to ionic, osmotic as well as oxidative stress; however, plants are able to synthesize osmolytes like proline, glycine betaine, etc., which can effectively protect the cell organelles from the stress-induced toxicity (Alsahli et al. 2020). Ahmad et al. (2020a) found that arsenic reacts with thiol groups of enzymes leading to inhibition of metabolism. As mediated growth inhibition has been proposed to be ascribed to cell cycle arrest and inhibition of DNA synthesis and repair mechanisms (Ghiani et al. 2014).

Nabi et al. (2019) reported that arsenic stress on mint plants caused increment of proline content of leaves which reached 55.7% at 100 mg As kg^{-1} of soil. Furthermore, the essential oil content was increased when As applied at 20 mg kg^{-1} of soil over the control. However, it was declined by 25.0% at 100 mg As kg^{-1}. In addition, menthol, the main constituent of mint oil decreased with the application of As, but the contents of menthone and menthyl acetate were significantly increased at 100 mg As kg^{-1} by 26.0% and 99.0% over the control. Ahmad et al. (2020a) observed high levels of proline and glycine betaine in *Vicia faba* plants under the effect of 10 and 20 μM. Alsahli et al. (2020) found that arsenic stress (20 mM) reduced relative water content by 46.56% in the pea plants with respect to the control. However, proline and glycine betaine contents increased under As stress by 82.81% and 71.83%, respectively, compared to the controls.

Other studies showed reduction in RWC under As stress in rice (Rahman et al. 2016) and soybean (Vezza et al. 2018). This effect most possibly results due to As-induced constraints to water uptake and damage to the root system architecture. Vezza et al. (2018) reported water absorption reduction of 25–39% in soybean under As stress. They demonstrated that As treatment applied could not decrease the osmotic potential, and suggested reduction in water absorption might be controlled instead of physiological or phenotypic changes. Armendariz et al. (2016) reported that As treatment to soybean plants resulted in thicker and lignified cell walls of root cells as well as deposition of huge quantity of dark deposits in the xylem vessels, and also observed a decline in root length and biomass. Kaya et al. (2020) represented that As stress reduced leaf RWC and leaf $_{\Psi l}$, and accumulation of proline content relative to the control was resultant. Such increase of proline content might help in maintaining the plant water status, analogous to what has been observed in soybean wherein reduction in leaf RWC and increase in proline content have been reported due to AsS (Vezza et al. 2018). However, kaya et al. (2020) found that As stress reduced endogenous NO content in maize plants, as has been found in *Vicia faba* (Xue and Yi 2017), macrophyte (da-Silva et al. 2018), and *Isatis cappadocica* (Souri et al. 2020). These reports suggest that NO might be involved as a signal molecule in many key physiological events of As-stressed plants. One of the key detrimental effects of As stress on plants is to disturb plant-water relations (Vezza et al. 2018). The reduced water status due to As in maize plants could be one of the strategies for lowering uptake of As and transferring it to the shoots, as has also been reported (Armendariz et al. 2016; Vezza et al. 2018). Reduced water uptake due to heavy metals or metalloids leads to root and leaf desiccation in several plant species (Weryszko-Chmielewska and Chwil 2005; Rucińska-Sobkowiak 2016).

Also, disturbance in hormonal homeostasis is one major toxic effect of As on plants. In this regard, in a rice cultivar T. hashemi, an increasing trend in endogenous JA content was observed in response to the increase in As concentration, and the highest increase was recorded under 50 μM As by 17.4% compared to control ones. However, the content of JA in Fajr reduced under arsenic treatment and the highest decrease was observed at 25 μM As (Mousavi et al. 2020). Other studies reported that arsenic stress and ABA accumulation are directly correlated; higher concentration of As in plants leads to accumulation of ABA in plant cells including guard cells (Huang et al. 2012), thereby causing stomatal closure and reduced transpiration (Stoeva et al. 2004). For example, Armendariz et al. (2016) reported that As stress induces ABA accumulation and stomatal closure in soybean, and reduces transpiration.

16.4.4 Reactive Oxygen Species and Membrane Deteriorations

Under un-stressful conditions, the reactive oxygen species (ROS) and the generation of free radicals were kept under tight control. But, the stresses accelerate the production of toxic ROS. These ROS causes potent damage to the nucleic acids, nucleotides, amino acids, and proteins and affects peroxidation of membrane lipids under acute stress conditions (Sallam et al. 2019). When plants are subjected to toxic levels of arsenic, they generate excessive quantities of reactive oxygen species (ROS) such as hydroxyl radical ($^{\cdot}$OH), superoxide anion ($O2^{\cdot-}$), and hydrogen peroxide (H2O2), which can cause oxidations and dysfunctions of critical macro- and micromolecule in plant cells (Mishra et al. 2019; Kohli et al. 2019). Lipid peroxidation affects the cellular processes and enhances the lipid-derived radicals (Dawood and Azooz 2019, 2020). Thus, the dangerous biochemical impact of As at the subcellular level is the production of reactive oxygen species (ROSs), such as hydroxyl radical (OH), superoxide radical ($O_2^{\bullet-}$) and hydrogen peroxide. ROSs are hazardous for plant metabolism and may lead to damage to macromolecules (Abbas et al. 2018). As also induces oxidative stress in rice plants by producing ROS (Nath et al. 2014), which can negatively affect rice growth and yield. The increased production of ROS and induction of oxidative stress are one of the major causes of the negative effects of As toxicity on plant growth (Mallick et al. 2011; Finnegan and Weihua 2012). As treatment increased H_2O_2 content and, as a result, induced oxidative stress as was obtained by Nath et al. (2014) in rice and Farooq et al. (2016) in *Brassica napus* L. As can induce the production and accumulation of MDA and H_2O_2 that in turn may increase electrolyte leakage has been reported by different authors (Ghosh et al. 2013; Rafiq et al. 2018). Induction of oxidative stress markers was the result of application of 10 and 20 μM of As in terms of hydrogen peroxide, electrolyte leakage, and MDA contents (Ahmad et al. 2020a). Mousavi et al. (2020) illustrated that As treatments caused a significant increase in MDA and H_2O_2 content in two rice varieties compared to the control treatment with the highest increase under 50 μM As concentration. In early growth of plants, AsV increases the formation of reactive oxygen species (ROS) that damaged the lipids of membrane and increase in malondialdehyde equivalent content noticed in *Pteris ensiformis* (Stoeva et al. 2005; Singh et al. 2006; Ahsan et al. 2008). Srivastava et al. (2005) and Singh et al. (2006) reported the induction of lipid peroxidation by AsV in *Pteris vittata* (hyperaccumulator) showing ROS production which is a common end point of aerobic life.

Arsenic stress also triggers higher accumulation of methylglyoxal, a key byproduct of glycolysis, which is found to be harmful for plant organelles. Methylglyoxal and ROS work together and hamper the normal functioning of plant organelles, and if they are not properly removed they may cause cell death (Dawood et al. 2021). In this regard, Alsahli et al. (2020) found that hydrogen peroxide, methyl glyoxal, electrolyte leakage, and MDA contents increased by 74.22%, 92.66%, from 10.77% to 69.05% and 63.09%, respectively, in the As-treated plants compared to the control. Kaya et al. (2020) reported that As treatment led to significant increases in H2O2, MDA content, and EL by 5.4-, 3.5- and 2.5-fold, respectively, with respect to non-stressed plants. It has similarly been reported that As stress induced oxidative stress, which was related with increased MDA and H2O2 (Praveen et al. 2017; Yadu et al. 2019). Moreover, As stress increased electrolyte leakage due to higher MDA and H2O2 contents as reported previously in soybean by Chandrakar et al. (2018). It is highly likely that these biomolecules may cause disruption of ion exchange capacity of cell membrane and all metabolic processes connected to cell membrane functioning (Zahra et al. 2018). The precise mechanism of the As-induced generation of ROS might be ascribed to As-detoxification mechanism, including the reduction of AsV to AsIII and the induction of phytochelatins (PC) synthesis (Meharg and Hartley-Whitaker 2002), plays an important role in ROS generation (Meharg and Hartley-Whitaker 2002).

16.4.5 Antioxidant System

When induced, the oxidative stress is mitigated by a complex series of defense systems, particularly enzymatic and/or non-enzymatic antioxidants. Such defenses improve plant's ability to counteract the over-accumulation of ROS, so as to protect the cells under oxidative stress (Dawood et al. 2021a). The enzymatic defense systems include among others ascorbate-glutathione cycle, peroxidases (POD), superoxide dismutases (SODs), peroxiredoxins, and catalase (CAT) (Sallam et al. 2019). This defensive role may also be accomplished by non-enzymatic antioxidants, including ascorbate, glutathione, and phenolic compounds (Abdel Latef et al. 2020). Furthermore, it has also been reported that the AsA-GSH cycle related enzymes are involved in As stress tolerance (Singh et al. 2015c). Plants can also synthesize methylglyoxal (MG) under stress conditions (Dawood et al. 2021). Over-accumulation of MG can harm cell membrane by degrading lipid and protein (Dawood et al. 2021). Plants upregulate the glyoxalase system relating enzymes, glyoxalase I (Gly I) and glyoxalase II (Gly II), to scavenge MG (Kaya et al. 2020).

Several studies stated that As induced oxidative stress was recorded to be associated with the redox mechanism of AsA and GSH as also suggested in the maize plants (Kaya et al. 2020) in pea seedlings (Singh et al., 2015b, d), Arabidopsis (Leterrier et al. 2012) or garlic plants (Ruíz-Torres et al. 2017). Kaya et al. (2020) stated that As stress decreased AsA and GSH contents of maize plants. However, the oxidized corresponding forms, DHA and GSSG, increased significantly by 4- and 5-fold, respectively, relative to the controls. This leads to significant declines in ratios of GSH/GSSG and AsA/DHA in the As-stressed plants. They also found that arsenic stress significantly decreased the activities of APX, GR, DHAR and MDHAR relative to the controls. In addition, the content of MG was triggered by 2.5-fold in response to As-stress, relative to that in control plants. The activities of the glyoxalase system enzymes such as that of glyoxalase I (Gly I) was significantly reduced, whereas the glyoxalase II (Gly II) was increased by 31% in the As-stressed plants compared with control plants.

SOD is characteristic enzyme responsible for dismutation of superoxide anion to water and oxygen and it acts as a first line of defense in managing oxidative stress. In response to As-stress, some studies showed increased SOD in maize plants (Kaya et al. 2020), in rice (Tripathi et al. 2013; Rahman et al. 2015; Dixit et al. 2015). Reports illustrated that the induction of SOD could be ascribed to the elevated level of H_2O_2 in plants treated with As (Mabrouk et al. 2019; Kaya et al. 2020). When As levels are very high, the extent of the damage caused by the pollutant decreases enzyme activity. The decrease in SOD activity at high heavy metal concentrations (Leão et al. 2016) may be attributed to the inactivation of the enzyme by H_2O_2 or to the existence of other ROS and the inactivation of other enzymes involved in the degradation of these compounds (Khan et al. 2009).

Hydrogen peroxide detoxification in plants has been done by multi-enzyme system including CAT, POD, and ascorbate-glutathione system. Peroxisomes inherently possess an efficient nitro-oxidative metabolism and CAT is the main peroxisomal antioxidant enzyme involved in scavenging of H_2O_2 (Kaya et al. 2020). Arsenic stress reduced the CAT activity in the maize plants, as has been reported in mungbean (Malik et al. 2012), and soybean (Chandrakar et al. 2016), but conflicting results were also reported in lupine (El-Shora and El-Gawad 2014), and rice (Dixit et al. 2016).

Moreover, POD is also one of the main enzymes involved in scavenging of H_2O_2, because of its different forms at cellular levels (Kaya et al. 2020). However, it was found to be decreased in maize plants subjected to As-stress (Kaya et al. 2020), as earlier recorded in Brassica napus (Farooq et al. 2016). In contrast, such a relationship of POD was not observed with H_2O_2 in *Vigna mungo* (Srivastava et al. 2017) and *Zea mays* (Anjum et al. 2016).

One of the important enzymes related to the AsA-GSH cycle is APX (ascorbate peroxidase), mostly existing in the plastid stroma and membrane to scavenge H_2O_2 through the ascorbate-glutathione cycle (Ghosh and Biswas 2017). Its activity was recorded to be decreased in the leaves of As-stressed maize plants, as was observed in pea seedlings (Singh et al. 2015b;), soybean (Chandrakar et al. 2016), rice (Singh et al. 2015d), and *Luffa* seedlings (Singh et al. 2015c). The decrement of APX activity can be accounted for over-accumulation of H_2O_2 in different subcellular compartments, consequently leading to damage of lipids and proteins as suggested by Singh et al. (2015b). However, some other works reported that As stress increased the APX enzyme activities as in wheat (Ghosh and Biswas 2017), *Isatis cappadocica* Desv. (Souri et al. 2018), oilseed rape (Farooq et al. 2018), and rice (Jung et al. 2019).

It is imperative to note that both DHAR and MDHAR are harmoniously critical to maintain the AsA concentrations and its redox status under stressful regimes (Wang et al. 2018). MDHAR plays a critical function in preserving the reduced pool of ascorbate redox and AsA states (Wang et al. 2018). MDHAR activity decreased in the As-stressed maize plants (Kaya et al. 2020) and in pea seedlings (Singh et al. 2015b). DHAR activity converts DHA to AsA (Suekawa et al. 2017). Arsenic stress reduced the activity of DHAR in maize (Kaya et al. 2020), in wheat (Hasanuzzaman and Fujita 2013), rice (Rahman et al. 2015; Jung et al. 2019), and maize (Ghosh et al. 2016).

To protect plants from oxidative stress, glutathione peroxidase (GPX) uses GSH as a substrate during scavenging of H2O2 and lipid hydroperoxides (Dawood and Azooz 2019). Rahman et al. (2015) reported that As exposure decreased GPX activity of rice, which might be due to increased H_2O_2 content, and the same conclusion was reported by Hasanuzzaman and Fujita (2013).

Stimulation of GST activity has been considered an important factor in stress tolerance (Bashandy et al. 2020) because it catalyzes the binding of different xenobiotics and their electrophilic metabolites to produce less toxic and water-soluble conjugates (Dawood and Azooz 2019). Rahman et al. (2015) demonstrated that GST activity decreased with increasing As stress.

The activity of GR which is related to the AsA-GSH cycle was found to be reduced in the leaves of As-stressed maize plants (Kaya et al. 2020), as earlier observed in pea (Singh et al. 2015b;). However, other plants responded to As-stress by increasing GR such as Arabidopsis (Leterrier et al. 2012), rice (Dixit et al., 2016), and wheat (Ghosh and Biswas 2017). Kaya et al. (2020), As-induced reduction in the GR activity might have reduced the GSH/GSSG ratio and the ascorbate pool, and had an effect on the ascorbate redox status with a reduced tolerance to stress. Another potential factor could be the NADPH availability, which is needed for the GR activity. It could be compromised in maize under As stress such as it has been described in the garlic plant (Ruíz-Torres et al. 2017).

Glyoxalase system composed of two enzymes (Gly I and Gly II) that reduce the exacerbation of methylglyoxal under stressful regimes (Li et al. 2017; Hasanuzzaman et al. 2019). Under As stress conditions, MG accumulation increased in maize (Kaya et al. 2020) and in rice (Rahman et al. 2015). The glyoxalase system is required to reverse the over-generated MG, so as to achieve improved stress tolerance. The up-regulation of the activities of both Gly I and Gly II enzymes may enhance abiotic stress tolerance in many species (Hossain et al. 2018; Zaid et al. 2019). Kaya et al. (2020) found that As stress elevated the MG content and Gly II activity, but reduced the Gly I activity. The enhanced MG content and reduced Gly I activity show that detoxification of MG is inadequate in the As-stressed plants, as observed in wheat by Hasanuzzaman and Fujita (2013) and in rice by Mostofa et al. (2014). Rahman et al. (2015) reported that Gly I activity was reduced and Gly II activity increased under metal stress.

The enzymatic activities showed dose response of plants to As, Leão et al. (2016) reported that SOD, CAT, POD and GR activities increased in *A. caroliniana* up to 1 mg L^{-1} As, however, the enzyme activity was negatively affected by exposure to As at 1.5 mg L^{-1}. However, the activity of APX showed a sharp decrease in plants exposed to As concentrations above 0.25 mg L^{-1}, indicating that even low pollutant concentrations are able to inactivate the enzyme. The same study also denoted that the content of non-enzymatic antioxidants total thiols, non-protein thiols and anthocyanins had increased whatever As concentration used. The response of non-enzymatic antioxidants did not decrease in the highest concentration of As, indicating lower sensitivity of these molecules to damage triggered by As. The toxic effects of As inactivated the enzymes were observed in rice (Shri et al. 2009) and aquatic plants (Farnese et al. 2014; Andrade et al. 2016) after exposure to high As.

For better growth performance under As stress, there is an equilibrium between production and elimination of ROS which decides the fate of a plant. To cope up with the damaging effects induced by As, plants either increase the formation of low molecular weight antioxidants such as GSH and AsA or up regulate the antioxidant enzymes such as SOD, CAT and APX or also synthesized the proline and phenolic that have antioxidant property (Mishra et al. 2021). Arsenite inhibits the function of enzymes and proteins through thiol interactions; while arsenate disrupts energy flow through phosphate replacement (Ma et al. 2008). Other studies reported that the enzyme activity shows elevated response at low level of As exposure in As-hyper-accumulator *P. vittata*, As-sensitive

Holcuslanatus and As-tolerant *Zea mays*, and the activity either remains the same or decreases at higher As levels (Cao et al. 2004; Sinha et al. 2010). Nabi et al. (2019) found that the different concentrations of As (20, 40, 60, 80, and 100 mg kg-1of soil) reduced carbonic anhydride for mint plants. Ahmad et al. (2020a) registered increased activities of CAT, SOD, as well as AsA-GSH cycle, whilst reduction of DHAR and MDHAR was also reported in faba bean plants. Arsenic applied as 20 μM enhanced GSH and GSSG by 62.25% and 123.10%, respectively, over the controls. However, Ascorbic acid (AsA) content decreased by 21.42% with 10 μM As and 48.57% with 20 μM As with respect to the control. Mousavi et al. (2020) showed that CAT, APX, and SOD activities were increased in two cultivars of rice under 25 and 50 μM As compared with control plants. Also, POD increased in response to 25 μM As treatment, while 50 μM As caused increment of POD by 100% in T. hashemi and reduced by 23.3% in the Fajr variety. Compared to the control treatment, ASA, DHA, and GSSG contents increased in response to As treatment and the highest increase was observed under high concentration of As (50 μM) in both rice varieties. On the other hand, As treatments decreased GSH content as well as the ASA/DHA and GSH/GSSG ratios in both rice varieties compared to the control treatment, with the highest reduction was noted at high As concentration. Similar effect of As on redox status of ASA-GSH in rice plants have also been reported by Singh et al. (2015) and Jung et al. (2019). It has been reported that Glutathione (GSH) plays an important role in As detoxification by chelating As (Ahmad et al. 2020a, b; Singh et al. 2015c).

Alsahli et al. (2020) registered that the activities of enzymatic antioxidants viz., SOD, CAT, APX, GR and GST, enhanced in the As-treated plants with reference to the controls. However arsenic stress decreased the ascorbate recycling enzymes MDHAR and DHAR by 49.16% and 49.06%, respectively, relative to the controls. Also, As stress decreased AsA and GSSG by 55.00% and 42.76%, respectively, but increased GSH by 75.60% in the As-treated plants relative to the controls. Arsenic stress applied to the pea plants differentially affected glyoxalase cycle by increasing Gly I by 63.23%, but decreased Gly II by 41.55% relative to the controls. Glyoxalase I and II are the main enzymatic components of the ubiquitous glyoxalase pathway involved in detoxifying increased MG levels (Ghosh et al. 2016).

Increased antioxidant functioning in response to As stress has been reported in fenugreek (Talukdar 2013), *Zea mays* (Yadav and Srivastava 2015), and *Ocimum tenuiflorum* (Siddiqui et al. 2015d). Treatment of As can impede plant growth by directly binding to the thiol groups of antioxidant enzymes, thereby hampered plant growth (Sharma 2012). Talukdar (2013) has demonstrated that protection against the As induced oxidative damage is due to increased expression of different isozymes of SOD and APX exhibiting coordinated functioning with CAT. Overall, plants have their own defense system, including enzymatic and non-enzymatic antioxidants, to counteract the effects of ROS. However, the antioxidant system of the plant may not play a sufficient role in decreasing the impacts of As-induced oxidative stress under severe toxicity.

16.4.6 Mineral Homeostasis

Being a phosphate analogue, AsV significantly suppresses the gene that is up regulated under phosphate starvation (Abercrombie et al. 2008). As a consequence, As in the nutrient solution decreases phosphate uptake, and anthocyanin synthesis is considered as a visible marker of As stress (Catarecha et al. 2007; Shaibur et al. 2013). Mousavi et al. (2020) denoted that the accumulation of As in the root was much higher (tenfold) than the leaf under As stress in two rice varieties. They also observed that the accumulation of Fe in roots and leaves of both rice varieties significantly increased under As treatment. However, most of the Fe absorbed by the plant accumulated in the root tissue.

Lsi2 is involved in the translocation of root arsenite to shoot in rice plant. Mousavi et al. (2020) showed that As treatment increased *Lsi2* gene expression in two rice varieties, which could explain As accumulation in leaves under As treatments. Increased *Lsi2* gene expression and As accumulation in rice plant leaves have also been reported by Singh et al. (2015). *Lsi1* is involved in the uptake of arsenite into the root from the external solution (Ma et al. 2008). Mousavi et al. (2020) reported that transcriptional levels of both *Lsi1* and *Lsi6* genes were up-regulated under arsenic treatment. Therefore, the increased As accumulation in the root of both rice varieties could be due to increased expression of *Lsi1* and *Lsi6* genes and consequently increased As uptake by root cells, which is in accordance with the results obtained by Singh et al. (2015) and Ma et al. (2008).

NRAMP1 and NRAMP5 are involved in the uptake of iron, manganese and cadmium in roots (Curie et al. 2000; Ishimaru et al. 2012). Tiwari et al. (2014) showed that *OsNRAMP1* is also involved in As uptake. Mousavi et al. (2020) showed that increasing As concentration up-regulated the expression of *NRAMP1* and *NRAMP5* genes and consequently increased As and Fe accumulation in the root of both rice varieties, which is in agreement with the results of Curie et al. (2000), Tiwari et al. (2014), and Singh et al. (2015). *OsFRDL1* transporter is involved in the xylem loading and the translocation of Fe to shoot in rice. The results showed that As treatments decreased *FRDL1* gene expression in the root of both rice varieties. Therefore, the decline in Fe accumulation in rice leaves under As treatments may be due to the decrease in *FRDL1* gene expression, which is involved in the translocation of Fe from root to shoot. Reduced *FRDL1* gene expression under As toxicity (Singh et al. 2015). Another Fe transporter, OsYSL2, is involved in long-distance transport in rice plants (Ishimaru et al. 2012). Mousavi et al. (2020) found that As significantly decreased *YSL2* expression in the root of both rice varieties. Ishimaru et al. (2012) indicated that the decreasing *OsYSL2* gene expression in rice plant reduces Fe translocation and, consequently, decreases Fe accumulation in shoot and grain. Therefore, the decline of Fe accumulation in the leaves under As treatments could be partly due to the decrease in *YSL2* gene expression. Thus, the accumulation of As inside root tissue change the overall profile of minerals uptake by plants which could affect the plant growth and productivity.

16.5 Mechanisms Induced by Plants for Arsenic Detoxification

1. From literature, the following strategies could be induced by plants under As stress as a trial to reduce its toxic effects: The synthesis of specific low molecular weight chelators plays an essential role to detoxify non-essential trace metalloids through binding and facilitate their transport into the vacuoles. Gamma-glutamylcysteine synthetase (γ-ECS) and glutathione synthetase (GS) enzymes synthesize the tripeptide glutathione (Glu-Cys-Gly) (Schulz et al. 2008). GSH has been found to bind with a number of metals and metalloids and is responsible for the cellular redox balance mechanism, which is generally the target of metalloid toxicity. The high level of metalloid binding capacity of GSH has been found to increase cellular defense against oxidative stress. On the other hand, GSH is the precursor of PC; Gasic and Korban (2007) have postulated that the increasing GSH and PC synthesis alone seems to be insufficient to achieve As tolerance or accumulation. PCs are synthesized from GSH and this reaction is catalyzed by PC synthase (PCS) and this PCS are responsible for the post-translational activation by metalloid As. Nonetheless, there are few reports on the PC synthesis, that they are important factors for the basic As tolerance (Clemens 2006), but not in the hyper-tolerant plants or hyper-accumulators (Ernst et al. 2008). Verbruggen et al. (2009) illustrated that the increase in GSH and PC synthesis in *A. thaliana* improved the tolerance and accumulation of both AsIII and AsV species. In *Arabidopsis thaliana*, PC deficiency leads to As hypersensitivity in the roots of non-hyper-tolerant plants; the major part of As is chelated by PC. In this regard, Kumari et al. (2017) observed that the expression of PCs increased by 2.4-fold and 1.6-fold in leaves and roots of *Artemisia* grown under As stress. It has been elucidated that As is a strong inducer of PC synthesis. Whenever rice is exposed to AsV, a number of genes and enzymes are up-regulated for the GSH synthesis, metabolism and transport (Ahsan et al. 2008); therefore, under As stress condition the plants have higher demand for GSH (Bleeker et al. 2006).
2. Arsenite enters in plants either directly by plant root or its concentration found to increased inside the cell when reduction of AsV into AsIII (reacts with metabolites having—SH groups) (Schulz et al. 2008). In As hyper-accumulator plants like *Pteris vittata*, there is gradient of AsV present toward the vacuole from the cytoplasm and the reason behind this gradient is due to involvement of energy-dependent transport of AsIII in tonoplast surrounding the vacuoles (Zhao et al. 2010). Arsenic detoxification is an important regulatory mechanism to minimize the As toxicity and a number of studies are conducted to prove that plants have inherent ability to reduce AsV into AsIII and then in less toxic methylated species via the action of an enzyme arsenate reductase (AR) (Xu et al. 2007). This process also mediates As efflux from cell to soil and maintains As bio-geochemical cycles (Shi et al. 2016). Gene encoding for AR is identified in different plants such as *Pteris vittata* (*PvACR2*), rice (*OsACR2; OsACR2*), *A. thaliana* (*ATQ1* and *HAC1*), and *H. lanatus* (*HlAsr*) by using sequence homology with *ACR2*; the arsenate reductase gene was found in *Saccharomyces cerevisiae*. Plants with mutant ACR2 gene showed higher As accumulation in stems, leaves or fruits leads the more susceptibility toward As stress (Dhankher et al. 2006). Further, overexpression of OsACR2;1 and OsACR2;2 in the rice plants shows higher resistance for As (Duan et al. 2007).
3. In plants, AsV is reduced to AsIII. AsIII efflux to the external medium is a vital way of As detoxification in plants. Inside plant cells, AsIII may be detoxified by complexation with phytochelatins (PCs), followed by the accumulation of AsIII–PC complexes in vacuoles through OsABCC1 transporters (Sun et al. 2018). OsABCC1 is one of the ATP-binding cassette (ABC) transporters. ABC transporter proteins play roles in the translocation of a broad range of substances within membranes using energy from ATP hydrolysis (Chang et al. 2018). Song et al. (2014) stated that OsABCC1 plays a vital role in the detoxification and decreasing As in rice grains. HAC1 contributes to the defence against As in plants (Meadows 2014) and is essential for the efflux of AsIII from roots for AsV detoxification (Salt 2017). In rice, OsHAC1;1, OsHAC1;2 and OsHAC4 function as AsV reductase. Moreover, glutaredoxin possesses AsV reductase enzyme activity in maintaining the glutathione (GSH) pool and assists in AsIII efflux (Shri et al. 2019). Brinke et al. (2018) stated that with rising AsIII concentrations, the importance of the term 'response to stress' is replaced by the detoxification ways 'glutathione biosynthesis', which is related to the term 'oxidation reduction'. GSH is applied as an electron donor by dehydroascorbate reductase to reconvert dehydroascorbate to ascorbate. GSH disulfide is the oxidized form of GSH, which may be reprocessed to GSH by glutathione reductase via reduced nicotinamide adenine dinucleotide phosphate. Hence, these different components of the ascorbate–GSH cycle may have a vital character in protecting cells against oxidative damage resulted by As toxicity (Jung et al. 2019). After the reduction of AsV to AsIII, further mechanisms of detoxification arise in the vacuole via vacuolar sequestration. AsIII chelates with sulfhydryl (–SH)-rich protein and arranges a complex that is separated by vacuolar transporters (PCs). In rice, two phytochelatin synthase enzymes have been testified, comprising OsPCS1 and OsPCS2 (Shri et al. 2019).

Plants modulate their defense system under As stress to reduce its toxicity, however, external application of some agents like mineral elements, chemicals, phytohormones, nanoparticles, etc. could strengthen the defense system. One of the distinguishing approaches for alleviation the toxicity of arsenic is the application of microbes. Thus, in the present chapter we will discuss extensively the recent approaches of phytobial remediation of arsenic toxicity.

16.6 Phytobial Remediation and Effects of As-Tolerant Microorganisms on Crop Plants

Phytobial remediation (bio-phyto-remediation) is an innovative strategy that aims to maximize the bioremediation efficiency of a certain pollutant through the use of both microorganisms and plants. However, the term phytoremediation alone have been described in recent studies as a process that uses soil microorganisms and green plants to provide a more sustainable clean-up method for large areas of contaminated soil (Alka et al. 2020). Interactions between plants and microbiota in the rhizosphere are complex phenomena and play a fundamental role in plant growth and development as well as in nutrient uptake and transport. Several reports have shown that plant's survival under different environmental stresses is directly related to microbiota present in their surroundings (Upadhyay et al. 2018). The rhizosphere environment is generally rich in different beneficial microbiota such as bacteria, fungi, and algae which is usually referred to as plant growth promoting microorganisms (PGPMs). The development of these microorganisms is dependent on the organic secretions of plant roots which include amino acids, organic acids, phenolics, sterols, sugars and vitamins. PGPMs that is characterized by As tolerance and plant growth promoting properties provide different direct or indirect traits to the plants (Table 16.1). Rhizospheric microbiota can indirectly influence plant growth and development through the prevention of phytopathogens. While direct effects include decreasing the bioavailability of As to plants (through secretion of protons, organic acids, redox reaction, metabolic reactions), As biotransformation (reduction, oxidation, methylation, demethylation), effect on As interactions with other elements like Fe, Si, etc., enhancing plant growth (by indole acetic acid (IAA), gibberellins, extracellular enzymes, nitrogen fixation, extracellular polysaccharides and siderophore production) (Alka et al. 2020; Lampis et al. 2015; Mesa et al. 2017; Wang et al. 2018; Zhang et al. 2020).

16.6.1 Role of Bacteria

An As hypertolerant bacterium namely *Staphylococcus arlettae* strain NBRIEAG-6 isolated from an As contaminated soil induced phytostabilization of As in the roots of Indian mustard (*Brassica juncea*) with a concurrent promotion of plant biomass, protein, chlorophyll and carotenoid contents (Srivastava et al. 2013). Similarly, *Acinetobacter* sp. strain nbri05 showed an ability to promote yield and phytostabilize As in the roots of chickpea plant (*Cicer arietinum*) (Srivastava and Singh 2014). As-resistant halophilic bacterial strains (*Kocuria flava* AB402 and *Bacillus vietnamensis* AB403) were isolated from the rhizosphere of mangrove plants and effectively colonized rice plants and promoted its growth and reduced its As uptake (Mallick et al. 2018).

Rhizospheric bacteria improve plant growth under a heavy metal stress by the production of siderophores and growth promoting hormones. Funes Pinter et al. (2017) showed that As resistant bacteria (ARB) such as *Micrococcus luteus* have the ability to produce siderophores, solubilize phosphates and fix atmospheric nitrogen in the presence of high NaAsO2 concentration. Accordingly, the inoculation of grapevine plants with *M. luteus* showed to increase plant biomass and protein content as well as mitigating As toxicity. Likewise, As-resistant purple nonsulfur bacteria (*Rhodopseudomonas palustris* C1 and *Rubrivivax benzoatilyticus* C31) were found to possess 5-aminolevulinic acid, indole acetic acid (IAA), exopolymeric substances, and siderophores production which significantly reduced As accumulation in treated rice plants with concomitant plant growth promotion (Nookongbut et al. 2018).

One of the most challenging strategy to help populations is to increase the yield of crop plants such as rice and reducing the internal exposure level of As. Arsenite oxidizing bacteria (*Achromobacter xylosoxidans* GD03) remarkedly accelerated arsenite oxidation in the paddy soil and produced IAA which supported the growth of rice plants (Wang et al. 2020). The bacterial inoculation resulted in a 34–69%, 43–74%, 24–76% and 35–57% decrease in arsenite concentration of the stems, leaves, bran and grain of rice, respectively with a concomitant 59–96% increase in the grain yield.

In the aquatic environment, microbial assemblages such as biofilms around aquatic plants play a crucial role in the As-cycling. This idea was recently investigated for the As-oxidizing, As-reducing and As-methylating bacteria coexisting in the phyllosphere of *Hydrilla verticillata* (Zhen et al. 2020). The authors showed that oxidizing bacteria can decrease the As accumulation (by approximately 64.44%) in *H. verticillata*, while reducing bacteria played a critical role in increasing As accumulation (by approximately 3.31-fold). Additionally, the existence of epiphytic bacteria enhanced As efflux by approximately 9-fold from the plant cells.

On the other hand, As-hyperaccumulating terrestrial plants and several types of rhizospheric bacteria have been reported as a promising type of phytobial remediation. Ghosh et al. (2011) isolated seven isolates of As resistant bacteria (ARB) belonging to the genera *Pseudomonas*, *Comamonas* and *Stenotrophomonas* from the rhizosphere of As-hyperaccumulator fern (*Pteris vittata*) and investigated their role in the phytobial remediation process. The ARB showed an effective role in converting insoluble As salts into soluble ones, which enhanced plant As uptake (from 18.1–21.9 to 35.3–236 mg kg^{-1} As in the fronds). The mechanism of As solubilization was related to the production of pyochelin-type siderophores by ARB as well as root exudate (dissolved organic C) by *P. vittata*. In another study, the production of catecholate-type siderophore by the rhizobacteria *Pseudomonas* PG12 strain was effective in promoting $FeAsO_4$ dissolution, and enhancing Fe and As uptake by *P. vittata* (Liu et al. 2005). Similarly, the siderophore and IAA-producing bacteria (*Bacillus* sp. MPV12, *Variovorax* sp. P4III4, and

TABLE 16.1

A Summary of Recent Studies on the Utilization of Microorganisms for the Amelioration of As Stress and for Regulating As Accumulation in Plants

Name	Targeted Plant	Processes	Reference
Bacteria			
Pseudomonas sp., *Comamonas* sp. and *Stenotrophomonas* sp.	*Pteris vittata* (As hyperaccumulator)	Solubilization of insoluble As species Enhanced plant As uptake Production of pyochelin-type siderophores	Ghosh et al. (2011)
Brevibacillus sp.	Chilli plant	Enhanced plant growth by colonizing the rhizosphere and reducing plant accumulation of As	Mallick et al. (2014)
Staphylococcus arlettae	*Brassica juncea*	Phystabilization of As in plant roots Enhancing plant biomass and biochemical composition	Srivastava et al. (2013)
Acinetobacter sp.	*Cicer aritenum*	Phytostabilization of As in plant roots Enhancing plant biomass and biochemical composition	Srivastava and Singh (2014)
Rhodopseudomonas palustris and *Rubrivivax benzoatilyticus*	Rice (*Oryza sativa*)	Reduction of both As(III) and As(V) toxicity Enhanced growth and enzymatic and non-enzymatic antioxidants Reduction of As uptake by rice	Nookongbut et al. (2018)
Epiphytic bacteria in the phyllosphere	*Hydrilla verticillate* (Aquatic plant)	The epiphytic bacteria possessed As(V) reduction, As(III) oxidation and As(III) methylation abilities As-oxidizing bacteria suppressed As uptake and accumulation in *H. verticillata* As-reducing bacteria induced As uptake and accumulation in *H. verticillata* Epiphytic bacteria increased As(III) efflux from plant cells	Zhen et al. (2020)
Ensifer sp.	Alfalfa (*Medicago sativa*)	Soil bioaugmentation with *Ensifer* sp. promoted the growth of alfalfa and stimulated the indigenous soil microbiota	Debiec-Andrzejewska et al. (2020)
Plant Growth Promoting Bacteria (two strains of *Stenotrophomonas maltophilia* sp. and one of *Agrobacterium* sp.)	*Arundo donax*	PGPB did not improve plant biomass or phytoremediation efficacy and increased the As concentration in the roots of *Arundo* plants	Guarino et al. (2020)
Pseudomonas sp.	soybean (*Glycine max*)	*Enhanced plant* growth under AsIII/AsV treatment. However, arsenic accumulation pattern in soybean was not modified by *Pseudomonas* sp. inoculation	Wevar Oller et al. (2020)
Bacillus licheniformis, Micrococcus luteus and *Pseudomonas fluorescens*	grapevine plants	*B. licheniformis* and *M. luteus* (both highly tolerant to As(III)) had the ability to produce siderophores in presence of high NaAsO2 concentration, but only M. luteus solubilized phosphates and fixed N2 under such condition *M. luteus* increased plant biomass and protein content, while *B. licheniformis* only increased plant biomass, and *P. fluorescens,* less tolerant to As(III), had no effect	Funes Pinter et al. (2017)
Iron oxide nanoparticles and *Bacillus subtilis*	*Cucurbita moschata*	Iron-NPs and *B. subtilis* improved plant growth attributes and proline contents and enhanced the biosynthesis of stress mitigating polyamines including spermidine and putrescine in As-stressed seedlings	Mushtaq et al. (2020)
Achromobacter xylosoxidans	Rice (*Oryza sativa*)	Inoculation of soil accelerated As(III) oxidation and reduced 35–57% of As accumulation in rice grains and led to 59–96% increase in grain yield	Wang et al. (2020)
Fungi			
Mycorrhiza *(Rhizoglomus intraradices)*	barley (*Hordeum vulgare*)	Down regulation of HvPht1;1 and HvPht1;2 genes (encoding high-affinity inorganic orthophosphate (Pi)-uptake systems in a direct pathway via root epidermis and root hairs) and upregulation of the AM-induced HvPht1;8 (encoding the Pi-uptake system responsible for transfer of Pi from the symbiotic interface to cortical cells) play a role in decreased As uptake and hence reduced As sensitivity in AM plants	Christophersen et al. (2009)
Mycorrhiza (*Glomus mosseae* and *Glomus intraradices*)	*Pteris vittata* (As hyperaccumulator) *Cynodon dactylon* (non-hyperaccumulator)	Enhanced plant biomass, P and N contents	Leung et al. (2013)

(continued)

TABLE 16.1 (Continued))

Name	Targeted Plant	Processes	Reference
Fungi			
Mycorrhiza (*Rhizophagus intraradices*)	Soybean (*Glycine max*)	Mycorrhizal inoculation improves plant biomass, height, and number of leaves, and limits the negative effect on root length The mycorrhizal inoculation decreased the As accumulation in the plant from 7.8 to 6.0 mg As kg^{-1}	Spagnoletti and Lavado (2015)
Mycorrhiza (*Rhizoglomus intraradices* and *Glomus etunicatum*)	*Triticum aestivum*	Enhanced plant growth Maintenance of P reduction of As translocation to grains Reduction of oxidative stress Enhancement of antioxidant molecules increasing cysteine, glutathione, non-protein thiols and glutathione-*S*-transferase	Sharma et al. (2017)
Mycorrhiza (*Rhizophagus Irregularis*)	*Medicago sativa*	Mycorrhizal inoculation improved plant P nutrition, restricted As uptake and retained more As in roots by upregulating the expression of the AM-induced P transporter gene MsPT4 and the metallothionein gene MsMT2	Li et al. (2018)
Aspergillus flavus (rhizospheric fungus)	Rice (*Oryza sativa*)	The fungus has the ability to bio-transform soluble arsenic into immobilized arsenic particles. Arsenic particles have reduced mobilization to plant and does not negatively impact beneficial plant-microbe interaction	Mohd et al. (2019)
Humicola sp.	*Bacopa monnieri*	Enhanced plant growth and bacoside-An under As stress in *B. monnieri*	Tripathi et al. (2020)
Mycorrhiza (*Rhizoglomus intraradices*)	black locust (*Robinia pseudoacacia*)	Improving plant growth, altering root morphology, regulating the concentrations and ratios of phytohormones, and increasing the concentration of soil glomalin	Zhang et al. (2020)
Mycorrhiza (*Rhizophagus intraradices*) amended with biochar, selenium, silica gel, and sulfur	Pea (*Pisum sativum*)	As in grains reduced by 77%, 71%, and 69% under AMF, Se, and Si-gel amendments, respectively	Alam et al. (2020)
Mycorrhiza (*Rhizophagus* and *Funelliformis* species)	*Pteris vittata* (As hyperaccumulator)	Enhanced plant growth and As accumulation. The authors concluded that the phytoremediation of the metallurgic site would require at least four years to reduce the concentration of As to 50 mg/kg	Cantamessa et al. (2020)
Piriformospora indica (endophytic fungus)	*Artemisia annua*	*P. indica* accumulated and restricted arsenic in the roots and enhanced plant antioxidant defense system and enzymatic activities (superoxide dismutase and peroxidase) Increasing levels of artemisinin biosynthesis genes, isoprenoids, terpenes, flavonoids biosynthetic pathway genes and signal molecules	Rahman et al. (2020)
Piriformospora indica (endophytic fungus)	Rice (*Oryza sativa*)	Improving plant tolerance immobilized As in the roots and reduced damage to photosynthetic organs. Regulating gene expression	Ghorbani et al. (2021)
Algae			
Chlorella vulgaris and *Nannochlropsis* sp.	Rice (*Oryza sativa*)	Increased plant growth, reduced As uptake and modulating antioxidant enzymes	Upadhyay et al. (2016)
Chlorella vulgaris and bacterium (*Pseudomonas putida*)	Rice (*Oryza sativa*)	Enhanced growth and imparts As and high P tolerance in rice. Reduction of As uptake, modulation of P trafficking and mineral nutrient uptake	Srivastava et al. (2018)
Anabaena sp.	Rice (*Oryza sativa*)	Effective reduction of As(III) accumulation than As(V) in rice Modulation of gene expression and antioxidant enzymes	Ranjan et al. (2018)
Two green microalgae and five cyanobacteria	Rice (*Oryza sativa*)	Increase soil pH, soil As speciation, and soil nutrient bioavailability. Enhanced rice growth, and nutrient uptake. Reduction of As and DMA accumulation. Reduction of As translocation to grains	Wang et al. (2018)
Chlorella sp.	Rice (*Oryza sativa*)	Reduction of As(III) and As(V) accumulation. Enhanced seedling growth. Enhanced S-adenosylmethionine dependent methyltransferases (SAMT) gene expression modulating glutathione mediated detoxification	Ranjan et al. (2021)

Pseudoxanthomonas sp. P4V6) showed to increase *P. vittata* biomass by up to 45% and promoted As removal efficiency from 13% without bacteria to 35% in the presence of the bacterial consortium (Lampis et al. 2015). Likewise, Wang et al. (2012) investigated the role of ASB from the rhizosphere of *P. vittata* and found a bacterium capable of As(III) oxidation (*Acinetobacter* sp. 33) while four bacterial strains belonging to the genera *Comamonas*, *Flavobacterium*, *Pseudomonas*, and *Staphylococcus* showed the ability for both As(V) reduction and As(III) oxidation.

The use of As-hyperaccumulator *Populus deltoides* inoculated by the ARB (*Agrobacterium radiobacter*) in the phytobial remediation of highly As-contaminated soil have been reported by Wang et al. (2011). *A. radiobacter* contributed to the increase in the As tolerance of *P. deltoides*, promoted its growth, increased As uptake efficiency and enhanced As translocation from roots to shoots. Mesa et al. (2017) investigated the potential use of rhizospheric and endophytic ARB in enhancing As phytoremediation by the autochthonous pseudometallophyte *Betula celtiberica* trees. Among the isolated bacteria, it was found that siderophore and IAA producing endophytic strains (*Variovorax paradoxus, Phyllobacterium myrsinacearum*) promoted As accumulation in the roots and shoots of *B. celtiberica*, while the rhizospheric strain *Ensifer adhaerens* enhanced the plant growth. However, in a recent study, the use of plant growth promoting bacteria (PGPB) (two strains of *Stenotrophomonas maltophilia* sp. and one of *Agrobacterium* sp.) did not improve plant biomass or phytoremediation efficacy of *Arundo donax* plants but the accumulation factor was doubled by increasing the As concentration in the plant roots (Guarino et al. 2020).

The efficiency of phytoremediation may be increased by reducing As phytotoxicity and stimulating the activity of indigenous soil microbiota through bioaugmenting of soil with ARB. Soil bioaugmentation using arsenite-oxidizing bacteria (*Ensifer* sp. M14) showed to promote alfalfa (*Medicago sativa*) biomass by about 60% compared to non-bioaugmented soil, while at the same time the As accumulation was doubled (Debiec-Andrzejewska et al. 2020). The soil bioaugmentation also induced an increase in the quantity of heterotrophic, denitrifying, nitrifying and cellulolytic bacteria as well as the activity of dehydrogenases and cellulases with no obvious disturbing of the natural microbial community structure. According to a recent study inoculation of *Cucurbita moschata* with *Bacillus subtilis* S4 in the presence of iron-nanoparticles have a synergistic ameliorative effect for As-stressed seedlings (Mushtaq et al. 2020). The combined bacterial and nanoparticle effects included an improvement in plant growth, proline contents and antioxidant enzymes' activities. Additionally, the biosynthesis of stress mitigating polyamines including spermidine and putrescine in As-stressed seedlings was promoted under bacterial and nanoparticle application.

An alternative strategy to increase the tolerance of plants against As phytotoxicity and to enhance the bioremediation efficiency is to develop genetically engineered plants. Li et al. (2005) engineered *Arabidopsis thaliana* to express the bacterial γ-glutamylcysteine synthetase (γ-*ECS*) gene. γ-ECS gene catalyzes the formation of the dipeptidethiol γ-glutamine-cysteine (γ-EC), the first step in the biosynthesis of phytochelatins (PCs). The levels of γ-EC and its derivatives, glutathione (GSH) and PCs, were increased in the transgenic plants to 3- to 20-fold higher than the wild plant under As, Cd and Hg stress. *A. thaliana* plants transformed with the arsenate reductase (*arsC*) gene and γ-glutamylcysteine synthetase (γ-*ECS*) gene also showed a promising potential for As accumulation (4- to 17-fold greater fresh shoot weight and 2- to 3-fold more arsenic per gram of tissue than wild plants) (Dhankher et al. 2002).

16.6.2 Role of Fungi

Several higher plants, which can grow in As-contaminated soils, are generally associated with arbuscular mycorrhizal fungi (AMF). Generally, more than 80% of terrestrial higher plants can establish symbiotic relationships with AMF. AMF play a crucial role as a "bridge" between plants and the rhizosphere, thus can help in the transfer of mineral nutrients (e.g., potassium, phosphorus, and nitrogen) from the rhizosphere soil to the host plant by their arbuscules and hyphal coils, and in return acquire carbohydrate compounds and lipids from their host (Cantamessa et al. 2020; Li et al. 2016). As stress can induce negative effects on mycorrhizal root colonization and hyphal density, however inoculated plants exhibited better growth and high As tolerance in comparison to non-inoculated plants (Li et al. 2018). *Hymenoscyphus ericae* (ericoid mycorrhizal fungus) showed to act as a filter to maintain low As levels in *Calluna vulgaris* through arsenite efflux while enhancing plant phosphorus uptake (Sharples et al. 2000). AMF symbioses ameliorate As toxicity in host plants due to lower As/P ratios in tissues of mycorrhizal plants, compared with those of non-mycorrhizal plants (Zhang et al. 2020). Spagnoletti and Lavado (2015) studied the response of mycorrhizal (*Rhizophagus intraradices*) soybean plants under As toxicity and found an enhancement of plant growth and biomass along with decreased As accumulation. The inoculation of six rice varieties with AMF (*R. intraradices*) reduced the inorganic/organic As ratio in grains, which was attributed to the transformation of inorganic As into a less toxic organic form (dimethylarsenic acid, DMA) (Li et al. 2016). Similarly, pea plants inoculated by AMF (*R. intraradices*) showed a remarkable reduction in As content in the seeds (Alam et al. 2020). Inoculation of Triticum aestivum with AMF (*Rhizoglomus intraradices* and *Glomus etunicatum*) indicated that AMF play a vital role in strengthening the antioxidant defense by increasing carotenoids, proline, and α-tocopherol concentrations along with increasing cysteine, glutathione, non-protein thiols, and activity of glutathione-*S*-transferase that facilitated sequestration of As into non-toxic complexes. In addition, the AMF inoculation showed to increase the activities of glutathione-ascorbate cycle enzymes and also augmented the glyoxalase system by increasing the activities of both glyoxalase I and glyoxalase II enzymes. In a recent study, Zhang et al. (2020) observed that *R. intraradices* inoculated *Robinia pseudoacacia* seedlings have improved root morphology and biomass under As stress in relation to the non-inoculated seedlings. Additionally, *R. intraradices* inoculation increased the concentrations of IAA and abscisic acid but decreased the concentrations of gibberellic acid and zeatin riboside.

Previous studies also verified that AMF could regulate the expression of some genes in host plants under As stress. In a study conducted by Pathare et al. (2016), AMF (*Rhizophagus irregularis*) induced upregulation of 14-3-3 protein genes, reflecting its involvement in ameliorating As toxicity in rice plants. The 14-3-3 proteins are phosphoserine-binding proteins that can regulate several plant activities through direct protein–protein interactions and can play crucial roles in regulating plant growth and its responses to different stresses. In another study, AMF (*Glomus intraradices*) inoculation to *Hordeum vulgare* downregulated HvPht1;1 and HvPht1;2 genes (encoding high-affinity inorganic orthophosphate (Pi)-uptake systems in a direct pathway via root epidermis and root hairs) and upregulated mycorrhizal-induced HvPht1;8 (encoding the Pi-uptake system responsible for transfer of Pi from the symbiotic interface to cortical cells), resulting in reduced As uptake and hence reduced As sensitivity in AMF plants (Christophersen et al. 2009). Induced expression of GiPT (High-affinity Pi/As transporter) gene in *G. intraradices* correlates with As uptake in AMF external hyphae (González-Chávez et al. 2011). Mycorrhizal inoculation of *Medicago sativa* improved plant phosphorus nutrition, restricted As uptake and retained more As in roots by upregulating the expression of the mycorrhizal-induced P transporter gene (MsPT4) and the metallothionein gene (MsMT2) (Li et al. 2018).

The phytobial remediation using As hyperaccumulating plants colonized by AMF is an emerging effective strategy, however it may take long period depending on the extent of pollution. In a recent field study, the phytobial remediation of a highly As polluted site using *Pteris vittata* (As hyperaccumulator fern) inoculated with AMF (*Rhizophagus* and *Funelliformis* species) take about four years (Cantamessa et al. 2020). Moreover, AMF colonization of *Pteris vittata* enhanced plant growth, As accumulation and cellular phosphorus and nitrogen contents (Cantamessa et al. 2020; Leung et al. 2013).

Rhizosphere fungi may also play a vital role in ameliorating As toxicity in plants. Mohd et al. (2019) observed that *Aspergillus flavus* in the rhizosphere of rice plants have the ability to transform soluble As into immobilized As particles, which cannot be readily absorbed by the plant cells as well as cannot negatively impact beneficial plant-microbe interactions. Recently, Tripathi et al. (2020) showed the ability of the soil fungus *Humicola* sp. to enhance the growth of *Bacopa monnieri* plant grown in an As contaminated soil. The fungus has the ability to bioaccumulate, bio-transform and bio-volatilize As from soil.

On the other hand, endophytic fungi are a versatile group of fungi that reside in the living plant without causing any disease symptoms and regarded as symbionts or mutualists. Endophytic fungi may increase the host defense system against various biotic and abiotic stresses. The endophytic fungus *Piriformospora indica* showed to accumulate and restrict As in the roots of *Artemisia annua* plant along with enhancing plant antioxidant defense system and enzymatic activities (superoxide dismutase and peroxidase) (Rahman et al. 2020). While at the same time, the levels of artemisinin biosynthesis genes, isoprenoids, terpenes, flavonoids biosynthetic pathway genes and signal molecules were enhanced. In a more recent study, *P. indica* exhibited the potential to immobilize As in the roots of rice plants, reduced damage to the photosynthetic organs and upregulated the genes responsible for Fe translocation to shoots (Ghorbani et al. 2021).

Genetic modification is a promising and practical approach to reduce As accumulation in rice grains. Verma et al. (2016) used fungal arsenic methyltransferase (*WaarsM*) gene of the soil fungus *Westerdykella aurantiaca* to develop transgenic rice. The expression improved arsenic resistance via volatilization and the plants were able to methylate As with a concomitant reduction of As concentration in the rice grains. Additionally, the yeast *Saccharomyces cerevisiae* cells expressing *WaarsM* showed an increased As methylation and volatilization potential (Verma et al. 2016).

16.6.3 Role of Algae

The use of algae or algal biomass in the removal of toxic pollutants is known as phycoremediation. The phycoremediation of heavy metals from a contaminated environment is efficient, ecococious, and cost-effective tool. Although As is toxic to several algae, a diverse group of prokaryotic and eukaryotic algae can accumulate As from a contaminated environment. Singh et al. (2016) investigated the potential role of soil microalgae in the bioaccumulation of As from soil and reported that diatoms can accumulate 760 mg As g^{-1} soil, followed by Cyanobacteria (*Oscillatoria* sp., 394 mg g^{-1}, *Pithophora*, 229 mg g^{-1} and *Phormidium* 144 mg g^{-1}). The pioneer steps for the bioaccumulation require biosorption of As on the surface of microalgae, followed by penetration into the cell. The presence of versatile functional groups such as carboxyl, hydroxyl, carbonyl, amine, and phosphate diester on the algal cell wall and plasma membrane help in the adsorption of As.

The inoculation of two green microalgae, *Chlorella vulgaris* and *Nannochloropsis* sp. was investigated under As toxicity in the hydroponic cultures of rice plants by Upadhyay et al. (2016). The phytotoxicity of As on rice seedling growth was ameliorated by algal inoculation with decreasing the antioxidant enzyme activities while at the same time the root, shoot length, biomass and protein were markedly enhanced. The existence of microalgae plays an important role for not only promoting soil fertility (e.g., nitrogen fixation or enhancing the mineralization of organic P compounds through the formation of organic acids), but also by modifying soil properties. Therefore soil microalgae can alter soil pH, soil As speciation, and soil nutrient availability which directly affects plant health (Wang et al. 2018).

The uptake and accumulation of As by plants can cause the transmission and biomagnification of As in the human food chain. Regarding this context, soil microalgae can reduce As translocation from roots to grains of rice grown in As-polluted paddy soil (Wang et al. 2018). Additionally, microalgae can affect plant growth and ameliorate As toxicity by modulating plant gene expression. The gene expression of lower silicon transporters (Lsi-1 and Lsi-2) in rice plant was found to be decreased in the presence of *Anabaena* sp. under As treatment, while S-adenosylmethionine dependent methyl transferase (SAMT) and nitrate reductase (NR) genes were enhanced simultaneously (Ranjan et al. 2018). The Lsi-1 and Lsi-2 transporters participate in the accumulation and transportation of

As in the plant cells, and their reduction in the presence of microalgae was attributed to the As bioaccumulation in the algal cells (Ranjan et al. 2018). S-adenosylmethionine is a well-known methyl donor, that is used by methyl transferases for the regulation of different biosynthetic pathways in the plant cells and its enhancement ensures better plant growth and development. In a recent study, the gene expression of four thiol dependent glutaredoxin (GRX) genes (GRX13950, GRX35340, GRX12190 and GRX07950) and two isomeric glutathione S-transferase (GST) genes (GST 38600 and GST 38610) in As treated rice plants were reduced upon co-culturing with *Chlorella* sp., which was related to a reduced oxidative stress (Ranjan et al. 2021).

In another study, Srivastava et al. (2018) studied the synchronized interaction of bacteria (*Pseudomonas putida*) and green algae (*Chlorella vulgaris*) in the uptake of As by rice plants. The authors found that the two microorganisms reduced As uptake by modulating phosphate trafficking. Therefore, most of the genes responsible for phosphate transport were upregulated in the presence of microorganisms. The balancing of reactive oxygen species (ROS), proline accumulation, hormone modulation, and As sequestration by bacterial and algal cells were elucidated as possible mechanisms of As detoxification.

Useful algal genes are also important and practical for the development of gene modified plants. Tang et al. (2016) genetically engineered two ecotypes of *Arabidopsis thaliana* with the AsIII S-adenosylmethyltransferase (arsM) gene from the chlorophyte *Chlamydomonas reinhardtii*. The transgenic *A. thaliana* plants showed a strong potential to biomethylate inorganic As into dimethylarsenate in the shoots with a simultaneous volatilization of small amounts of volatile As.

REFERENCES

Abbas, G., Murtaza, B., Bibi, I., Shahid, M., Niazi, N. K., Khan, M. I., Amjad, M., Hussain, M. 2018. Arsenic uptake, toxicity, detoxification, and speciation in plants: physiological, biochemical, and molecular aspects. *Int J Environ. Res Public Health* 15: 59.

Abdel Latef, A. A. H., Dawood, M. F. A., Hassanpour, H., Rezayian, M., Younes, N. A. 2020. Impact of the static magnetic field on growth, pigments, osmolytes, nitric oxide, hydrogen sulfide, phenylalanine ammonia-lyase activity, antioxidant defense system, and yield in lettuce. *Biology* 9(7): 172.

Abedi T., Mojiri A. 2020. Arsenic uptake and accumulation mechanisms in rice species. *Plants* 9: 129.

Abedin, M. J., Feldmann, J., Meharg, A. A. 2002. Uptake kinetics of arsenic species in rice plants. *Plant Physiol* 128: 1120–1128.

Abercrombie, J., Halfhill, M., Ranjan, P., Rao, M., Saxton, A., Yuan, J., Stewart, C. N. 2008. Transcriptional responses of *Arabidopsis thaliana* plants to As (V) stress. *BMC Plant Biol* 8: 87.

Ahmad, P., Alam, P., Balawi, T. H., Atalayan, F., Ahanger, A. M., Ashraf, M. 2020a. Sodium nitroprusside (SNP) improves tolerance to arsenic (As) toxicity in *Vicia faba* through the modifications of biochemical attributes, antioxidants, ascorbate-glutathione cycle and glyoxalase cycle. *Chemosphere* 244: 125480.

Ahmad, P., Alyemeni, M. N., Al-Huquail, A. A., Alqahtahi, M. A., Wijaya, L., Asharaf, M., Kaya, C., Bajguz, A. 2020b. Zinc oxide nanoparticle application alleviates arsenite toxicity in Soybean plants by restricting the uptake of As and modulating key biochemical attributes, antioxidant enzymes ascorbate-glutathione cycle and glyoxalase system. *Plants*9: 825.

Ahsan, N., Lee, D. G., Alam, I., Kim, P. J., Lee, J. J., Ahn, Y. O., Kwak, S. S., Lee, I. J., Bahk, J. D., Kang, K. Y., Renaut, J., Komatsu, S., Lee, B. H. 2008. Comparative proteomic study of arsenic-induced differentially expressed proteins in rice roots reveals glutathione plays a central role during As stress. *Proteomics* 8: 3561–3576.

Alam, M. Z., Hoque, M. A., Ahammed, G. J., Carpenter-Boggs, L. 2020. Effects of arbuscular mycorrhizal fungi, biochar, selenium, silica gel, and sulfur on arsenic uptake and biomass growth in *Pisum sativum* L. *Emerg Contam* 6: 312–322.

Alka, S., Shahir, S., Ibrahim, N., Chai, T. T., Mohd Bahari, Z., Abd Manan, F., 2020. The role of plant growth promoting bacteria on arsenic removal: A review of existing perspectives. *Environ Technol Innov* 17: 100602.

Alsahli, A. A., Bhat, J. A., Alyemeni, M. N., Ashraf, M., Ahmad, P. 2020. Hydrogen sulfide (H_2S) mitigates arsenic (As)-induced toxicity in pea (*Pisum sativum* L.) plants by regulating osmoregulation, antioxidant defense system, ascorbate glutathione cycle and glyoxalase system. *J. Plant Growth Reg.*2020:1–20.

Amaral, C. D. B., Nobrega, J. A., Nogueira, A. R. A. 2014. Investigation of arsenic species stability by HPLC-ICP-MS in plants stored under different conditions for 12 months. *Microchem J* 117: 122–126.

Andrade, H. M., Oliveira, J. A., Farnese, F. S., Ribeiro, C., Silva, A. A., Campos, F. V., Neto, J. L. 2016. Arsenic toxicity: cell signalling and the attenuating effect of nitric oxide in *Eichhornia crassipes*. *Biologia Plant* 60: 173–180.

Anjum, S. A., Tanveer, M., Hussain, S., Shahzad, B., Ashraf, U., Fahad, S., Hassan, W., Jan, S., Khan, I., Saleem, M. F., Bajwa, A. A., Wang, L., Mahmood, A., Samad, R.A., Tung, S. A. 2016. Osmoregulation and antioxidant production in maize under combined cadmium and arsenic stress. *Environ Sci Pollut Res* 23: 11864–11875.

Anjum, S. A., Xie, X.-y., Wang, L.-c., Saleem, M. F., Man, C., Lei, W. 2011. Morphological, physiological and biochemical responses of plants to drought stress. *Afr J Agric Res* 6: 2026–2032.

Anwar, H. M., Rengel, Z., Damon, P., Tibbett, M. 2018. Arsenic-phosphorus interactions in the soil-plant-microbe system: dynamics of uptake, suppression and toxicity to plants. *Environ Pollut* 233: 1003–1012.

Armendariz, A. L., Talano, M. A., Travaglia, C., Reinoso, H., Wevar Oller, A. L., Agostini, E. 2016. Arsenic toxicity in soybean seedlings and their attenuation mechanisms. *Plant Physiol Biochem* 98: 119–127.

Bakhat, H. F., Zia, Z., Abbas, S., Hammad, H. M., Shah, G. M., Khalid, S., Shahid, N., Sajjad, M., Fahad, S. 2019. Factors controlling arsenic contamination and potential remediation measures in soil-plant systems. *Groundw Sustain Dev* 9: 100263.

Bashandy, S. R., Abd-Alla, M. H., Dawood, M. F. A. 2020. Alleviation of the toxicity of oily wastewater to canola plants by the N2-fixing, aromatic hydrocarbon biodegrading bacterium *Stenotrophomonas maltophilia* -SR1. *Appl Soil Ecol* 154: 103654.

Basu, A., Saha, D., Saha, R., Ghosh, T., Saha, B. A. 2014. Review on sources, toxicity and remediation technologies for removing arsenic from drinking water. *Res Chem Intermed* 40 (2): 447–485.

Bhattacharyya, R., Chatterjee, D., Nath, B., Jana, J., Jacks, G., Vahter, M. 2003. High arsenic groundwater: Mobilization, metabolism and mitigation – an overview in the Bengal Delta Plain. *Mol Cell Biochem* 253: 347–355.

Bienert, G. P., Thorsen, M., Schüssler, M. D., Nilsson, H. R., Wagner, A., Tamás, M. J., Jahn, T. P. 2008. A subgroup of plant aquaporins facilitate the bi-directional diffusion of As(OH)3 and Sb(OH)3 across membranes. *BMC Biol* 6: 26.

Bleeker, P. M., Hakvoort, H. W. J., Bliek, M., Souer, E., Schat, H. 2006. Enhanced arsenate reduction by a CDC25-like tyrosine phosphatase explains increased phytochelatin accumulation in arsenate- tolerant Holcuslanatus. *Plant J* 45: 917–929.

Boye, K., Lezama-Pacheco, J., Fendorf, S. 2017. Relevance of Reactive Fe:S Ratios for Sulfur Impacts on Arsenic Uptake by Rice. *Soils* 1: 1.

Brinke, A., Reifferscheid, G., Klein, R., Feiler, U., Buchinger, S. 2018. Transcriptional changes measured in rice roots after exposure to arsenite-contaminated sediments. *Environ Sci Pollut Res* 25: 2707–2717.

Calvo, C., Bolado, S., Alvarez-Benedí, J., Andrade, M. A. 2006. Arsenic uptake and accumulation in curly endives (*Cichorium endivia* L.) irrigated with contaminated water. *J Environ Sci Health Part B* 41: 459–470.

Cantamessa, S., Massa, N., Gamalero, E., Berta, G. 2020. Phytoremediation of a highly arsenic polluted site, using Pteris vittata L. And arbuscular mycorrhizal fungi. *Plants* 9: 1–13.

Cao, X., Ma, L. Q., Tub, C. 2004. Antioxidative responses to arsenic in the arsenic-hyperaccumulator Chinese brake fern (*Pteris vittata* L.). *Environ Pollut* 128: 317–325.

Cao, Y., Sun, D., Mei, H., Liu, X., Sun, S., Xu, G., Liu, Y., Chen, Y. 2017. Knocking out OsPT4 gene decreases arsenate uptake by rice plants and inorganic arsenic accumulation in rice grains. *Environ Sci Technol* 51: 12131–12138.

Catarecha, P., Segura, M. D., Franco-Zorrilla, J. M., García-Ponce, B., Lanza, M., Solano, R., Paz-Ares, J., Leyva, A. 2007. A mutant of the Arabidopsis phosphate transporter PHT1,1 displays enhanced arsenic accumulation. *Plant Cell* 19: 1123–1133.

Chakraborty, S., Alam, M. O., Bhattacharya, T., Singh, Y. N. 2014. Arsenic accumulation in food crops: a potential threat in Bengal Delta Plain. *Water Qual Exposure Health* 6: 233–246.

Chandrakar, V., Dubey, A., Keshavkant, S. 2016. Modulation of antioxidant enzymes by salicylic acid in arsenic exposed Glycine max L. *J Soil Sci Plant Nut* 16: 662–676.

Chandrakar, V., Dubey, A., Keshavkant, S. 2018. Modulation of arsenic-induced oxidative stress and protein metabolism by diphenyleneiodonium, 24-epibrassinolide and proline in Glycine max L. *Acta Bot Croat* 77: 51–61.

Chang, Z., Jin, M., Yan, W., Chen, H., Qiu, S., Fu, S., Xia, J., Liu, Y., Chen, Z., Wu, J., Tang, X.2018. The ATP-binding cassette (ABC) transporter OsABCG3 is essential for pollen development in rice. *Rice* 11: 58.

Chen, Y., Wu, F., Liu, M., Parvez, F., Slavkovich, V., Eunus, M., Ahmed, A., Argos, M., Islam, T., Rakibuz-Zaman, M., Hasan, R., Sarwar, G., Levy, D., Graziano, J., Ahsan, H. 2013. A prospective study of arsenic exposure, arsenic methylation capacity, and risk of cardiovascular disease in Bangladesh. *Environ Health Perspect* 121 (7): 832–838.

Christophersen, H. M., Smith, F. A., Smith, S. E. 2009. Arbuscular mycorrhizal colonization reduces arsenate uptake in barley via downregulation of transporters in the direct epidermal phosphate uptake pathway. *New Phytol* 184: 962–974.

Clemens, S. 2006. Toxic metal accumulation, responses to exposure and mechanisms of tolerance in plants. *Biochimie* 88: 1707–1719.

Cozzolino, V., Pigna, M, Di Meo, V., Caporale, A. G., Violante, A. 2010. Effects of arbuscular mycorrhizal inoculation and phosphorus supply on the growth of *Lactuca sativa* L. and arsenic and phosphorus availability in an arsenic polluted soil under nonsterile conditions. *Appl Soil Ecol* 45: 262–268.

Curie, C., Alonso, J. M., Le Jean, M., Ecker, J. R., Briat, J. F. (2000) Involvement of NRAMP1 from *Arabidopsis thaliana* in iron transport. *Biochem J* 347: 749–755.

Da-Silva, C. J., Canatto, R. A., Cardoso, A. A., Ribeiro, C., de Oliveira, J. A. 2018. Oxidative stress triggered by arsenic in a tropical macrophyte is alleviated by endogenous and exogenous nitric oxide. *Braz J Bot* 41: 21–28.

Dawood, M. F. A., Azooz, M. M. 2019. Concentration-dependent effects of tungstate on germination, growth, lignification-related enzymes, antioxidants, and reactive oxygen species in broccoli (*Brassica oleracea* var. *italica* L.). *Environ Sci Pollut Res* 26: 36441–36457.

Dawood, M. F. A., Azooz, M. M. 2020. Insights into the oxidative status and antioxidative responses of germinating broccoli (*Brassica oleracea* var. *italica* L.) seeds in tungstate contaminated water. *Chemosphere* 261: 127585.

Dawood M. F.A., Sohag A. A., Tahjib-Ul-Arif, M., Abdel Latef, A. A. 2021a. Hydrogen sulfide priming can enhance the tolerance of artichoke seedlings to individual and combined saline-alkaline and aniline stresses. *Plant Physiol Biochem* 159: 347–362.

Dawood, M. F. A., Tahjib-Ul-Arif, M., Sohag, A. A. M., Abdel Latef, A. A. H., Ragaey, M. M. 2021b. Mechanistic Insight of Allantoin in Protecting Tomato Plants Against Ultraviolet C Stress. *Plants* 10: 11.

Debiec-Andrzejewska, K., Krucon, T., Piatkowska, K., Drewniak, L. 2020. Enhancing the plants growth and arsenic uptake from soil using arsenite-oxidizing bacteria. *Environ Pollut* 264: 114692.

Dhankher, O. P., Li, Y., Rosen, B. P., Shi, J., Salt, D., Senecoff, J. F., Sashti, N. A., Meagher, R. B. 2002. Engineering tolerance and hyperaccumulation of arsenic in plants by combining arsenate reductase and γ-glutamylcysteine synthetase expression. *Nat Biotechnol* 20: 1140–1145.

Dhankher, O. P., Rosen, B. P., McKinney, E. C., Meagher, R. B. 2006. Hyperaccumulation of arsenic in the shoots of Arabidopsis silenced for arsenate reductase (ACR2). *PNAS USA* 103: 5413–5418.

Dixit, G., Singh, A.P., Kumar, A., Mishra, S., Dwivedi, S., Kumar, S., Tripathi, R.D. 2016. Reduced arsenic accumulation in rice (*Oryza sativa* L.) shoot involves sulfur mediated improved thiol metabolism, antioxidant system and altered arsenic transporters. *Plant Physiol Biochem* 99: 86–96.

Dixit, G., Singh, A. P., Kumar, A., Singh, P. K., Kumar, S., Dwivedi, S., Tripathi, R. D. 2015. Sulfur mediated reduction of arsenic toxicity involves efficient thiol metabolism and the antioxidant defense system in rice. *J Hazard Mater* 298: 241–251.

Duan, G. L., Zhou, Y., Tong, Y. P., Mukhopadhyay, R., Rosen, B. P., Zhu, Y. G. 2007. A CDC25 homologue from rice functions as an arsenate reductase. *New Phytol* 174: 311–321.

Dwivedi, S. Mishra, A., Tripathi, P., Dave, R., Kumar, A., Srivastava, S., Chakrabarty, D., Trivedi, P. K., Adhikari, B., Norton, G. J., Tripathi, R. D., Nautiyal, C. S. 2012. Arsenic affects essential and non-essential amino acids differentially in rice grains: inadequacy of amino acids in rice based diet. *Environ Int* 46: 16–22.

El-Shora, H. M., El-Gawad, A. M. A. 2014. Environmental toxicity of arsenic on Lupine (*Lupinus termis* L.) as C3 crop plant and possible alleviation. *Inter J Agric Crop Sci* 7: 687.

Emamverdian, A., Ding, Y., Mokhberdoran, F., Xie, Y. 2015. Heavy metal stress and some mechanisms of plant defense response. *Sci World J* 2015:1:15.

Ernst, W. H., Krauss, G. J., Verkleij, J. A., Wesenberg, D. 2008. Interaction of heavy metals with the sulphur metabolism in angiosperms from an ecological point of view. *Plant Cell Environ* 31: 123–143.

Farnese, F. S., Oliveira, J. A., Lima, F. S., Leao, G. A., Gusman, G. S., Silva, L. C. 2014. Evaluation of the potential of *Pistia stratiotes* L. (water lettuce) for bioindication and phytoremediation of aquatic environments contaminated with arsenic. *Brazilian J Biol* 74: S103–S112.

Farooq, M. A., Gill, R. A., Islam, F., Ali, B., Liu, H., Xu, J., He, S., Zhou, W. 2016. Methyl jasmonate regulates antioxidant defense and suppresses arsenic uptake in *Brassica napus* L. *Front Plant Sci* 7: 468.

Farooq, M. A., Islam, F., Yang, C., Nawaz, A., Gill, R. A., Ali, B., Zhou, W. 2018. Methyl jasmonate alleviates arsenic-induced oxidative damage and modulates the ascorbate–glutathione cycle in oilseed rape roots. *Plant Growth Regul* 84: 135–148.

Finnegan, P. M., Chen, W. 2012. Arsenic toxicity: the effects on plant metabolism. *Front Physiol* 3: 1–18.

Finnegan, P., Weihua, C. 2012. Arsenic toxicity: the effects on plant metabolism. *Front Physiol* 3: 182.

Funes Pinter, I., Salomon, M. V., Berli, F., Bottini, R., Piccoli, P. 2017. Characterization of the As(III) tolerance conferred by plant growth promoting rhizobacteria to in vitro-grown grapevine. *Appl Soil Ecol* 109: 60–68.

Gasic, K., Korban, S. S. 2007. Transgenic Indian mustard (Brassica juncea) plants expressing an Arabidopsis phytochelatin synthase (AtPCS1) exhibit enhanced As and Cd tolerance. *Plant Mol Biol* 64: 361–369.

Ghiani, A., Fumagalli, P., Nguyen Van, T., Gentili, R., Citterio, S. 2014. The Combined Toxic 662 and Genotoxic Effects of Cd and As to Plant Bioindicator *Trifolium repens* L. *PLOS ONE* 9: e99239.

Ghorbani, A., Tafteh, M., Roudbari, N., Pishkar, L., Zhang, W., Wu, C., 2021. Piriformospora indica augments arsenic tolerance in rice (Oryza sativa) by immobilizing arsenic in roots and improving iron translocation to shoots. *Ecotoxicol Environ Saf* 209: 111793.

Ghosh, P., Rathinasabapathi, B., Ma, L. Q. 2011. Arsenic-resistant bacteria solubilized arsenic in the growth media and increased growth of arsenic hyperaccumulator *Pteris vittata* L. *Bioresour Technol* 102: 8756–8761.

Ghosh, S., Biswas, A. K. 2017. Selenium modulates growth and thiol metabolism in wheat (*Triticum aestivum* L.) during arsenic stress. *Am J Plant Sci* 8: 363.

Ghosh, S., Derle, A., Ahire, M., More, P., Jagtap, S., Phadatare, S. D., Patil, A. B., Jabgunde, A. M., Sharma, G. K., Shinde, V. S., Pardesi, K., Dhavale, D. D., Chopade, B. A. 2013. Phytochemical analysis and free radical scavenging activity of medicinal plants *Gnidia glauca* and *Dioscorea bulbifera*. *PLoS ONE* 8: e82529.

Ghosh, S., Shaw, A.K., Azahar, I., Adhikari, S., Jana, S., Roy, S., Hossain, Z. 2016. Arsenate (AsV) stress response in maize (*Zea mays* L.). *Environ Exp Bot* 130: 53–67.

González-Chávez, M. del C. A., Ortega-Larrocea, M. del P., Carrillo-González, R., López-Meyer, M., Xoconostle-Cázares, B., Gomez, S. K., Harrison, M. J., Figueroa-López, A. M., Maldonado-Mendoza, I. E. 2011. Arsenate induces the expression of fungal genes involved in As transport in arbuscular mycorrhiza. *Fungal Biol* 115: 1197–1209.

Guarino, F., Miranda, A., Castiglione, S., Cicatelli, A. 2020. Arsenic phytovolatilization and epigenetic modifications in *Arundo donax* L. assisted by a PGPR consortium. *Chemosphere* 251: 126310.

Guillod-Magnin, R., Brüschweiler, B. J., Aubert, R., Haldimann, M. 2018. Arsenic species in rice and rice-based products consumed by toddlers in Switzerland. *Food Addit Contam Part A* 35: 1164–1178.

Gupta, D. K., Tiwari, S., Razafindrabe, B. H. N., Chatterjee, S. 2017. Arsenic contamination from historical aspects till present situation. In: Gupta D. K., Chatterjee S. (Eds.), Arsenic Contamination in the Environment: The Issues and Solutions. Springer International Publishing AG, Cham, Switzerland, pp 1–12.

Gupta, M., Ahmad, M. A. 2014. Arsenate induced differential response in rice genotypes. *Ecotoxicol Environ Saf* 107: 46–54.

Hasanuzzaman, M., Alam, M. M., Nahar, K., Mohsin, S. M., Bhuyan, M. B., Parvin, K., Fujita, M. 2019. Silicon-induced antioxidant defense and methylglyoxal detoxification works coordinately in alleviating nickel toxicity in *Oryza sativa* L. *Ecotoxicol* 28: 261–276.

Hasanuzzaman, M. Nahar, K., Hossain, M., Mahmud, J. A., Rahman, A., Inafuku, M., Oku, H., Fujita, M. 2017. Coordinated actions of glyoxalase and antioxidant defense systems in conferring abiotic stress tolerance in plants. *Int J Mol Sci* 18: 200.

Hasanuzzaman, M., Fujita, M. 2013. Exogenous sodium nitroprusside alleviates arsenic–induced oxidative stress in wheat (*Triticum aestivum* L.) seedlings by enhancing antioxidant defense and glyoxalase system. *Ecotoxicol* 22: 584–596.

Herbert, K. J., Holloway, A., Cook, A. L., Chin, S. P., Snow, E. T. 2014. Arsenic exposure disrupts epigenetic regulation of SIRT1 in human keratinocytes. *Toxicol Appl Pharmacol* 281: 136–145.

Hettick, B. E., Cañas-Carrell, J. E., French, A. D., Klein, D. M. 2015. Arsenic: A Review of the Element's Toxicity, Plant Interactions, and Potential Methods of Remediation. *J Agric Food Chem* 63: 7097–7107.

Hossain, K., Quaik, S., Pant, G., Yada, S., Maruthi, Y. A., Rafatullah, M., Nasir, M., Ismail, N. 2015. Arsenic fate in the ground water and its effect on soil-crop systems. *Res J Environ Toxicol* 9: 231–240.

Hossain, M. A., Li, Z. G., Hoque, T. S., Burritt, D. J., Fujita, M., Munne-Bosch, S. 2018. Heat or cold priming-induced cross-tolerance to abiotic stresses in plants: key regulators and possible mechanisms. *Protoplasma* 255: 399–412.

Huang, H., Jia, Y., Sun, G.-X., Zhu, Y.-G. 2012. Arsenic speciation and volatilization from flooded paddy soils amended with different organic matters. *Environ Sci Technol* 46: 2163–2168.

Hüvely, A., Petö, J., Hoyk, E. 2014. The arsenic accumulating capacity of lettuce growing in aggregate hydroponics under the influence of arsenic polluted nutrient solution. *Acta Tech Corvininesis–Bull Eng* 7 (2): 89–92.

Ishimaru, Y., Takahashi, R., Bashir, K., Shimo, H., Senoura, T., Sugimoto, K. 2012. Characterizing the role of rice NRAMP5 in manganese, iron and cadmium transport. *Sci Rep* 2: 286.

Islam, M. N., Das, B. K., Huque, M. E. 2012. Arsenic accumulation in common vegetables from irrigation. *J Sci Res* 4: 675–688.

Islam, S., Rahman, M. M., Islam, M. R., Naidu, R. 2017. Geographical variation and age-related dietary exposure to arsenic in rice from Bangladesh. *Sci Total Environ* 601: 122–131.

Jia, Y., Sun, G. -X., Huang, H., Zhu, Y. -G. 2013. Biogas slurry application elevated arsenic accumulation in rice plant through increased arsenic release and methylation in paddy soil. *Plant Soil* 365: 387–396.

Jung, H. I., Kong, M. S., Lee, B. R., Kim, T. H., Cae, M. J., Lee, E. J., Jung, G. B., Lee, C. H., Sung, J. K., Kim, Y. H. 2019. Exogenous glutathione increases arsenic translocation into shoots and alleviates arsenic-induced oxidative stress by sustaining ascorbate–glutathione homeostasis in rice seedlings. *Front Plant Sci* 10: 1089.

Kaya, C., Ashraf, M., Alyemeni, M. N., Corpas, F. J., Ahmad, P. 2020. Salicylic acid-induced nitric oxide enhances arsenic toxicity tolerance in maize plants by up regulating the ascorbate-glutathione cycle and glyoxalase system. *J Hazard Mater* 399: 123020.

Khalid, S., Shahid, M., Niazi, N. K., Rafiq, M., Bakhat, H. F., Imran, M., Abbas, T., Bibi, I., Dumat, C. 2017. Arsenic behaviour in soil-plant system: biogeochemical reactions and chemical speciation influences. In: *Enhancing Cleanup of Environmental Pollutants*. Springer, New York, pp. 97–140.

Khan, I., Ahmad, A., Iqbal, M. 2009. Modulation of antioxidant defense system for arsenic detoxification in Indian mustard. *Ecotoxicol Environ. Safety* 72: 626–634.

Khan, E., Gupta, M. 2018. Arsenic–silicon priming of rice (*Oryza sativa* L.) seeds influence mineral nutrient uptake and biochemical responses through modulation of Lsi-1, Lsi-2, Lsi-6 and nutrient transporter genes. *Sci Rep* 8: 10301.

Kobya, M., Soltani, R. D. C., Omwene, P. I., Khataee, A. 2020. A review on decontamination of arsenic-contained water by electrocoagulation: reactor configurations and operating cost along with removal mechanisms. *Environ Technol Innov* 17: 100519.

Kohli, S. K., Khanna, K., Bhardwaj, R., Allah, E. F. A., Ahmad, P., Corpas, F. J. 2019. Assessment of subcellular ROS and NO metabolism in higher plants: multifunctional signaling molecules. *Antioxidants* 8: 641.

Kumar, A., Dwivedi, S., Singh, R. P., Chakrabarty, D., Mallick, S., Trivedi, P. K., Adhikari, B., Tripathi, R. D. 2014. Evaluation of amino acid profile in contrasting arsenic accumulating rice genotypes under arsenic stress. *Biol Plant* 58: 733–742.

Kumarathilaka, P., Seneweera, S., Meharg, A., Bundschuh, J. 2018. Arsenic accumulation in rice (*Oryza sativa* L.) is influenced by environment and genetic factors. *Sci. Total Environ.* 642: 485–496.

Kumari, A., Pandey, N., Pandey-Rai, S. 2017. Protection of *Artemisia annua* roots and leaves against oxidative stress induced by arsenic. *Biol Planta* 61: 367–377.

Lampis, S., Santi, C., Ciurli, A., Andreolli, M., Vallini, G. 2015. Promotion of arsenic phytoextraction efficiency in the fern Pteris vittata by the inoculation of As-resistant bacteria: a soil bioremediation perspective. *Front Plant Sci* 6: 80.

Leão, G.A., de Oliveira, J. A., Felipe, R. T. A., Farnese, F. S. 2016. Phytoremediation of arsenic-contaminated water: the role of antioxidant metabolism of *Azolla caroliniana* Willd. (Salviniales). *Acta Botanica Brasilica* 31(2): 161–168.

Lee, C. H., Hsieh, Y. C., Lin, T. H., Lee, D. Y. 2013. Iron plaque formation and its effect on arsenic uptake by different genotypes of paddy rice. *Plant Soil* 363: 231–241.

Leterrier, M., Airaki, M., Palma, J. M., Chaki, M., Barroso, J. B., Corpas, F. J. 2012. Arsenic triggers the nitric oxide (NO) and S-nitrosoglutathione (GSNO) metabolism in Arabidopsis. *Environ Pollut* 166: 136–143.

Leung, H. M., Leung, A. O. W., Ye, Z. H., Cheung, K. C., Yung, K. K. L. 2013. Mixed arbuscular mycorrhizal (AM) fungal application to improve growth and arsenic accumulation of Pteris vittata (As hyperaccumulator) grown in As-contaminated soil. *Chemosphere* 92: 1367–1374.

Li, H., Chen, X.W., Wong, M.H., 2016. Arbuscular mycorrhizal fungi reduced the ratios of inorganic/organic arsenic in rice grains. *Chemosphere* 145, 224–230.

Li, J., Sun, Y., Jiang, X., Chen, B., Zhang, X., 2018. Arbuscular mycorrhizal fungi alleviate arsenic toxicity to Medicago sativa by influencing arsenic speciation and partitioning. *Ecotoxicol Environ Saf* 157, 235–243.

Li, Y., Dhankher, O. P., Carreira, L., Balish, R. S., Meagher, R. B., 2005. Arsenic and mercury tolerance and cadmium sensitivity in Arabidopsis plants expressing bacterial γ-glutamylcysteine synthetase. *Environ. Toxicol. Chem.* 24, 1376.

Li, Z.G., Duan, X. Q., X, Min, Zhou, Z. H. 2017. Methylglyoxal as a novel signal molecule induces the salt tolerance of wheat by regulating the glyoxalase system, the antioxidant system, and osmolytes. *Protoplasma* 254: 1995–2006.

Liu, C. W., Chen, Y. Y., Kao, Y. H., Maji, S. K. 2014. Bioaccumulation and translocation of arsenic in the ecosystem of the Guandu Wetland, Taiwan. *Wetlands* 34: 129–140.

Liu, W. L., Zhu, Y. G., Smith, F. A. 2005. Effects of iron and manganese plaques on arsenic uptake by rice seedlings (*Oryza sativa* L.) grown in solution culture supplied with arsenate and arsenite. *Plant Soil* 277: 127–138.

Liu, X., Yang, G. M., Guan, D. X., Ghosh, P., Ma, L. Q. 2015. Catecholate-siderophore produced by As-resistant bacterium effectively dissolved FeAsO4 and promoted Pteris vittata growth. *Environ Pollu* 206: 376–381.

Luan, M., Liu, J., Liu, Y., Han, X., Sun, G., Lan, W., Luan, S. 2018. Vacuolar phosphate transporter 1 (vpt1) affects arsenate tolerance by regulating phosphate homeostasis in Arabidopsis. *Plant Cell Physiol* 59: 1345–1352.

Ma, R., Shen, J. L., Wu, J. S., Tang, Z., Shen, Q. R., Zhao, F. J. 2014. Impact of agronomic practices on arsenic accumulation and speciation in rice grain. *Environ Pollut* 194: 217–223.

Ma, J. F., Yamaji, N., Mitani, N., Xu, X. Y., Su, Y. H., McGrath, S. P., Zhao, F. J. 2008. Transporters of arsenite in rice and their role in arsenic accumulation in rice grain. *Proc Natl Acad Sci USA* 105: 9931–9935.

Mabrouk, B., Kaab, S. B., Rezgui, M., Majdoub, N., da Silva, J. T., Kaab, L. B. B. 2019. Salicylic acid alleviates arsenic and zinc toxicity in the process of reserve mobilization in germinating fenugreek (*Trigonella foenum-graecum* L.) seeds. *South Afr J Bot* 124: 235–243.

Malik, J.A., Goel, S., Kaur, N., Sharma, S., Singh, I., Nayyar, H. 2012. Selenium antagonizes the toxic effects of arsenic on mungbean (*Phaseolus aureus* Roxb.) plants by restricting its uptake and enhancing the antioxidative and detoxification mechanisms. *Environ Exp Bot* 77: 242–248.

Mallick, I., Bhattacharyya, C., Mukherji, S., Dey, D., Sarkar, S. C., Mukhopadhyay, U. K., Ghosh, A. 2018. Effective rhizoinoculation and biofilm formation by arsenic immobilizing halophilic plant growth promoting bacteria (PGPB) isolated from mangrove rhizosphere: A step towards arsenic rhizoremediation. *Sci. Total Environ.* 610–611, 1239–1250.

Mallick, I., Hossain, S. T., Sinha, S., Mukherjee, S. K. 2014. Brevibacillus sp. KUMAs2, a bacterial isolate for possible bioremediation of arsenic in rhizosphere. *Ecotoxicol. Environ Saf* 107: 236–244.

Mallick, S., Sinam, G., Sinha, S. 2011. Study on arsenate tolerant and sensitive cultivars of Zea mays L.: differential detoxification mechanism and effect on nutrients status. *Ecotoxicol Environ Saf* 74: 1316–1324.

Martinez, V. D., Vucic, E. A., Becker-Santos, D. D., Gil, L., Lam, W. L. 2011. Arsenic exposure and the induction of human cancers. *J Toxicol*, 2011: 1–13.

Meadows, R. 2014. How Plants Control Arsenic Accumulation. *PLoS Biol* 12: e1002008.

Meharg, A. A., Hartley-Whitaker, J. 2002. Arsenic uptake and metabolism in arsenic resistant and non-resistant plant species. *New Phytol* 154: 29–43.

Mesa, V., Navazas, A., González-Gil, R., González, A., Weyens, N., Lauga, B., Gallego, J. L. R., Sánchez, J., Peláez, A. I. 2017. Use of endophytic and rhizosphere bacteria to improve phytoremediation of arsenic-contaminated industrial soils by autochthonous Betula celtiberica. *Appl Environ Microbiol* 83.

Milivojevic, D. B., Nikolic, B. R., Drinic, G. 2006. Effects of arsenic on phosphorous content in different organs and chlorophyll fluorescence in primary leaves of soybean. *Biol Plant* 1: 149–151.

Mishra, B. K., Dubey, C. S., Shukla, D. P., Bhattacharya, P., Usham, A. L. 2014. Concentration of arsenic by selected vegetables cultivated in the Yamuna flood plains (YFP) of Delhi, India. *Environ Earth Sci* 72 (9): 3281–3291.

Mishra R. K., Tiwari S., Patel A., Prasad S. M. 2021. Arsenic contamination, speciation, toxicity and defense strategies in plants. *Brazilian J Bot* 44: 1–10.

Mishra, S., Dwivedi, S., Mallick, S., Tripathi, R. D. 2019. Redox homeostasis in plants under arsenic stress. In: Panda, K.S., Yamamoto, Y.Y. (Eds.), *Redox Homeostasis in Plants*. Springer, Cham., pp. 179–198.

Mishra, S., Mattusch, J., Wennrich, R. 2017. Accumulation and transformation of inorganic and organic arsenic in rice and role of thiol-complexation to restrict their translocation to shoot. *Sci Rep* 7: 40522.

Mitani, N., Yamaji, N., Ma, J. F. 2008. Characterization of substrate specificity of a rice silicon transporter, Lsi1. *Pflüg. Arch* 456: 679–686.

Miteva, E. 2002. Accumulation and effect of arsenic on tomatoes. *Comm Soil Sci Plant Anal* 33: 1917–1926.

Mitra, A., Chatterjee, S., Moogouei, R., Gupta, D. K. 2017. Arsenic Accumulation in Rice and Probable Mitigation Approaches: A Review. *Agronomy* 7: 67.

Mohd, S., Kushwaha, A. S., Shukla, J., Mandrah, K., Shankar, J., Arjaria, N., Saxena, P. N., Khare, P., Narayan, R., Dixit, S., Siddiqui, M. H., Tuteja, N., Das, M., Roy, S. K., Kumar, M., 2019. Fungal mediated biotransformation reduces toxicity of arsenic to soil dwelling microorganism and plant. *Ecotoxicol Environ Saf* 176: 108–118.

Moreno-Jiménez, E., Esteban, E., Peñalosa, J. M. 2015. The Fate of Arsenic in Soil-Plant Systems. In *Reviews of Environmental Contamination and Toxicology*, Whitacre, D. M. (Ed.), Springer, Berlin, Germany.

Mostofa, M. G., Seraj, Z. I., Fujita, M. 2014. Exogenous sodium nitroprusside and glutathione alleviate copper toxicity by reducing copper uptake and oxidative damage in rice (*Oryza sativa* L.) seedlings. *Protoplasma* 251: 1373–1386.

Mousavi, S. R., Niknejad, Y., Fallah, H., Tari, D. B. 2020. Methyl jasmonate alleviates arsenic toxicity in rice. *Plant Cell Reports* 39:1041–1060.

Muehe, E. M., Wang, T., Kerl, C. F., Planer-Friedrich, B., Fendorf, S. 2019. Rice production threatened by coupled stresses of climate and soil arsenic. *Nat Commun* 10: 4985.

Mushtaq, T., Shah, A. A., Akram, W., Yasin, N. A. 2020. Synergistic ameliorative effect of iron oxide nanoparticles and Bacillus subtilis S4 against arsenic toxicity in *Cucurbita moschata*: polyamines, antioxidants, and physiochemical studies. *Int. J. Phytoremediation* 22: 1408–1419.

Nabi, A., Naeem, M., Aftab, T., Masroor, M., Khan, A. 2019. Arsenic Toxicity Induced Changes in Growth, Photosynthetic Pigments, Antioxidant Machinery, Essential Oil, Menthol and Other Active Constituents of Menthol Mint (*Mentha arvensis* L.), *J. Essential Oil Bearing Plants* 22: 1333–1348.

Nath, S., Panda, P., Mishra, S., Dey, M., Choudhury, S., Sahoo, L., Panda, S. K. 2014. Arsenic stress in rice: redox consequences and regulation by iron. *Plant Physiol Biochem* 80: 203–210.

Nookongbut, P., Kantachote, D., Megharaj, M., Naidu, R., 2018. Reduction in arsenic toxicity and uptake in rice (*Oryza sativa* L.) by As-resistant purple nonsulfur bacteria. *Environ Sci Pollut Res* 25: 36530–36544.

Norton, G. J., Adomako, E. E., Deacon, C. M., Carey, A. M., Price, A. H., Mehang, A. A. 2013. Effect of organic matter amendment, arsenic amendment and water management regime on rice grain arsenic species. *Environ Pollut* 177: 38–47.

Pathare, V., Srivastava, S., Sonawane, B. V., Suprasanna, P. 2016. Arsenic stress affects the expression profile of genes of 14-3-3 proteins in the shoot of mycorrhiza colonized rice. *Physiol Mol Biol Plants* 22: 515–522.

Pena-Fernandez, A., Gonzalez-Munoz, M. J., Lobo-Bedmar, M. C. 2014. Establishing the importance of human health risk assessment for metals and metalloids in urban environments. *Environ Int* 72: 176–185.

Pigna, M., Cozzolino, V., Caporale, A. G., Mora, M. L., Meo, V. D., Jara, A. A., Violante, A. 2010. Effect of phosphorus fertilization on arsenic uptake by while grown in polluted soils. *J Soil Sci Plant Nutr* 10: 428–442.

Pigna, M., Cozzolina, V., Violante, A., Meharg, A. A. 2008. Influence of phosphate on the arsenic uptake by wheat (Triticum durum L.) irrigated with arsenic solutions at three different concentrations. *Water Air Soil Pollut* 197: 371–380.

Pommerrenig, B., Diehn, T.A., Bernhardt, N., Bienert, D., Mitani-Ueno, N., Fuge, J., Bieber, A., Spitzer, C., Bräutigam, A., Ma, J. F. Chaumont, F., Bienert, G. P. 2019. Functional evolution of nodulin 26-like intrinsic proteins: From bacterial arsenic detoxification to plant nutrient transport. *New Phytol* 225: 1383–1396.

Pommerrenig, B., Diehn, T. A., Bienert, G. P. 2015. Metalloidoporins: Essentiality of Nodulin 26-like intrinsic proteins in metalloid transport. *Plant Sci* 238: 212–227.

Praveen, A., Pandey, C., Khan, E., Panthri, M., Gupta, M. 2017. Silicon mediated genotoxic alterations in Brassica juncea under arsenic stress: comparative study of biochemical and molecular markers. *Pedosphere* 30: 517–527.

Qiao, J. T., Li, X. M., Hu, M., Li, F., Young, L. Y., Sun, W., Huang, W., Cui, J. 2018.Transcriptional activity of arsenic-reducing bacteria and genes regulated by lactate and biochar during arsenic transformation in flooded paddy soil. *Environ Sci Technol* 51: 61–70.

Quazi, S., Datta, R., Sarkar, D. 2011. Effects of soil types and forms of arsenical pesticide on rice growth and development. *Int J Environ Technol* 8: 445–460.

Rafiq, M., Shahid, M., Shamshad, S., Khalid, S., Niazi, N.K., Abbas, G., Saeed, M.F., Ali, M. and Murtaza, B., 2018. A comparative study to evaluate efficiency of EDTA and calcium in alleviating arsenic toxicity to germinating and young Vicia faba L. seedlings. J Soils and Sediments, 18(6), pp. 2271–2281.

Rafiq, M., Shahid, M., Abbas, G., Shamshad, S., Khalid, S., Niazi, N. K., Dumat, C. 2017. Comparative effect of calcium and EDTA on arsenic uptake and physiological attributes of *Pisum sativum*. *Int J Phytoremediation* 19:662–669.

Rahman, A., Mostofa, M. G., Alam, M., Nahar, K., Hasanuzzaman, M., Fujita, M. 2015. Calcium mitigates arsenic toxicity in rice seedlings by reducing arsenic uptake and modulating the antioxidant defense and glyoxalase systems and stress markers. *Biomed Res Int* 2015: 340812.

Rahman, H., Ramanathan, V., Nallathambi, J., Duraialagaraja, S., Muthurajan, R. 2016. Over-expression of a NAC 67 transcription factor from finger millet (*Eleusine coracana* L.) confers tolerance against salinity and drought stress in rice. *BMC Biotechnol* 16: 35–35.

Rahman, M.A., Hasegawa, H., Rahman, M. M., Miah, M. A. M., Tasmin, A. 2008. Straighthead disease of rice (*Oryza sativa* L.) induced by arsenic toxicity. *Environ Exp Bot* 62:54–59.

Rahman, M. A., Hasegawa, H., Rahman, M. M., Rahman, M. A., Miah, M. A. M. 2007. Accumulation of arsenic in tissues of rice plant (*Oryza sativa* L.) and its distribution in fractions of rice grain. *Chemosphere* 69: 942–948.

Rahaman, M. S., Akter, M., Rahman, M. M., Sikder, M. T., Hosokawa, T., Saito, T., Kurasaki, M. 2020. Investigating the protective actions of D-pinitol against arsenic-induced toxicity in PC12 cells and the underlying mechanism. *Environ Toxicol Pharm* 74: 103302.

Rahaman, S., Sinha, A. C., Mukhopadhyay, D. 2011. Effect of water regimes and organic matters on transport of arsenic in summer rice (*Oryza sativa* L.). *J Environ Sci* 23: 633–639.

Rahman, S. ur, Khalid, M., Kayani, S. I., Tang, K. 2020. The ameliorative effects of exogenous inoculation of Piriformospora indica on molecular, biochemical and physiological parameters of Artemisia annua L. under arsenic stress condition. *Ecotoxicol Environ Saf* 206: 111202.

Ranjan, R., Kumar, N., Dubey, A. K., Gautam, A., Pandey, S. N., Mallick, S. 2018. Diminution of arsenic accumulation in rice seedlings co-cultured with *Anabaena* sp.: Modulation in the expression of lower silicon transporters, two nitrogen dependent genes and lowering of antioxidants activity. *Ecotoxicol Environ Saf* 151: 109–117.

Ranjan, R., Kumar, N., Gautam, A., Kumar Dubey, A., Pandey, S. N., Mallick, S. 2021. Chlorella sp. modulates the glutathione mediated detoxification and S-adenosylmethionine dependent methyltransferase to counter arsenic toxicity in *Oryza sativa* L. *Ecotoxicol Environ Saf* 208: 111418.

Rodriguez-Iruretagoiena, A., Trebolazabala, J., Martinez-Arkarazo, I., De Diego, A., Madariaga, J. M. 2015. Metals and metalloids in fruits of tomatoes (*Solanum lycopersicum*) and their cultivation soils in the Basque Country: concentrations and accumulation trends. *Food Chem* 173: 1083–1089.

Rodríguez-Ruiz, M., Aparicio-Chacon, M. V., Palma, J. M., Corpas, F. J. 2019. Arsenate disrupts ion balance, sulfur and nitric oxide metabolisms in roots and leaves of pea (*Pisum sativum* L.) plants. *Environ Exp Bot* 161: 143–156.

Rucińska-Sobkowiak, R. 2016. Water relations in plants subjected to heavy metal stresses. *Acta Physiol Plant* 38: 257.

Ruíz-Torres, C., Feriche-Linares, R., Rodríguez-Ruíz, M., Palma, J. M., Corpas, F. J. 2017. Arsenic-induced stress activates sulfur metabolism in different organs of garlic (*Allium sativum* L.) plants accompanied by a general decline of the NADPH-generating systems in roots. *J Plant Physiol* 211: 27–35.

Sallam, A., Alqudah, A. M., Dawood, M. F. A., Baenziger, P. S., Börner, A. 2019. Drought stress tolerance in wheat and barley: Advances in physiology, breeding and genetics research. *Int J Mol Sci* 20(13): 3137.

Salt, D. E. 2017. Would the real arsenate reductase please stand up? *New Phytol* 215: 1090–1101.

Schmidt, C. W. 2014. Low-dose arsenic. *Environ Health Perspect* 122 (5): A130–A134.

Schulz, H., Hortling, S., Tanneberg, H. 2008. The identification and quantification of arsenic-induced phytochelatins—comparison between plants with varying As sensitivities. *Plant Soil* 303: 275–287.

Shaibur, M. R., Adjadeh, T. A., Kawai, S. 2013. Effect of phosphorus on the concentrations of arsenic, iron and some other elements in barley grown hydroponically. *J Soil Sci Plant Nut* 13: 87–98.

Shaibur, M. R., Kawai, S. 2009. Effect of arsenic on visible symptom and arsenic concentration in hydroponic Japanese mustard spinach. *Environ Exp Bot* 67: 65–70.

Shakoor, M. B., Riaz, M., Niazi, N. K., Ali, S., Rizwan, M., Arif, M. S., Arif, M. 2019. Chapter 18—Recent Advances in Arsenic Accumulation in Rice. In: Hasanuzzaman, M., Fujita, M., Nahar, K., Biswas, J. K. (Eds.), *Advances in Rice Research for Abiotic Stress Tolerance*. Elsevier, Amsterdam, The Netherlands.

Sharma, A. K., Tjell, J. C., Sloth, J. J., Holm, P. E. 2014. Review of arsenic contamination, exposure through water and food and low cost mitigation options for rural areas. *Appl Geochem* 41:11–33.

Sharma, I. 2012. Arsenic induced oxidative stress in plants. *Biologia* 67: 447–53.

Sharma, S., Anand, G., Singh, N., Kapoor, R. 2017a. Arbuscular mycorrhiza augments arsenic tolerance in wheat (*Triticum aestivum* L.) by strengthening antioxidant defense system and thiol metabolism. *Front Plant Sci* 8: 906.

Sharma, S., Anand, G., Singh, N., Kapoor, R. 2017b. Arbuscular mycorrhiza augments arsenic tolerance in wheat (*Triticum aestivum* L.) by strengthening antioxidant defense system and thiol metabolism. *Front Plant Sci* 8: 1–21.

Sharples, J. M., Meharg, A. A., Chambers, S. M., Cairney, J. W. G. 2000. Symbiotic solution to arsenic contamination. *Nature* 404: 951–952.

Shi, S., Wang, T., Chen, Z., Tang, Z., Wu, Z., Salt, D. E., Chao, D. Y., Zhao, F. 2016. OsHAC1, 1 and OsHAC1, 2 function as arsenate reductases and regulate arsenic accumulation. *Plant Physiol* 172: 1708–1719.

Shri, M., Kumar, S., Chakrabarty, D., et al. 2009. Effect of arsenic on growth, oxidative stress, and antioxidant system in rice seedlings. *Ecotoxicol Environm Safety* 72: 1102–1110.

Shri, M., Singh, P. K., Kidiwai, M., Gautam, N., Dubey, S., Verma, G., Chakrabarty, D. 2019. Recent advances in arsenic metabolism in plants: Current status, challenges and highlighted biotechnological intervention to reduce grain arsenic in rice. *Metallomics* 11: 519–532.

Siddiqui, F., Tandon, P.K., Srivastava, S. 2015c. Arsenite and arsenate impact the oxidative status and antioxidant responses in *Ocimum tenuiflorum* L. *Physiol Mol Biol Plants* 21: 453–458.

Singh, A. P., Dixit, G., Mishra, S., Dwivedi, S., Tiwari, M., Mallick, S., Pandey, V., Trivedi, P. K., Chakrabarty, D., Tripathi, R. D. 2015a. Salicylic acid modulates arsenic toxicity by reducing its root to shoot translocation in rice (*Oryza sativa* L.). *Front Plant Sci* 6: 340.

Singh, M., Kumar, J., Singh, S., Singh, V. P., Prasad, S. M. et al. 2015b. Adaptation strategies of plants against heavy metal toxicity: a short review. *Biochem Pharmacol* 4: 161.

Singh, N., Ma, L. Q., Srivastava, M., Rathinasabapathi, B. 2006. Metabolic adaptations to arsenic induced oxidative stress in *Pteris vittata* L. and *Pteris ensiformis* L. *Plant Sci* 170: 274–282.

Singh, N. K., Raghubanshi, A. S., Upadhyay, A. K., Rai, U. N. 2016. Arsenic and other heavy metal accumulation in plants and algae growing naturally in contaminated area of West Bengal, India. *Ecotoxicol Environ Saf* 130: 224–233. https://doi.org/10.1016/j.ecoenv.2016.04.024

Singh, R., Parihar, P., Prasad, P. 2018. Simultaneous exposure of sulphur and calcium hinder As toxicity: up-regulation of growth, mineral nutrients uptake and antioxidant system. *Ecotoxicol Environ Saf* 161: 318–331.

Singh, R., Parihar, P., Prasad, S. M. 2020. Interplay of calcium and nitric oxide in improvement of growth and arsenic induced toxicity in mustard seedlings. *Sci Rep* 10: 1–12.

Singh, R., Singh, S., Parihar, P., Singh, V. P., Prasad, S. M. 2015d. Arsenic contamination, consequences and remediation techniques: a review. *Ecotoxicol Environ Saf* 112: 247–270.

Singh, V. P., Singh, S., Kumar, J., Prasad, S. M. 2015c. Investigating the roles of ascorbate-glutathione cycle and thiol metabolism in arsenate tolerance in ridged Luffa seedlings. *Protoplasma* 252: 1217–1229.

Sinha, S., Sinam, G., Mishra, R. K., Mallick, S. 2010. Metal accumulation, growth, antioxidants and oil yield of *Brassica juncea* L. exposed to different metals. *Ecotoxicol Environ Saf* 73: 1352–1361.

Sodhi, K.K., Kumar, M., Agrawal, P.K., Singh, D.K. 2019. Perspectives on arsenic toxicity, carcinogenicity and its systemic remediation strategies. *Environ Technol Innov* 16: 100462.

Song, W. Y., Yamaki, T., Yamaji, N., Ko, D., Jung, K. H., Fujii-Kashino, M., An, G., Martinoia, E., Lee, Y., Ma, F. J. 2014. A rice ABC transporter, OsABCC1, reduces arsenic accumulation in the grain. *PNAS* 111: 15699–15704.

Souri, Z., Karimi, N., de Oliveira, L. M. 2018. Antioxidant enzymes responses in shoots of arsenic hyperaccumulator, Isatis cappadocica Desv., under interaction of arsenate and phosphate. *Environ Technol* 39: 1316–1327.

Souri, Z., Karimi, N., Farooq, M. A., Sandalio, L. M. 2020. Nitric oxide improves tolerance to arsenic stress in *Isatis cappadocica* desv. shoots by enhancing antioxidant defenses. *Chemosphere* 239: 124523.

Spagnoletti, F., Lavado, R. S. 2015. The Negative Effects of Arsenic on Soybeanplants. *Agronomy* 5(2): 188–199.

Srivastava, M., Ma, L. Q., Singh, N., Singh, S. 2005. Antioxidant responses of hyper-accumulator and sensitive fern species to arsenic. *J Exp Bot* 56: 1335–1342.

Srivastava, S., Akkarakaran, J. J., Sounderajan, S., Shrivastava, M., Suprasanna, P. 2016. Arsenic toxicity in rice (*Oryza sativa* L.) is influenced by sulfur supply: Impact on the expression of transporters and thiol metabolism. *Geoderma* 270: 33–42.

Srivastava, S., Shrivastava, M., Suprasanna, P., D'Souza, S. F. 2011. Phytofiltration of arsenic from simulated contaminated water using Hydrilla verticillata in field conditions. *Ecol Eng* 37: 1937–1941.

Srivastava, S., Singh, N. 2014. Mitigation approach of arsenic toxicity in chickpea grown in arsenic amended soil with arsenic tolerant plant growth promoting Acinetobacter sp. *Ecol Eng* 70: 146–153.

Srivastava, S., Sinha, P., Sharma, Y. K. 2017. Status of photosynthetic pigments, lipid peroxidation and anti-oxidative enzymes in Vigna mungo in presence of arsenic. *J Plant Nutr* 40: 298–306.

Srivastava, S., Srivastava, S., Bist, V., Awasthi, S., Chauhan, R., Chaudhry, V., Singh, P. C., Dwivedi, S., Niranjan, A., Agrawal, L., Chauhan, P. S., Tripathi, R. D., Nautiyal, C. S. 2018. *Chlorella vulgaris* and *Pseudomonas putida* interaction modulates phosphate trafficking for reduced arsenic uptake in rice (*Oryza sativa* L.). *J Hazard Mater* 351: 177–187.

Srivastava, S., Verma, P. C., Chaudhry, V., Singh, Namrata, Abhilash, P. C., Kumar, K. V., Sharma, N., Singh, Nandita 2013. Influence of inoculation of arsenic-resistant Staphylococcus arlettae on growth and arsenic uptake in *Brassica juncea* (L.) Czern. Var. R-46. *J Hazard Mater* 262: 1039–1047.

Stoeva, N., Berova, M., Vassilev, A., Zlatev, Z. 2005. Effect of arsenic on some physiological parameters in bean plants. *Biol Planta* 49: 293.

Stoeva, N., Berova, M., Zlatev, Z. 2004. Effect of arsenic on some physiological parameters in bean plants. *Biol Plant* 49: 293–296.

Stoeva, N., Bineva, T. Z. 2003. Oxidative changes and photosynthesis in Oat plants grown in As-contaminated soil. *Bulg J Plant Physiol* 29: 87–95.

Suda, A., Makino, T. 2016. Effect of organic amendments on arsenic solubilization in soils during long-term flooded incubation. *Int J Environ Sci Technol* 13: 2375–2382.

Suekawa, M., Fujikawa, Y., Esaka, M. 2017. Physiological role of ascorbic acid recycling enzymes in plants. In: Hossain, M.A., Munné-Bosch, S., Burritt, D. J., Diaz-Vivancos, P., Fujita, M., Lorence, A. (Eds.), *Ascorbic Acid in Plant Growth, Development and Stress Tolerance*. Springer, Cham, pp. 355–373.

Sultana, R., Kobayashi, K., Kim, K. H. 2015. Comparison of arsenic uptake ability of barnyard grass and rice species for arsenic phytoremediation. *Environ Monit Assess* 187(1): 4101.

Sun, S. K., Chen, Y., Che, J., Konishi, N., Tang, Z., Miller, A. J., Ma, J. F., Zhao, F. J. 2018. Decreasing arsenic accumulation in rice by overexpressing OsNIP1,1 and OsNIP3,3 through disrupting arsenite radial transport in roots. *New Phytol* 219: 641–653.

Suriyagoda, L. D. B., Ditteret, K., Lambers, H. 2018. Mechanism of arsenic uptake, translocation and plant resistance to accumulate arsenic in rice grains. *Agri Ecosyst Environ* 253: 23–37.

Syu, C. H., Wu, P. R., Lee, C. H., Juang, L. W., Lee, D. Y. 2019. Arsenic phytotoxicity and accumulation in rice seedlings grown in arsenic-contaminated soils as influenced by the characteristics of organic matter amendments and soils. *J Plant Nutr Soil Sci* 182: 60–71.

Talukdar, D. 2013. Arsenic-induced changes in growth and antioxidant metabolism of fenugreek. *Russian J. Plant Physiol.* 60: 652–660.

Tang, Z., Lv, Y., Chen, F., Zhang, W., Rosen, B. P., Zhao, F. J. 2016. Arsenic Methylation in *Arabidopsis thaliana* expressing an algal arsenite methyltransferase gene increases arsenic phytotoxicity. *J Agric Food Chem* 64: 2674–2681.

Tiwari, M., Sharma, D., Dwivedi, S., Singh, M., Tripathi, R. D., Trivedi, P. K. 2014. Expression in Arabidopsis and cellular localization reveal involvement of rice NRAMP, OsNRAMP1, in arsenic transport and tolerance. *Plant Cell Environ* 37: 140–152.

Tripathi, P., Khare, P., Barnawal, D., Shanker, K., Srivastava, P. K., Tripathi, R. D., Kalra, A. 2020. Bioremediation of arsenic by soil methylating fungi: role of Humicola sp. strain 2WS1 in amelioration of arsenic phytotoxicity in Bacopa monnieri L. *Sci Total Environ* 716: 136758.

Tripathi, P., Tripathi, R. D., Singh, R. P., Dwivedi, S., Goutam, D., Shri, M., Chakrabarty, D. 2013. Silicon mediates arsenic tolerance in rice (*Oryza sativa* L.) through lowering of arsenic uptake and improved antioxidant defence system. *Ecol Eng* 52: 96–103.

Tu, S., Ma, L. Q. 2003. Interactive effects of pH, arsenic and phosphorus on uptake of As and P and growth of the arsenic hyperaccumulator *Pteris vittata* L. under hydroponic conditions. *Environ Exp Bot* 50: 243–251.

Ultra, V. U., Nakayama, A., Tanaka, S., Kang, Y., Sakurai, K., Iwasaki, K. 2009. Potential for the alleviation of arsenic toxicity in paddy rice using amorphous iron-(hydr)oxide amendments. *Soil Sci Plant Nutr* 55: 160–169.

Upadhyay, A. K., Singh, N. K., Singh, R., Rai, U. N. 2016. Amelioration of arsenic toxicity in rice: Comparative effect of inoculation of *Chlorella vulgaris* and *Nannochloropsis* sp. on growth, biochemical changes and arsenic uptake. *Ecotoxicol Environ Saf* 124: 68–73.

Upadhyay, M. K., Yadav, P., Shukla, A., Srivastava, S. 2018. Utilizing the potential of microorganisms for managing arsenic contamination: A feasible and sustainable approach. *Front Environ Sci* 6: 1–11.

Verbruggen, N., Hermans, C., Schat, H. 2009. Mechanisms to cope with arsenic or cadmium excess in plants. *Curr Opin Plant Biol* 12: 364–372.

Verma, S., Verma, P. K., Pande, V., Tripathi, R. D., Chakrabarty, D. 2016. Transgenic Arabidopsis thaliana expressing fungal arsenic methyltransferase gene (WaarsM) showed enhanced arsenic tolerance via volatilization. *Environ Exp Bot* 132: 113–120.

Vezza, M. E., Llanes, A., Travaglia, C., Agostini, E., Talano, M. A. 2018. Arsenic stress effects on root water absorption in soybean plants: physiological and morphological aspects. *Plant Physiol Biochem* 123: 8–17.

Wang, K. T., Li, Y. P., Wu, Y. C., Qiu, Z. Q., Ding, Z. X., Wang, X. J., Chen, W., Wang, R. J., Fu, F. F., Rensing, C., Yang, G. D. 2020. Improved grain yield and lowered arsenic accumulation in rice plants by inoculation with arsenite-oxidizing Achromobacter xylosoxidans GD03. *Ecotoxicol Environ Saf* 206: 111229.

Wang, Q., Xiong, D., Zhao, P., Yu, X., Tu, B., Wang, G. 2011. Effect of applying an arsenic-resistant and plant growth-promoting rhizobacterium to enhance soil arsenic phytoremediation by Populus deltoides LH05-17. *J Appl Microbiol* 111: 1065–1074.

Wang, X., Rathinasabapathi, B., Oliveira, L.M. De, Guilherme, L.R.G., Ma, L.Q. 2012. Bacteria-mediated arsenic oxidation and reduction in the growth media of arsenic hyperaccumulator *Pteris vittata. Environ Sci Technol* 46: 11259–11266.

Wang, X., Xia, H., Ni, Z., Liang, D. 2018. Comparison of enzyme activities involved in AsA-GSH cycle in red-flesh kiwifruit varieties. *IOP Publishing. In IOP Conference Series: Materials Science and Engineering* 392: 052015.

Wang, Y., Li, Y. Q., Lv, K., Cheng, J. J., Chen, X. L., Ge, Y., Yu, X. Y. 2018. Soil microalgae modulate grain arsenic accumulation by reducing dimethylarsinic acid and enhancing nutrient uptake in rice (*Oryza sativa* L.). *Plant Soil* 430: 99–111.

Weryszko-Chmielewska, E., Chwil, M. 2005. Lead-induced histological and ultrastructural changes in the leaves of soybean (*Glycine max* (L.) merr.). *Soil Sci Plant Nut* 51: 203–212.

Wevar Oller, A. L., Regis, S., Armendariz, A. L., Talano, M. A., Agostini, E. 2020. Improving soybean growth under arsenic stress by inoculation with native arsenic-resistant bacteria. *Plant Physiol Biochem* 155: 85–92.

WHO 2001. The World Health Report 2001: Mental health: new understanding, new hope.

Xu, J. Y., Li, H. B., Liang, S., Luo, J., Ma, L. N. Q. 2014. Arsenic enhanced plant growth and altered rhizosphere. Characteristics of hyperacaimulator *Pteris vittata*. *Environ Pollut* 194: 105–111.

Xu, X. Y., McGrath, S. P., Zhao, F. J. 2007. Rapid reduction of arsenate in the medium mediated by plant roots. *New Phytol* 176: 590–599.

Xue, M., Yi, H. 2017. Nitric oxide signaling is involved in arsenic-induced guard cell death in *Vicia faba* L.(Fabaceae). *Braz J Bot* 40: 635–642.

Yadav, R. K., Srivastava, S. K. 2015. Effect of arsenite and arsenate on lipid peroxidation, enzymatic and non-enzymatic antioxidants in *Zea mays* Linn. *Biochem Physiol* 4: 81.

Yadu, B., Chandrakar, V., Tamboli, R., Keshavkant, S. 2019. Dimethylthiourea antagonizes oxidative responses by up-regulating expressions of pyrroline-5-carboxylate synthetase and antioxidant genes under arsenic stress. *Int J Environ Sci Technol* 16: 8401.

Ye, Y., Yuan, J., Chang, X., Yang, M., Zhang, L., Lian, X. 2015. The phosphate transporter gene OsPht1,4 is involved in phosphate homeostasis in rice. *PLoS ONE* 10: e0126186.

Zahra, S., Mahmood, S., Noreen, S., Akrem, A. 2018. Independent and combined nickel and cadmium induced lipid peroxidation of biological membranes and its mitigation through antioxidant enzymes in *Grewia asiatica* L. *Pak J Life Soc Sci* 16: 48–54.

Zaid, A., Mohammad, F., Wani, S. H., Siddique, K. M. 2019. Salicylic acid enhances nickel stress tolerance by up-regulating antioxidant defense and glyoxalase systems in mustard plants. *Ecotoxicol Environ Safety* 180: 575–587.

Zhang, H., Liu, N., J, Zhao, F, Ge, Y, Xu, Chen, Y., 2019. Disturbance of photosystem II oxygen evolution complex induced the oxidative damage in *Chlorella vulgaris* under the stress of cetyltrimethylammonium chloride. *Chemosphere* 223: 659–667.

Zhang, J., Zhao, Q.Z., Duan, G.L., Huang, Y.C. 2011 Influence of sulphur on arsenic accumulation and metabolism in rice seedlings. *Environ. Exp. Bot.* 72: 34–40.

Zhang, Q.M., Gong, M., Liu, K., Chen, Y., Yuan, J., Chang, Q., 2020. Rhizoglomus intraradices improves plant growth, root morphology and phytohormone balance of Robinia pseudoacacia in arsenic-contaminated soils. *Front Microbiol* 11: 1–13.

Zhao, F. J., Ago, Y., Mitani, N., Li, R. Y., Su, Y. H., Yamaji, N., McGrath. S. P., Ma, J. F. 2010. The role of the rice aquaporin Lsi1 in arsenite efflux from roots. *New Phytol* 186: 392–399.

Zhao, F. J., Zhu, Y. G., Meharg, A. A. 2013. Methylated arsenic species in rice: geographical variation, origin, and uptake mechanisms. *Environ Sci Technol* 47: 3957–3966.

Zhen, Z., Yan, C., Zhao, Y. 2020. Influence of epiphytic bacteria on arsenic metabolism in Hydrilla verticillata. *Environ Pollut* 261: 114232.

Zhu, F. K., Yang, S. K., Fan, W. X., Wang, A. R., Hao, H. L., Yao, S. W. 2014. Heavy metals in jujubes and their potential health risks to the adult consumers in Xinjiang province, *China. Environ Monit Assess* 186(10): 6039–6046.

Zvobgo, G., Hu, H., Shang, S., Shamsi, I. H., Zhang, G. 2015. The effects of phosphate on arsenic uptake and toxicity alleviation in tobacco genotypes with differing arsenic tolerances. *Environ Toxicol Chem* 34(1): 45–52.

17

Proteomic and Genomic Approaches to Study Plant Physiological Responses under Heavy Metal Stress

Abhishek Joshi, Bhanupriya Kanthaliya, and Supriya Meena
Laboratory of Biomolecular Technology, Department of Botany, M. L. Sukhadia University, Udaipur, Rajasthan, India

Vishnu D. Rajput and Tatiana Minkina
Academy of Biology and Biotechnology, Southern Federal University, Stachki, Rostov-on-Don, Russia

Jaya Arora
Laboratory of Biomolecular Technology, Department of Botany, M. L. Sukhadia University, Udaipur, Rajasthan, India

CONTENTS

17.1 Introduction

Nowadays, contamination of ecosystems by heavy metals (HMs) or metalloids is a major concern worldwide (Masindi and Muedi 2018). It has been estimated that more than 10 million sites being contaminated with HMs and similar toxic substances all over the world (Liu et al. 2018; Li et al. 2019). This issue has mainly arisen from the rapid industrialization, increased anthropogenic activities, and modern agriculture practices during the last few decades (Singh et al. 2018). Usually, HMs are present as essential and non-essential in the ecosystem, but their excess concentration has exacerbated several toxic and hazardous impacts on the living organism, including plants (Singh et al. 2011). For plants, some of the HMs (i.e. Cu, Zn, Mn, Fe, Mo, B, Ni, Co, Cl, and Si) considered as essential nutrients and they have been reported to be involved in various process related to plant growth and development (Hänsch and Mendel 2009; Vatansever et al. 2017). At elevated concentrations, both the essential and non-essential metals (i.e. As, Cd, Cr, Pb, Hg, etc.) possess serious threats to the basic functioning of plants through either direct or indirect actions. The direct actions include inhibition of cytoplasmic enzymes and damage to cellular structures (Minkina et al. 2019) because of oxidative stress, while indirect actions lead to the replacement of crucial cation from their specific binding sites and disturb plant homeostasis via generation of reactive oxygen species (ROS) and oxidation of major biomolecules such as to lipids, proteins and nucleic acids (Bielen et al. 2013; Küpper and Andresen 2016). Here, it is important to understand why these metals can go from being essential to being highly toxic at a slightly higher concentration, and how plants established their homeostatic equilibrium. These understandings come from the organizing systematic analysis of plants' physiological, biochemical, and molecular machinery (Viehweger 2014; Joshi et al. 2020). In the last two decades, substantial research on plant response to HMs as well as metal-containing nanoparticles has been conducted to unravel the distinct biological questions related to their stress tolerance mechanism (Nagajyoti et al. 2010; Rastogi et al. 2017; Rajput et al. 2018a, b; Ghori et al. 2019; Seneviratne et al. 2019).

Currently, high-throughput OMICS techniques, namely genomics, transcriptomics, proteomics and metabolomics

DOI: 10.1201/9781003155089-17

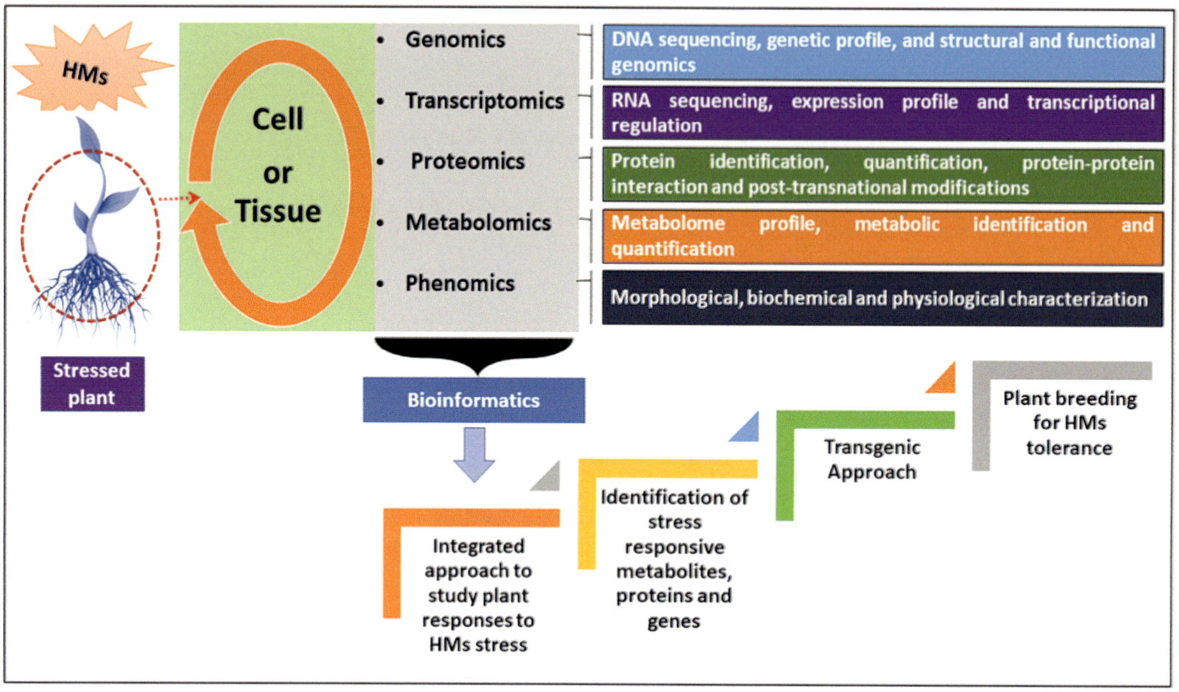

FIGURE 17.1 Integrated omics approach to study plant physiological responses under HMs stress.

(Figure 17.1) are being extensively performed to dissect plants strategies of HMs stress tolerance (Singh et al. 2016; Mehta et al. 2019). Among them, both proteomics and genomics are the imperative techniques not only for scanning, identification, characterization, and assessment of target genes/proteins related to HMs stress but also for their introgression to develop transgenic with multiple adaptive traits (Jha and Bohra 2016; Mosa et al. 2017; Nguyen et al. 2018). The advancement of these OMICS techniques will provide meticulous data set for the identification of the genetic determinants of the hyperaccumulation and more biological system analysis (Peng et al. 2021; Anwar and Kim 2020).

Therefore, the present chapter summarizes the response of plants under HM stress covering the role of stress-inducible proteins, transcription factors, and genes. Special emphasis is given to the state of the art of recent development in these systems biology approaches so far for better understanding the complex mechanism of plant HM stress acclimatization and development of multiple stress resistance transgenic plants.

17.2 An Overview on Plant Responses to HMs Stress

Heavy metals are habitually non-biodegradable, have steady persistence in both the soil and water ecosystem, and easily get accumulated in plants (Ali et al. 2013; Chaplygin et al. 2020). These metals enter a plant from the rhizosphere via roots through either passive or active transport. After entering, such metal ions can be translocated to the aerial parts (leaves, stem, etc.) with the help of the xylem stream (Page and Feller 2015). Although plants require many of HMs like Cu, Zn, Mn, B, Ni, Co, and Si for their growth and development, excessive concentration can become toxic to plants. Plant responses to essential metals involve distinct phases from deficiency to sufficiency and tolerance to toxicity.

In the case of non-essential metals, only the tolerance and toxicity phases occur (Hänsch and Mendel 2009; Arif et al. 2016). The term critical or threshold concentration is widely used to explain their toxicity initiating concentration, which can be varied with plant species to species, their stage of development, and the chemical nature as well as the concentration of the metal (Baderna et al. 2015; Chowdhury et al. 2017; Uchimiya et al. 2020). The toxicity caused by both the essential and non-essential metals are broadly categorized into the following processes: (1) competence with the essential metal cations: for instance, the lack of specificity of a transporter involved in the uptake of essential metals such as P, Zn, Fe, and Ca allow the entry of As, Cd, and Pb, respectively, which resulted in resentment for absorption at the root surface; (2) direct interaction with sulfhydryl groups of function proteins; (3) displacement of essential cations in specific binding sites; and (4) generation of oxidative stress through distorting macromolecules and cellular enzymes (DalCorso et al. 2013; Dubey et al. 2018). The accumulation of HMs in plants implicates negative impacts on inadequate functioning. Heavy metals induced toxic effects on plants have been well versed like overall growth inhibition, reduction in photosynthesis, oxidation of essential protein, and enzymes, production of ROS, promoting cell death and nucleic acid damage through different mode of action (Asati et al. 2016; Edelstein and Ben-Hur 2018; Singh et al. 2020).

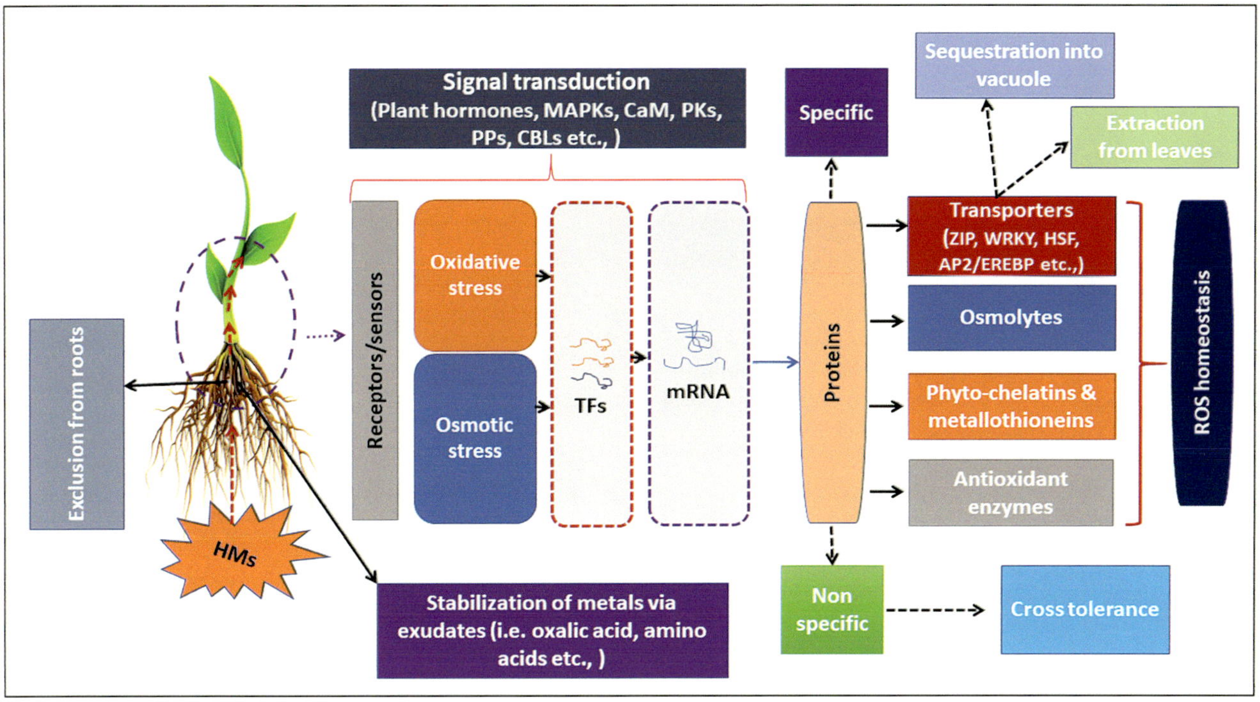

FIGURE 17.2 Integrative view of mechanism of HMs tolerance and cross-tolerance in plants. **HMs** = heavy metals, **MAPKs** = mitogen-activated protein kinase, **CaM** = Ca^{2+}/Calmodulin, **PKs** = ROS-modulated protein kinases, **PPs** = protein phosphatases, **CBLs** = calcineurin-B-like protein, **TFs** = transcription factor, ZIP = leucine zipper.

To counterbalancing the HM's stress, plants have evolved a wide range of defense mechanisms to regulate the uptake, mobilization, and intracellular concentration of HM ions (Figure 17.2). Structural barriers like the thick cuticle, trichomes, as well as dense cell wall and mycorrhizal symbiosis are the first line of defense in these mechanisms (Shi and Cai 2009; Kumar et al. 2016). It has also been reported that the plant roots barricading the entry of HMs through the synthesis and deposition of callose as well as transform in gown structural plasticity (O'Lexy et al. 2018). Apart from this, the most common defense mechanism includes the synthesis of metal transporter (for metal uptakes and vacuolar transport), thiol-containing chelating compounds (for metal detoxification and tolerance), and molecular chaperones for delivery and trafficking of metal ions (Chaudhary et al. 2018; Çelik et al. 2020). Furthermore, increased abundance of defense proteins, osmolytes and both the enzymatic or non-enzymatic antioxidant components help HMs stressed plants to effective ROS scavenging and re-establishing their redox homeostasis (Singh et al. 2015; Maleki et al. 2017; Nikalje and Suprasanna 2018; Joshi et al. 2020; Kosakivska et al. 2021). It has also been reported that plants modulate their most crucial metabolic pathways like photosynthesis, mitochondrial respiration, and nitrogen metabolism to compensate for the high-energy demands of HMs challenged cells (Khan et al. 2016; Singh et al. 2018; Hussain et al. 2020).

In addition, plant hormones (phytohormones) play an essential role in plant HM tolerance as they coordinate signaling mechanisms and stimulate distinct adaptive responses. Abscisic acid (ABA) concentration is known to increase after HM exposure and several studies have confirmed their increased abundance in the different tissue of plants like rice (Kim et al. 2014), potato tubers (Stroiński et al. 2010), cucumbers (Wang et al. 2014), and chickpea (Atici et al. 2005). Also, transcriptomic studies of rice roots under vanadium (V) and arsenic (As) stress has revealed strong expression of genes associated with signaling and biosynthesis of ABA (Huang et al. 2012; Lin et al. 2013). Similarly, Xu et al. (2019) perform transcriptomic analysis of wild paper mulberry plant (*Broussonetia papyrifera* rice) exposed cadmium (Cd) showed that this metal triggered the expression of genes associated with signaling and biosynthesis of Auxin and ABA.

17.3 Approaches to Study Plant Responses to HMs Stress

Over decades, several classical biochemical and physiological methodologies have been used to understand the stress tolerance mechanism in plants, and OMICS approaches are the very modest strategies that have been applied to gain insight into complete biological mechanisms of the plant during abiotic stresses such as HMs stress. These approaches mainly include genomics, transcriptomics, proteomics, metabolomics, and phenomics where the key emphasis was given to the mapping of information objects (mainly the proteins, transcript, and genes), finding the interactive relationship among the objects, and engineering the networks as well as objects to modulate the regulatory mechanisms. In the past decade, a paradigm shift has been brought in research by the genomics and proteomics approaches as they collectively open up

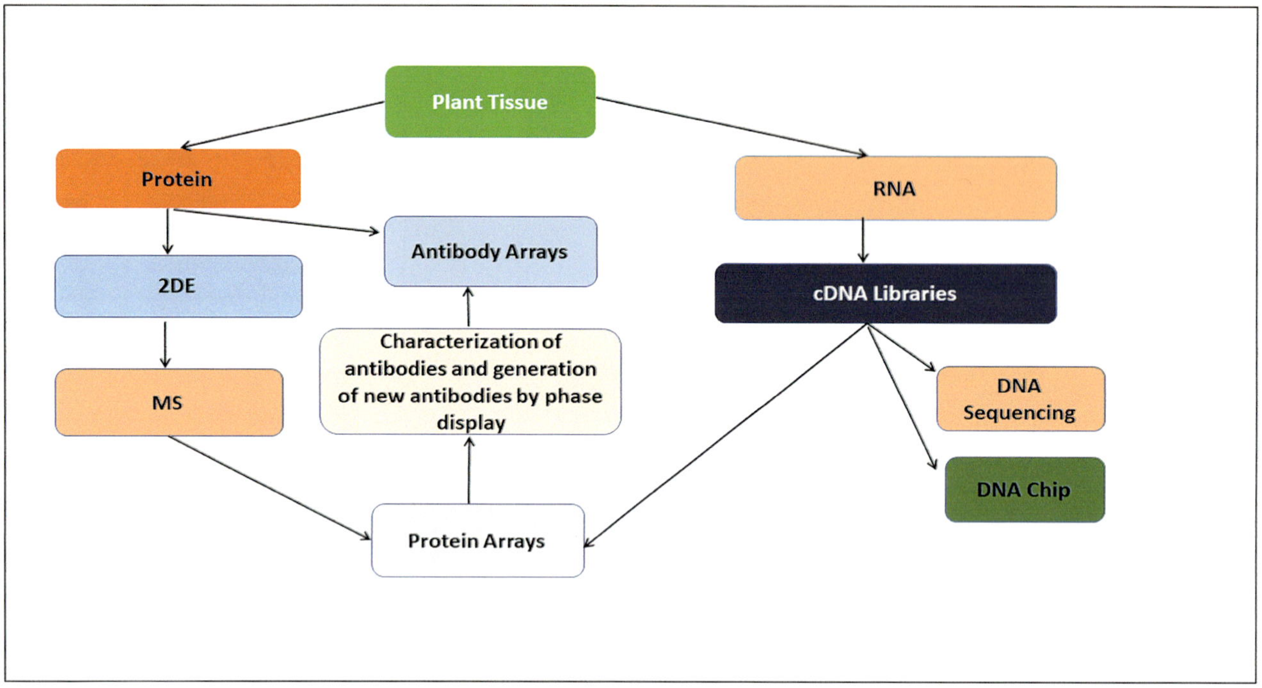

FIGURE 17.3 Correlation between genomics and proteomics approaches, **2DE** = two-dimensional gel electrophoresis, **MS** = Mass spectrometry.

new horizons for enhanced stress tolerance via reverse genetics (Figure 17.3). Therefore, in the sub-sections, we have illustrated the comprehensive scenario of these two consistent methodologies to study plant physiological responses under HM stress.

17.3.1 Proteomic Approaches to Study Plant Responses to HMs Stress

17.3.1.1 Overview on Proteomics Tools

Protein extraction and purification from the plant tissue is the most crucial step to initiate any proteomic experiment. One reason is the plants have low protein content and produce a broad spectrum of interfering substances (i.e. phenolic compounds, proteolytic and oxidative enzymes; organic acids, pigments, and carbohydrates) that can contaminate protein extracts, and distract the subsequent protein fractionation and downstream analysis (Pavoković et al. 2012; Joshi et al. 2018). Many differential precipitation protocols (using polyethylene glycol, trichloroacetic acid, and cold acetone, extraction buffer containing MgSO4, urea, and Tris) have been developed to overcome the difficulties associated with plant protein extraction and purification. Nonetheless, choosing an appropriate method largely influenced by the nature of the extracted tissue and the downstream applications (Wang et al. 2016; Andrés-Colás and Van Der Straeten 2017; Joshi et al. 2019; Meena et al. 2020). Apart from the isolation, the key objective of any proteomic investigation is to qualitative and quantitative information about the constituent protein. In most plant proteomic studies, including those focused on HMs toxicity, pre-fractionation of the protein sample prior to mass spectrometry (MS) analysis is carried out via either the gel-based or by certain gel-free technique (Figure 17.4). As compared to conventional staining procedure of both the 1-DE and two-dimensional gel electrophoresis (2-DE) gel using coomassie brilliant blue (CBB) and silver staining, differential fluorescence labeling (using Cy2, Cy3, and Cy5 dyes) of two-dimensional difference gel electrophoresis (2-DDIGE) based proteomic approach is more convenient to compare deferential expressions of proteins from both the control and HM stressed tissue within the same gel (Abdallah et al. 2012; Wu et al. 2014; Borges et al. 2019). In blue native (BN) and sodium dodecyl sulfate (SDS) gels, protein complexes are solubilized directly from the tissue and mixed with CBB dye that binds to the isolated complexes before polyacrylamide gel electrophoresis (PAGE) based separation. The combination of BN-PAGE and denaturing SDS-PAGE has been used to study metal stress-induced protein shift in plants (Qureshi et al. 2010; Nowakowski et al. 2014). With the advancement of technology multiplexed isotope-coded affinity tag (ICAT), stable-isotope labeling by amino acids in cell culture (SILAC), isobaric tags for relative and absolute quantitation (iTRAQ), isotope-coded protein label (ICPL), N-terminal labeling, and even label-free methods of peptides have allowed comparative as well as quantitative analysis of multiple samples (Belwal et al. 2018; Jorrin-Novo et al. 2019; Nishioka et al. 2021). It has also been reported that the reproducibility of liquid chromatography (LC) based separation is much better than that achieved by 2D-GE. Most of the gel-based and gel-free techniques are coupled with either the MS or matrix-assisted laser desorption ionization-time-of-flight (MALDI-TOF) mass spectrometer to allowing sequencing of both proteins as well as peptides and providing more confident proteome maps (Angel et al. 2012; Gemperline et al. 2016). Surface-enhanced laser desorption/ionization-time off-light (SELDI-TOF) analysis has also been used to perceive protein expression under metal exposer (Moon et al. 2014; Scebba et al. 2016).

Furthermore, image analysis software are the crucial tool to generate proteome maps in both gel-based and gel-free

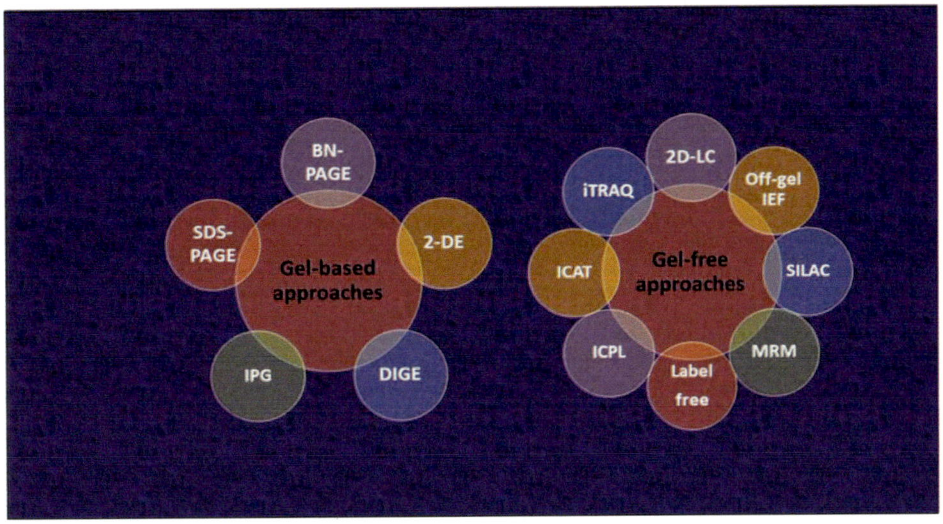

FIGURE 17.4 Potential approaches for studying plant response to HM toxicity by MS based proteomics. **2-DE** = two-dimensional gel electrophoresis, **2-DDIGE** = two-dimensional difference gel electrophoresis, **BN-PAGE** = blue native polyacrylamide gel electrophoresis, **SDS-PAGE** = sodium dodecyl sulfate polyacrylamide gel electrophoresis, **ICAT** = multiplexed isotope-coded affinity tag, **SILAC** = stable-isotope, **iTRAQ** = isobaric tags for relative and absolute quantitation, **ICPL** = isotope-coded protein label, **LC** = liquid chromatography.

proteomics approaches (Lemeer et al. 2012; Schmidt et al. 2014). Nowadays, distinct types of software and tools are available to unravel the proteome profile through database searches using MS data for protein identification and gel image analysis (Table 17.1). These resources are either freely or commercially assailable and their operational proficiency depends on many computational and graphical user interfaces (Martinez 2016; Pirzadah et al. 2019; Zhang et al. 2020).

17.3.1.2 Applications of Proteomics Methodologies to Elucidate Plant Stress Response

Proteomics is the promising tool for analysis of complete proteome or sum of all proteins from a plant cell or tissue under the HMs stress condition (Hossain and Komatsu 2013). Over the past decade, thorough researches on proteomic analyses of plants under HMs stress have been performed to unravel the tolerance mechanism (Cvjetko et al. 2014; Hasan et al. 2017; Jain et al. 2018). Most of the proteomic analyses confirmed that the proteins related to basic metabolic pathways (i.e. photosynthesis, respiration, nitrogen metabolism, and sulfur metabolism) have been differentially regulated in response to metal toxicity (Singh et al. 2016; Enot et al. 2020; Zhang et al. 2020). It has been reported that the many proteins linked to transport and cellular trafficking has shown differential expression under the HMs stress (Yang et al. 2016; Wang et al. 2020). Several proteomic studies have also shown that proteins involved in antioxidant defense like catalase (CAT), superoxide dismutase (SOD) glutathione (GSH), and allene oxide synthase (AOS), are differentially regulated under metal stresses (Xue et al. 2015; Wang et al. 2016; Baig et al. 2018; Chauhan et al. 2020).

In terms of transcript amount, it has also been reported that the expression of differentially expressed proteins (DEPs) varies among different tissue as well as developmental stages of plants. For instance, under Mn stress, a total of 356 differentially expressed proteins (DEPs) linked to defense response, photosynthesis, metabolism, cell wall modulation as well as signal transduction were identified from stylo (*Stylosanthes guianensis* of which 206 proteins are expressed in leaves while only 150 in roots (Liu et al. 2019). Something similar happened with proteins involved in oxide reduction processes, stress response, and photosynthetic processes of cacao tree (*Theobroma cacao*) under Cd stress, these proteins are greatly expressed in the plant roots than the leaves (Oliveira et al. 2020). Several reports have shown how the same plants exposed to different concentrations of the same metal may provide differential proteomic responses (Chen et al. 2015; Alotaibi et al. 2019; Lyu et al. 2020). Table 17.2 summarizes most of the investigations carried out to evaluate plant proteomes under HMs stress to date; these findings could be utilized for the development of HM tolerant crops and transgenic with multiple traits.

17.3.2 Genomics Approaches to Study Plant Response under HM Stress

17.3.2.1 Overview on Genomic Tools

Genomics tools are imperative tools to decipher the entire genetic make-up of plants. These tools provide valuable insights not only to the plant physiological and molecular responses but also to the specific strategies for crop improvement against abiotic stress (Akpınar et al. 2013). The innovation of different types of molecular markers such as restriction fragment length polymorphisms (RFLPs), random amplification of polymorphic DNAs (RAPDs), cleaved amplified polymorphic sequence (CAPS), simple sequence repeats (SSRs), amplified fragment length polymorphisms (AFLPs), single nucleotide polymorphisms (SNPs) and single feature polymorphisms (SFPs) has enabled consideration of quantitative traits into their single genetic modules and helped in selection as well as pyramiding QTL alleles through the marker-assisted

TABLE 17.1

Summary of Functional Proteomic Analyses in Response to HMs Stress (2010 to 2020)

HMs	Plant (Tissue)	Used Proteomics Methodologies/ Techniques	Major Findings	References
Arsenic (As)	*Brassica napus* (leaf)	Isobaric tag for relative and absolute quantitation (iTRAQ) based proteomic analysis	Increased abundance of proteins linked to primary metabolism, scavenging of ROS, and signal transduction network	Farooq et al. (2021)
Antimony (Sb)	*Miscanthus sinensis* (leaves and root)	2-DE,MALDI-TOF-MS and MALDI-TOF-TOF-MS	Increased expression of protein relative to antioxidant defense and stress response	Xue et al. (2015)
Mn	*Stylosanthes guianensis* (leaves and root)	Label-free proteomic approach, LC-MS/MS analysis	A total of 356 differentially expressed proteins (DEPs) were identified (206 proteins from leaves and 150 proteins from roots), which were mainly involved in defense response, photosynthesis, carbon fixation, metabolism, cell wall modulation and signaling	Liu et al. (2019)
Cadmium (Cd)	*Oryza sativa* (seedling)	HPLC-MS/MS, SDS-PAGE and iTRAQ	Enhanced expression of several proteins associated with carbohydrate metabolism, redox reaction and signal transduction	Liu et al. (2020)
	Theobroma cacao (leaves and roots)	2D SDS-PAGE and MS/MS	A greater expression of differentially accumulated proteins in the roots than in the leaves, especially proteins involved in oxidoreduction processes, defense and stress response, and photosynthetic processes	Oliveira et al. (2020)
	Sorghum bicolor (leaves)	2-DE coupled MALDI TOF/TOF-MS	Significant increases in the levels of proteins known to be involved in carbohydrate metabolism, transcriptional regulation, translation and stress responses	Roy et al. (2016)
	Spinacia oleracea (leaves)	2DE-MALDI TOF/TOF MS	Up-regulation of proteins associated with cell signaling, curtailment of oxygen species, suppression of proteolytic pathways, as well as cellular infrastructure	Bagheri et al. (2015)
	Helianthus annuus (leaves)	2D-DIGE and nESI-LC–MS/MS	Abundant expression of proteins those are involved in energy and disease/defense(including stress), the ribulose bisphosphate carboxylase large chain, transketolase, and heat shock proteins	Júnior et al. (2015)
	Populus yunnanensis (leaves)	SDS-PAGE, 2-DE coupled with MALDI-TOF/TOF-MS	Transiently induced photosynthesis- and energy-associated proteins, antioxidant enzymes and heat shock proteins (HSPs) accumulated to enhance protein stability and establish a new cellular homeostasis	Yang et al. (2015
	Oryza sativa (leaves)	2-DE coupled MALDI-TOF/TOF-MS	Induced up-regulation of protein like class III peroxidase, thaumatin-like protein irrespective, disulfide isomerase, a HSP70 homologue, a NADH-ubiquinone oxidoreductase, and a putative phosphogluconate dehydrogenase	Nwugo and Huerta (2011)
Copper (Cu)	*Phragmites australis*	LC-MS/MS and iTRAQ	Up regulation of proteins related to stress defense	Wu et al. (2020)
	Ocimum basilicum	SDS-PAGE,LC-MS-MS/MS	Increased accumulation of specific proteins related to transpiration and photosynthetic processes	Georgiadou et al. (2018)
	Oenothera glazioviana (root)	Label-free proteomic analysis	Increased expression of 58 DAPs, of which most of the identified proteins could be annotated to signal transduction, detoxification, stress defence, carbohydrate, energy, and protein metabolism, development, and oxidoreduction	Wang et al. (2017)
	Sorghum bicolor (root)	2D-PAGE combined with MALDI-TOF/TOF MS	21 proteins associated with the stress-related, metabolism and regulatory proteins showed increased abundance and provide valuable insights to establish a homeostasis under Cu stress	Roy et al. (2017)
	Oryza sativa (root)	2-DE and MALDI-TOF/TOF MS	Increased expression of responsive proteins which play an important role in the detoxification of excess Cu and in maintaining cellular homeostasis	Chen et al. (2015)
Chromium (Cr)	*Zea mays* (shoots and roots)	2-DE and MALDI-TOF-MS	Expression of DEPs associated with stress response (41.3%), energy and carbohydrate metabolism (21.7%), protein metabolism (6.5%), amino acid metabolism (6.5%), and others of unknown functions	Terzi and Yıldız (2020)
	Brassica napus (leaves)	2-DEand MALDI- TOF/TOF-MS	Increased accumulation of proteins associated with photosynthesis, energy metabolism, stress defense and redox regulation	Yıldız and Terzi (2016)

Lead (Pb)	*Allium cepa* (root)	2D-PAGE, TOF/TOF MS	A total of 17 proteins were expressed differentially, of which increased expression of NDPK indicated improved transcription and protein synthesis while the increased expression of PR1 and CHI1 indicated improved pathogen invasion	Lyu et al. (2020)
	Rhoeo discolour (leaves)	2D-DIGE,LC–MS/MS and label-free quantitation	Significant abundance of proteins relative to photosynthetic apparatus	Enot et al. (2020)
	Glycine max (root nodules)	SDS-PAGE, 2D, MALDI TOF MS/MS	Increased abundance of defence, development and repair related proteins. it has also induced aunique protein, sulfotransferase which controls the activity of several bioactive compounds involved in plant stress response	Baig et al. (2018)
	Raphanus sativus (roots)	iTRAQLC–MS/MS proteomic analysis	Increased abundance of antioxidative defense, detoxification, carbohydrate and energy metabolism related proteins	Wang et al. (2016)
Mn	*Stylosanthes guianensis* (leaves and root)	Label-free proteomic approach, LC-MS/MS analysis	A total of 356 differentially expressed proteins (DEPs) were identified (206 proteins from leaves and 150 proteins from roots), which were mainly involved in defense response, photosynthesis, carbon fixation, metabolism, cell wall modulation and signaling	Liu et al. (2019)
Mercury (Hg)	*Paspalum distichum* (root)	LC-MS/MS	A total 49 DEPs involving photosynthesis and energy metabolism (31%), oxidative stress (14%), protein folding (16%), sulfur compound metabolism (10%), metal binding, and ion transport (29%) are differently regulated	Ding et al. (2019)
Zinc (Zn)	*Lactuca sativa* (leaves)	SDS, Hybrid quadrupole-time-of flight(Q-TOF) MS	Enhanced synthesis of proteins relative to photosynthetic apparatus and oxidative defense	Lucini and Bernardo (2015)

TABLE 17.2

Some Commercially Available Resources Used in Proteomics Research

Application of Resources	Name of Software/Tools/Databases	Source
Gel image analysis	DeCyder™ 2D analysis	www.apbiotech.com
	ImageMasterTM 2D Elite	www.apbiotech.com
	Investigator™ HT PC Analyser	www.genomicsolutions.com
	PDQuest™ 2-D	www.bio-rad.com
	Phoretix™ 2D	www.nonlinear.com
	Progenesis™	www.nonlinear.com
	TotalLab	www.totallab.com
	Quantity One	www.bio-rad.com
	Z3 2D-Gel Analysis System	www.2dgels.com
Mass spectrometry analysis	Mascot	www.matrixscience.com
	MS-FIT	https://prospector.ucsf.edu/
	MS-Seq	https://prospector.ucsf.edu/
	PeptideSearch	www.narrador.embl-heidelberg.de
	PeptIdent	www.expasy.ch
	PepFrag	http://prowl.rockefeller.edu/
	Sequest	http://fields.scripps.edu/
	Sonars MS/MS	www.proteometrics.com
	TagIdent	www.expasy.ch
Tools and databases for protein sequences	BLAST	http://www.ncbi.nlm.nih.gov/Tools/BLAST
	Sequence Manipulation Suite	http://wire.ndsu.nodak.edu/DEALING/DMtools/SMS
	Bioinformatic.Net	http://www.bioinformatics.vg/ A
	Molecular Biology Database Collection	http://www3.oup.co.uk/nar/database/c/
	SwissProt	http://us.expasy.org/sprot/
	UniProt	http://www.uniprot.org/

selection (MAS) approach (Nadeem et al. 2018; Gantait et al. 2019; Ramesh et al. 2020; Younis et al. 2020). High-throughput sequencing techniques like next-generation sequencing (NGS) and genome-wide association studies (GWAS) have made them more feasible and provide remarkable insights to predict abiotic stress-related genes (Challa and Neelapu 2018; Saidi and Hajibarat 2020).

Several technologies are available for the analysis of gene expression, which can be grouped into two major categories such as open system and closed system. The open system such as cDNA-AFLP, MPSS (massively parallel signature sequencing), and SAGE (serial analysis of gene expression) does not necessitate any previous knowledge about the genome or transcriptome of plants. A closed system such as microarray, quantitative real-time PCR (qRT-PCR) requires somewhat knowledge about the genome that being studied. Both types of systems have been successfully used to elucidate the expression of thousands of genes in control and stressed tissues at various developmental stages of plants (Bansal et al., 2014; Hou et al. 2015; Bagati et al. 2018). The expressed sequence tags (ESTs), digital gene expression (DGE), and some other parallel strategies like 454 sequencings, complementation and promoter trapping, random and targeted mutagenesis have also been used to an expression of thousands of genes in control and abiotic stress-treated tissues at different developmental stages (Ambardar et al. 2016; Yuan et al. 2017; Sui et al. 2018). The CRISPR-Cas9 (clustered regularly interspaced short palindromic repeats) system has emerged as a tool with high-throughput applications like gene editing, gene regulation, epigenetic modification, and chromosomal imaging (Zhang et al. 2019; Bhat et al. 2021).

The biological data generated during the functional genomic studies contains a vast amount of information that cannot be grasped without the help of computational methods including bioinformatics (Martinez 2016). Furthermore, the free availability of functional genomic data related to not just model plants but also the other species provides fertile ground for biologists to advancing and accelerating more biological discovery (Baxevanis and Bateman 2015; Hickey et al. 2017). Several important tools and databases currently exist to make the relevant biological information available to the researchers and plant breeders (Table 17.3).

TABLE 17.3

List of Some Important Tools and Databases Available for Plant Genomic Research

Type of Resources	Name	Description/Applications	Source
Tools	MISA	Used for the identification as well as localization of perfect microsatellites and compound microsatellites in sequences	http://pgrc.ipkgatersleben.de/misa/
	AutoSNP	For identify putative polymorphisms between orthologous and parologous sequences from expressed sequence databases (ESTs)	http://www.cerealsdb.uk.net/discover.htm
	SNP2CAPS	For computational conversion of single nucleotide polymorphisms (SNPs) into cleaved amplified polymorphic sequence(CAPS) markers	http://pgrc.ipkgatersleben.de/snp2caps/
	MicroArray Software	A software package for microarray data analysis, datamining and data visualization	https://www.cs.tcd.ie/Nadia.Bolshakova/softwaretotal.html
	TASSEL	Software package to evaluate trait associations, evolutionary patterns as well as linkage disequilibrium	http://www.maizegenetics.net/bioinformatics/tasselindex.htm
	ORFfinder	For identification of all possible open reading frames in a DNA sequence	http://www.ncbi.nlm.nih.gov/Tools/ORFfinder
	e-PCR	To search DNA sequence for sequence tagged sites of a genome	http://www.ncbi.nlm.nih.gov/Tools/e-PCR
	MapMan	User-driven tool for displays large datasets onto diagrams of metabolic pathways or more processes	http://gabi.rzpd.de/projects/MapMan/
Databases	Genbank	Public sequence repository	http://www.ncbi.nlm.nih.gov/genbank/
	EMBL	General public sequence repository	http://www.ebi.ac.uk/embl/
	NCBI	Public databases with software tools for storing, disseminating and analyzing genome data	http://www.ncbi.nih.gov/
	GRAMENE	Curated and web-accessible data resource for comparative genome analysis especially for grasses	http://www.gramene.org/
	PlantMarkers	Database of predicted plant molecular markers	http://markers.btk.fi/
	PEDANT	Genome database that provides exhaustive automatic analysis of genomic sequences using bioinformatics tools	http://pedant.gsf.de/
	Plant GDB	A web resource devoted to develop robust genome annotation methods and tools for plant genomes	http://www.plantgdb.org/
	SALAD	Comparative genomics database constructed from plant-genome-based proteome data sets	http://salad.dna.affrc.go.jp/
	PlantTribes	Gene family resource for comparative genomics	http://fgp.bio.psu.edu/tribedb/
	PlantGenIE.org	Collection of web resources for searching, visualizing and analyzing genomics and transcriptomics data	http://plantgenie.org/
	GOLD	Genomes online database	https://gold.jgi.doe.gov/

17.3.2.2 Applications of Genomic Methodologies to Elucidate Plant Stress Responses

Plant responses to HMs stress is quantitative and complex traits that controlled by either the set of a single gene or multiple genes, commonly called the quantitative trait loci (QTLs). Therefore, genomics studies focusing on plant response to HMs toxicity have been largely summarized into the following groups: structural genomics (aiming to identify, locate, and order genomic features along chromosomes), functional genomics (functions of genes and gene networks), and comparative genomics (utilizing the conservation between closely related or distantly related species). Together, these genomics approaches can characterize a genome to its full extent, and provide promising perspectives inbreeding as well as transgenic improved varieties against HMs stress and more agronomic traits (Akpınar et al. 2013; Panda et al. 2015; Jha and Bohra 2016; Gallo-Franco et al. 2020; Anwar and Kim, 2020, Figure 17.5).

Genetic mapping of QTLs by DNA marker assays is the first phase for discovering the genetic framework of a quantitative or complex trait. Several studies have been conducted over the last two decades to the identification of QTLs in both the model and non-model plants under HMs stress (Table 17.4). It has also been reported that increasing implementation of genome-scale techniques like high-throughput GWAS dramatically improves the scope for genomics-assisted breeding for HM resistance crops (Korte and Farlow 2013; Qin et al. 2015; Shrestha et al. 2018). Zhang et al. (2017) have been performed a GWAS to QTL mapping and candidate gene analysis across Fe and Zn toxicity at the seedling stage in rice. The authors find out a total of 31 putative QTL affecting various morphological features like shoot height, root length, shoot fresh weight, shoot dry weight, root dry weight, shoot water content across metal exposure. Five toxicity tolerances QTL and 22 candidate genes for 10 important QTL regions were also detected in the same genomic regions. Recently, the implementation of GWAS in rapeseed (*Brassica napus* L.) reveals four genetic loci of lead (Pb) tolerance during seedling establishment (Zhang et al. 2020).

One of the greatest advantages of functional genomics compared to other approaches is the prospectus of analyzing the expression of a number of genes simultaneously; resulting in a more comprehensive picture of genome or transcriptome reforms under stress conditions (Bohnert et al. 2006). Plant stress-responsive genes have been largely grouped as early responsive genes (induced quickly for a short time period) and delayed responsive genes (induced gradually with long-lasting expression). It has been observed that most of the metal stress-responsive genes involve complex regulation mechanisms at both the transcriptional and post-transcriptional levels (Rascio and Navari-Izzo 2011; Sun et al. 2014). The regulation mechanism primarily governs by the transcriptional factors (TFs) and several families of TFs such as bHLH (basic helix-loop-helix), basic leucine zipper (bZIP), myeloblastosis protein (MYB), ethylene-responsive factor(ERF), WRKY, and many others have been recognized to influence metal stress response in plants(Hussain et al. 2011; Thapa et al. 2012; Singh et al. 2016). Table 17.5 summarizes recent investigations carried out to evaluate the involvement of TFs under HMs stress; these findings could be utilized to generate genetically improved crops for extreme environmental circumstances like HM stress. The miRNA-based approaches have also been used to deciphering transcriptional and post-translational as well as epigenetic regulation of plant adaptions toward metal stress conditions (Mendoza-Soto et al. 2012; Noman and Aqeel 2017). Several reports on metal stress-responsive miRNAs and their targets are available in cereals (Budak et al. 2015; Wu et al. 2018; Ding et al. 2018; Yu et al. 2019; Zhai et al. 2020) and

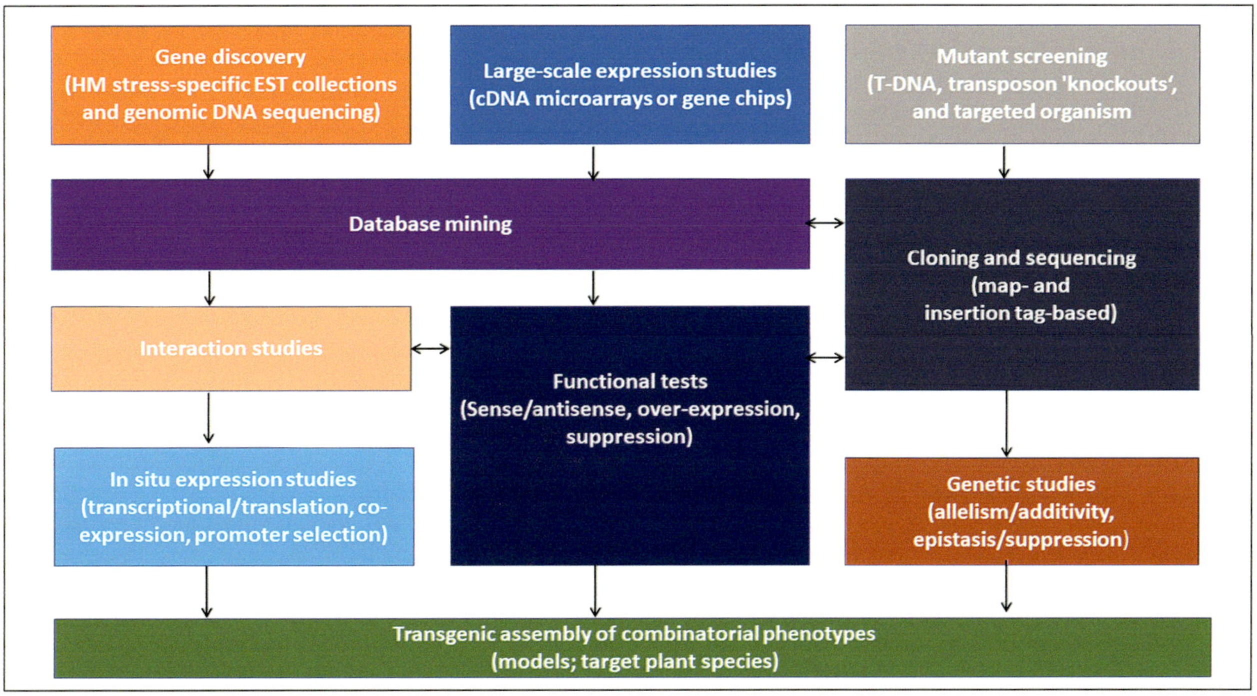

FIGURE 17.5 Integrated genomics strategy from gene discovery to evaluation of metal stress tolerance phenotypes.

TABLE 17.4

Selected List of Experiments Performed for the Identification of QTLs in Plants under HMs Stress

Plant	Mapping Population	QTL/Loci	Genotyping/ Linkage Map	Phenotypic Variance (PV) %	Chromosomal Location	Significance/ applications	Reference
Arabidopsis thaliana	RIL, Col-0 × Bur-0	3QTLs	CAPS	50	5	Cadmium(Cd) tolerance	Fischer et al. (2017)
Arabidopsis halleri	BC1, *A. halleri* × *A. l. petraea*	3 QTLs (Cdtol-1*)	–	16–43	3	Cd and Zn (Zinc) tolerance	Courbot et al. (2007)
Maize	RIL, B73 × Mo17	1	–	20	2	Cd accumulation	Zdunić et al. (2014)
	RIL, kernels × 9782	qPC1 and qPC4	SSR	5.15–11.13	1and 4	Lead tolerance	Zhao et al. (2014)
Populus	sBC, *P. trichocarpa* Torr. and Gray × *P. deltoides* Bart	16 QTLs	Whole-genome microarray	5.9–11.6	2	Cd tolerance, identification of nine Cd-responsive genes	Induri et al. (2012)
Raphanus sativus L	F2,NAU-Dys9 × NAU-Yh	6 QTLs (qRCd9*)	RAPD,SSR,ISSR and RAMP	9.86–48.64	1, 4, 6, and 9	Cd accumulation in root	Xu et al. (2012)
Rice	Japonica, Yuefu × IRAT109	3 QTLs	–	35.7	1, 2 and 5	Mercury tolerance, relative root length	Wang et al. (2013)
	RIL, Suwon490 × SNU-SG1	scc10, gcc3, gcc9, gcc11, sgr5	SSR	43.5–59.9	10, 3, 9, 11, and 5	Cd accumulation in shoot	Yan et al. (2013)
Thlaspi caerulescens	F3, Lellingen (LE) × La Calamine (LC)	2QTL	AFLP	16.6–21.7	–	Zn accumulation in root	Assunção et al. (2006)
Wheat	RIL, Ch 9 × Sh	26 QTLs	–	7.97–60.16	–	Cd tolerance and accumulation in root	Ci et al. (2012)

RIL = recombinant inbred line, sBC = pseudo-backcross, BC1 = first-generation backcross population, *= Predominant QTLs.

other species like *Ricinus communis* (Çelik and Akdaş 2019), *Raphanus sativus* (Liu et al. 2015), and *Brassica napus* (Jian et al. 2018; Fu et al. 2019).

17.3.2.3 An Overview of Modern Technologies for Developing Metal-Resilient Plants and Phytoremediation

Nowadays, genomes of many metal hyperaccumulators, like the model hyper accumulator *Thlaspic aerulescens* (now *Noccaeac aerulescens*, Cd, Ni, and Zn), *Arabidopsis halleri*(Cd and Zn), and non-model *Pterisvittata* (As, Pb), *Brassica juncea* (Pb, Zn, Cd), *Hirschfeldia incana* (Pb) and several other species have been either fully or partially sequenced (Briskine et al. 2017; Auguy et al. 2016; Yang et al. 2016; Mandáková et al. 2015; Xie et al. 2009). Rajput et al. (2021) has also illustrated phylogenetic clustered for revealing the phylogenetic relationship of several metal hyperaccumulators. Manipulation of genomic sequences of these metal hyperaccumulators may facilitate the identification and characterization of key genetic determinants for the development of metal-resilient plants and exploration of phytoremediation strategies. Sequence data information of these plants can be utilized to establish ZFNs (zinc finger nucleases), TALENs (transcription activator-like effector nucleases), and a high-throughput CRISPR-Cas9 system for the targeted engineering of mechanisms involved in the tolerance, uptake, and hyperaccumulation of HMs or metalloids (Estrela and Cate 2016).

Areas of focus for both purposes may include CRISPR-mediated increased expression genes that confer HM tolerance and hyperaccumulation like metal ligands, defense molecules, metal transporters proteins, plant growth hormones, root exudates, and TFs (Basharat et al. 2018). In fact, CRISPR-Cas9 systems have already been used to modify the genome of many species for increased agronomic traits (like plant biomass, growth rate, yield, and productivity) and disease and climate resistance (Miglani 2017). Recently, it has applied to modify the genome of species like poplar and maize, known for their applicability in phytoremediation processes (Fan et al. 2015; Agarwal et al. 2018).

Conclusion and Future Prospectus

Increasing contamination of ecosystem by the HMs is a grim concern, as they are non-biodegradable and easily get accumulated in plant tissues. Plant responses to HM stress is genetically complex traits, having involvement of multiple genes and signaling components and difficult to get controlled or engineered. Over decades, several efforts have been made to elucidate the osmolytes, stress related proteins, TFs, signaling molecules and signaling cascades that confer the plant HMs stress tolerance. As results, considerable amount of data is now available at the genome as well as proteome levels on plant response to HMs stress. Further advancement is required in various functional tools and resources for both the genomics and proteomics strategies to intricate HMs stress regulatory

TABLE 17.5
Some Examples of TFs Whose Overexpression Confers HMs Stress Tolerance in Plants

Name of TFs	Family of TFs	Source Species	Targeted Transgenic	Functional Response	References
PoWRKY	WRKY	*Polygonatum odoratum*	*Arabidopsis*	Improved seed germination, root growth and oxidative defense by improved super oxide dismutase (SOD) activity and proline content	Wei et al. (2021)
OsMYB-R1	MYBs	*Oryza sativa* L.	*Arabidopsis*	Cd tolerance, promoting growth and defense response	Tiwari et al. (2020)
GmWRKY142	WRKY	*Glycine max*	*Arabidopsis thaliana*	Enhanced Cd tolerance and lead to extensive transcriptional reprogramming of stress-responsive genes	Cai et al. (2020)
NAC	NAC	*Aegilops markgrafii*	Wheat	Improved Cd tolerance	Du et al. (2020)
ZmbHLH105	basic helix-loop-helix (bHLH)	*Zea mays*	*Nicotiana tabacum*	Improved Mn stress tolerance by regulating antioxidant machinery-mediated ROS scavenging and expression of Mn/Fe-related transporters	Sun et al. (2019)
RsMYB1	MYBs	*Raphanus sativus*	*Petunia hybrida*	Improves tolerance to Cu, Zn, and Mn by enhanced expression of glutathione S-transferase (GST), phytochelatin synthase (PCS) and antioxidants like SOD, catalase (CAT), and peroxidase (POX)	Ai et al. (2018)
StDREB1 and StDREB2	Drought-responsive element binding (DREB)	*Solanum tuberosum*	*Solanum tuberosum*	Improves tolerance to Cd by improving plant growth, proline and antioxidant production leading to low oxidative stress damage	Charfeddine et al. (2017)
BnbZIP3	bZIP	*Boehmeria nivea*	*Arabidopsis*	Improved Cd tolerance and roots growth	Huang et al. (2016)
SbMT-2		*Salicornia brachiate*	*Nicotiana tabacum*	Enhanced accumulation of Cu, Cd, and Zn and improved tolerance by modulating ROS scavenging	Chaturvedi et al. (2014)
PtsrMYB	MYB	*Poncirus trifoliata*	*Nicotiana tabacum*	Enhanced dehydration tolerance and elevated polyamine biosynthesis	Sun et al. (2014)
WRKY6	WRKY	–	*Arabidopsis thaliana*	Modulates arsenate As(V) uptake, transposon expression and provide coordinated strategy for As(V) tolerance	Castrillo et al. (2013)
bHLH100	bHLH	*Sedum alfredii*	*Sedum alfredii*	Improved Cd hyper-accumulation and tolerance	Gao et al. (2013)

networks and their downstream genes. However, the genome sequences are available for some of metal hyperaccumulators, more genome sequencing and *in silico* analysis of metal hyperaccumulators are required to decipher more novel genes, to develop metal-resilient varieties through genetic engineering (CRISPR-Cas9) and genome assistant breeding for sustainable agriculture and phytoremediation.

REFERENCES

Abdallah, C., Dumas-Gaudot, E., Renaut, J. and Sergeant, K. 2012. Gel-based and gel-free quantitative proteomics approaches at a glance. *Int J Plant Genom*, 2012, Article ID 494572, 1–17.

Agarwal, A., Yadava, P., Kumar, K., Singh, I., Kaul, T., Pattanayak, A. and Agrawal, P. K. 2018. Insights into maize genome editing via CRISPR/Cas9. *PMBP*, 24, 175–183.

Ai, T. N., Naing, A. H., Yun, B.W., Lim, S. H. and Kim, C. K. 2018. Overexpression of RsMYB1 enhances anthocyanin accumulation and heavy metal stress tolerance in transgenic petunia. *Front Plant Sci*, 9, 1388.

Akpınar, B.A., Lucas, S. J. and Budak, H. 2013. Genomics approaches for crop improvement against abiotic stress. *Sci World J*, 2013, Article ID 361921.

Ali, H., Khan, E. and Sajad, M. A. 2013. Phytoremediation of heavy metals—concepts and applications. *Chemosphere*, 91, 869–881.

Alotaibi, M. O., Mohammed, A. E., Almutairi, T. A. and Elobeid, M. M. 2019. Morpho-physiological and proteomic analyses of *Eucalyptus camaldulensis* as a bioremediator in copper-polluted soil in Saudi Arabia. *Plants*, 8, 43.

Ambardar, S., Gupta, R., Trakroo, D., Lal, R. and Vakhlu, J. 2016. High throughput sequencing: an overview of sequencing chemistry. *Indian J Microbiol*, 56, 394–404.

Andrés-Colás, N. and Van Der Straeten, D. 2017. Optimization of non-denaturing protein extraction conditions for plant PPR proteins. *PloS One*, 12, e0187753.

Angel, T. E., Aryal, U. K., Hengel, S. M., Baker, E. S., Kelly, R. T., Robinson, E. W. and Smith, R. D. 2012. Mass spectrometry-based proteomics: existing capabilities and future directions. *Chem Soc Rev*, 41, 3912–3928.

Anwar, A. and Kim, J. K. 2020. Transgenic breeding approaches for improving abiotic stress tolerance: recent progress and future perspectives. *Int J Mol Sci*, 21, 2695.

Arif, N., Yadav, V., Singh, S., Singh, S., Ahmad, P., Mishra, R. K., Sharma, S., Tripathi, D. K., Dubey, N. K. and Chauhan, D. K. 2016. Influence of high and low levels of plant-beneficial heavy metal ions on plant growth and development. *Front Environ Sci*, 4, 69.

Asati, A., Pichhode, M. and Nikhil, K. 2016. Effect of heavy metals on plants: an overview. *IJAIEM*, 5, 56–66.

Assunção, A. G., Pieper, B., Vromans, J., Lindhout, P., Aarts, M. G. and Schat, H. 2006. Construction of a genetic linkage map of *Thlaspic aerulescens* and quantitative trait loci analysis of zinc accumulation. *New Phytologist*, 170, 21–32.

Atici, Ö., Ağar, G. and Battal, P. E. Y. A. M. İ. 2005. Changes in phytohormone contents in chickpea seeds germinating under lead or zinc stress. *Biol Plant*, 49, 215–222.

Auguy, F., Fahr, M., Moulin, P., El Mzibri, M., Smouni, A., Filali-Maltouf, A., Béna, G. and Doumas, P. 2016. Transcriptome changes in *Hirschfeldia incana* in response to lead exposure. *Front Plant Sci*, 6,1231.

Baderna, D., Lomazzi, E., Pogliaghi, A., Ciaccia, G., Lodi, M. and Benfenati, E. 2015. Acute phytotoxicity of seven metals alone and in mixture: are Italian soil threshold concentrations suitable for plant protection?. *Environ Res*, 140, 102–111.

Bagati, S., Mahajan, R., Nazir, M., Dar, A. A. And Zargar, S. M. 2018. "Omics": A gateway towards abiotic stress tolerance. In S.M, Zargar; M.Y, Zargar (Eds.) *Abiotic stress-mediated sensing and signaling in plants: An omics perspective (pp 1-45)*. Singapore, Springer.

Bagheri, R., Bashir, H., Ahmad, J., Iqbal, M. And Qureshi, M. I. 2015. Spinach (*Spinacia oleracea* L.) modulates its proteome differentially in response to salinity, cadmium and their combination stress. *Plant Physiol Biochem*, 97, 235–245.

Baig, M. A., Ahmad, J., Bagheri, R., Ali, A. A., Al-Huqail, A. A., Ibrahim, M. M. and Qureshi, M. I. (2018). Proteomic and ecophysiological responses of soybean (*Glycine max* L.) root nodules to Pb and hg stress. *BMC Plant Biol*, 18, 1–21.

Bansal, K. C., Lenka, S. K. and Mondal, T. K. 2014. Genomic resources for breeding crops with enhanced abiotic stress tolerance. *Plant Breed*, 133, 1–11.

Basharat, Z., Novo, L. A. and Yasmin, A. 2018. Genome editing weds CRISPR: what is in it for phytoremediation? *Plants*, 7, 51.

Baxevanis, A. D. and Bateman, A. 2015. The importance of biological databases in biological discovery. *Curr Protoc Bioinf*, 50, 1–1.

Belwal, T., Ezzat, S. M., Rastrelli, L., Bhatt, I. D., Daglia, M., Baldi, A., Devkota, H. P., Orhan, I. E., Patra, J. K., Das, G., Anandharamakrishnan, C., Gomez-Gomez, L., Nabavi, S. F., Nabavi, S. M., and Atanasov, A. G. 2018. A critical analysis of extraction techniques used for botanicals: trends, priorities, industrial uses and optimization strategies. *TrAC*, 100, 82–102.

Bhat, M. A., Mir, R. A., Kumar, V., Shah, A. A., Zargar, S. M., Rahman, S. and Jan, A. T. 2021. Mechanistic insights of CRISPR/Cas mediated genome editing towards enhancing abiotic stress tolerance in plants. *Physiol Plant*, 172:1255–1268

Bielen, A., Remans, T., Vangronsveld, J. and Cuypers, A. 2013. The influence of metal stress on the availability and redox state of ascorbate, and possible interference with its cellular functions. *Int J Mol Sci*, 14(3), 6382–6413.

Bohnert, H. J., Gong, Q., Li, P. and Ma, S. 2006. Unraveling abiotic stress tolerance mechanisms–getting genomics going. *Curropin Plant Biol*, 9, 180–188.

Borges, K. L. R., Salvato, F., Loziuk, P. L., Muddiman, D. C. and Azevedo, R. A. 2019. Quantitative proteomic analysis of tomato genotypes with differential cadmium tolerance. *ESPR*, 26, 26039–26051.

Briskine, R. V., Paape, T., Shimizu-Inatsugi, R., Nishiyama, T., Akama, S., Sese, J. and Shimizu, K. K. 2017. Genome assembly and annotation of *Arabidopsis halleri*, a model for heavy metal hyperaccumulation and evolutionary ecology. *Mol Ecol Resour,* 17, 1025–1036.

Budak, H., Kantar, M., Bulut, R. and Akpinar, B. A. 2015. Stress responsive miRNAs and isomiRs in cereals. *Plant Sci*, 235, 1–13.

Cai, Z., Xian, P., Wang, H., Lin, R., Lian, T., Cheng, Y., Ma, Q. and Nian, H. 2020. Transcription factor GmWRKY142 confers cadmium resistance by up-regulating the cadmium tolerance 1-like genes. *Front Plant Sci*, 11, 724.

Castrillo, G., Sánchez-Bermejo, E., de Lorenzo, L., Crevillén, P., Fraile-Escanciano, A., Mohan, T. C., Mouriz, A., Catarecha, P., Sobrino-Plata, J., Olsson, S. and Del Puerto, Y. L. 2013. WRKY6 transcription factor restricts arsenate uptake and transposon activation in *Arabidopsis*. *The Plant Cell*, 25, 2944–2957.

Çelik, Ö. and Akdaş, E. Y. 2019. Tissue-specific transcriptional regulation of seven heavy metal stress-responsive miRNAs and their putative targets in nickel indicator castor bean (*R. communis* L.) plants. *Ecotoxicol Environ Saf*, 170, 682–690.

Çelik, Ö., Ayan, A., Meriç, S. and Atak, Ç. 2020. Heavy metal stress-responsive phyto-miRNAs. *Cellular and molecular phytotoxicity of heavy metals*. Springer, Cham.

Challa, S. and Neelapu, N. R. 2018. Genome-wide association studies (GWAS) for abiotic stress tolerance in plants. *Biochemical, physiological and molecular avenues for combating abiotic stress tolerance in plants*. Elsevier Inc, Academic Press.

Chaplygin, V. A., Rajput, V. D., Mandzhieva, S. S., Minkina, T. M., Nevidomskaya, D. G., Nazarenko, O. G., Kalinitchenko, V. P., Singh, R., Maksimov, A. Y., and Popova, V. A., 2020. Comparison of heavy metal content in *Artemisia austriaca* in various impact zones. *ACS Omega*, 5, 23393–23400.

Charfeddine, M., Charfeddine, S., Bouaziz, D., Messaoud, R. B. and Bouzid, R. G., 2017. The effect of cadmium on transgenic potato (*Solanum tuberosum*) plants overexpressing the StDREB transcription factors. *PCTOC*, 128, 521–541.

Chaturvedi, A. K., Patel, M. K., Mishra, A., Tiwari, V. and Jha, B., 2014. The SbMT-2 gene from a halophyte confers abiotic stress tolerance and modulates ROS scavenging in transgenic *tobacco*. *PloS One*, 9, e111379.

Chaudhary, K., Agarwal, S. and Khan, S. 2018. Role of phytochelatins (PCs), metallothioneins (MTs), and heavy metal ATPase (HMA) genes in heavy metal tolerance. *Mycoremediation and environmental sustainability*. Springer, Cham.

Chauhan, R., Awasthi, S., Indoliya, Y., Chauhan, A. S., Mishra, S., Agrawal, L., Srivastava, S., Dwivedi, S., Singh P. C., Mallick, S., Chauhan, P. S., Pande, V., Chakrabarty, D. and Tripathi, R. D. 2020. Transcriptome and proteome analyses reveal selenium mediated amelioration of arsenic toxicity in rice (*Oryza sativa* L.). *J Hazard Mater*, 390, 122122.

Chen, C., Song, Y., Zhuang, K., Li, L., Xia, Y. and Shen, Z. 2015. Proteomic analysis of copper-binding proteins in excess copper-stressed roots of two rice (*Oryza sativa* L.) varieties with different Cu tolerances. *PLoS One*, 10, e0125367.

Chowdhury, R., Favas, P. J., Jonathan, M. P., Venkatachalam, P., Raja, P. and Sarkar, S. K. 2017. Bioremoval of trace metals from rhizosediment by mangrove plants in Indian Sundarban Wetland. *Mar Pollut Bull*, 124, 1078–1088.

Ci, D., Jiang, D., Li, S., Wollenweber, B., Dai, T. and Cao, W. 2012. Identification of quantitative trait loci for cadmium tolerance and accumulation in Wheat. *Acta Physiol. Plant*, 34, 191–202.

Courbot, M., Willems, G., Motte, P., Arvidsson, S., Roosens, N., Saumitou-Laprade, P. and Verbruggen, N. 2007. A major quantitative trait locus for cadmium tolerance in *Arabidopsis hallericolocalizes* with HMA4, a gene encoding a heavy metal ATPase. *Plant Physiol*, 144, 1052–1065.

Cvjetko, P., Zovko, M. and Balen, B. 2014. Proteomics of heavy metal toxicity in plants. *Arh Hig Rada Toksikol*, 65, 1–18.

DalCorso, G., Manara, A. and Furini, A. 2013. An overview of heavy metal challenge in plants: from roots to shoots. *Metallomics*, 5, 1117–1132.

Ding, W., Zhang, J., Wu, S. C., Zhang, S., Christie, P. and Liang, P. 2019. Responses of the grass *Paspalum distichum* L. to Hg stress: a proteomic study. *Ecotoxicol Environ Saf*, 183, 109549.

Ding, Y., Gong, S., Wang, Y., Wang, F., Bao, H., Sun, J., Cai, C., Yi, K., Chen, Z. and Zhu, C. 2018. MicroRNA166 modulates cadmium tolerance and accumulation in rice. *Plant Physiol*, 177, 1691–1703.

Du, X., He, F., Zhu, B., Ren, M. and Tang, H. 2020. NAC transcription factors from *Aegilops markgrafii* reduce cadmium concentration in transgenic wheat. *Plant and Soil*, 449, 39–50

Dubey, S., Shri, M., Gupta, A., Rani, V. and Chakrabarty, D. 2018. Toxicity and detoxification of heavy metals during plant growth and metabolism. *Environ Chem Lett*, 16, 1169–1192.

Edelstein, M. and Ben-Hur, M. 2018. Heavy metals and metalloids: Sources, risks and strategies to reduce their accumulation in horticultural crops. *Sci Hortic*, 234, 431–444.

Enot, M. M., Weiland, F., Mittal, P., Hoffmann, P., Sillero-Mahinay, M. and Pukala, T. 2020. Differential proteome analysis of the leaves of lead hyperaccumulator, *Rhoeodiscolor* (L. Her.) Hance. *J Mass Spectrom*, e4689. https://doi.org/10.1002/jms.4689

Estrela, R. and Cate, J. H. D. 2016. Energy biotechnology in the CRISPR-Cas9 era. *Curr Opin Biotechnol*, 38, 79–84.

Fan, D., Liu, T., Li, C., Jiao, B., Li, S., Hou, Y. and Luo, K. 2015. Efficient CRISPR/Cas9-mediated targeted mutagenesis in *Populus* in the first generation. *Sci Rep*, 5(1), 1–7.

Farooq, M. A., Hong, Z., Islam, F., Noor, Y., Hannan, F., Zhang, Y., Ayyaz, A., Mwamba, T. M., Zhou, W. and Song, W. 2021. Comprehensive proteomic analysis of arsenic induced toxicity reveals the mechanism of multilevel coordination of efficient defense and energy metabolism in two *Brassica napus* cultivars. *Ecotoxicol Environ Saf*, 208, 111744.

Fischer, S., Spielau, T. and Clemens, S. 2017. Natural variation in *Arabidopsis thaliana* Cd responses and the detection of quantitative trait loci affecting Cd tolerance. *Sci Rep*, 7, 1–14.

Fu, Y., Mason, A.S., Zhang, Y., Lin, B., Xiao, M., Fu, D. and Yu, H. 2019. Micro RNA-mRNA expression profiles and their potential role in cadmium stress response in *Brassica napus*. *BMC Plant Biol*, 19, 1–20.

Gallo-Franco, J. J., Sosa, C. C., Ghneim-Herrera, T. and Quimbaya, M. 2020. Epigenetic control of plant response to heavy metal stress: a new view on aluminum tolerance. *Front Plant Sci*, 11, 602625.

Gantait, S., Sarkar, S. and Verma, S. K. 2019. Marker-assisted Selection for Abiotic Stress Tolerance in Crop Plants. *Molecular plant abiotic stress: biology and biotechnology*. John Wiley & Sons Ltd, Hoboken, NJ.

Gao, J., Sun, L., Yang, X. and Liu, J. X. 2013. Transcriptomic analysis of cadmium stress response in the heavy metal hyperaccumulator *Sedum alfredii* Hance. *PloS One*, 8, e64643.

Gemperline, E., Keller, C. and Li, L. 2016. Mass spectrometry in plant-omics. *Anal Chem*, 88, 3422–3434.

Georgiadou, E. C., Kowalska, E., Patla, K., Kulbat, K., Smolińska, B., Leszczyńska, J. and Fotopoulos, V. 2018. Influence of heavy metals (Ni, Cu, and Zn) on nitro-oxidative stress responses, proteome regulation and allergen production in basil (*Ocimum basilicum* L.) plants. *Front Plant Sci*, 9, 862.

Ghori, N. H., Ghori, T., Hayat, M. Q., Imadi, S. R., Gul, A., Altay, V. and Ozturk, M. 2019. Heavy metal stress and responses in plants. *IJEST*, 16, 1807–1828.

Hänsch, R. and Mendel, R. R. 2009. Physiological functions of mineral micronutrients (Cu, Zn, Mn, Fe, Ni, Mo, B, Cl). *Curr Opin Plant Biol*, 12, 259–266.

Hasan, M., Cheng, Y., Kanwar, M. K., Chu, X. Y., Ahammed, G. J. and Qi, Z. Y. 2017. Responses of plant proteins to heavy metal stress—a review. *Front Plant Sci*, 8, 1492.

Hickey, J. M., Chiurugwi, T., Mackay, I., Powell, W., Eggen, A., Kilian, A., Jones, C., Canales, C., Grattapaglia, D., Bassi, F. and Atlin, G. 2017. Genomic prediction unifies animal and plant breeding programs to form platforms for biological discovery. *Nat Genet*, 49, 1297.

Hossain, Z. and Komatsu, S. 2013. Contribution of proteomic studies towards understanding plant heavy metal stress response. *Front Plant Sci*, 3, 310.

Hou, J., Liu, X., Wang, J., Zhao, S. and Cui, B. 2015. Microarray-based analysis of gene expression in *Lycopersicon esculentum* seedling roots in response to cadmium, chromium, mercury, and lead. *Environ Sci Technol*, 49, 1834–1841.

Huang, T. L., Nguyen, Q. T. T., Fu, S. F., Lin, C. Y., Chen, Y. C. and Huang, H. J. 2012. Transcriptomic changes and signalling pathways induced by arsenic stress in rice roots. *Plant Mol Biol*, 80, 587–608.

Huang, C., Zhou, J., Jie, Y., Xing, H., Zhong, Y., She, W., Wei, G., Yu, W. and Ma, Y. 2016. A ramie (*Boehmeria nivea*) bZIP transcription factor BnbZIP3 positively regulates drought, salinity and heavy metal tolerance. *Mol Breed*, 36, 1–15.

Hussain, S., Khaliq, A., Noor, M. A., Tanveer, M., Hussain, H. A., Hussain, S., Shah, T. and Mehmood, T. 2020. Metal toxicity and nitrogen metabolism in plants: an overview. *Carbon and nitrogen cycling in soil*. Springer, Singapore.

Hussain, S. S., Kayani, M. A. and Amjad, M. 2011. Transcription factors as tools to engineer enhanced drought stress tolerance in plants. *Biotechnol Prog*, 27, 297–306.

Induri, B. R., Ellis, D. R., Slavov, G. T., Yin, T., Zhang, X., Muchero, W., Tuskan, G. A. and DiFazio, S. P. 2012. Identification of quantitative trait loci and candidate genes for cadmium tolerance in *Populus*. *Tree Physiol*, 32, 626–638.

Jain, S., Muneer, S., Guerriero, G., Liu, S., Vishwakarma, K., Chauhan, D. K., Dubey, N. K., Tripathi, D. K. and Sharma, S. 2018. Tracing the role of plant proteins in the response to metal toxicity: a comprehensive review. *Plant Signal Behav*, 13, e1507401.

Jian, H., Yang, B. O., Zhang, A., Ma, J., Ding, Y., Chen, Z., Li, J., Xu, X. and Liu, L. 2018. Genome-wide identification of microRNAs in response to cadmium stress in oilseed rape (*Brassica napus* L.) using high-throughput sequencing. *Int J Mol Sci*, 19, 1431.

Jorrin-Novo, J. V., Komatsu, S., Sanchez-Lucas, R. and de Francisco, L. E. R. 2019. Gel electrophoresis-based plant proteomics: past, present, and future. Happy 10th anniversary Journal of Proteomics! *J Proteom*, 198, 1–10.

Joshi, A., Kanthaliya, B. and Arora, J. 2018. Halophytes of Thar Desert: potential source of nutrition and feedstuff. *Int J Bioassays*, 8, 5674–5683.

Joshi, A., Kanthaliya, B. and Arora, J. 2019. Evaluation of growth and antioxidant activity in *Suaeda monoica* and *Suaeda nudiflora* callus cultures under sequential exposure to saline conditions. *Curr Biotechnol*, 8, 42–52.

Joshi, A., Kanthaliya, B. and Arora, J. 2020. Halophytic plant existence in Indian salt flats: biodiversity, biology, and uses. *Handbook of halophytes: from molecules to ecosystems towards biosaline agriculture*. Springer, Switzerland AG,1–22.

Joshi, A., Kanthaliya, B., Rajput, V., Minkina, T. and Arora, J. 2020. Assessment of phytoremediation capacity of three halophytes: *Suaedamonoica, Tamarixindica* and *Cressa critica*. *Biol Futur*, 71, 301–312.

Júnior, C. A. L., de Sousa Barbosa, H., Galazzi, R. M., Koolen, H. H. F., Gozzo, F. C. and Arruda, M. A. Z. 2015. Evaluation of proteome alterations induced by cadmium stress in sunflower (*Helianthus annuus* L.) cultures. *Ecotoxicol Environ Saf*, 119, 170–177.

Khan, M. I. R., Iqbal, N., Masood, A., Mobin, M., Anjum, N. A. and Khan, N. A. 2016. Modulation and significance of nitrogen and sulfur metabolism in cadmium challenged plants. *Plant Growth Regul*, 78, 1–11.

Kim, Y. H., Khan, A. L., Kim, D. H., Lee, S. Y., Kim, K. M., Waqas, M., Jung, H. Y., Shin, J. H., Kim, J. G. and Lee, I. J. 2014. Silicon mitigates heavy metal stress by regulating P-type heavy metal ATPases, *Oryza sativa* low silicon genes, and endogenous phytohormones. *BMC Plant Biol*, 14, 1–13.

Korte, A. and Farlow, A. 2013. The advantages and limitations of trait analysis with GWAS: a review. *Plant Methods*, 9, 1–9.

Kosakivska, I. V., Babenko, L. M., Romanenko, K. O., Korotka, I. Y. and Potters, G. 2021. Molecular mechanisms of plant adaptive responses to heavy metals stress. *Cell Biol Int*, 45, 258–272.

Kumar, R., Mishra, R. K., Mishra, V., Qidwai, A., Pandey, A., Shukla, S. K., Pandey, M., Pathak, A. and Dikshit, A. 2016. Detoxification and tolerance of heavy metals in plants. *Plant metal interaction*. Elsevier, Amsterdam, Netherlands.

Küpper, H. and Andresen, E. 2016. Mechanisms of metal toxicity in plants. *Metallomics*, 8, 269–285.

Lemeer, S., Hahne, H., Pachl, F. and Kuster, B. 2012. Software tools for MS-based quantitative proteomics: a brief overview. *Quantitative methods in proteomics*. Humana Press, Totowa, NJ.

Li, C., Zhou, K., Qin, W., Tian, C., Qi, M., Yan, X., and Han, W. 2019. A review on heavy metals contamination in soil: effects, sources, and remediation techniques. *Soil Sediment Contam*, 28; 380–394.

Lin, C. Y., Trinh, N. N., Lin, C. W. and Huang, H. J. 2013. Transcriptome analysis of phytohormone, transporters and signaling pathways in response to vanadium stress in rice roots. *Plant Physiol Biochem*, 66, 98–104.

Liu, L., Li, W., Song, W. and Guo, M. 2018. Remediation techniques for heavy metal-contaminated soils: principles and applicability. *Sci Total Environ*, 633, 206–219.

Liu, P., Huang, R., Hu, X., Jia, Y., Li, J., Luo, J., Liu, Q., Luo, L., Liu, G. and Chen, Z. 2019. Physiological responses and proteomic changes reveal insights into Stylosanthes response to manganese toxicity. *BMC Plant Biol*, 19, 1–21.

Liu, S., Li, Y., Liu, L., Min, J., Liu, W., Li, X., Pan, X., Lu, X. and Deng, Q. 2020. Comparative proteomics in rice seedlings to characterize the resistance to cadmium stress by high-performance liquid chromatography–tandem mass spectrometry (HPLC-MS/MS) with isobaric tag for relative and absolute quantitation (iTRAQ). *Anal Lett*, 53, 807–820.

Liu, W., Xu, L., Wang, Y., Shen, H., Zhu, X., Zhang, K., Chen, Y., Yu, R., Limera, C. and Liu, L. 2015. Transcriptome-wide analysis of chromium-stress responsive microRNAs to explore miRNA-mediated regulatory networks in radish (*Raphanus sativus* L.). *Sci Rep*, 5, 1–17.

Lucini, L. and Bernardo, L. 2015. Comparison of proteome response to saline and zinc stress in lettuce. *Front Plant Sci*, 6, 240.

Lyu, G., Li, D., Li, S., Ning, C. and Qin, R. 2020. Genotoxic effects and proteomic analysis on *Allium cepa var. agrogarum* L. root cells under Pb stress. *Ecotoxicol*, 29, 959–972.

Maleki, M., Ghorbanpour, M. and Kariman, K. 2017. Physiological and antioxidative responses of medicinal plants exposed to heavy metals stress. *Plant Gene*, 11, 247–254.

Mandáková, T., Singh, V., Krämer, U. and Lysak, M. A., 2015. Genome structure of the heavy metal hyperaccumulator *Noccaea caerulescens* and its stability on metalliferous and nonmetalliferous soils. *Plant Physiol*, 169, 674–689.

Martinez, M. 2016. Computational tools for genomic studies in plants. *Curr Genomics*, 17, 509–514.

Masindi, V. and Muedi, K. L. 2018. Environmental contamination by heavy metals. In H., Saleh, R., Aglan (Eds.) *Heavy metals*. IntechOpen. London, United Kingdom.

Meena, S., Kanthaliya, B., Joshi, A., Khan, F. and Arora, J. 2020. Biologia futura: medicinal plants-derived bioactive peptides in functional perspective—a review. *Biologia Futur*, 71, 195–208.

Mehta, S., James, D. and Reddy, M. K. 2019. Omics technologies for abiotic stress tolerance in plants: current status and prospects. *Recent approaches in omics for plant resilience to climate change*. Springer, Cham.

Mendoza-Soto, A.B., Sánchez, F. and Hernández, G. 2012. MicroRNAs as regulators in plant metal toxicity response. *Front Plant Sci*, 3, 105.

Miglani, G. S. 2017. Genome editing in crop improvement: present scenario and future prospects. *J Crop Improv*, 31, 453–559.

Minkina, T., Rajput, V., Fedorenko, G., Fedorenko, A., Mandzhieva, S., Sushkova, S., Morin, T., Yao, J. 2019. Anatomical and ultrastructural responses of Hordeum sativum to the soil spiked by copper. *Environ Geochem Health*, 42, 45–58.

Moon, Y. S., Park, E. S., Kim, T. O., Lee, H. S. and Lee, S. E. 2014. SELDI-TOF MS-based discovery of a biomarker in *Cucumis sativus* seeds exposed to CuO nanoparticles. *Environ Toxicol Pharmacol*, 38, 922–931.

Mosa, K. A., Ismail, A. and Helmy, M. 2017. Functional genomics combined with other omics approaches for better understanding abiotic stress tolerance in plants. *Plant stress tolerance*. Springer, Cham.

Nadeem, M. A., Nawaz, M. A., Shahid, M. Q., Doğan, Y., Comertpay, G., Yıldız, M., Hatipoğlu, R., Ahmad, F., Alsaleh, A., Labhane, N., Özkan, H., Chung G. and Baloch, F. S. 2018. DNA molecular markers in plant breeding: current status and recent advancements in genomic selection and genome editing. *Biotechnol Biotechnol Equip*, 32, 261–285.

Nagajyoti, P. C., Lee, K. D. and Sreekanth, T. V. M. 2010. Heavy metals, occurrence and toxicity for plants: a review. *Environ Chem Lett*, 8(3), 199–216.

Nguyen, H. C., Lin, K. H., Ho, S. L., Chiang, C. M. and Yang, C. M. 2018. Enhancing the abiotic stress tolerance of plants: from chemical treatment to biotechnological approaches. *Physiol Plant*, 164, 452–466.

Nikalje, G. C. and Suprasanna, P. 2018. Coping with metal toxicity–cues from halophytes. *Front Plant Sci*, 9, 777.

Nishioka, K., Kato, Y., Ozawa, S. I., Takahashi, Y. and Sakamoto, W. 2021. Phos-tag-based approach to study protein phosphorylation in the thylakoid membrane. *Photosynth Res*, 147, 107–124.

Noman, A. and Aqeel, M., 2017. miRNA-based heavy metal homeostasis and plant growth. *Environmental Science and Pollution Research*, 24,10068–10082.

Nowakowski, A. B., Wobig, W. J. and Petering, D. H. 2014. Native SDS-PAGE: high resolution electrophoretic separation of proteins with retention of native properties including bound metal ions. *Metallomics*, 6, 1068–1078.

Nwugo, C. C. and Huerta, A. J. 2011. The effect of silicon on the leaf proteome of rice (*Oryza sativa* L.) plants under cadmium-stress. *J Proteome Res*, 10, 518–528.

O'Lexy, R., Kasai, K., Clark, N., Fujiwara, T., Sozzani, R. and Gallagher, K. L. 2018. Exposure to heavy metal stress triggers changes in plasmodesmatal permeability via deposition and breakdown of callose. *J Exp Bot*, 69, 3715–3728.

Oliveira, B. R. M., de Almeida, A. A. F., Pirovani, C. P., Barroso, J. P., Neto, C. H. D. C., Santos, N. A., Ahnert, D., Baligar, V. C. and Mangabeira, P. A. O. 2020. Mitigation of Cd toxicity by Mn in young plants of cacao, evaluated by the proteomic profiles of leaves and roots. *Ecotoxicol*, 29, 340–358.

Page, V. and Feller, U. 2015. Heavy metals in crop plants: transport and redistribution processes on the whole plant level. *Agron J*, 5, 447–463.

Panda, P., Sahoo, L. and Panda, S. K. 2015. Heavy metal and metalloid stress in plants: the genomics perspective. *Abiotic Stresses in Crop Plants*, 164-177.

Pavoković, D., Križnik, B. and Krsnik-Rasol, M. 2012. Evaluation of protein extraction methods for proteomic analysis of non-model recalcitrant plant tissues. *Croat Chem Acta*, 85, 177–183.

Peng, J. S., Guan, Y. H., Lin, X. J., Xu, X. J., Xiao, L., Wang, H. H. and Meng, S. 2021 Comparative understanding of metal hyper accumulation in plants: a mini-review. *Environ Geochem Health*, 43:1599–1607

Pirzadah, T. B., Malik, B. and Hakeem, K. R. 2019. Integration of "Omic" *approaches to unravel the heavy metal tolerance in plants. Essentials of bioinformatics Volume III.* Springer, Cham.

Qin, P., Wang, L., Liu, K., Mao, S., Li, Z., Gao, S., Shi, H. and Liu, Y. 2015. Genomewide association study of *Aegilops tauschii* traits under seedling-stage cadmium stress. *Crop J*, 3, 405–415.

Qureshi, M. I., D'Amici, G. M., Fagioni, M., Rinalducci, S. and Zolla, L. 2010. Iron stabilizes thylakoid protein–pigment complexes in Indian mustard during Cd-phytoremediation as revealed by BN-SDS-PAGE and ESI-MS/MS. *J Plant Physiol*, 167, 761–770.

Rajput, V., Minkina, T., Fedorenko, A., Sushkova, S., Mandzhieva, S., Lysenko, V., Duplii, N., Fedorenko, G., Dvadnenko, K. and Ghazaryan, K. 2018a. Toxicity of copper oxide nanoparticles on spring barley (*Hordeum sativum distichum*). *Sci Total Environ*, 645, 1103–1113.

Rajput, V. D., Minkina, T., Suskova, S., Mandzhieva, S., Tsitsuashvili, V., Chapligin, V. and Fedorenko, A. 2018b. Effects of copper nanoparticles (CuO NPs) on crop plants: a mini review. *BioNanoSci*, 8, 36–42.

Rajput, V., Minkina, T., Semenkov, I., Klink, G., Tarigholizadeh, S. and Sushkova, S. 2021. Phylogenetic analysis of hyperaccumulator plant species for heavy metals and polycyclic aromatic hydrocarbons. *Environ Geochem Health*, 43:1629–1654.

Ramesh, P., Mallikarjuna, G., Sameena, S., Kumar, A., Gurulakshmi, K., Reddy, B. V., Reddy, P.C. and Sekhar, A. C. 2020. Advancements in molecular marker technologies and their applications in diversity studies. *J Biosci*, 45, 1–15.

Rascio, N. and Navari-Izzo, F. 2011. Heavy metal hyperaccumulating plants: how and why do they do it? And what makes them so interesting?. *Plant Sci*, 180, 169–181.

Rastogi, A., Zivcak, M., Sytar, O., Kalaji, H. M., He, X., Mbarki, S. and Brestic, M. 2017. Impact of metal and metal oxide nanoparticles on plant: a critical review. *Front chem*, 5, 78.

Roy, S. K., Cho, S. W., Kwon, S. J., Kamal, A. H. M., Kim, S. W., Oh, M. W., Lee, M. S., Chung, K. Y., Xin, Z. and Woo, S. H. 2016. Morpho-physiological and proteome level responses to cadmium stress in sorghum. *PLoS One*, 11, e0150431.

Roy, S. K., Cho, S. W., Kwon, S. J., Kamal, A. H. M., Lee, D. G., Sarker, K., Lee, M. S., Xin, Z. and Woo, S. H. 2017. Proteome characterization of copper stress responses in the roots of sorghum. *Biometals*, 30, 765–785.

Saidi, A. and Hajibarat, Z. 2020. Application of Next Generation Sequencing, GWAS, RNA seq, WGRS, for genetic improvement of potato (*Solanum tuberosum* L.) under drought stress. *Biocatal Agric Biotechnol*, 29:101801.

Scebba, F., Tognotti, D., Presciuttini, G., Gabellieri, E., Cioni, P., Angeloni, D., Basso, B. and Morelli, E. 2016. A SELDI-TOF approach to ecotoxicology: comparative profiling of low molecular weight proteins from a marine diatom exposed to CdSe/ZnS quantum dots. *Ecotoxicol Environ Saf*, 123, 45–52.

Schmidt, A., Forne, I. and Imhof, A. 2014. Bioinformatic analysis of proteomics data. *BMC Syst Biol*, 8, 1–7.

Seneviratne, M., Rajakaruna, N., Rizwan, M., Madawala, H. M. S. P., Ok, Y. S. and Vithanage, M. 2019. Heavy metal-induced oxidative stress on seed germination and seedling development: a critical review. *Environ Geochem Health*, 41, 1813–1831.

Shi, G. and Cai, Q. 2009. Leaf plasticity in peanut (*Arachishypogaea* L.) in response to heavy metal stress. *Environ. Exp. Bot*, 67, 112–117.

Shrestha, A., Dziwornu, A. K., Ueda, Y., Wu, L. B., Mathew, B. and Frei, M. 2018. Genome-wide association study to identify candidate loci and genes for Mn toxicity tolerance in rice. *PLoS One*, 13, e0192116.

Singh, M., Kumar, J., Singh, S., Singh, V. P., Prasad, S. M. and Singh, M. P. V. V. B. 2015. Adaptation strategies of plants against heavy metal toxicity: a short review. *Biochem Pharmacol (Los Angel)*, 4, 2167–0501.

Singh, R., Ahirwar, N. K., Tiwari, J. and Pathak, J. 2018. Review on sources and effect of heavy metal in soil: its bioremediation. *IMPACT: IJRANSS*, 6, 1–22.

Singh, R., Gautam, N., Mishra, A., and Gupta, R. 2011.Heavy metals and living systems: An overview. *Indian J Pharmacol*, 43:246–253.

Singh, R., Parihar, P. and Prasad, S. M. 2018. Sulfur and calcium simultaneously regulate photosynthetic performance and nitrogen metabolism status in As-challenged *Brassica juncea* L. seedlings. *Front Plant Sci*, 9, 772.

Singh, S., Kumar, V., Datta, S., Dhanjal, D. S., Singh, S., Kumar, S., Kapoor, D., Prasad, R. and Singh, J. 2020. Physiological responses, tolerance, and remediation strategies in plants exposed to metalloids *ESPR*, https://doi.org/10.1007/s11356-020-10293-2.

Singh, S., Parihar, P., Singh, R., Singh, V. P. and Prasad, S. M. 2016. Heavy metal tolerance in plants: role of transcriptomics, proteomics, metabolomics, and ionomics. *Front Plant Sci*, 6, 1143.

Stroiński, A., Chadzinikolau, T., Giżewska, K. and Zielezińska, M. 2010. ABA or cadmium induced phytochelatin synthesis in potato tubers. *Biol Plant*, 54, 117–120.

Sui, J., Jiang, P., Qin, G., Gai, S., Zhu, D., Qiao, L. and Wang, J. 2018. Transcriptome profiling and digital gene expression analysis of genes associated with salinity resistance in peanut. *Electron J Biotechn*, 32, 19–25.

Sun, K., Wang, H. and Xia, Z. 2019. The maize bHLH transcription factor bHLH105 confers manganese tolerance in transgenic tobacco. *Plant Sci*, 280, 97–109.

Sun, P., Zhu, X., Huang, X. and Liu, J. H. 2014. Overexpression of a stress-responsive MYB transcription factor of *Poncirus trifoliata* confers enhanced dehydration tolerance and increases polyamine biosynthesis. *Plant Physiol Biochem*, 78,71–79.

Terzi, H. and Yıldız, M. 2020. Proteomic analysis reveals the role of exogenous cysteine in alleviating chromium stress in maize seedlings. *Ecotoxicol Environ Saf*, 209, 111784.

Thapa, G., Sadhukhan, A., Panda, S. K. and Sahoo, L. 2012. Molecular mechanistic model of plant heavy metal tolerance. *Biometals*, 25, 489–505.

Tiwari, P., Indoliya, Y., Chauhan, A. S., Pande, V. and Chakrabarty, D. 2020. Over-expression of rice R1-type MYB transcription factor confers different abiotic stress tolerance in transgenic *Arabidopsis*. *Ecotoxicol Environ Saf*, 206, 111361.

Uchimiya, M., Bannon, D., Nakanishi, H., McBride, M. B., Williams, M. A. and Yoshihara, T. 2020. Chemical speciation, plant uptake, and toxicity of heavy metals in agricultural soils. *J. Agric. Food Chem*, 68, 12856–12869.

Vatansever, R., Ozyigit, I. I. and Filiz, E. 2017. Essential and beneficial trace elements in plants, and their transport in roots: a review. *Appl Biochem Biotechnol*, 181, 464–482.

Viehweger, K. 2014. How plants cope with heavy metals. *Bot Stud*, 55, 1–12.

Wang, C. Q., Tao, W. A. N. G., Ping, M. U., LI, Z. C. and Ling, Y. A. N. G. 2013. Quantitative trait loci for mercury tolerance in rice seedlings. *Rice Sci*, 20, 238–242.

Wang, Y., Wang, Y., Kai, W., Zhao, B., Chen, P., Sun, L., Ji, K., Li, Q., Dai, S., Sun, Y., Wang, Y., Pei Y. and Leng, P. 2014. Transcriptional regulation of abscisic acid signal core components during cucumber seed germination and under Cu2+, Zn2+, NaCl and simulated acid rain stresses. *Plant Physiol. Biochem*, 76, 67–76.

Wang, C., Wang, J., Wang, X., Xia, Y., Chen, C., Shen, Z. and Chen, Y. 2017. Proteomic analysis on roots of *Oenothera glazioviana* under copper-stress conditions. *Sci Rep*, 7, 1–12.

Wang, N., Wu, X., Ku, L., Chen, Y. and Wang, W. 2016. Evaluation of three protein-extraction methods for proteome analysis of maize leaf midrib, a compound tissue rich in sclerenchyma cells. *Front Plant Sci*, 7, 856.

Wang, X., Xu, M., Gao, C., Zeng, Y., Cui, Y., Shen, W. and Jiang, L. 2020. The roles of endomembrane trafficking in plant abiotic stress responses. *J Integr Plant Biol*, 62, 55–69.

Wang, Y., Xu, L., Tang, M., Jiang, H., Chen, W., Zhang, W., Wang, R. and Liu, L. 2016. Functional and integrative analysis of the proteomic profile of radish root under Pb exposure. *Front Plant Sci*, 7, 1871.

Wei, Z., Ye, J., Zhou, Z., Chen, G., Meng, F. and Liu, Y. 2021. Isolation and characterization of PoWRKY, an abiotic stress-related WRKY transcription factor from *Polygonatum odoratum*. *Physiol Mol Biol Plants,27*:1–9.

Wu, X., Gong, F. and Wang, W. 2014. Protein extraction from plant tissues for 2DE and its application in proteomic analysis. *Proteomics*, 14, 645–658.

Wu, J., Hu, J., Wang, L., Zhao, L. and Ma, F. 2020. Responses of *Phragmites australis* to copper stress: a combined analysis of plant morphology, physiology and proteomics. *Plant Biol* 23:351–362.

Wu, L., Yu, J., Shen, Q., Huang, L., Wu, D. and Zhang, G. 2018. Identification of microRNAs in response to aluminum stress in the roots of Tibetan wild barley and cultivated barley. *BMC Genomics*, 19,1–14.

Xie, Q. E., Yan, X. L., Liao, X. Y. and Li, X. 2009. The arsenic hyperaccumulator fern *Pteris vittata* L. *Environ Sci Technol*, 43, 8488–8495.

Xu, Z., Dong, M., Peng, X., Ku, W., Zhao, Y. and Yang, G. 2019. New insight into the molecular basis of cadmium stress responses of wild paper mulberry plant by transcriptome analysis. *Ecotoxicol Environ Saf*, 171, 301–312.

Xu, L., Wang, L., Gong, Y., Dai, W., Wang, Y., Zhu, X., Wen, T. and Liu, L. 2012. Genetic linkage map construction and QTL mapping of cadmium accumulation in radish (*Raphanus sativus* L.). *Theor Appl Genet*, 125, 659–670.

Xue, L., Ren, H., Li, S., Gao, M., Shi, S., Chang, E., Wei, Y., Yao, X., Jiang, Z. and Liu, J. 2015. Comparative proteomic analysis in *Miscanthus sinensis* exposed to antimony stress. *Environ Pollut*, 201, 150–160.

Yan, Y. F., Lestari, P., Lee, K. J., Kim, M. Y., Lee, S. H. and Lee, B. W. 2013. Identification of quantitative trait loci for cadmium accumulation and distribution in rice (*Oryza sativa*). *Genome*, 56, 227–232.

Yang, L., Ji, J., Harris-Shultz, K. R., Wang, H., Wang, H., Abd-Allah, E. F., Luo, Y. and Hu, X. 2016. The dynamic changes of the plasma membrane proteins and the protective roles of nitric oxide in rice subjected to heavy metal cadmium stress. *Front Plant Sci*, 7, 190.

Yang, Y., Li, X., Yang, S., Zhou, Y., Dong, C., Ren, J., Sun, X. and Yang, Y. 2015. Comparative physiological and proteomic analysis reveals the leaf response to cadmium-induced stress in poplar (*Populus yunnanensis*). *PLoS One*, 10, e0137396.

Yang, J., Liu, D., Wang, X., Ji, C., Cheng, F., Liu, B., Hu, Z., Chen, S., Pental, D., Ju, Y. and Yao, P. 2016. The genome sequence of allopolyploid *Brassica juncea* and analysis of differential homoeolog gene expression influencing selection. *Nat Genetics*, 48, 1225–1232.

Yıldız, M. and Terzi, H. 2016. Proteomic analysis of chromium stress and sulfur deficiency responses in leaves of two canola (*Brassica napus* L.) cultivars differing in Cr (VI) tolerance. *Ecotoxicol Environ Saf*, 124, 255–266.

Younis, A., Ramzan, F., Ramzan, Y., Zulfiqar, F., Ahsan, M. and Lim, K. B. 2020. Molecular markers improve abiotic stress tolerance in crops: a review. *Plants*, 9, 1374.

Yu, J., Wu, L., Fu, L., Shen, Q., Kuang, L., Wu, D. and Zhang, G. 2019. Genotypic difference of cadmium tolerance and the associated microRNAs in wild and cultivated barley. *Plant Growth Regul*, 87, 389–401.

Yuan, C., Mo, J., Yuang, L., Yu, R., Liu, Q., Zhong, J. and Liu, K. 2017. Digital gene expression profiling analysis of *Avicennia marina* under heavy metal stress treatment. *For Res, Beijing*, 30, 206–213.

Zdunić, Z., Grljušić, S., Ledenčan, T., Duvnjak, T. and Šimić, D. 2014. Quantitative trait loci mapping of metal concentrations in leaves of the maize IBM population. *Hereditas*, 151, 55–60.

Zhai, R., Ye, S., Zhu, G., Lu, Y., Ye, J., Yu, F., Chu, Q. and Zhang, X. 2020. Identification and integrated analysis of glyphosate stress-responsive microRNAs, lncRNAs, and mRNAs in rice using genome-wide high-throughput sequencing. *BMC genomics*, 21, 1–15.

Zhang, J., Chen, K., Pang, Y., Naveed, S.A., Zhao, X., Wang, X., Wang, Y., Dingkuhn, M., Pasuquin, J., Li, Z. and Xu, J. 2017. QTL mapping and candidate gene analysis of ferrous iron and zinc toxicity tolerance at seedling stage in rice by genome-wide association study. *BMC Genomics*, 18, 828.

Zhang, F., Ge, W., Ruan, G., Cai, X. and Guo, T. 2020. Data-independent acquisition mass spectrometry-based proteomics and software tools: a glimpse in 2020. *Proteomics*, 20, 1900276.

Zhang, Y., Malzahn, A. A., Sretenovic, S. and Qi, Y. 2019. The emerging and uncultivated potential of CRISPR technology in plant science. *Nat Plants*, 5, 778–794.

Zhang, F., Xiao, X., Xu, K., Cheng, X., Xie, T., Hu, J. and Wu, X. 2020. Genome-wide association study (GWAS) reveals genetic loci of lead (Pb) tolerance during seedling establishment in rapeseed (*Brassica napus* L.). *BMC genomics*, 21, 1–12.

Zhang, H., Xu, Z., Guo, K., Huo, Y., He, G., Sun, H., Guan, Y., Xu, N., Yang, W. and Sun, G. 2020. Toxic effects of heavy metal Cd and Zn on chlorophyll, carotenoid metabolism and photosynthetic function in tobacco leaves revealed by physiological and proteomics analysis. *Ecotoxicol Environ Saf*, 202, 110856.

Zhao, X., Lin, H., Zhang, Z., Shen, Y. and Pan, G. 2014. Mapping of QTLs controlling Pb (2+) content in maize kernels under Pb (2+) stress. *Yi Chuan= Hereditas*, 36(8), 821–826. 10.3724/sp.j.1005.2014.0821

18

Reactive Oxygen and Nitrogen Species and Antioxidant Defense Studies in Plants

Sharad K. Tripathi and Afifa Qidwai
Biological Product Laboratory, Department of Botany, University of Allahabad, Prayagraj, India

Rajesh Kumar
P.G. Department of Botany, Mahatma Gandhi Government Arts College, Mahe, Puducherry, India

CONTENTS

18.1 Introduction

Plants face various environmental stresses such as salinity, drought, extreme temperature, toxic metals, flooding, etc., due to drastic and harsh climate change (Pereira 2016). These diverse abiotic stresses have become a major threat to sustainable crop production. On the other hand, numerous toxic effects lead to oxidative stress through the over accumulation of reactive oxygen species (ROS) including free radicals (superoxide anion, O_2^{-}; hydroperoxyl radical, HO_2; alkoxy radical and hydroxyl radical, OH^{-}) and non-radical molecules (hydrogen peroxide, H_2O_2 and singlet oxygen, $1O_2$) (Hasanuzzaman et al. 2019). The primary cellular ROS generation sites are chloroplasts, mitochondria, peroxisomes,

DOI: 10.1201/9781003155089-18

apoplast, and plasma membranes. Although ROS are formed in the plant as part of normal cellular metabolism, but over accumulation due to stress severely damages necessary cellular ingredients and functions including carbohydrates, proteins, lipids, DNA, etc. (Raja et al. 2017). Moreover, ROS are well known as secondary messengers or signaling molecules and transport the signal to the nucleus through redox reactions using mitogen-activated protein kinase (MAPK) pathway in a variety of cellular mechanisms to increase tolerance against diverse abiotic stresses and participate as major molecules in the acclimation process of plants under environmental stimuli. They actually act as signal transduction molecules, which control diverse pathways throughout the acclimation of the plant under stress conditions. Other studies demonstrated that ROS are essential for the success of numerous fundamental natural processes, including cellular proliferation (Mittler 2017). Apart from ROS, H_2O_2 is a critical component of stress response regulation in crop plants such as rice, wheat, maize, mung bean, soybean cucumber, sour orange, strawberry, basil, and rapeseed. Additionally, it is established along with ROS, reactive nitrogen species (RNS), reactive sulfur species (RSS), and reactive carbonyl species (RCS) also play a key signaling role and are all involved in a cross-talk in plant abiotic stress tolerance (Yamasaki et al. 2019). Therefore, these species play a crucial, dual role in plant biology, representing a fascinating area of research for plant biologists.

18.1.1 Types of RNS

The RNS term introduced to add on some of the NO related molecules like peroxynitrite and S-nitroglutathione which play an important role in different aspects of physiological role in plants and animals (Halliwell and Gutteridge 2007). With the aim of promoting interest of plant biologists in NO, the following chapter will endeavor to summarize current recognition of both properties and physiological roles of NO and RNS in plants.

18.1.2 Nitric Oxide

18.1.2.1 Properties of Nitric Oxide

Nitrogen monoxide gain more attention due to its stability comes by gaining or losing an electron which favors energetically more favorable electron structure. It was represented by major three forms – nitroxyl anion (NO^-) nitric oxide radical (NO) and nitrosonium cation (NO^+) (Stamler et al. 1992; Wojtaszek 2000). NO freely diffuse not only in the cytoplasm but also in lipid regions of plasma membrane due to its lipophilic character (Arasimowicz and Floryszak-Wieczorek 2007).

18.1.2.2 Various Roles of NO in Plant Physiology

NO firstly came under consideration for regulation during plant pathogen interactions (Delledonne et al. 1998; Durner et al. 1998). Now, it is well known that NO is involved in the stimulation of seed (Beligni and Lamattina 2000) pollen germination (Šírová et al. 2011) inflection of plant growth and development (Durner and Klessig 1999), maturation of plant and their senescence (Leshem et al. 1998; Guo and Crawford 2005; Wilhelmová et al. 2006), floral regulation (He et al. 2004), involvement in stomatal movement (García- Mata and Lamattina 2001; Neill et al. 2002; Guo et al. 2003; Desikan et al. 2004; Bright et al. 2006), gravitropism (Hu et al. 2005), some functions of mitochondria (Zottini et al. 2002), regulation of photosynthesis (Takahashi and Yamasaki 2002) or promotes light-mediated greening (Zhang et al. 2006a). Additionally, NO is involved in responses to various stresses, such as drought, salt, and heat stresses, risk element stress, disease resistance and apoptosis (Durner and Klessig 1999; García-Mata and Lamattina 2002; Zhao et al. 2004, 2007; Zhang et al. 2006b; Procházková et al. 2012). In addition, NO plays an important role in symbiotic organisms, particularly between legumes and *Sinorhizobium* (Baudouin et al. 2006). NO act as a transcription regulator to regulate the gene expression in disease resistance process (Polverari et al. 2003) and the expression of stress-related transcription factors and signaling-related kinases (Parani et al. 2004), and by the interaction with other signaling molecules such as salicylic acid and jasmonic acid (Grün et al. 2006; Lozano-Juste et al. 2011).

In peroxisomes, NO nitrosylates proteins such as catalase, glyoxylate oxidase and malate dehydrogenase are involved in photorespiration, β-oxidation and the detoxification of ROS (Ortega-Galisteo et al. 2012).

18.1.3 Peroxynitrite

18.1.3.1 Properties of Peroxynitrite

Peroxynitrite spontaneously formed by reaction between NO and O_2^- by diffusion limited reaction (Huie and Padmaja 1993). This reaction does not require any enzyme to produce $OONO^-$ (peroxynitrtite) because no enzyme can catalyze as fast as it happen spontaneously. NO reacts faster with superoxide which compete out the function of Superoxide dis mutase (SOD). Subsequently, the formation of the $OONO^-$ in vivo always forward direction due to kinetics and thermodynamics of the reaction (Pacher et al. 2007). As far as peroxynitrtite can also be prepared by the enzyme nitrate reductase (NR) with reactants oxygen and NAD(P)H (Bethke et al. 2004). There is a base level of $OONO^-$ is always present in photosynthesizing chloroplasts in plants but it gets higher in the stress conditions which promotes the production of both NO and ROS (Vandelle and Delledonne 2011). $OONO^-$ is reactive nitrogen species, it is short-lived at physiological pH and temperature range which may voluntarily move through the membrane and react with target substrate not only the host cell but in the surrounding cells in the radius between (~5–20 μm) (Szabó et al. 2007). This rare stability of $OONO^-$ because it bends to form a stable cis-configuration where the negative charge restricted over the whole molecule (Tsai et al. 1994). In plants, at first, the generation of $OONO^-$ was confirmed during plant response to biotic stress. Saito et al. (2006) detected intracellular time-dependent $OONO^-$ production in tobacco BY-2 cells treated with INF1 elicitin secreted by *Phytophthora infestans*. Corpas et al. (2009) found that production of $OONO^-$ in coarse of abiotic stress: they noticed that the production of $OONO^-$ in Arabidopsis during salinity stress.

18.1.4 Generation and Function of Nitric Oxide in Plants

In plants there are numerous possible sources of NO including enzymatic and non-enzymatic systems (Wilson et al. 2008; Mur et al. 2012; Hancock, 2012; del Río et al. 2014). A concise list of different substrates used in the production of NO in plant cells. There were different reports of L-arginine-dependent NOS activity in different extracts in different plant species. (Cueto et al. 1996; Jasid et al. 2006; Zhao et al. 2007; reviewed by Corpas et al. 2009, and by del Río 2011). Firstly, the NOS activity in plant tissue was reported by (Barroso et al. 1999; Corpas et al. 2004; del Río et al. 2006, 2014; del Río 2011). Not only in peroxisomes some other organelles also had been reported to produce NO like in mitochondria (Gupta and Kaiser 2010), chloroplasts (Jasid et al. 2006). It was found that NO also regulates some signaling process likewise it regulates different processes by inducing gene transcription or activating secondary messengers (Besson-Bard et al. 2008; Gaupels et al. 2011). NO play different physiological role like in seed germination pollen tube growth, lignification on cell wall, root organogenesis, flowering, fruit ripening and senescence and biotic and abiotic stress (Wendehenne and Hancock 2011; Puppo et al. 2013; Nasir Khan et al. 2014).

18.2 Mechanism in Defense

Like ROS, RNS also play an important role in the signaling in plants by the stimulation of abiotic stress (Wilson et al. 2008; del Río 2013; Corpas et al. 2013a; Nasir Khan et al. 2014; Yu et al. 2014). Likewise, NO is an important facilitator, in co-operation with ROS, in the defense response to pathogen attacks in plants (Bellin et al. 2013; Trapet et al. 2014; Yu et al. 2014). Commonly, it has been noticed that, when plants are exposed to biotic and environmental stresses, a rapid overproduction of ROS and RNS takes place (del Río and Puppo, 2009; Mittler et al. 2011; Airaki et al. 2012; Sandalio et al. 2012; del Río 2013; Baxter et al. 2014; Nasir Khan et al. 2014; Yu et al. 2014). Different types of ROS and RNS has been grouped in a given. Normally the rapid production of NO and ROS leads to programmed cell death (PCD) process. PCD helps in maintaining different aspects of growth and development and also remove infected cells during response to environmental stresses and pathogen infection (Wang et al. 2013). It is known that NO in the presence of O_2 can react with reduced glutathione (GSH), by an S-nitrosylation reaction, to form S-nitrosoglutathione (GSNO) which is an important mobile container of NO bioactivity which have been noticed in different plant species (Ortega-Galisteo et al. 2012; Barroso et al. 2013; Corpas et al. 2013c; Xu et al. 2013; Yu et al. 2014). On the other hand, the RNS peroxynitrite ($ONOO^-$) is a powerful oxidant/nitrating species which is formed by the rapid reaction between O_2^- and NO (Radi 2013), and its occurrence in plant organelles, like peroxisomes, has been reported. It brings to conclusion that NO present in plant tissue and generation of $ONOO^-$ conferred important posttranslational modifications in natural as well as stress condition *viz.*, nitration of proteins and nitrosylation (Romero-Puertas et al. 2013; Corpas et al. 2013a). In peroxisomes, catalase and glycolate oxidase activity are inhibited by *S*-nitrosylation and this could regulate the cellular level of key signaling molecules like H_2O_2 (Ortega-Galisteo et al. 2012). In contrast $ONOO^-$ can cause tyrosine nitration of plant proteins although at some level of endogenous nitration have some regulatory function. Some recent results EM immunogold-labeling in pea plants concluded that nitrated proteins occurs in different sub-cellular compartments like peroxisomes, chloroplasts, mitochondria and in cytosol of the leaf cells (Barroso et al. 2013).

Among ROS, O_2^- predominantly acts as a reducing agent forming strong oxidants. Moreover, O_2^- reacts with nitric oxide (NO) producing RNSs, RSSs, and RCSs. These compounds also promote oxidative stress and are involved in "shaping" the intra- and extracellular redox signal.

18.3 ROS: Site of Generation and Effects

ROS are a group of free radicals, reactive molecules, and ions that are derived from O_2. It has been estimated that about 1% of O_2 consumed by plants is diverted to produce ROS in various subcellular loci such as chloroplasts, mitochondria, peroxisomes (Asada and Takahashi, 1987). ROS are well recognized for playing a dual role as both deleterious and beneficial species depending on their concentration in plants. At high concentration ROS cause damage to biomolecules, whereas at low/moderate concentration it acts as second messenger in intracellular signaling cascades that mediate several responses in plant cells.

18.3.1 Types of ROS

The most common ROS include superoxide (O_2), hydroxyl (-OH), peroxyl (RO_2), hydroperoxyl (HO_2), alkoxyl (RO*), carbonate (CO_3^-) as radicals and hypobromous acid (HOBr), hypochlorous acid (HOCl), singlet oxygen ($1O_2$), ozone (O_3), hydroperoxide (ROOH) as non-radicals (Table 18.1). Oxygen itself is a totally harmless molecule as in its ground state it has two unpaired electrons with parallel spin which makes it paramagnetic and, hence, unlikely to participate in reactions with organic molecules unless it is activated (Apel and Hirt 2004). Activation of O_2 may occur by two different mechanisms: (i) absorption of sufficient energy to reverse the spin on one of the unpaired electrons and (ii) stepwise monovalent reduction.

TABLE 18.1
Types of ROS in Plants

Types of ROS	
Radicals	**Non-radicals**
Superoxide (O_2), hydroxyl (-OH), peroxyl (RO_2), hydroperoxyl (HO_2), alkoxyl (RO*), carbonate (CO_3^-)	Hypobromous acid (HOBr), hypochlorous acid (HOCl), singlet oxygen ($1O_2$), ozone (O_3), hydroperoxide (ROOH)

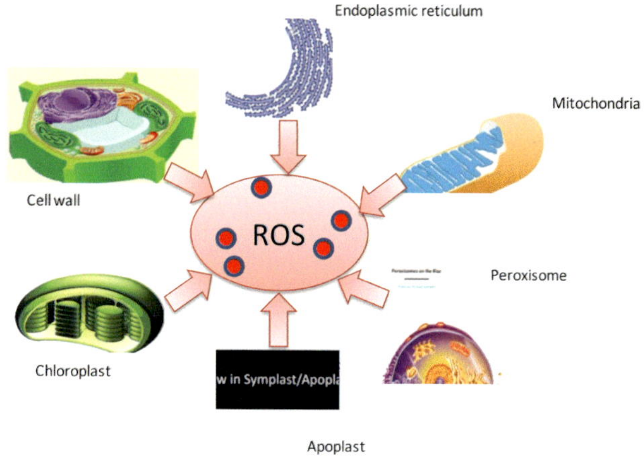

FIGURE 18.1 Site of production of ROS and RNS in plants.

18.3.2 Sites of Production of ROS

ROS are produced in both unstressed and stressed cells at several locations in chloroplasts, mitochondria, plasma membranes, peroxisomes, apoplast, endoplasmic reticulum, and cell walls (Figure 18.1). ROS are always formed by the inevitable leakage of electrons onto O_2 from the electron transport activities of chloroplasts, mitochondria, and plasma membranes or as a by-product of various metabolic pathways localized in different cellular compartments.

18.3.2.1 Chloroplasts

In chloroplasts, various forms of ROS are generated from several locations. ETCs in PSI and PSII are the main sources of ROS in chloroplasts. Production of ROS by these sources is enhanced in plants by conditions limiting CO_2 fixation, such as drought, salt, and temperature stresses, as well as by the combination of these conditions with high-light stress. Under normal conditions, the electron flow from the excited PS centers to NADP which is reduced to NADPH which, then, enters the Calvin cycle and reduces the final electron acceptor, CO_2. In case of overloading of the ETC, due to decreased NADP supply resulting from stress conditions, there is leakage of electron from ferredoxin to O_2, reducing it to O_2•– (Elstner 1991).

18.3.2.2 Mitochondria

Mitochondria, powerhouse of cell, produce ROS in several sites of ETC. In mitochondria direct reduction of oxygen to O_2 •– occurs in the flavoprotein region of NADH dehydrogenase segment (complex I) of the respiratory chain (Arora et al. 2002). When NAD+-linked substrates for complex I are limited, electron transport can occur from complex II to complex I (reverse electron flow). This process has been shown to increase ROS production at complex I and is regulated by ATP hydrolysis (Turrens 2003; Noctor et al. 2007). Ubiquinone-cytochrome region (complex III) of the ETC also produces O_2 •– from oxygen. In plants, under normal aerobic conditions, ETC and ATP syntheses are tightly coupled; however, various stress factors lead to inhibition and modification of its component, leading to over reduction of electron carriers and, hence, formation of ROS and RNS (Murphy 2009).

18.3.2.3 Endoplasmic Reticulum

In endoplasmic reticulum, NAD(P)H-dependent electron transport involving Cyt P450 produces O_2 •– Organic substrate, RH, reacts first with Cyt P450 and then is reduced by a flavoprotein to form a radical intermediate (Cyt P450R–). Triplet oxygen can readily react with this radical intermediate as each has one unpaired electron. This oxygenated complex (Cyt P450-ROO–) may be reduced by cytochrome b or occasionally the complexes may decompose releasing O_2 •.

18.3.2.4 Peroxisomes

Peroxisomes are probably the major sites of intracellular H_2O_2 production, as a result of their essentially oxidative type of metabolism called photorespiration. The main metabolic processes responsible for the generation of H_2O_2 in different types of peroxisomes are the glycolate oxidase reaction, the fatty acid β-oxidation, the enzymatic reaction of flavin oxidases, and the disproportionation of O_2 •– radicals (Baker and Graham 2002). During photorespiration, the oxidation of glycolate by glycolate oxidase in peroxisomes accounts for the majority of H_2O_2 production (Noctor et al. 2002).

18.3.2.5 Plasma Membranes

Electron transporting oxidoreductases are ubiquitous at plasma membranes and lead to generation of ROS at plasma membrane. Production of ROS was studied using EPR spin-trapping techniques and specific dyes in isolated plasma membranes from the growing and the nongrowing zones of hypocotyls and roots of etiolated soybean seedlings as well as coleoptiles and roots of etiolated maize seedlings. NAD(P)H mediated the production of $O_2{}^{\bullet-}$ in all plasma membrane samples. It was suggested that in soybean plasma membranes, $O_2{}^{\bullet-}$ production could be attributed to the action of at least two enzymes, an NADPH oxidase, and, in the presence of menadione, a quinone reductase. NADPH oxidase has been proposed to play a key role in the production and accumulation of ROS in plants under stress conditions (Kwak et al. 2003).

18.3.2.6 Cell Walls

Cell walls are also regarded as active sites for ROS production. Role of cell-wall-associated peroxidase in H_2O_2 generation has been shown. In horseradish, peroxidase associated with isolated cell walls catalyzes the formation of H_2O_2 in the presence of NADH. The reaction is stimulated by various monophenols, especially of coniferyl alcohol. Malate dehydrogenase was found to be the sole candidate for providing NADH. The generation of ROS by cell-wall located peroxidases has been shown during hypersensitive response (HR) triggered in cotton by the bacterium *Xanthomonas campestris* pv. malvacearum and potassium (K) deficiency stress in Arabidopsis (Kim et al. 2010).

18.3.2.7 Apoplast

Cell-wall-located enzymes have been proved to be responsible for apoplastic ROS production. The cell-wall-associated oxalate oxidase, also known as germin, releases H_2O_2 and CO_2 from oxalic acid (Wojtaszek 1997). This enzyme was reported to be involved in apoplastic hydrogen peroxide accumulation during interactions between different cereals species and fungi (Lane 2002). Amine oxidase-like enzymes may contribute to defense responses occurring in the apoplast following biotic stress, mainly through H_2O_2 production (Cona et al. 2006).

18.3.3 Role of ROS as Messengers

At low/moderate concentration, ROS have been implicated as second messengers in intracellular signaling cascades that mediate several plant responses in plant cells, including stomatal closure, programmed cell death, gravitropism and acquisition of tolerance to both biotic and abiotic stresses (Bethke and Jones 2001, Joo et al. 2001, Miller et al. 2008). The role of ROS as second messenger in hormone mediated cellular responses in plants which involves sense, transduction and translation of ROS signal into appropriate cellular responses with the help of some redox-sensitive proteins, calcium mobilization, protein phosphorylation, and gene expression. ROS can be sensed directly also by key signaling proteins such as a tyrosine phosphatase through oxidation of conserved cysteine residues (Xiong et al. 2002). ROS can also modulate the activities of many components in signaling, such as protein phosphatases, protein kinases and transcription factors and communicate with other signal molecules and the pathway forming part of the signaling network that controls response downstream of ROS.

18.3.4 ROS and Oxidative Damage to Biomolecules

When the level of ROS exceeds the defense mechanisms, a cell is said to be in a state of "oxidative stress". However, the equilibrium between production and scavenging of ROS is perturbed under a number of stressful conditions such as salinity, drought, high light, toxicity due to metals, pathogens, and so forth. Enhanced level of ROS can cause damage to biomolecules such as lipids, proteins and DNA (Figure 18.2). These reactions can alter intrinsic membrane properties like fluidity, ion transport, loss of enzyme activity, protein cross-linking, inhibition of protein synthesis, DNA damage, and so forth ultimately resulting in cell death (Xiong et al. 2002).

18.4 Antioxidative Defense System in Plants

Plants possess complex antioxidative defense system comprising of nonenzymatic and enzymatic components to scavenge ROS. In plant cells, specific ROS producing and scavenging systems are found in different organelles such as chloroplasts, mitochondria, and peroxisomes. ROS scavenging pathways from different cellular compartments are coordinated (Pang and Wang, 2008). Under normal conditions, potentially toxic oxygen metabolites are generated at a low level and there is an appropriate balance between production and quenching of ROS. The balance between production and quenching of ROS may be perturbed by a number of adverse environmental factors, giving rise to rapid increases in intracellular ROS levels, which can induce oxidative damage to lipids, proteins, and nucleic acids. In order to avoid the oxidative damage, higher plants raise the level of endogenous antioxidant defense (Sharma et al. 2010). Various components of antioxidative defense system involved in ROS scavenging have been manipulated, overexpressed or downregulated to add to the present knowledge and understanding the role of the antioxidant systems (Figure 18.3).

18.4.1 Nonenzymatic Components of Antioxidative Defense System

Nonenzymic components of the antioxidative defense system include the major cellular redox buffers ascorbate (AsA) and glutathione (γ-glutamyl-cysteinyl-glycine, GSH) as well as tocopherol, carotenoids, and phenolic compounds. They interact with numerous cellular components and in addition to crucial roles in defense and as enzyme cofactors, these antioxidants influence plant growth and development by modulating processes from mitosis and cell elongation to senescence and cell death (Semchuk et al. 2009).

18.4.1.1 Ascorbate

Ascorbate (AsA) is the most abundant, low molecular weight antioxidant that has a key role in defense against oxidative stress caused by enhanced level of ROS. AsA is considered powerful antioxidant because of its ability to donate electrons in a number of enzymatic and nonenzymatic reactions (Isherwood

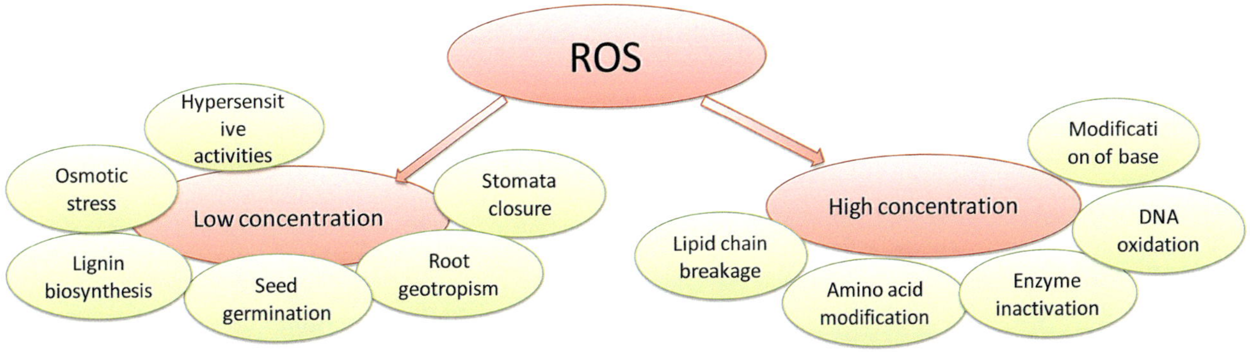

FIGURE 18.2 Effect of ROS in plants in low and high concentration.

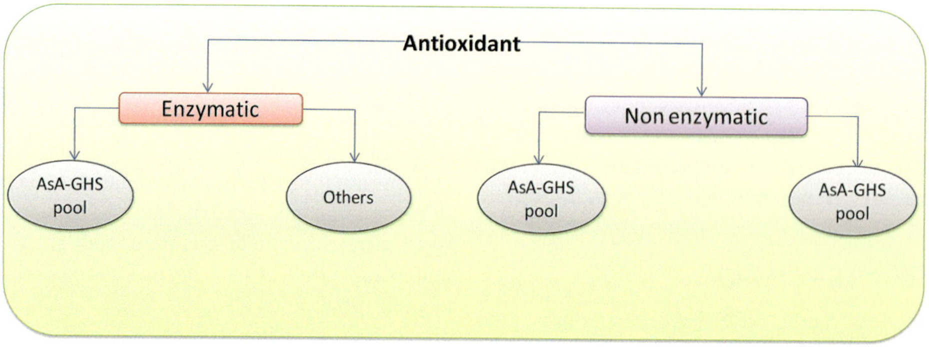

FIGURE 18.3 Overview of plant antioxidant system.

et al. 1954). AsA has been shown to play important role in several physiological processes in plants, including growth, differentiation, and metabolism. The majority of the AsA pool in plants is contributed by D-mannose/L-galactose commonly called Smirnoff-Wheeler pathway which proceeds via GDP-D-mannose, GDP-L-galactose, L-galactose, and L-galactono-1,4-lactone (Wheeler et al. 1998). AsA is also synthesized via uronic acid intermediates such as D-galacturonic acid. In this pathway D-galacturonic acid is reduced to L-galactonic acid by galacturonic acid reductase, which is subsequently converted to L-galactono-1,4-lactone. The L-galactono-1,4-lactone is further oxidized to AsA by Lgalactono-1,4-lactone dehydrogenase (GALDH) enzyme. It is synthesized in the mitochondria by L-galactono-γ-lactone dehydrogenase and is transported to the other cell components through a proton-electrochemical gradient or through facilitated diffusion. It is detected in the majority of plant cell types, organelles and apoplast in plants and is found to be particularly abundant in photosynthetic tissues (Shao et al. 2008). AsA protects critical macromolecules from oxidative damage. Under normal physiological condition, AsA mostly exists in reduced state in chloroplast where it also acts as a cofactor of violaxanthin de-epoxidase, thus, sustaining dissipation of excess excitation energy (Smirnoff 2000).

18.4.1.2 Glutathione

Tripeptide glutathione (γ-glutamylcysteinyl-glycine, GSH) is one of the crucial low molecular weight nonprotein thiol that plays an important role in intracellular defense against ROS-induced oxidative damage. It has been detected virtually in all cell compartments such as cytosol, chloroplasts, endoplasmic reticulum, vacuoles, and mitochondria (Foyer and Noctor 2003). GSH is synthesized in the cytosol and chloroplasts of plant cells by compartment specific isoforms of γ-glutamyl-cysteinyl synthetase (γ-ECS) and glutathione synthetase (GS). The balance between the GSH and glutathione disulfide (GSSG) is a central component in maintaining cellular redox state. Due to its reducing power, GSH plays an important role in diverse biological processes, including cell growth/division, regulation of sulfate transport, signal transduction, conjugation of metabolites, enzymatic regulation, synthesis of proteins and nucleic acids.

18.4.1.3 Tocopherols

Tocopherols (α, β, γ, and δ) represent a group of lipophilic antioxidants involved in scavenging of oxygen free radicals, lipid peroxy radicals. Relative antioxidant activity of the tocopherol isomers in vivo is $\alpha>\beta>\gamma>\delta$ which is due to the methylation pattern and the amount of methyl groups attached to the phenolic ring of the polar head structure (Fukuzawa et al. 1982). Hence, α-tocopherol with its three methyl substituents has the highest antioxidant activity of tocopherols (Kamal-Eldin and Appelqvist 1996). Tocopherols are synthesized only by photosynthetic organisms and are present in only green parts of plants. The tocopherol biosynthetic pathway utilizes two compounds homogentisic acid (HGA) and phytyl diphosphate (PDP) as precursors. Tocopherols are known to protect lipids and other membrane components by physically quenching and chemically reacting with O_2 in chloroplasts, thus protecting the structure and function of PSII (Ivanov and Khorobrykh 2003). Tocopherols prevent the chain propagation step in lipid autooxidation which makes it an effective free radical trap.

18.4.1.4 Carotenoids

Carotenoids also belong to the group of lipophilic antioxidants and are able to detoxify various forms of ROS as depicted by J. Young (1991). Carotenoids are found in plants as well as micro-organisms. In plants, carotenoids absorb light in the region between 400 and 550 nm of the visible spectrum and pass the captured energy to the chlorophyll (Sieferman-Harms 1987). As an antioxidant, they scavenge $1O_2$ to inhibit oxidative damage and quench triplet sensitizer (3Chl*) and excited chlorophyll (Chl*) molecule to prevent the formation of $1O_2$ to protect the photosynthetic apparatus. Carotenoids also serve as precursors to signaling molecules that influence plant development and biotic/abiotic stress responses (Li et al. 2008). The ability of carotenoids to scavenge, prevent or minimize the production of triplet chlorophyll may be accounted for by their chemical specificity and high carotenoids content favors better adaptation of sugarcane plants under saline condition (Gomathi and Rakkiyapan 2011).

18.4.1.5 Phenolic Compounds

Phenolics are diverse secondary metabolites (flavonoids, tannins, hydroxycinnamate esters, and lignin) which possess antioxidant properties. They are abundantly found in plant tissues (Grace and Logan, 2000). Polyphenols contain an aromatic ring with –OH or OCH_3 substituents which together contribute to their biological activity, including antioxidant action. They have been shown to outperform well-known antioxidants, AsA and α-tocopherol, in in vitro antioxidant assays because of their strong capacity to donate electrons or hydrogen atoms. Polyphenols can chelate transition metal ions, can directly scavenge molecular species of active oxygen, and can inhibit lipid peroxidation by trapping the lipid alkoxyl radical. They also modify lipid packing order and decrease fluidity of the membranes (Arora et al. 2000). These changes could strictly hinder diffusion of free radicals and restrict peroxidative reactions. There is some evidence of induction of phenolic metabolism in plants as a response to multiple stresses (Michalak 2006).

18.4.2 Enzymatic Components

The enzymatic components of the antioxidative defense system comprise of several antioxidant enzymes such as superoxide dismutase (SOD), catalase (CAT), guaiacol peroxidase (GPX), enzymes of ascorbate glutathione (AsA-GSH) cycle ascorbate peroxidase (APX), monodehydroascorbate reductase (MDHAR), dehydroascorbate reductase (DHAR), and glutathione reductase (GR). These enzymes operate in different subcellular compartments and respond in concert when cells are exposed to oxidative stress.

Conclusion

ROS and RNS are unavoidable by products of normal cell metabolism. They are generated by electron transport activities of chloroplast, mitochondria, and plasma membrane or as a byproduct of various metabolic pathways localized in different cellular compartments. Under normal growth condition, ROS production in various cell compartments is low. However, various environmental stresses such as drought, salinity, chilling, metal toxicity, and UV-B, if prolonged over to a certain extent, disrupt the cellular homeostasis and enhance the production of ROS and RNS. They play two divergent roles in plants; in low concentrations they act as signaling molecules that mediate several plant responses in plant cells, including responses under stresses, whereas in high concentrations they cause exacerbating damage to cellular components. Enhanced level of these species causes oxidative damage to lipid, protein, and DNA leading to altered intrinsic membrane properties like fluidity, ion transport, loss of enzyme activity, protein crosslinking, inhibition of protein synthesis, DNA damage, ultimately resulting in cell death. In order to avoid the oxidative damage, higher plants possess a complex antioxidative defense system comprising of nonenzymatic and enzymatic components. The metabolism of ROS is crucial in crop growth, development, adaptation, and existence under stressful environments. The production and scavenging of ROS are essential factors of plant defense processes, and modulation and overexpression of candidate genes encoding ROS detoxifying enzymes are widely used to enhance tolerance against several abiotic stresses. However, the balance among the detoxification and generation of ROS is maintained by both enzymatic and nonenzymatic antioxidant systems under stressful conditions. Notably, ROS are known to perform a dual role in plant biology due to molecular cross-talk with other signaling molecules such as RNS, RSS, and ROS. ROS is very important for various biological mechanisms, such as cellular proliferation is known to exert a signaling role at low concentrations. Both ROS and RNS can create oxidative and nitrosative stress solely or together nitro-oxidative stress; however, they are also involved in signaling process of higher plants especially under adverse environmental situations. Mechanism study of formation and fate of these species using advanced analytical techniques will help in developing broader view of the role in plants. Future progress in genomics, metabolomics, and proteomics will help in clear understanding of biochemical networks involved in cellular responses to oxidative stress. Improved understanding of these will be helpful in producing in-built capacity of enhanced levels of tolerance of plant using biotechnological approach.

REFERENCES

Airaki M, Leterrier M, Mateos RM, Valderrama R, Chaki M, Barroso JB, del Río LA, Palma JM, Corpas FJ. 2012. Metabolism of reactive oxygen species and reactive nitrogen species in pepper (*Capsicum annuum* L.) plants under low temperature stress. *Plant and Cell Physiology* 35, 281–295.

Apel K, Hirt H. 2004. Reactive oxygen species: metabolism, oxidative stress, and signal transduction. *Annual Review of Plant Biology* 55, 373–399.

Arasimowicz M, Floryszak-Wieczorek J. 2007. Nitric oxide as a bioactive signalling molecule in plant stress responses. *Plant Science* 172, 876–887.

Arora A, Byrem TM, Nair MG, Strasburg GM. 2000. Modulation of liposomal membrane fluidity by flavonoids and isoflavonoids. *Archives of Biochemistry and Biophysics* 373(1), 102–109.

Arora A, Sairam RK, Srivastava GC. 2002. Oxidative stress and antioxidative system in plants. *Current Science* 82(10), 1227–1238.

Asada K, Takahashi M. (1987) Production and Scavenging of Active Oxygen in Chloroplasts. In: Kyle, D.J., Osmond, C.B. and Arntzen, C.J., Eds., Photoinhibition, Elsevier, Amsterdam, 227–287.

Baker A and Graham AI, *Plant peroxisomes: biochemistry, cell biology and biotechnological applications*. Dordrecht, The Netherlands: Kluwer Academic Publishers, 2002.

Barroso JB, Corpas FJ, Carreras A, Sandalio LM, Valderrama R, Palma JM, Lupiáñez JA, del Río LA. 1999. Localization of nitric-oxide synthase in plant peroxisomes. *Journal of Biological Chemistry* 274, 36729–36733.

Barroso JB, Valderrama R, Corpas FJ. 2013. Immunolocalization of *S*-nitrosoglutathione, *S*-nitrosoglutathione reductase and tyrosine nitration in pea leaf organelles. *Acta Physiologia Plantarum* 35, 2635–2640.

Baudouin E, Pieuchot L, Engler G *et al.* 2006. Nitric oxide is formed in Medicago truncatula-*Sinorhizobium meliloti* functional nodules. *Molecular Plant–Microbe Interactions* 19, 970–975.

Baxter A, Mittler R, Suzuki N. 2014. ROS as key players in plant stress signalling. *Journal of Experimental Botany* 65, 1229–1240.

Beligni MV, Lamattina L. 2000. Nitric oxide stimulates seed germination and de-etiolation, and inhibits hypocotyl elongation, three light-inducible responses in plants. *Planta* 210, 215–221.

Bellin D, Asai S, Delledonne M, Yoshioka H. 2013. Nitric oxide as a mediator for defence responses. *Molecular Plant–Microbe Interactions* 26, 271–277.

Besson-Bard A, Pugin A, Wendehenne D. 2008. New insights into nitric oxide signaling in plants. *Annual Review of Plant Biology* 59, 21–39.

Bethke PC, Badger MR, Jones RL. 2004. Apoplastic synthesis of nitric oxide by plant tissues. *Plant Cell* 16, 332–341.

Bethke PC, Jones RL. 2001. Cell death of barley aleurone protoplasts is mediated by reactive oxygen species. *Plant Journal* 25(1), 19–29.

Bright J, Desikan R, Hancock JT *et al.* 2006. ABA-induced NO generation and stomatal closure in *Arabidopsis* are dependent on H2O2 synthesis. *The Plant Journal* 45, 113–122.

Cona A, Rea G, Angelini R, Federico R, Tavladoraki P. 2006. Functions of amine oxidases in plant development and defence. *Trends in Plant Science* 11(2), 80–88.

Corpas FJ, Alché JD, Barroso JB. 2013c. Current overview of *S*-nitrosoglutathione (GSNO) in higher plants. *Frontiers in Plant Science* 4, 126.

Corpas FJ, Barroso JB, Carreras A, *et al.* 2004. Cellular and subcellular localization of endogenous nitric oxide in young and senescent pea plants. *Plant Physiology* 136, 2722–2733.

Corpas FJ, del Río LA, Barroso JB. 2013a. Protein tyrosine nitration in higher plants under natural and stress conditions. *Frontiers in Plant Science* 4, 29.

Corpas FJ, Leterrier M, Begara-Morales JC, *et al.* 2013b. Inhibition of peroxisomal hydroxypyruvate reductase (HPR1) by tyrosine nitration. *Biochimica et Biophysica Acta* 1830, 4981–4989.

Corpas FJ, Palma JM, del Río LA, Barroso JB. 2009. Evidence supporting the existence of L-arginine-dependent nitric oxide synthase activity in plants. *New Phytologist* 184, 9–14.

Cueto M, Hernández-Perea O, Martín R, Ventura ML, Rodrigo J, Lamas S, Golvano PM. 1996. Presence of nitric oxide synthase activity in roots and nodules of *Lupinus albus*. *FEBS Letters* 398, 159–164.

del Río LA. 2011. Peroxisomes as a source of reactive nitrogen species signal molecules. *Archives of Biochemistry and Biophysics* 506, 1–11.

del Río LA (ed.). 2013. *Peroxisomes and their key role in cellular signaling and metabolism*. Berlin, Heidelberg: Springer-Verlag.

del Río LA, Corpas FJ, Barroso JB, López-Huertas E, Palma JM. 2014. Function of peroxisomes as a cellular source of nitric oxide and other reactive nitrogen species. In: Nasir Khan M, Mobin M, Mohammad F, Corpas FJ, eds. *Nitric oxide in plants: metabolism and role in stress physiology*. Berlin, Heidelberg: Springer-Verlag, 33–55.

del Río LA, Puppo A (eds.). 2009. *Reactive oxygen species in plant signaling*. Berlin, Heidelberg: Springer-Verlag.

del Río LA, Sandalio LM, Corpas FJ, Palma JM, Barroso JB. 2006. Reactive oxygen species and reactive nitrogen species in peroxisomes. Production, scavenging, and role in cell signalling. *Plant Physiology* 141, 330–335.

Delledonne M, Xia Y, Dixon RA, Lamb C. 1998. Nitric oxide functions as a signal in plant disease resistance. *Nature* 394, 585–588.

Desikan R, Cheung MK, Bright J *et al.* 2004. ABA, hydrogen peroxide and nitric oxide signalling in stomatal guard cells. *Journal of Experimental Botany* 55, 205–212.

Durner J, Klessig D. 1999. Nitric oxide as a signal in plants. *Current Opinion in Plant Biology* 2:369–374.

Durner J, Wendehenne D, Klessig DF. 1998. Defence gene induction in tobacco by nitric oxide, cyclic GMP, and cyclic ADP-ribose. *Proceedings of the National Academy of Sciences, USA* 95, 10328–10333.

Elstner EF. 1991. Mechanisms of oxygen activation in different compartments of plant cells. In: Pell EJ, Steffen, KL, eds. *Active oxygen/oxidative stress and plant metabolism*. Rockville, MD: American Society of Plant Physiologists, 13–25.

Foyer CH, Noctor G. 2003. Redox sensing and signaling associated with reactive oxygen in chloroplasts, peroxisomes and mitochondria. *Physiologia Plantarum* 119(3).

Fukuzawa K, Tokumura A, Ouchi S, Tsukatani H. 1982. Antioxidant activities of tocopherols on Fe2+-ascorbate induced lipid peroxidation in lecithin liposomes. *Lipids* 17(7), 511–514.

García-Mata C, Lamattina L. 2001. Nitric oxide induces stomatal closure and enhances the adaptive plant responses against drought stress. *Plant Physiology* 126, 1196–1204.

García-Mata C, Lamattina L. 2002. Nitric oxide and abscisic acid cross talk in guard cells. *Plant Physiology* 128, 790–792.

Gaupels F, Kuruthukulangarakoola GT, Durner J. 2011. Upstream and downstream signals of nitric oxide in pathogen defence. *Current Opinion in Plant Biology* 14, 707–714.

Gomathi R, Rakkiyapan P. 2011. Comparative lipid peroxidation, leaf membrane thermostability, and antioxidant system in four sugarcane genotypes differing in salt tolerance. *International Journal of Plant Physiology and Biochemistry* 3(4), 67–74.

Grace SG, Logan BA. 2000. Energy dissipation and radical scavenging by the plant phenylpropanoid pathway. *Philosophical Transactions of the Royal Society B* 355.

Grün S, Lindermayr C, Sell S, Durner J. 2006. Nitric oxide and gene regulation in plants. *Journal of Experimental Botany* 57, 507–516.

Guo FQ, Crawford NM. 2005. Arabidopsis nitric oxide synthase1 is targeted to mitochondria and protects against oxidative damage and dark-induced senescence. *The Plant Cell*, 17(12), 3436–3450.

Guo FQ, Okamoto M, Crawford NM. 2003. Identification of a plant nitric oxide synthase gene involved in hormone signalling. *Science* 302, 100–103.

Gupta KJ, Kaiser WM. 2010. Production and scavenging of nitric oxide by barley root mitochondria. *Plant and Cell Physiology* 51, 576–584.

Halliwell B, Gutteridge JMC. 2007. *Free radicals in biology and medicine*. Oxford University Press, USA.

Hancock JT. 2012. NO synthase? Generation of nitric oxide in plants. *Periodicum Biologorum* 114, 19–24.

Hasanuzzaman M, Bhuyan M, Anee TI, Parvin K, Nahar K, Mahmud JA, Fujita M. 2019. Regulation of ascorbate-glutathione pathway in mitigating oxidative damage in plants under abiotic stress. *Antioxidants* 8, 384.

He YK, Tang RH, Yi H *et al.* 2004. Nitric oxide represses the *Arabidopsis* floral transition *Science* 305:1968–1971.

Hu X, Neill SJ, Tang Z, Cai W. 2005. Nitric oxide mediates gravitropic bending in soybean roots. *Plant Physiology.* 137, 663–670.

Huie RE, Padmaja S. 1993. The reaction rate of nitric oxide with superoxide. Free Rad Res in pea (*Pisum sativum* L.) leaf peroxisomes: changes under abiotic stress. *Journal of Experimental Botany* 18, 195–198.

Isherwood FA, Chen YT, Mapson LW. 1954. Synthesis of L-ascorbic acid in plants and animals. *The Biochemical Journal* 56(1), 1–15.

Ivanov BN, Khorobrykh S. 2003. Participation of photosynthetic electron transport in production and scavenging of reactive oxygen species. *Antioxidants and Redox Signaling* 5(1), 43–53.

Jasid S, Simontacchi M, Bartoli CG, Puntarulo S. 2006. Chloroplasts as a nitric oxide cellular source. Effect of reactive nitrogen species on chloroplastic lipids and proteins. *Plant Physiology* 142, 1246–1255.

Joo JH, Bae YS, Lee JS. 2001. Role of auxin-induced reactive oxygen species in root gravitropism. *Plant Physiology* 126(3), 1055–1060.

Kamal-Eldin A, Appelqvist LA. 1996. The chemistry and antioxidant properties of tocopherols and tocotrienols. *Lipids* 31(7), 671–701.

Kim MJ, Ciani S, Schachtman DP. 2010. A peroxidase contributes to ROS production during *Arabidopsis* root response to potassium deficiency. *Molecular Plant* 3(2), 420–427.

Kwak JM, Mori IC, Pei ZM *et al.* 2003. NADPH oxidase AtrbohD and AtrbohF genes function in ROS-dependent ABA signaling in *Arabidopsis*. *EMBO Journal* 22(11), 2623–2633.

Lane BG. 2002. Oxalate, germins, and higher-plant pathogens. *IUBMB Life* 53(2), 67–75.

Leshem YY, Wills RBH, Ku VVV. 1998. Evidence for the function of the free radical gas nitric oxide (NO) as an endogenous maturation and senescence regulating factor in higher plants. *Plant Physiology and Biochemistry* 36, 825–833.

Li F, Vallabhaneni R, Yu J, Rocheford T, Wurtzel ET. 2008. The maize phytoene synthase gene family: overlapping roles for carotenogenesis in endosperm, photomorphogenesis, forms. *Science* 258, 1898–1902.

Lozano-Juste J, Colom-Moreno R, León J. 2011. In vivo protein tyrosine nitration in Arabidopsis thaliana. *Journal of Experimental Botany*, 62(10), 3501–3517.

Michalak A. 2006. Phenolic compounds and their antioxidant activity in plants growing under heavy metal stress. *Polish Journal of Environmental Studies* 15(4), 523–530.

Miller G, Shulaev V, Mittler R. 2008. Reactive oxygen signaling and abiotic stress. *Physiologia Plantarum* 133(3), 481–489.

Mittler, R. 2017. ROS are good. *Trends in Plant Science* 22, 11–19.

Mittler R, Vanderauwera S, Suzuki N, Miller G, Tognetti VB, Vandepoele K, Gollery M, Shulaev V, van Breusegem F. 2011. ROS signaling: the new wave? *Trends in Plant Science* 16, 300–309.

Mur LA, Mandon J, Persijn S *et al.* 2012. Nitric oxide in plants: an assessment of the current state of knowledge. *AoB PLANTS*, 5:pls052. doi: 10.1093/aobpla/pls052.

Murphy MP. 2009. How mitochondria produce reactive oxygen species. *Biochemical Journal* 417(1), 1–13.

Nasir Khan M, Mobin M, Mohammad F, Corpas FJ (eds.). 2014. *Nitric oxide in plants: metabolism and role in stress physiology.* Berlin, Heidelberg: Springer-Verlag.

Neill SJ, Desikan R, Clarke A, Hancock JT. 2002. Nitric oxide is a novel component of abscisic acid signaling in stomatal guard cells. *Plant Physiology* 128, 13–16.

Noctor G, De Paepe R, and Foyer CH. 2007. Mitochondrial redox biology and homeostasis in plants. *Trends in Plant Science* 12(3), 125–134.

Noctor G, Veljovic-Jovanovic S, Driscoll S, Novitskaya L, Foyer CH. 2002. Drought and oxidative load in the leaves of C3 plants: a predominant role for photorespiration? *Annals of Botany* 89(7), 841–850.

Ortega-Galisteo AP, Rodríguez-Serrano M, Pazmiño DM, Gupta DK, Sandalio LM, Romero-Puertas MC. 2012. *S*-Nitrosylated proteins in pea (*Pisum sativum* L.) leaf peroxisomes: changes under abiotic stress. *Journal of Experimental Botany* 63, 2089–2103.

Pacher P, Beckman JS, Liaudet L. 2007. Nitric oxide and peroxynitrite in health and disease *Physiological Reviews.* 87(1): 315–424.

Pang CH, Wang BS. 2008. Oxidative stress and salt tolerance in plants. In: Lüttge, U, Beyschlag, W, Murata, J, eds. *Progress in botany.* Berlin: Springer, 231–245.

Parani M, Rudrabhatla S, Myers R *et al.* 2004. Microarray analysis of nitric oxide responsive *Plant Biotechnology.* 2(4), 359–366.

Pereira A. 2016. Plant abiotic stress challenges from the changing environment. *Frontiers in Plant Science* 7, 1123.

Polverari A, Molesini B, Pezzotti M *et al.* 2003. Nitric oxide-mediated transcriptional changes in *Arabidopsis thaliana. Molecular Plant Microbe Interactions* 16, 1094–1105.

Procházková D, Haisel D, Pavlíková D *et al.* 2012. The effect of risk elements in soil to nitric oxide metabolism in tobacco plants. *Plant Soil Environment*, 58 (10), 435–440.

Puppo A, Pauly N, Boscari A, Mandon K, Brouquisse R. 2013. Hydrogen peroxide and nitric oxide: key regulators of the legume– *Rhizobium* and mycorrhizal symbioses. *Antioxidants & Redox Signaling* 18, 2202–2219.

Radi R. 2013. Protein tyrosine nitration: biochemical mechanisms and structural basis of functional effects. *Accounts of Chemical Research* 46, 550–559.

Raja V, Majeed U, Kang H, Andrabi KI, John R. 2017. Abiotic stress: interplay between ROS, hormones and MAPKs. *Environmental and Experimental Botany* 137, 142–157.

Romero-Puertas MC, Rodríguez-Serrano M, Sandalio LM. 2013. Protein *S*-nytrosylation in plants under abiotic stress: an overview. *Frontiers in Plant Science* 4, 373.

Saito S, Yamamoto-Katou A, Yoshioka H *et al.* 2006. Peroxynitrite generation and tyrosine nitration in defence responses in tobacco BY-2 cells. *Plant and Cell Physiology* 47, 689–697.

Sandalio LM, Rodríguez-Serrano M, Gupta DK, Archilla A, Romero-Puertas MC, del Río LA. 2012. Reactive oxygen species and nitric oxide in plants under cadmium stress: from toxicity to signaling. In: Ahmad P, Prasad MNV, eds. *Environmental adaptations and stress tolerance of plants in the era of climate change.* Berlin, Heidelberg: Springer-Verlag, 199–215.

Semchuk NM, Lushchak OV, Falk J, Krupinska K, and Lushchak VI. 2009. Inactivation of genes, encoding tocopherol biosynthetic pathway enzymes, results in oxidative stress in outdoor grown Arabidopsis thaliana. *Plant Physiology and Biochemistry*, 47(5), 384–390.

Shao HB, Chu LY, Lu ZH, Kang CM. 2008. Primary antioxidant free radical scavenging and redox signaling pathways in higher plant cells. *International Journal of Biological Sciences* 4(1), 8–14.

Sharma, P, Jha, AB, Dubey, RS. 2010. Oxidative stress and antioxidative defence system in plants growing under abiotic Stresses. In: Pessarakli, M, ed. *Handbook of plant and crop stress*. Boca Raton, FL: CRC Press, Taylor and Francis Publishing Company, 3rd edition, 89–138.

Sieferman-Harms D. 1987. The light harvesting function of carotenoids in photosynthetic membrane. *Plant Physiology* 69(3), 561–568.

Smirnoff N. 2000. Ascorbic acid: metabolism and functions of a multi-facetted molecule. *Current Opinion in Plant Biology* 3(3), 229–235.

Stamler JS, Singel DJ, Loscalzo J. 1992. Biochemistry of nitric oxide and its redox-activated state of knowledge. *AOB Plants* 5, 1–17.

Szabó C, Ischiropoulos H, Radi R. 2007. Peroxynitrite: biochemistry, pathophysiology and development of therapeutics. *Nature Reviews Drug Discovery*, 6(8), 662–80.

Takahashi S, Yamasaki H. 2002. Reversible inhibition of photophosphorylation in chloroplasts by nitric oxide. *FEBS Letters* 512, 145–148.

Trapet P, Kulik K, Lamotte O, Jeandroz S, Bourque S, Nicolas-Francès V, Rosnoblet C, Besson-Bard A, Wendehenne D. 2014. NO signaling in plant immunity: a tale of messengers. *Phytochemistry* 112, 72–79.

Tsai JHM, Harrison JG, Martin JC *et al.* 1994. Role of conformation of peroxynitrite anion (ONOO-) with Its Stability and Toxicity. *Journal of the American Chemical Society*, 116, 4115–4116.

Turrens JF. 2003. Mitochondrial formation of reactive oxygen species. *Journal of Physiology* 552(2), 335–344.

Vandelle E, Delledonne M. 2011. Peroxynitrite formation and function in plants. *Plant Science* 5(1), 43–53.

Wang Y, Loake GJ, Chu C. 2013. Cross-talk of nitric oxide and reactive oxygen species in plant programmed cell death. *Frontiers in Plant Science* 4, 314.

Wendehenne D, Hancock JT. 2011. New frontiers in nitric oxide biology in plants. *Plant Science* 181, 507–508.

Wheeler GL, Jones MA, Smirnoff N. 1998. The biosynthetic pathway of vitamin C in higher plants. *Nature* 393(6683), 365–369.

Wilhelmová, N., Fuksova, H., Srbova, M., Mikova, D., Mýtinová, Z., Prochazkova, D.& Wilhelm, J. (2006). The effect of plant cytokinin hormones on the production of ethylene, nitric oxide, and protein nitrotyrosine in ageing tobacco leaves. *Biofactors*, 27(1–4), 203–211.

Wilson ID, Neill SJ, Hancock JT. 2008. Nitric oxide synthesis and signaling in plants. *Plant, Cell and Environment* 31, 622–631.

Wojtaszek P. 1997. Oxidative burst: an early plant response to pathogen infection. *Biochemical Journal* 322(3), 681–692.

Wojtaszek P. 2000. Nitric oxide in plants: to NO or not to NO. *Phytochemistry* 54, 1–4.

Xiong L, Schumaker KS, Zhu JK. 2002. Cell signaling during cold, drought, and salt stress, *Plant Cell* 14, S165–S183.

Xu S, Guerra D, Lee U, Vierling E. 2013. *S*-nitrosoglutathione reductases are low-copy number, cysteine-rich proteins in plants that control multiple developmental and defence responses in *Arabidopsis*. *Frontiers in Plant Science* 4, 430.

Yamasaki H, Ogura MP, Kingjoe KA, Cohen MF. 2019. D-Cysteine-induced rapid root abscission in the water fern Azolla pinnata: implications for the linkage between d-amino acid and reactive sulfur species (RSS) in plant environmental responses. *Antioxidants* 8, 411.

Young J. 1991. The photoprotective role of carotenoids in higher plants. *Physiologia Plantarum* 83(4), 702–708.

Yu M, Lamattina L, Spoel SH, Loake GJ. 2014. Nitric oxide function in plant biology: a redox cue in deconvolution. *New Phytologist* 202, 1142–1156.

Zhang L, Wang Y, Zhao L *et al.* 2006a. Involvement of nitric oxide in light mediated greening of barley seedlings. *Journal of Plant Physiology* 163, 818–826.

Zhang YY, Wang LL, Liu YL *et al.* 2006b. Nitric oxide enhances salt tolerance in maize seedlings through increasing activities of proton-pump and Na+/H+ antiport in the tonoplast *Planta* 224(3), 545–555.

Zhao L, Zhang F, Guo J *et al.* 2004. Nitric oxide functions as a signal in salt resistance in the calluses from two ecotypes of reed. *Plant Physiology* 134, 849–857.

Zhao MG, Tian QY, Zhang WH. 2007. Nitric oxide synthase-dependent nitric oxide production is associated with salt tolerance in *Arabidopsis*. *Plant Physiology* 144, 206–217.

Zottini M, Formentin E, Scattolin M *et al.* 2002. Nitric oxide affects plant mitochondrial functionality in vivo. *FEBS Letters* 515, 75–78.

Index

Note: Locators in *italics* represent figures and **bold** indicate tables in the text.